Monographs in

Volume 4

Number Theory

An Elementary Introduction Through
Diophantine Problems

Monographs in Number Theory

ISSN 1793-8341

Monographs in

Volume 4

Number Theory

An Elementary Introduction Through Diophantine Problems

Daniel Duverney

Baggio Engineering School, France

World Scientific

NEW JERSEY · LONDON · SINGAPORE · BEIJING · SHANGHAI · HONG KONG · TAIPEI · CHENNAI

Published by

World Scientific Publishing Co. Pte. Ltd.

5 Toh Tuck Link, Singapore 596224

USA office: 27 Warren Street, Suite 401-402, Hackensack, NJ 07601

UK office: 57 Shelton Street, Covent Garden, London WC2H 9HE

British Library Cataloguing-in-Publication Data
A catalogue record for this book is available from the British Library.

Originally published in French as *Théorie des nombres* © Dunod, 2007, 2nd edition, Paris

Monographs in Number Theory — Vol. 4
NUMBER THEORY
An Elementary Introduction Through Diophantine Problems

Copyright © 2010 by World Scientific Publishing Co. Pte. Ltd.

ISBN-13 978-981-4307-45-1
ISBN-10 981-4307-45-9
ISBN-13 978-981-4307-46-8 (pbk)
ISBN-10 981-4307-46-7 (pbk)

Printed in Singapore.

Preface

This textbook aims at introducing the reader to number theory.

Number theory (or higher arithmetic) was considered the "Queen of Mathematics" by Carl Friedrich Gauss (1777-1855). He wrote in particular: "The higher arithmetic presents us with an inexhaustible storehouse of interesting truths – of truths, too, which are not isolated but stand in the closest relation to one another, and between which, with every successive advance of the science, we continually discover new and sometimes wholly unexpected points of contact."

One hundred years later, Louis Mordell (1872-1952) added: "The theory of numbers is unrivalled for the number and variety of its results and for the beauty and wealth of its demonstrations. The higher arithmetic seems to include most of the romance of mathematics."

A great part of the beauty, charm and romance of number theory seems to result, as already observed by Gauss, from the contrast between the simplicity of many statements and the difficulty we encounter in proving them. Overcoming these difficulties has been often the historical source of deep and fruitful theories, which have shown themselves powerful and efficient in many other branches of mathematics.

For instance, its is known from ancient Babylonia (1900-1600 b.c.) that the equation $x^2 + y^2 = z^2$ has solutions for integers x, y, z. For example $3^2 + 4^2 = 5^2$. However, not until the seventieth century did Pierre de Fermat (1601-1665) prove that $x^4 + y^4 = z^4$ has no solution for non-zero integers x, y, z.

In a most famous conjecture, he even asserted that, in fact, the equation $x^n + y^n = z^n$ has no solution for non-zero integers x, y, z and $n \geq 3$.

Despite of the simplicity of its statement, *Fermat's last theorem* has shown incredibly difficult to prove. It was not until the end of the twentieth century that a complete proof was given by Andrew Wiles. This proof goes far beyond the scope of this book and rests on deep and complicated theories. We will content ourselves with proving Fermat's last theorem in the cases $n = 3$, 4, 5. However, even studying these "elementary" cases will convince the reader of the difficulty of the general problem.

Thus, this book aims at introducing the reader to the charm and beauty of number theory. It is intended for graduate and advanced undergraduate students, as well as for professional mathematicians, maths teachers, and mathematically educated laymen.

The prerequisites are an undergraduate course in algebra (groups, rings, fields and vector spaces) and calculus of one or two variables. A few parts of the book will make use of basic knowledge in complex analysis.

Evidently, it was not possible to cover, in one single book, all subjects in number theory. We have focused our attention on *diophantine problems*, named after Diophantus of Alexandria (around 250 b.c.), who wrote the first known book devoted exclusively to number theory. Diophantine problems can be roughly divided into two categories:

- Diophantine equations, that is equations whose unknowns are integers. This was the main subject of Diophantus book.

- Irrationality and transcendence problems, which go back to the irrationality of $\sqrt{2}$ (Pythagorean school, around 500 b.c.). The main tool for proving irrationality results consists in using *diophantine approximations*, that is approximations of real numbers by rational numbers.

From this entry point, the book offers the reader to discover many fundamental and instructive subjects in number theory, deliberately from an elementary point of view:

- Expansions of real numbers in series, infinite products and continued fractions
- Representations of integers as sums of squares
- Arithmetical functions (with an insight into prime number theory)
- Algebraic theory of numbers
- Diophantine equations
- Irrationality and transcendence methods...

The book is completely self-contained. Some proofs have been left to the reader as solved exercises. The reason for this is to promote a more "active" reading and also, sometimes, to highlight the main ideas of selected proofs or parts of the text.

However, a greater part of the 209 solved exercises consists in applications and further study of the concepts developed in the theory. They will hopefully help the reader understanding and mastering the powerful tools forged by mathematicians, all along the centuries, in order to handle and solve, in "honor of the human mind" (Carl Gustav Jacobi, 1804-1851), number-theoretical problems.

I am grateful to Michel Garcia for his kind assistance in checking thoroughly the first and second french editions of this book and pointing out a number of misprints and mathematical mistakes. Many thanks also to Jordan Taren, who helped me improving this translation into english by correcting a number of grammatical mistakes.

Daniel Duverney,
Lille, France
May 2010

Structure of the book

An arrow linking chapter p to chapter q means that it is necessary to have read chapter p before studying chapter q. If the arrow is dotted, some parts of chapter q use results from chapter p, but studying chapter p is not necessary for the understanding of chapter q.

If there is no arrow between two chapters, these chapters are independent from each other.

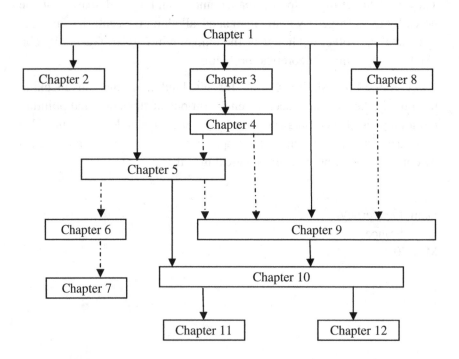

Contents

Chapter 4. Regular continued fractions

Chapter 5. Quadratic fields and diophantine equations

Chapter 6. Squares and sums of squares

Chapter 7. Arithmetical functions

Solutions to the exercises

Bibliography   331

Index   333

Chapter 1

Irrationality and diophantine approximation

In this introductory chapter, we prove that the numbers $\sqrt{d}$, e and π are irrational, as well as the values of the Tschakaloff function (section 1.4). These four proofs allow us to identify the notion of diophantine approximation and to link it with irrationality problems (section 1.5). We conclude with some methodological remarks (section 1.6).

1.1   Irrationality of $\sqrt{d}$

Theorem 1.1   *Let $d \in \mathbb{N}$. Assume that d is not a perfect square. Then $\sqrt{d}$ is irrational.*

Proof   Assume on the contrary that $\sqrt{d}$ is a rational number. Then $\sqrt{d} = a/b$, where a and b are coprime integers. Squaring this equality yields $b^2 d = a^2$. As d is not a perfect square, there exists a prime p and an integer k such that $d = p^{2k+1}\delta$, with $p \nmid \delta$. Then $p^{2k+1} \mid a^2$ and therefore $p^{k+1} \mid a$, so that $a = p^{k+1}\alpha$. Hence $b^2\delta = p\alpha^2$. The prime number p divides $b^2\delta$ but does not divide δ. Therefore $p \mid b^2$, whence $p \mid b$. So $p \mid a$ and $p \mid b$, contradiction because a and b are coprime. Theorem 1.1 is proved.

Corollary 1.1 *If* $d \in \mathbb{N}$ *is not a perfect square, the numbers* 1 *and* $\sqrt{d}$ *are linearly independent over* $\mathbb{Q}$. *In other words, if* $p, q \in \mathbb{Q}$ *and* $p + q\sqrt{d} = 0$, *then* $p = q = 0$.

Proof If q was different from zero, the equality $p + q\sqrt{d} = 0$ would imply $\sqrt{d} = -p/q \in \mathbb{Q}$, contradiction with theorem 1.1. Therefore $q = 0$ and $p = 0$.

We will use corollary 1.1 repeatedly in chapters 4 and 5.

1.2   Irrationality of *e*

Theorem 1.2 *e is irrational.*

Proof For every natural integer n, we can write

$$e = \sum_{k=0}^{n} \frac{1}{k!} + R_n, \qquad \text{with } R_n = \sum_{k=n+1}^{+\infty} \frac{1}{k!}. \tag{1.1}$$

It is clear that $R_n > 0$. Moreover,

$$R_n = \frac{1}{(n+1)!} + \frac{1}{(n+2)!} + \frac{1}{(n+3)!} + \cdots$$

$$R_n = \frac{1}{(n+1)!}\left(1 + \frac{1}{n+2} + \frac{1}{(n+2)(n+3)} + \frac{1}{(n+2)(n+3)(n+4)} + \cdots\right)$$

$$R_n < \frac{1}{(n+1)!}\left(1 + \frac{1}{2} + \left(\frac{1}{2}\right)^2 + \left(\frac{1}{2}\right)^3 + \cdots\right) = \frac{2}{(n+1)!}.$$

Therefore (1.1) yields

$$0 < n!e - n!\sum_{k=0}^{n} \frac{1}{k!} < \frac{2}{n+1}. \tag{1.2}$$

We argue again by *reductio ad absurdum*. Assume that $e = a/b$, where a and b are natural integers. Denote $\alpha_n = n!\sum_{k=0}^{n} \frac{1}{k!}$. Then $\alpha_n \in \mathbb{N}$, and (1.2)

yields $0 < n!a - b\alpha_n < \dfrac{2b}{n+1}$.

This implies that the integer $\beta_n = n!a - b\alpha_n$ is not zero and vanishes when n tends to infinity. This is impossible, because *the absolute value of a non zero integer is always greater than* 1. This proves theorem 1.2.

1.3　Irrationality of π

Theorem 1.3　π *is irrational.*

Proof　Let $P(x)$ be any polynomial of degree $2n$. Put
$$F(x) = P(x) - P''(x) + P^{(4)}(x) - \cdots + (-1)^n P^{(2n)}(x).$$
We observe that $P(x)\sin x = (F'(x)\sin x - F(x)\cos x)'$, which yields at once *Hermite's formula*:
$$\int_0^\pi P(x)\sin x\,dx = F(0) + F(\pi).　　　　(1.3)$$
Assume that $\pi = a/b$, $a, b \in \mathbb{N}$, and apply Hermite formula with
$$P(x) = \frac{1}{n!}x^n(a - bx)^n.$$
Denote $I_n = \int_0^\pi P(x)\sin x\,dx$.

Then $I_n > 0$ because $P(x)\sin x$ is a continuous, non negative and non identically zero function on $[0, \pi]$. Moreover, $x(a - bx) \le a^2/4b$ on $[0, \pi]$, whence $I_n \le \dfrac{1}{n!}\pi\left(\dfrac{a^2}{4b}\right)^n$. Therefore $\lim_{n\to+\infty} I_n = 0$. But it can be proved (exercise 1.1) that $F(0) \in \mathbb{Z}$ and $F(\pi) \in \mathbb{Z}$ for every $n \in \mathbb{N}$. Thus $I_n \in \mathbb{Z}$ for every $n \in \mathbb{N}$. Hence the *positive* sequence I_n vanishes at infinity, which is impossible. Therefore π is irrational.

1.4.　Irrationality of the values of the Tschakaloff function

Let $q \in \mathbb{C}$, $q > 1$. The *Tschakaloff function* is defined by
$$T_q(x) = \sum_{n=0}^{+\infty} \frac{x^n}{q^{\frac{n(n+1)}{2}}},　　\forall x \in \mathbb{C}.　　　(1.4)$$
It satisfies the functional equation
$$T_q(qx) = 1 + xT_q(x).　　　　(1.5)$$

We will prove the following result.

Theorem 1.4   *Let* $q \in \mathbb{Z}$, $|q| \geq 2$. *Then* $T_q(x)$ *is irrational for every*
$x \in \mathbb{Q}^*$.

We need the following lemma, which will be also useful in the study of
Padé approximants (chapter 8).

Lemma 1.1   *Let* $\mathbb{K}$ *be any subfield of* $\mathbb{C}$, *and let* $f(x) = \sum_{n=0}^{+\infty} a_n x^n$,
with $a_n \in \mathbb{K}$ *for every* $n \in \mathbb{N}$ *and radius of convergence* $R > 0$. *Assume*
p, q, r are 3 natural integers satisfying $p < r \leq p + q + 1$. *Then there exist*
$P, Q \in \mathbb{K}[x]$, $Q \neq 0$, *and a series* $g(x) = \sum_{n=0}^{+\infty} b_n x^n$, $|x| < R$, *such that*
$\deg P \leq p, \deg Q \leq q$ *and*

$$Q(x)f(x) + P(x) = x^r g(x). \tag{1.6}$$

The proof of lemma 1.1 is left as an exercise (exercise 1.2).

We now prove theorem 1.4. Let $x = \alpha/\beta$, $(\alpha, \beta) \in \mathbb{Z}^2$. Assume that
$T_q(x) = \mu/\nu, (\mu, \nu) \in \mathbb{Z}^2$. An easy induction using the functional equation

(1.5) shows that $T_q\left(\dfrac{x}{q^n}\right) = \dfrac{A_n}{\nu\alpha^n}$ for every $n \in \mathbb{N}$, where $A_n \in \mathbb{Z}$.

Let ρ be a *fixed integer*, such that $|\alpha/q^\rho| < 1$.

We use lemma 1.1, with $\mathbb{K} = \mathbb{Q}$, $f(x) = T_q(x)$, $p = q = 2\rho, r = 3\rho$. Then
the polynomials P and Q have *rational* coefficients. However, if we
multiply (1.6) by the LCM of these coefficients, we see that we may
assume that P and Q have *integer* coefficients. Hence we can write

$$Q(x)T_q(x) + P(x) = x^{3\rho} g(x), \tag{1.7}$$

where $P, Q \in \mathbb{Z}[x]$, $\deg Q \leq 2\rho$, $\deg P \leq 2\rho$, $Q \neq 0$.

But g is not identically zero : if it was the case, T_q would be a rational
fraction because of (1.7), and therefore a polynomial because it is
defined on $\mathbb{C}$, which is impossible considering its Taylor expansion
(1.4). Therefore at least one of the Taylor coefficients of g is not zero.
Hence there exist an integer $\sigma \geq 0$ and a function h such that

$$Q(x)T_q(x) + P(x) = x^{3\rho+\sigma} h(x), \qquad h(0) \neq 0. \tag{1.8}$$

Replace $x = \alpha/\beta$ by x/q^n in (1.8), multiply by $v\alpha^n \beta^{2\rho} q^{2\rho n}$, and denote by B_n the common value of both sides of the equation

$$B_n = \frac{v\alpha^{3\rho+\sigma}}{\beta^{\rho+\sigma}} \left(\frac{\alpha}{q^{\rho+\sigma}}\right)^n h\left(\frac{\alpha}{\beta q^n}\right)$$

$$= \left(\beta^{2\rho} q^{2\rho n} Q\left(\frac{\alpha}{\beta q^n}\right)\right)\left(v\alpha^n T_q\left(\frac{\alpha}{\beta q^n}\right)\right) + v\alpha^n\left(\beta^{2\rho} q^{2\rho n} P\left(\frac{\alpha}{\beta q^n}\right)\right). \tag{1.9}$$

As $\left(\beta^{2\rho} q^{2\rho n} P\left(\dfrac{\alpha}{\beta q^n}\right)\right)$, $v\alpha^n T_q\left(\dfrac{\alpha}{\beta q^n}\right)$, and $\left(\beta^{2\rho} q^{2\rho n} Q\left(\dfrac{\alpha}{\beta q^n}\right)\right) \in \mathbb{Z}$, we see

that $B_n \in \mathbb{Z}$. Moreover, $B_n \underset{\infty}{\sim} \dfrac{v\alpha^{3\rho+\sigma}}{\beta^{\rho+\sigma}} \left(\dfrac{\alpha}{q^{\rho+\sigma}}\right)^n h(0)$, which implies that

$\lim_{n \to +\infty} B_n = 0$ because $\left|\alpha/q^\rho\right| < 1$, and also that $B_n \neq 0$ as $\alpha \neq 0$ and $h(0) \neq 0$. Again, we have constructed a sequence of non zero integers which vanishes at infinity. This proves theorem 1.4.

1.5 Diophantine approximation

Constructing a *diophantine approximation* of a given real number α means finding a sequence of rational numbers P_n/Q_n and a function f vanishing at infinity, such that

$$\left|\alpha - \frac{P_n}{Q_n}\right| \leq f(Q_n), \qquad \forall n \in \mathbb{N}. \tag{1.10}$$

Observe that we have proved the irrationality of the numbers e and $T_q(\alpha/\beta)$ by using diophantine approximations.

Indeed, in the case of e, we can write (1.2) as

$$0 < e - \frac{P_n}{Q_n} < \frac{2}{(n+1)Q_n} \leq \frac{2}{Q_n}, \tag{1.11}$$

where $P_n = n! \displaystyle\sum_{k=0}^{n} \frac{1}{k!}$ and $Q_n = n!$.

In the case of $T_q(\alpha/\beta)$, it is not so clear. One has to observe that

$$T_q\left(\frac{\alpha}{\beta q^n}\right) = \frac{k_n}{\alpha^n} T_q\left(\frac{\alpha}{\beta}\right) + \frac{\ell_n}{\alpha^n},$$

where $(k_n, \ell_n) \in \mathbb{Z}^2$ (induction by using (1.5)). Hence (1.9) yields

$$T_q\left(\frac{\alpha}{\beta}\right) + \frac{P_n}{Q_n} \le \frac{c}{Q_n}\left(\frac{\alpha}{q^{\rho+\sigma}}\right)^n \le \frac{c}{Q_n} \tag{1.12}$$

with $\quad Q_n = k_n \beta^{2\rho} q^{2\rho n} Q(\alpha/\beta q^n)$, $\quad P_n = \alpha^n \beta^{2\rho} q^{2\rho n} P(\alpha/\beta q^n) + \ell_n Q_n/k_n$,

and $c = \max_{n \in \mathbb{N}} \alpha^{3\rho+\sigma} \beta^{-\rho-\sigma} h\left(\alpha\beta^{-1}q^{-n}\right)$.

The difference between e and $T_q(\alpha/\beta)$ consists in the fact that the diophantine approximation given by par (1.11) is an *explicit* one (one can compute explicitly P_n and Q_n as functions of n), whereas the one given by (1.12) is not. In this last case, P_n and Q_n are expressed by using polynomials P and Q, and lemma 1.1 asserts the *existence* of these polynomials, but gives no way how to compute them.

However, in both cases, we get an irrationality result, because after multiplying by Q_n, the product $Q_n f(Q_n)$ vanishes at infinity. We say we have obtained a *good diophantine approximation*, and can state the following result.

Theorem 1.5 *Let $\alpha \in \mathbb{R}$. Assume there exists a sequence P_n/Q_n of rational numbers satisfying*

$$\forall n \in \mathbb{N}, \quad 0 < \left|\alpha - \frac{P_n}{Q_n}\right| \le \frac{\varepsilon(n)}{Q_n}, \quad \text{with } \lim_{n \to +\infty} \varepsilon(n) = 0.$$

Then α is irrational.

Proof See exercise 1.3.

The following theorem has been proved by Dirichlet (1805-1859). It shows that, for a given irrational α, there exist good diophantine approximations.

Theorem 1.6 *Let $\alpha \in \mathbb{R}$. Assume α to be irrational. Then there exists an infinite sequence of rational numbers P_n/Q_n satisfying*

$$0 < \left|\alpha - \frac{P_n}{Q_n}\right| \le \frac{1}{Q_n^2}, \quad \forall n \in \mathbb{N}.$$

For the proof of theorem 1.6, we will need the following lemma.

Lemma 1.2   *Assume α is irrational. Then, for every integer $Q > 1$, one can find $p/q \in \mathbb{Q}$ such that $1 \leq q < Q$ and $0 < |q\alpha - p| \leq \dfrac{1}{Q}$.*

Proof of lemma 1.2 Denote by $[\alpha]$ the integral part of α, and consider the $Q+1$ numbers 0, 1, $\alpha - [\alpha]$, $2\alpha - [2\alpha]$, ..., $(Q-1)\alpha - [(Q-1)\alpha]$.
All these numbers belong to the interval $[0,1]$, and all are of the form $a\alpha + b$, a and b integers, $0 \leq a \leq Q-1$. We now use the *pigeon-hole principle*. We divide the interval $[0,1]$ into Q sub-intervals, namely $[0, 1/Q]$, $[1/Q, 2/Q]$, ..., $[(Q-1)/Q, 1]$. Then at least two of the above $Q+1$ numbers belong to the *same* sub-interval (try to put the $Q+1$ numbers into the Q intervals). Denote these two numbers by $\xi_1 = a_1\alpha + b_1$ and $\xi_2 = a_2\alpha + b_2$, with $0 \leq a_1 \leq Q-1$, $0 \leq a_2 \leq Q-1$, and $a_1 \neq a_2$ (because $a_1 = a_2$ implies $a_1 = a_2 = 0$, that is $\xi_1 = 0$ and $\xi_2 = 1$, which is impossible). We can assume that $a_1 > a_2$. By the choice of ξ_1 and ξ_2, $|\xi_1 - \xi_2| = |(a_1 - a_2)\alpha + b_1 - b_2| \leq 1/Q$, with $0 < a_1 - a_2 \leq Q-1$, which proves lemma 1.2.

We now prove theorem 1.6 by induction. Choose an arbitrary integer $Q > 1$. By lemma 1.2, we can find a rational number P_1/Q_1 such that $0 < |\alpha - P_1/Q_1| \leq 1/QQ_1$ and $1 \leq Q_1 < Q$. Hence $0 < |\alpha - P_1/Q_1| < 1/Q_1^2$.
Now we use again lemma 1.2, by choosing Q such that $Q^{-1} < |\alpha - P_1/Q_1|$. We can find a rational number P_2/Q_2 satisfying $1 \leq Q_2 < Q$ and $0 < |\alpha - P_2/Q_2| \leq 1/QQ_2 < 1/Q_2^2$. Moreover,

$$\left| \frac{P_1}{Q_1} - \frac{P_2}{Q_2} \right| \geq \left\| \alpha - \frac{P_2}{Q_2} \right| - \left| \alpha - \frac{P_1}{Q_1} \right\| \neq 0$$

since $|\alpha - P_1/Q_1| > 1/Q$ and $|\alpha - P_2/Q_2| \leq 1/QQ_2 \leq 1/Q$.
By induction we obtain an infinite sequence of rational numbers P_n/Q_n satisfying $|\alpha - P_n/Q_n| < 1/Q_n^2$, which proves theorem 1.6.

Remark 1.1 Theorem 1.6 asserts the existence of the sequence P_n/Q_n, but gives no way how to compute it. The theory of regular continued fractions (formula (4.8)) will enable us to get an explicit result.

1.6   Methodological remarks

We have used, in this chapter, three methods worth describing.

For proving the irrationality of $\sqrt{d}$, we have argued by *reductio ad absurdum*. Assuming that $\sqrt{d} = a\,/\,b$, we have proved that $\sqrt{d} = a'\,/\,b'$, with $a' < a$, $b' < b$. From here, there are two possibilities:

    a.   Use the fact that the starting point was minimal. In our proof, we have assumed $a\,/\,b$ to be irreducible.

    b.   Alternatively, consider that, starting from a given value a, we construct a decreasing sequence $a > a' > a'' > \cdots$ of *positive integers*, which is clearly impossible. This last process is called the *descent method*.

For proving the irrationality of e, π and $T_q(\alpha/\beta)$, we have constructed a *sequence of positive integers vanishing at infinity*, which is impossible. Generally, the most difficult part in this sort of proof, which forms the basis of transcendence proofs (chapter 12), consists in proving that no term of this sequence can be zero.

Finally, we have used the *pigeon-hole principle*: if $n+1$ pigeons nest in n pigeon-holes, then at least 2 pigeons have to nest in the same hole. A consequence of this principle is the following result.

If $(u_n)_{n \in \mathbb{N}}$ is a bounded sequence of integers, there exist two integers n and p, $n \neq p$, such that $u_n = u_p$.

Exercises

1.1   With the notations of theorem 1.3, prove that $P(0) = P'(0) = \cdots$
$\cdots = P^{(n-1)}(0) = 0$. Deduce that $F(0) \in \mathbb{Z}$ and $F(\pi) \in \mathbb{Z}$.

1.2   Prove lemma 1.1.

1.3   Prove theorem 1.5.

1.4   Prove that $\alpha = \sqrt[3]{2} + \sqrt{5}$ is irrational.

1.5   Prove that $\beta = \log_{10} 2$ is irrational.

1.6   Let $P(x) = a_0 + a_1 x + \cdots + a_n x^n$, with $a_n \neq 0$, be a polynomial with integer coefficients. Let $r = p/q$, p and q coprimes and not zero, a rational root with multiplicity d of $P(x)$. Prove that q^d divides a_n.

Application: Prove that $\cos\dfrac{\pi}{n}$ is irrational for every $n \geq 4$.

1.7   Let $m \in \mathbb{Z}$, $m \geq 2$. Prove that $\displaystyle\sum_{n=0}^{+\infty} \frac{1}{m^{n^2}}$ is irrational.
Give two different proofs.

1.8   An irrational number α is said to be *quadratic* if there exist integers a, b, c with $ac \neq 0$, such that $a\alpha^2 + b\alpha + c = 0$.
1) Prove that the *golden number* $\Phi = (1 + \sqrt{5})/2$ is quadratic.
2) Prove that e is not quadratic.

1.9   The *Fermat numbers* F_n are defined by $F_n = 2^{2^n} + 1$. Our aim is to prove that $\chi = \displaystyle\sum_{n=0}^{+\infty} \frac{1}{F_n}$ is irrational. We define, for $|x| < 1$,

$$f(x) = \sum_{n=0}^{+\infty} \frac{x^{2^n}}{1 - x^{2^n}}, \quad g(x) = \sum_{n=0}^{+\infty} \frac{x^{2^n}}{1 + x^{2^n}}.$$

1) Prove that $f(x) - g(x) = 2\left(f(x) - \dfrac{x}{1-x}\right)$.
2) Prove that $f(1/2)$ is irrational.
3) Deduce that χ is irrational.

1.10   For $a \in \mathbb{N}$, let $f_a(x) = \displaystyle\sum_{n=0}^{+\infty} (1+a)(1+aq)\ldots(1+aq^{n-1})x^n q^{-\frac{n(n+1)}{2}}$,

with $q \in \mathbb{Z}$, $|q| \geq 2$. Prove that, if $x \in \mathbb{Q}^*$ and $|x| < \dfrac{|q|}{a}$, then $f_a(x) \notin \mathbb{Q}$.

1.11 Pell's equation

Let $d \in \mathbb{N}$. Assume d is not a square. Our aim is to prove that the diophantine equation $x^2 - dy^2 = 1$ has at least one solution $(x, y) \neq (1, 0)$.

1) Prove there exist infinitely many pairs of non-zero positive integers (x, y) such that $0 < |x^2 - dy^2| < 1 + 2\sqrt{d}$.

2) Prove there exist an integer k ($|k| < 1 + 2\sqrt{d}$), and integers m and n such that the system $x^2 - dy^2 = k$, $x \equiv m \pmod{k}$, $y \equiv n \pmod{k}$ admits infinitely many solutions.

3) Let (x', y') and (x'', y'') be two pairs solutions of this system. Prove there exists $(\xi, \eta) \in \mathbb{Z}^2$ such that

$$\begin{cases} (x' - y'\sqrt{d})(x'' + y''\sqrt{d}) = k(\xi + \eta\sqrt{d}) \\ (x' + y'\sqrt{d})(x'' - y''\sqrt{d}) = k(\xi - \eta\sqrt{d}) \end{cases}$$

4) Conclude.

1.12 An example of transcendental number

Let $\alpha \in \mathbb{R}$. We say that α is algebraic of degree d if there exist an integer $d \geq 1$ and rational integers a_0, a_1, $\cdots$, a_d, with $a_d \neq 0$, such that $a_0 + a_1\alpha + \cdots + a_d\alpha^d = 0$. For example, rational numbers are algebraic of degree 1, quadratic numbers are algebraic of degree 2. We say that α is transcendental if it is not algebraic. The purpose of this exercise is to prove that, for every $q \in \mathbb{Z}$ satisfying $|q| \geq 2$, the number $\alpha = \sum_{n=0}^{+\infty} q^{-2^n}$ is transcendental.

1) Prove that, for every $h \in \mathbb{N} - \{0\}$, $\alpha^h = \sum_{n=0}^{+\infty} b_h(n)q^{-n}$, where $b_h(n) = 0$ if the expression of n in base 2 counts at least $h+1$ digits 1, and $b_h(n) = h!$ if it counts exactly h digits 1.

2) For every integer $k \geq 2$, let $n_k = \left(1 + 2 + \cdots + 2^{d-1}\right)2^k$. For every $h \in \{1, 2, \cdots, d\}$, compute a) $b_h(n_k)$, b) $b_h(n_k + 1)$, $b_h(n_k + 2)$, $\cdots$, $b_h\left(n_k + 2^{k-1}\right)$, c) $b_h(n_k - 1)$, $b_h(n_k - 2)$, ..., $b_h\left(n_k - 2^{k-2}\right)$.

3) Prove that α is transcendental.

Chapter 2

Representations of real numbers by infinite series and products

Here, we give some classical representations of real numbers: p-adic expansion, Engel series, Cantor infinite product. For each of these representations, there is an elementary irrationality criterion.

2.1 p-adic expansion of a real number

2.1.1 For every real number x, we denote by $[x]$ the integral part of x. Let p be a natural integer, $p \geq 2$, and let γ_0 be a positive real number. We define the natural integers $c_0, c_1, c_2, \ldots$, and the real numbers $\gamma_1, \gamma_2, \gamma_3, \ldots$ by the following algorithm.

$$\begin{cases} \gamma_0 = c_0 + \dfrac{\gamma_1}{p} & \text{with } c_0 = [\gamma_0], \quad 0 \leq \gamma_1 < p \\[2mm] \gamma_1 = c_1 + \dfrac{\gamma_2}{p} & \text{with } c_1 = [\gamma_1], \quad 0 \leq \gamma_2 < p \\[1mm] \vdots & \quad \vdots \qquad\qquad \vdots \\[1mm] \gamma_n = c_n + \dfrac{\gamma_{n+1}}{p} & \text{with } c_n = [\gamma_n], \quad 0 \leq \gamma_{n+1} < p \end{cases} \tag{2.1}$$

An easy induction shows that

$$\gamma_0 = \sum_{k=0}^{n} \frac{c_k}{p^k} + \frac{\gamma_{n+1}}{p^{n+1}}. \tag{2.2}$$

As $0 \leq \gamma_{n+1} < p$, we see that $\lim_{n \to +\infty} \dfrac{\gamma_{n+1}}{p^{n+1}} = 0$, whence

$$\gamma_0 = \sum_{k=0}^{+\infty} \frac{c_k}{p^k}. \tag{2.3}$$

We have obtained *the p-adic expansion* of γ_0.

2.1.2 By construction, $c_n = [\gamma_n] \le \gamma_n < p$ for every $n \ge 1$. As c_n is an *integer*, we see that $c_n \le p-1$ for every $n \ge 1$. We now prove that there exist infinitely many n such that $c_n \le p-2$. Assume, on the contrary, that we can find an N such that $c_n = p-1$ for every $n \ge N$. Then (2.1) yields $\gamma_n = p-1+\gamma_{n+1}/p$ for every $n \ge N$, whence $p-\gamma_n = (p-\gamma_{n+1})/p$ for every $n \ge N$. By induction, we see that $p-\gamma_N = (p-\gamma_n)/p^{n-N}$ for every $n \ge N$. Now let n tend to infinity. As $0 < p-\gamma_n \le p$, we get $p-\gamma_N = 0$, a contradiction because $\gamma_N < p$. Therefore there exist infinitely many n such that $c_n \le p-2$.

2.1.3 Last, we prove that the *p-adic expansion* of γ_0 in the form (2.2), with $c_n \le p-1$ for $n \ge 1$ and $c_n \le p-2$ for infinitely many n, is unique. To do this, assume that

$$\gamma_0 = \sum_{k=0}^{+\infty} \frac{d_k}{p^k}, \tag{2.4}$$

with $d_n \le p-1$ for $n \ge 1$ and $d_n \le p-2$ for infinitely many n. Then we have $\gamma_0 = d_0 + \sum_{k=1}^{+\infty} d_k/p^k$, and

$$0 \le \sum_{k=1}^{+\infty} \frac{d_k}{p^k} < \sum_{k=1}^{+\infty} \frac{p-1}{p^k},$$

because at least one of the d_k's is less than $p-1$.

Hence $0 \le \sum_{k=1}^{+\infty} d_k/p^k < (p-1)\sum_{k=1}^{+\infty} 1/p^k = 1$.

This means that $d_0 = [\gamma_0] = c_0$. By induction, assume now we have proved that $d_k = c_k$ for all $k \le n$. We proceed to prove that $d_{n+1} = c_{n+1}$. By (2.4) and (2.2), we have

$$\gamma_0 - \sum_{k=0}^{n} \frac{d_k}{p^k} = \sum_{k=n+1}^{+\infty} \frac{d_k}{p^k} = \frac{\gamma_{n+1}}{p^{n+1}},$$

whence $\gamma_{n+1} = d_{n+1} + p^{n+1}\sum_{k=n+2}^{+\infty} d_k/p^k$.

But $0 \le p^{n+1}\sum_{k=n+2}^{+\infty} d_k/p^k < p^{n+1}\sum_{k=n+2}^{+\infty}(p-1)/p^k = 1$.

Therefore $d_{n+1} = [\gamma_{n+1}] = c_{n+1}$, and we have proved by induction that $d_n = c_n$ for every natural integer n. We can summarize 2.1.1, 2.1.2, 2.1.3 by the following theorem.

Theorem 2.1   *Every positive real number γ_0 can be written, in a unique way, in the form*

$$\gamma_0 = \sum_{n=0}^{+\infty} \frac{c_n}{p^n} = c_0 + 0,c_1c_2c_3...,$$

where $c_0 = [\gamma_0]$, $0 \le c_n \le p-1$ for every $n \ge 1$ and $c_n \le p-2$ for infinitely many n. The integers c_0, c_1, c_2, ... can be computed by the algorithm (2.1). We call this expression of γ_0 its p-adic expansion.

2.1.4.   It is easy to know if γ_0 is rational or not by looking at its *p*-adic expansion.

Theorem 2.2   *A real positive number γ_0 is rational if, and only if, its p-adic expansion is ultimately periodic, that is if there exist two integers $N \ge 0$ and $k > 0$ such that $c_{n+k} = c_n$ for every $n \ge N$.*

Proof   Assume first that γ_0 is rational, $\gamma_0 = a/b$, $(a,b) \in \mathbb{N}^2$. Then $\gamma_0 = c_0 + \gamma_1/p$ implies that $\gamma_1 = p(a-c_0b)/b$. By induction, we see that $\gamma_n = pa_n/b$, with $a_n \in \mathbb{N}$, $0 \le \gamma_n < p$, and $0 \le a_n < b$. Hence the integers a_n can take only a *finite* number of values, namely 0, 1, 2, ..., $b-1$. Therefore there exist two *different* subscripts N and $N+k$ such that $a_N = a_{N+k}$ and, consequently, $\gamma_N = \gamma_{N+k}$. As these two numbers are equal, they have the same integral and fractional parts, whence $c_N = c_{N+k}$ and $\gamma_{N+1} = \gamma_{N+k+1}$. An easy induction shows now that $c_{N+n} = c_{N+k+n}$ for every $n \ge 0$. This proves that the *p*-adic expansion of γ_0 is periodic from rank N.

Conversely, assume that, for some $k > 0$, $c_{n+k} = c_n$ for every $n \ge N$. Then

$$\gamma_0 = \sum_{m=0}^{N-1} \frac{c_m}{p^m} + \sum_{m=0}^{+\infty} \left(\frac{c_N}{p^N} + \frac{c_{N+1}}{p^{N+1}} + \cdots + \frac{c_{N+k-1}}{p^{N+k-1}} \right) \frac{1}{p^{mk}}$$

$$= \sum_{m=0}^{N-1} \frac{c_m}{p^m} + \left(\frac{c_N}{p^N} + \frac{c_{N+1}}{p^{N+1}} + \cdots + \frac{c_{N+k-1}}{p^{N+k-1}} \right) \frac{p^k}{p^k - 1}.$$

Therefore $\gamma_0 \in \mathbb{Q}$, which completes the proof of theorem 2.2.

Example 2.1 Let $p \geq 2$ be an integer, and let $\theta = \sum_{n=0}^{+\infty} 1/p^{n^2}$. Then θ is irrational, because its p-adic expansion is defined by $c_n = 1$ if n is a square and $c_n = 0$ if not. The sequence c_n is not ultimately periodic, therefore θ is irrational by theorem 2.2. For another proof of the irrationality of θ, see exercise 1.7.

2.2 Expansion of a real number in Engel series

Let $x_0 \in \,]0,1]$. Consider the following algorithm:

$$\begin{cases} u_0 = \left[\dfrac{1}{x_0} \right] + 1 \\ x_1 = u_0 x_0 - 1 \end{cases} \begin{cases} u_1 = \left[\dfrac{1}{x_1} \right] + 1 \\ x_2 = u_1 x_1 - 1 \end{cases} \cdots \begin{cases} u_n = \left[\dfrac{1}{x_n} \right] + 1 \\ x_{n+1} = u_n x_n - 1 \end{cases} \tag{2.5}$$

One can prove the following theorem (see exercise 2.1 for details).

Theorem 2.3 *Every real number $x_0 \in \,]0,1]$ can be expanded, in a unique way, by using (2.5), as an Engel series*

$$x_0 = \frac{1}{u_0} + \frac{1}{u_0 u_1} + \frac{1}{u_0 u_1 u_2} + \cdots + \frac{1}{u_0 u_1 \cdots u_n} + \cdots,$$

where $u_n \in \mathbb{N}$, $u_n \geq 2$, u_n non decreasing.

Example 2.2 We know that $e - 1 = \dfrac{1}{1} + \dfrac{1}{1 \cdot 2} + \dfrac{1}{1 \cdot 2 \cdot 3} + \cdots + \dfrac{1}{n!} + \cdots$. This is the Engel series expansion of $e - 1$. Here $u_n = n + 1$.

Example 2.3 Let $q \in \mathbb{N}$, $q \geq 2$, $\alpha = \sum_{n=0}^{+\infty} 1/q^{n^2}$. Then $\alpha - 1 = \sum_{n=1}^{+\infty} 1/q^{n^2}$ is an Engel series, with $u_n = q^{2n+1}$.

Theorem 2.4   *Let $x_0 \in]0,1]$. Then x_0 is rational if, and only if, the sequence u_n of its expansion in Engel series is ultimately constant.*

We give here a proof of this theorem. A second one is proposed in the exercise 2.2. First assume that u_n is a constant from rank N. Then

$$x_0 = \sum_{n=0}^{N-1} \frac{1}{u_0 u_1 \ldots u_n} + \sum_{n=N}^{+\infty} \frac{1}{u_0 u_1 \ldots u_n} = \sum_{n=0}^{N-1} \frac{1}{u_0 u_1 \ldots u_n} + \frac{1}{u_0 u_1 \ldots u_N} \sum_{k=0}^{+\infty} \frac{1}{u_N^{\,k}}$$

$$= \sum_{n=0}^{N-1} \frac{1}{u_0 u_1 \ldots u_n} + \frac{1}{u_0 u_1 \ldots u_N} \times \frac{u_N}{u_N - 1} \in \mathbb{Q}.$$

Conversely, assume that x_0 is rational, $x_0 = A_0 / B_0$ with $A_0 \in \mathbb{N}$, $B_0 \in \mathbb{N}$. Dividing B_0 by A_0 yields $B_0 = A_0 Q_0 + R_0$, with $0 \le R_0 < A_0$. By using (2.6), we get

$$x_1 = u_0 x_0 - 1 = \left(\left[\frac{B_0}{A_0} \right] + 1 \right) \frac{A_0}{B_0} - 1 = \left(\left[Q_0 + \frac{R_0}{A_0} \right] + 1 \right) \frac{A_0}{B_0} - 1.$$

As $0 \le R_0 < A_0$, we have $0 \le R_0 / A_0 < 1$, so that

$$x_1 = (Q_0 + 1) \frac{A_0}{B_0} - 1 = \frac{Q_0 A_0 + A_0 - B_0}{B_0} = \frac{A_0 - R_0}{B_0}.$$

Therefore $x_1 = A_1 / B_0$, with $1 \le A_1 \le A_0$. Arguing the same way shows now that $x_n = A_n / B_0$, with $1 \le A_n \le A_{n-1}$ for every $n \ge 1$. But *a non increasing sequence of natural integers must be ultimately constant.* Therefore there exists $N \in \mathbb{N}$ such that $A_n = A_{n+1}$ for every $n \ge N$. Hence $x_{n+1} = x_n = x_N$ for $n \ge N$. By (2.5), this implies $x_N = u_{n+1} x_N - 1$ and $x_N = u_n x_N - 1$, whence $u_{n+1} = u_n$ for all $n \ge N$.

Example 2.4   The numbers $e = \sum_{n=0}^{+\infty} 1/n!$ and $\alpha = \sum_{n=0}^{+\infty} q^{-n^2}$ ($q \in \mathbb{N}$, $q \ge 2$) are irrational because the sequences of their expansions in Engel series tend to infinity.

2.3   Cantor infinite product

Let α_0 be a real number greater than 1. We put

$$q_0 = \left[\frac{\alpha_0}{\alpha_0 - 1} \right], \quad \alpha_1 = \frac{\alpha_0}{1 + 1/q_0}. \qquad (2.6)$$

By definition of q_0, we see that $q_0 \le \dfrac{\alpha_0}{\alpha_0 - 1} < q_0 + 1$.

Therefore $\alpha_0 < (q_0 + 1)(\alpha_0 - 1)$, whence $\alpha_0 > 1 + 1/q_0$. This implies that $\alpha_1 > 1$, and we can repeat the same process with α_1. Then we define the sequences (q_n) and (α_n) by the following algorithm.

$$q_n = \left[\frac{\alpha_n}{\alpha_n - 1} \right], \quad \alpha_{n+1} = \frac{\alpha_n}{1 + 1/q_n}. \tag{2.7}$$

An easy induction shows that

$$\alpha_0 = \left(1 + \frac{1}{q_0} \right)\left(1 + \frac{1}{q_1} \right) \dots \left(1 + \frac{1}{q_n} \right) \alpha_{n+1}. \tag{2.8}$$

Now, it can be proved (see exercice 2.3) that $\lim_{n \to \infty} \alpha_n = 1$, which yields the following result.

Theorem 2.5 *Every real number $\alpha_0 > 1$ can be expanded, in a unique way, as a Cantor infinite product*

$$\alpha_0 = \prod_{n=0}^{+\infty} \left(1 + \frac{1}{q_n} \right),$$

with $q_n \in \mathbb{N}$, $q_{n+1} \ge q_n^2$ for every $n \in \mathbb{N}$, $q_n \ge 2$ for n sufficiently large.

Example 2.5 Let $q_0 \in \mathbb{N} - \{0, 1\}$. We define the sequence (q_n) by $q_{n+1} = 2q_n^2 - 1$ for every $n \in \mathbb{N}$. As $q_{n+1} \ge q_n^2$ for every integer n, the infinite product $\prod_{n=0}^{+\infty} (1 + 1/q_n)$ is a Cantor infinite product, whose value can be computed. Indeed, we first check easily that

$$\left(1 + \frac{1}{q_n} \right)\left(1 + \frac{1}{q_{n+1}} \right) = \frac{1 - 1/q_{n+1}}{1 - 1/q_n}.$$

Therefore $\displaystyle\prod_{n=0}^{+\infty} \left(1 + \frac{1}{q_n} \right) \prod_{n=0}^{+\infty} \left(1 + \frac{1}{q_{n+1}} \right) = \frac{1}{1 - 1/q_0} = \frac{q_0}{q_0 - 1}$.

This can be written $\displaystyle\left(\prod_{n=0}^{+\infty} \left(1 + \frac{1}{q_n} \right) \right)^2 = \frac{q_0}{q_0 - 1}\left(1 + \frac{1}{q_0} \right) = \frac{q_0 + 1}{q_0 - 1}$.

Hence the expansion of $\sqrt{(q_0 + 1)/(q_0 - 1)}$ in Cantor infinite product is

$$\sqrt{\frac{q_0+1}{q_0-1}} = \prod_{n=0}^{+\infty}\left(1+\frac{1}{q_n}\right), \quad q_{n+1} = 2q_n^2 - 1. \qquad (2.9)$$

Theorem 2.6 Let $\alpha_0 \in \mathbb{R}$, $\alpha_0 > 1$. Then α_0 is rational if, and only if, the sequence (q_n) of its expansion in Cantor infinite product satisfies $q_{n+1} = q_n^2$ for every large n.

Proof See exercise 2.4.

Exemple 2.6 Choose $q_0 = 2$ and $q_0 = 3$ in (2.9). We obtain the expansions in Cantor infinite products of $\sqrt{3}$ and $\sqrt{2}$:

$$\sqrt{3} = \prod_{n=0}^{+\infty}\left(1+\frac{1}{q_n}\right), \quad q_0 = 2, \quad q_{n+1} = 2q_n^2 - 1.$$

$$\sqrt{2} = \prod_{n=0}^{+\infty}\left(1+\frac{1}{q_n}\right), \quad q_0 = 3, \quad q_{n+1} = 2q_n^2 - 1.$$

Hence $\sqrt{2}$ and $\sqrt{3}$ are irrational by theorem 2.6. Evidently, there is a much simpler proof (see section 1.1). These expansions are however interesting because they converge very fast.

Exercises

2.1   Proof of theorem 2.3

1) Prove that $x_n > 0$ for every $n \in \mathbb{N}$, and that (x_n) is not increasing.

2) Prove that (u_n) is not decreasing and that $u_n \geq 2$ for every $n \in \mathbb{N}$.

3) Prove that $x_0 = \sum_{n=0}^{+\infty} 1/(u_0 u_1 ... u_n)$.

4) Prove this expansion to be unique when $u_n \geq 2$, u_n non decreasing.

2.2   Let the Engel series $x_0 = \sum_{n=0}^{+\infty} 1/(u_0 u_1 ... u_n)$, with $u_n \in \mathbb{N} - \{0,1\}$, $\lim_{n \to +\infty} u_n = +\infty$. Prove that x_0 is irrational by using a diophantine approximation (theorem 1.5).

2.3   Proof of theorem 2.5

1) Prove that $q_{n+1} \geq q_n^2$ for every $n \in \mathbb{N}$.

2) Prove that there exists an integer N such that $q_N \geq 2$.

3) Deduce that $\alpha_0 = \prod_{n=0}^{+\infty} (1 + 1/q_n)$.

4) Prove the uniqueness of the expansion in Cantor infinite product, that is: if the sequence q_n' of integers satisfies $q_{n+1}' \geq q_n'^2$ for every $n \in \mathbb{N}$ and if $\alpha_0 = \prod_{n=0}^{+\infty} (1 + 1/q_n')$, then $q_n' = q_n$ for every $n \in \mathbb{N}$.

2.4   Proof of theorem 2.6

1. Prove that, for all $x \in \mathbb{R}$, $(1-x) \prod_{k=0}^{n-1} (1 + x^{2^k}) = 1 - x^{2^n}$.

2. Deduce that $\prod_{k=0}^{+\infty} (1 + x^{2^k}) = 1/(1-x)$ if $x < 1$, then check that α_0 is rational if $q_{n+1} = q_n^2$ for n sufficiently large.

3. Assume that $\alpha_0 \in \mathbb{Q}$. Prove that $q_{n+1} = q_n^2$ for n sufficiently large.

2.5   Compute the p-adic expansion of $\alpha = \dfrac{1}{p-1}$.

2.6   Compute the 7-adic expansion of $\beta = \dfrac{35}{9}$.

2.7   By using an expansion in Engel series, prove that $e^{\sqrt{2}} \notin \mathbb{Q}$.

2.8   Stratemeyer's formula

Let the sequence (t_n) be defined by its first term $t_0 \in \mathbb{N} - \{0,1\}$ and by the recurrence relation $t_{n+1} = 2t_n^2 - 1$.

1. Prove that $t_n - \sqrt{t_n^2 - 1} = \dfrac{1}{2t_n} + \dfrac{1}{2t_n}\left(t_{n+1} - \sqrt{t_{n+1}^2 - 1}\right)$.

2. Prove *Stratemeyer's formula*, which gives the expansion in Engel series of $t_0 - \sqrt{t_0^2 - 1}$:

$$t_0 - \sqrt{t_0^2 - 1} = \sum_{n=0}^{+\infty} \frac{1}{2t_0 2t_1 \ldots 2t_n}.$$

3) Deduce expressions of $\sqrt{2}, \sqrt{3}, \sqrt{5}$ and $\sqrt{7}$ by means of Engel series, and the irrationality of these numbers.

2.9   Expansion in Sylvester series

Let $\gamma \in]0,1]$. First we define $\alpha_0 = \dfrac{1}{\gamma}$, $q_0 = [\alpha_0] + 1$, $\dfrac{1}{\alpha_0} = \dfrac{1}{q_0} + \dfrac{1}{\alpha_1}$.

Then $q_1 = [\alpha_1] + 1$, $\dfrac{1}{\alpha_1} = \dfrac{1}{q_1} + \dfrac{1}{\alpha_2}$, and so on.

1) Prove that $q_{n+1} \geq q_n^2 - q_n + 1$.

2) Prove that every real number $\gamma \in]0,1]$ can be expanded, in a unique way, as a *Sylvester series*

$$\gamma = \frac{1}{q_0} + \frac{1}{q_1} + \cdots + \frac{1}{q_n} + \cdots,$$

with $q_{n+1} \geq q_n^2 - q_n + 1$ for every $n \in \mathbb{N}$.

3) Prove that γ is rational if, and only if, the sequence q_n of its expansion in Sylvester series satisfies $q_{n+1} = q_n^2 - q_n + 1$ for any n sufficiently large.

4) Prove that the number

$$\sum_{n=0}^{+\infty} \frac{1}{2^{2^n} - a}$$

is irrational for every integer a which is not of the form 2^{2^n}.

2.10   A striking number

With the exception of rational numbers, there are few numbers α whose decimal expansion is known, together with the decimal expansion of their inverse $1/\alpha$. The purpose of this exercise is to provide an example of such a number (Blanchard and Mendès-France, 1982).

Let $E = \{0, 1, 4, 5, 16, 17, 20, 21, 64, \ldots\}$ be the set of all natural numbers whose expression in base 4 uses only the digits 0 and 1.

Let $2E = \{0, 2, 8, 10, 32, 34, 40, 42, \ldots\}$ be the set of all natural numbers whose expression in base 4 uses only the digits 0 and 2.

Define $\alpha = 3 \sum_{n \in E} 10^{-n}$.

1) Compute the first 25 digits of the decimal expansion of α.

2) Prove that α is irrational.

3) Let $f(x) = \displaystyle\prod_{n=0}^{+\infty} \left(1 + x^{4^n}\right)$, which exists if $|x| < 1$.

Find the representation of $f(x)$ as a power series.

4) Let $g(x) = \prod_{n=0}^{+\infty} \left(1 + x^{2.4^n}\right)$, which exists if $|x| < 1$.

Compute $f(x) g(x)$.

5) Compute the decimal expansion of $\dfrac{1}{\alpha}$.

Chapter 3

Continued fractions

This chapter is devoted to the study of general continued fractions. The important special case of regular continued fractions will be considered later in chapter 4. In section 3.1, we first define continued fractions and present their basic properties. This section is sufficient for the reader who would like to go directly to chapter 4. Sections 3.2, 3.3, 3.4 are devoted to the proof of general theorems about the convergence of continued fractions in $\mathbb{C}$ and in the Riemann sphere $\Sigma = \mathbb{C} \cup \{\infty\}$, which is the natural frame for studying this problem. These theorems are then used in section 3.5, in order to expand a quotient $J_\nu / J_{\nu-1}$ of contiguous Bessel functions into a continued fraction. At last, in 3.6, we prove an irrationality criterion for continued fractions with integer coefficients, and apply it to the quotient $J_\nu / J_{\nu-1}$ and the irrationality of π.

3.1 Introduction

Let $(a_n)_{n\geq 0}$ and $(b_n)_{n\geq 1}$ be two sequences of complex numbers, with $b_n \neq 0$ for every $n \geq 1$. We consider the following fractions:

$$R_0 = a_0, \quad R_1 = a_0 + \frac{a_1}{b_1}, \quad R_2 = a_0 + \cfrac{a_1}{b_1 + \cfrac{a_2}{b_2}}, \quad R_3 = a_0 + \cfrac{a_1}{b_1 + \cfrac{a_2}{b_2 + \cfrac{a_3}{b_3}}}, \quad \dots$$

More generally, we get R_n by replacing, in R_{n-1}, b_{n-1} by $b_{n-1} + a_n / b_n$. For the sake of simplicity, we will make use of the following notation.

$$R_1 = a_0 + \frac{a_1}{b_1}, \quad R_2 = a_0 + \frac{a_1}{b_1} + \frac{a_2}{b_2}, \quad R_3 = a_0 + \frac{a_1}{b_1} + \frac{a_2}{b_2} + \frac{a_3}{b_3}, \quad \dots$$

Hence, for every integer n,

$$R_n = a_0 + \frac{a_1}{b_1} \frac{a_2}{+ b_2} + \cdots \frac{a_{n-1}}{b_{n-1}} \frac{a_n}{+ b_n} = a_0 + \frac{a_1}{b_1} \frac{a_2}{+ b_2} + \cdots \frac{a_{n-1}}{b_{n-1} + \dfrac{a_n}{b_n}}. \qquad (3.1)$$

Due to the divisions, it can occur that some terms in the sequence R_n are not defined. However, we will assume here that there exists an integer N such that R_n exists for every $n \geq N$. Then the sequence $(R_n)_{n \geq N}$ defines a *continued fraction*. The numbers R_n are called the *convergents* to this continued fraction. Indeed, we will say that the continued fraction is *convergent* if the sequence R_n is convergent and *divergent* if it is not. In case of convergence, the limit R of R_n is called the *value* of the continued fraction, and we will write

$$R = a_0 + \frac{a_1}{b_1} \frac{a_2}{+ b_2} + \cdots \frac{a_n}{b_n} + \cdots \qquad (3.2)$$

The following theorem is fundamental.

Theorem 3.1 *For every $n \geq 0$, we have $R_n = P_n / Q_n$, where P_n and Q_n are defined for $n \geq -1$ and satisfy*

$$P_{-1} = 1, \quad Q_{-1} = 0, \quad P_0 = a_0, \quad Q_0 = 1,$$

and for every $n \geq 1$

$$\begin{cases} P_n = b_n P_{n-1} + a_n P_{n-2} \\ Q_n = b_n Q_{n-1} + a_n Q_{n-2} \end{cases} \qquad (3.3)$$

Proof Observe the first terms of sequence R_n. First $R_0 = a_0 = P_0 / Q_0$, with $P_0 = a_0$, $Q_0 = 1$. Next $R_1 = a_0 + a_1 / b_1 = P_1 / Q_1$, with $P_1 = a_0 b_1 + a_1$ and $Q_1 = b_1$. Hence (3.3) is true for $n = 1$.

Now replace, in R_1, b_1 by $b_1 + a_2 / b_2$. We obtain

$$R_2 = \frac{a_0 \left(b_1 + \dfrac{a_2}{b_2} \right) + a_1}{b_1 + a_2 b_2} = \frac{a_0 b_1 b_2 + a_0 a_2 + a_1 b_2}{b_1 b_2 + a_2} = \frac{P_2}{Q_2},$$

where $P_2 = b_2 (a_0 b_1 + a_1) + a_2 a_0 = b_2 P_1 + a_2 P_0$, $Q_2 = b_2 b_1 + a_2 = b_2 Q_1 + a_2 Q_0$. Hence (3.3) is also true for $n = 2$. Now we prove (3.3) by induction. We can write

$$R_{n+1} = a_0 + \frac{a_1}{b_1} \frac{a_2}{+ b_2} \frac{a_n}{+ \cdots b_n} \frac{a_{n+1}}{+ b_{n+1}} = a_0 + \frac{a_1}{b_1} \frac{a_{n-1}}{+ \cdots b_{n-1}} \frac{a_n}{+ b_n + \frac{a_{n+1}}{b_{n+1}}} = \frac{P'_n}{Q'_n}.$$

Here R_{n+1} can be seen as a convergent of order n, where b_n is replaced by $b_n + a_{n+1} / b_{n+1}$. Hence, by induction hypothesis,

$$\begin{cases} P'_n = (b_n + a_{n+1}/b_{n+1}) P_{n-1} + a_n P_{n-2} \\ Q'_n = (b_n + a_{n+1}/b_{n+1}) Q_{n-1} + a_n Q_{n-2} \end{cases}.$$

This implies that

$$\begin{cases} b_{n+1} P'_n = b_{n+1} (b_n P_{n-1} + a_n P_{n-2}) + a_{n+1} P_{n-1} \\ b_{n+1} Q'_n = b_{n+1} (b_n Q_{n-1} + a_n Q_{n-2}) + a_{n+1} Q_{n-1} \end{cases}.$$

By using again the induction hypothesis, we obtain

$$\begin{cases} b_{n+1} P'_n = b_{n+1} P_n + a_{n+1} P_{n-1} \\ b_{n+1} Q'_n = b_{n+1} Q_n + a_{n+1} P_{n-1} \end{cases}$$

Now define $P_{n+1} = b_{n+1} P'_n$ and $Q_{n+1} = b_{n+1} Q'_n$. We get $R_{n+1} = P_{n+1}/Q_{n+1}$, which proves theorem 3.1.

Remark 3.1 The n-th convergent R_n exists only if $Q_n \neq 0$.

Theorem 3.2 *For every $n \geq 1$, we have*

$$P_n Q_{n-1} - P_{n-1} Q_n = (-1)^{n-1} a_1 a_2 \cdots a_n \qquad (3.4)$$

$$R_n - R_{n-1} = \frac{(-1)^{n-1} a_1 a_2 \cdots a_n}{Q_n Q_{n-1}} \quad \text{if } Q_n Q_{n-1} \neq 0 \qquad (3.5)$$

The proof of theorem 3.2 is left as an exercise (see exercise 3.2). As a consequence, we get

Theorem 3.3 *The continued fraction* $a_0 + \dfrac{a_1}{b_1} \dfrac{a_2}{+ b_2} \dfrac{a_n}{+ \cdots b_n}$ *is convergent*

if, and only if, the series $\sum_{n=1}^{+\infty} (-1)^{n-1} a_1 a_2 \ldots a_n / Q_n Q_{n-1}$ *is convergent. In this case,*

$$a_0 + \frac{a_1}{b_1} \frac{a_2}{+ b_2} \frac{a_n}{+ \cdots b_n +\cdots} = a_0 + \sum_{n=1}^{+\infty} (-1)^{n-1} \frac{a_1 a_2 \ldots a_n}{Q_n Q_{n-1}}.$$

Proof   See exercise 3.2.

An interesting consequence of theorem 3.3 is the following result, which allows transformation of a series into a continued fraction.

Theorem 3.4   *Assume that the series* $\sum_{n=0}^{+\infty} a_0 a_1 \ldots a_n / b_0 b_1 \ldots b_n$ *is convergent, and that* $a_n + b_n \neq 0$ *for every* $n \geq 1$. *Then the continued fraction* $\dfrac{a_0}{b_0} \dfrac{a_1 b_0}{-a_1 + b_1} \dfrac{a_2 b_1}{-a_2 + b_2} - \cdots$ *is convergent and*

$$\sum_{n=0}^{+\infty} \frac{a_0 a_1 \ldots a_n}{b_0 b_1 \ldots b_n} = \frac{a_0}{b_0} \frac{a_1 b_0}{-a_1 + b_1} \frac{a_2 b_1}{-a_2 + b_2} - \cdots \frac{a_n b_{n-1}}{a_n + b_n} - \cdots.$$

Proof   See exercise 3.3.

Example 3.1   We have $e^{-1} = \sum_{n=0}^{+\infty} (-1)^n / n! = \sum_{n=0}^{+\infty} a_0 a_1 \ldots a_n / b_0 b_1 \ldots b_n$, with $a_0 = 1$, $a_n = -1$ for $n \geq 1$, and $b_n = n + 2$. Hence

$$\begin{cases} \dfrac{1}{e} = \dfrac{1}{2} \dfrac{1}{+2} \dfrac{2}{+3} \dfrac{3}{+\cdots} \dfrac{n}{n} + \cdots \\ e = 2 + \dfrac{2}{2} \dfrac{3}{+3} \dfrac{n}{+\cdots n} + \cdots \end{cases} \tag{3.6}$$

Example 3.2   By Leibniz formula we know that

$$\frac{\pi}{4} = \sum_{n=0}^{+\infty} \frac{(-1)^n}{2n+1} = 1 + \sum_{n=1}^{+\infty} \frac{(-1)(-3)(-5) \cdots (1 - 2n)}{3 \cdot 5 \cdots (2n+1)} \tag{3.7}$$

Therefore we can apply theorem 3.4 with $a_0 = 1$, $b_0 = 1$, and $a_n = 1 - 2n$, $b_n = 2n + 1$ for $n \geq 1$. We obtain *Brouncker's formula*:

$$\frac{\pi}{4} = \frac{1}{1} + \frac{1^2}{2} + \frac{3^2}{2} + \frac{5^2}{2} + \cdots \tag{3.8}$$

3.2   A criterion of convergence

Theorem 3.3 is not very useful, in practice, for studying the convergence of a continued fraction. Indeed, the sequence Q_n is not explicitly known, which makes it difficult to study the convergence of the series $\sum_{n=1}^{+\infty} (-1)^{n-1} a_1 a_2 \ldots a_n / Q_n Q_{n-1}$.

We now look for a more practical criterion of convergence. We assume, without loss of generality, that $a_0 = 0$, and observe that any fraction

$$\frac{a_1}{b_1} + \frac{a_2}{b_2} + \cdots \frac{a_n}{b_n} + \cdots \quad \text{can also be written in the form} \quad \frac{1}{c_1} \cdot \frac{1}{c_2} + \cdots + \frac{1}{c_n} + \cdots.$$

Indeed, we first have $R_1 = a_1/b_1 = 1/c_1$, with $c_1 = b_1/a_1$. Then we see that

$$R_2 = \frac{a_1}{b_1 + a_2/b_2} = \frac{1}{b_1/a_1 + a_2/(a_1 b_2)} = \frac{1}{c_1} \frac{1}{+ c_2}, \quad \text{with } c_2 = \frac{a_1 b_2}{a_2}.$$

We obtain R_3 by replacing, in R_2, b_2 by $b_2 + a_3/b_3$, whence

$$R_3 = \frac{1}{c_1 + \dfrac{a_1}{a_2}\left(b_2 + \dfrac{a_3}{b_3}\right)} = \frac{1}{c_1} + \frac{1}{c_2 + \dfrac{a_1 a_3}{a_2 b_3}} = \frac{1}{c_1} \frac{1}{+ c_2} \frac{1}{+ c_3}, \quad \text{with } c_3 = \frac{a_2 b_3}{a_1 a_3}.$$

More generally, it is easy to prove by induction the following

Theorem 3.5 *Let $(a_n)_{n\geq 1}$ and $(b_n)_{n\geq 1}$ be two sequences of non-zero complex numbers. Then the continued fractions*

$$\frac{a_1}{b_1} + \frac{a_2}{b_2} + \cdots \frac{a_n}{b_n} + \cdots \quad \text{and} \quad \frac{1}{c_1} \frac{1}{+ c_2} + \cdots \frac{1}{c_n} + \cdots,$$

where $c_{2n} = \dfrac{a_1 a_3 \ldots a_{2n-1}}{a_2 a_4 \ldots a_{2n}} b_{2n}$ *and* $c_{2n+1} = \dfrac{a_2 a_4 \ldots a_{2n}}{a_1 a_3 \ldots a_{2n+1}} b_{2n+1}$, *have the same convergents R_n. We say that they are equivalent. Therefore they are both convergent or divergent and, if convergent, have the same value.*

However, for the second type of continued fraction, we have the following elementary convergence criterion.

Theorem 3.6 *Let $(c_n)_{n\geq 1}$ be a sequence of complex numbers such that $|c_n| \geq 2$ for every $n \geq 1$, and such that the series $\sum_{n=1}^{+\infty} 1/|c_n c_{n+1}|$ is convergent. Then the continued fraction $\dfrac{1}{c_1} \dfrac{1}{+ c_2} \dfrac{1}{+ c_n} + \cdots$ is convergent.*

Proof We know by theorem 3.1 that $P_0 = 0$, $P_1 = 1$, $P_n = c_n P_{n-1} + P_{n-2}$. We first prove by induction that $|P_n| \geq |P_{n-1}|$. This is true for $n = 1$. Now $|P_{n+1}| = |c_{n+1} P_n + P_{n-1}| \geq ||c_{n+1} P_n| - |P_{n-1}||$. By the induction hypothesis,

$|P_n| \geq |P_{n-1}|$. As $|c_{n+1}| \geq 2$, we get $|P_{n+1}| \geq 2|P_n| - |P_n|$, whence $|P_{n+1}| \geq |P_n|$. In particular, $|P_{n+1}| \geq |P_1| = 1$, whence $P_n \neq 0$ for every $n \geq 1$. By arguing the same way, we see that $Q_n \neq 0$ for every $n \geq 1$. This proves that the convergents R_n are well defined for every $n \geq 1$.

Now we put, for every $n \geq 1$,

$$u_n = \frac{c_n P_{n-1}}{P_n}, \quad v_n = \frac{c_n Q_{n-1}}{Q_n}. \tag{3.9}$$

As $P_n = c_n P_{n-1} + P_{n-2}$ and $Q_n = c_n Q_{n-1} + Q_{n-2}$, we can write, for $n \geq 2$,

$$P_n = c_n \left(1 + \frac{u_{n-1}}{c_{n-1} c_n} \right) P_{n-1}, \quad Q_n = c_n \left(1 + \frac{v_{n-1}}{c_{n-1} c_n} \right) Q_{n-1}. \tag{3.10}$$

Hence, as $P_1 = 1$ and $Q_1 = c_1$,

$$\begin{cases} P_n = c_2 c_3 \ldots c_n \prod_{k=1}^{n-1} \left(1 + \frac{u_k}{c_k c_{k+1}} \right) & (n \geq 2) \\ \\ Q_n = c_1 c_2 \ldots c_n \prod_{k=1}^{n-1} \left(1 + \frac{v_k}{c_k c_{k+1}} \right) & (n \geq 2) \end{cases} \tag{3.11}$$

But by (3.9) and (3.10) we have

$$u_n = \frac{1}{1 + \dfrac{u_{n-1}}{c_{n-1} c_n}}, \quad v_n = \frac{1}{1 + \dfrac{v_{n-1}}{c_{n-1} c_n}}. \tag{3.12}$$

As $u_1 = 0$ and $v_1 = 1$, an easy induction shows that $|u_n| \leq 2$ and $|v_n| \leq 2$ for every $n \geq 1$. Therefore the series

$$\sum_{k=1}^{+\infty} \left| \frac{u_k}{c_k c_{k+1}} \right| \leq 2 \sum_{k=1}^{+\infty} \frac{1}{|c_k c_{k+1}|}, \quad \sum_{k=1}^{+\infty} \left| \frac{v_k}{c_k c_{k+1}} \right| \leq 2 \sum_{k=1}^{+\infty} \frac{1}{|c_k c_{k+1}|}$$

are convergent, and so are the infinite products

$$\beta = \prod_{k=1}^{+\infty} \left(1 + \frac{u_k}{c_k c_{k+1}} \right), \quad \gamma = \prod_{k=1}^{+\infty} \left(1 + \frac{v_k}{c_k c_{k+1}} \right)$$

(see exercise 3.4). Moreover, $\beta \neq 0$ and $\gamma \neq 0$ since $1 + u_k/(c_k c_{k+1}) \neq 0$ and $1 + v_k/(c_k c_{k+1}) \neq 0$ for every $k \geq 1$. Indeed, $|u_k/(c_k c_{k+1})| \leq 1/2$ and $|v_k/(c_k c_{k+1})| \leq 1/2$ (see exercise 3.4, question 2). Finally, we see by (3.11) that P_n/Q_n converges to $\beta/(c_1 \gamma)$, which proves theorem 3.6.

Example 3.3 The continued fraction $\dfrac{1}{1+}\dfrac{1}{2+}\cdots\dfrac{1}{n+}\cdots$ is convergent by theorem 3.6. Its value can be expressed in terms of Bessel functions (see section 3.5).

3.3 How to divide by zero

We consider a continued fraction

$$\frac{a_1}{b_1 +}\frac{a_2}{b_2 +}\frac{a_3}{b_3 +}\cdots\frac{a_n}{b_n +}\cdots, \qquad (3.13)$$

where $a_n \in \mathbb{C}^*$ and $b_n \in \mathbb{C}^*$ for every $n \geq 1$. Assume this continued fraction to be convergent, with value 0. Then the continued fraction

$$b_1 + \frac{a_2}{b_2 +}\frac{a_3}{b_3 +}\cdots\frac{a_n}{b_n +}\cdots \qquad (3.14)$$

is *divergent*, and its convergents R_n satisfy $\lim_{n \to +\infty}|R_n| = +\infty$. Hence the continued fraction

$$b_2 + \frac{a_3}{b_3 +}\cdots\frac{a_n}{b_n +}\cdots \qquad (3.15)$$

converges and its value is 0. And so on...

Taking this into account, we will add to $\mathbb{C}$ an element, denoted by ∞, satisfying the following properties:

$$\begin{cases} \forall a \in \mathbb{C}, \quad a + \infty = \infty. \\[2mm] \forall a \in \mathbb{C}, \quad \dfrac{a}{\infty} = 0. \\[2mm] \forall a \in \mathbb{C}^*, \quad \dfrac{a}{0} = \infty. \end{cases} \qquad (3.16)$$

It is evidently forbidden to divide zero by zero.

We denote $\Sigma = \mathbb{C} \cup \{\infty\}$. It is convenient to visualize Σ as the *Riemann sphere* (figure 3.1, page 28). The straight line which links the north pole of Σ to the complex number z defines the representant Z of z in Σ. The north pole of Σ is ∞. If $|z| \to +\infty$, its representant $Z \to \infty$ in Σ. The south pole of Σ is 0.

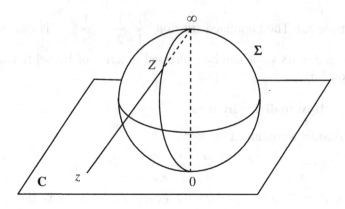

Figure 3.1

We will say that a continued fraction *is convergent in* Σ if its convergents R_n have a limit in Σ. Hence, the continued fractions (3.13), (3.14) and (3.15) are convergent in Σ and

$$\begin{cases} \cfrac{a_1}{b_1} + \cfrac{a_2}{b_2} + \cfrac{a_3}{b_3} + \cdots \cfrac{a_n}{b_n} + \cdots = 0 \\[2mm] b_1 + \cfrac{a_2}{b_2} + \cfrac{a_3}{b_3} + \cdots \cfrac{a_n}{b_n} + \cdots = \infty \\[2mm] b_2 + \cfrac{a_3}{b_3} + \cdots \cfrac{a_n}{b_n} + \cdots = 0 \; .... \end{cases}$$

It should be noted that, if $a_n \neq 0$ for every $n \geq 1$, the convergents R_n are well defined in Σ. Indeed, as $P_n Q_{n-1} - P_{n-1} Q_n = (-1)^{n-1} a_1 a_2 \ldots a_n$ (theorem 3.2), P_n and Q_n cannot both be zero. Hence, if $Q_n = 0$, then $R_n = \infty$.

3.4   Expansion in continued fraction in Σ

Consider first a continued fraction $\alpha_1 = \cfrac{1}{c_1} + \cfrac{1}{c_2} + \cdots \cfrac{1}{c_n} + \cdots$ ($c_n \in \mathbb{C}^*$ for

$n \geq 1$), convergent in Σ. If we put $\alpha_n = \cfrac{1}{c_n} + \cfrac{1}{c_{n+1}} + \cdots$, we see that

$$\alpha_n = \frac{1}{c_n + \alpha_{n+1}} \quad \text{for every } n \geq 1. \tag{3.17}$$

Conversely, consider a sequence $(\alpha_n)_{n\geq 1}$ in Σ satisfying (3.17), where the c_n's are non zero complex numbers. Then

$$\alpha_1 = \frac{1}{c_1+\alpha_2} = \frac{1}{c_1} \frac{1}{+c_2+\alpha_3} = \frac{1}{c_1} \frac{1}{+c_2} \frac{1}{+c_3+\alpha_4} = \frac{1}{c_1} \frac{1}{+c_2} \frac{1}{+\cdots c_{n-1}} \frac{1}{+c_n+\alpha_{n+1}}.$$

Hence it is natural to ask if $\alpha_1 = \dfrac{1}{c_1} \dfrac{1}{+c_2} \dfrac{1}{+\cdots c_n} + \cdots$.

Theorem 3.7 *Let $(c_n)_{n\geq 1}$ be a sequence of non zero complex numbers and let $(\alpha_n)_{n\geq 1}$ be a sequence in Σ satisfying (3.17). Assume there exists an integer N such that $|c_n| \geq 2$ and $\alpha_n \in \mathbb{C}$ for every $n \geq N$. Moreover, assume that $\lim_{n\to+\infty} \alpha_{n+1}/c_n = 0$ and that the series $\sum_{n=1}^{+\infty} 1/|c_n c_{n+1}|$ is convergent. Then $\alpha_1 = \dfrac{1}{c_1} \dfrac{1}{+c_2} \dfrac{1}{+\cdots c_n} + \cdots$.*

Proof By (3.17) we have

$$\alpha_1 = \frac{1}{c_1} \frac{1}{+c_2} \frac{1}{+\cdots c_{N-1}} \frac{1}{+c_N+\alpha_{N+1}}. \tag{3.18}$$

But by (3.17) we have also

$$\alpha_{N+1} = \frac{1}{c_{N+1}} \frac{1}{+c_{N+2}} \frac{1}{+\cdots c_{N+n}} \frac{1}{+\alpha_{N+n+1}} = \frac{1}{c_{N+1}} \frac{1}{+\cdots c_{N+n}} \frac{\alpha_{N+n+1}}{+\quad 1} = \frac{P'_n}{Q'_n}.$$

Hence α_{N+1} can be considered a convergent of order $n+1$. By theorem 3.1, we can write

$$\begin{cases} P'_n = P_n + \alpha_{N+n+1}P_{n-1} \\ Q'_n = Q_n + \alpha_{N+n+1}Q_{n-1} \end{cases}$$

where $R_n = P_n/Q_n$ is the convergent of order n of de $\dfrac{1}{c_{N+1}} \dfrac{1}{+c_{N+2}} + \cdots$.

Therefore $\alpha_{N+1} = \dfrac{P_n}{Q_n} \cdot \dfrac{1+\alpha_{n+N+1}\cdot\dfrac{P_{n-1}}{P_n}}{1+\alpha_{n+N+1}\cdot\dfrac{Q_{n-1}}{Q_n}}.$

But relations (3.10) and the proof of theorem 3.6 show that $\dfrac{P_{n-1}}{P_n} \underset{+\infty}{\sim} \dfrac{1}{c_{n+N}}$

and $\dfrac{Q_{n-1}}{Q_n} \underset{+\infty}{\sim} \dfrac{1}{c_{n+N}}$. Hence $\alpha_{N+1} = \lim\limits_{n\to+\infty} \dfrac{P_n}{Q_n} = \dfrac{1}{c_{N+1}} + \dfrac{1}{c_{N+2}} + \cdots$.

Replacing in (3.18) proves theorem 3.7.

3.5   A quotient of Bessel functions

In what follows, we will assume for the sake of simplicity that v is a real number. We first recall some basic results about Bessel functions. For every $v > 0$, the *gamma function* is defined by

$$\Gamma(v) = \int_0^{+\infty} t^{v-1} e^{-t}\, dt. \tag{3.19}$$

The gamma function satisfies the functional relation

$$\Gamma(v+1) = v\Gamma(v). \tag{3.20}$$

This can be easily seen by using an integration by parts in (3.19). In particular, for every $n \in \mathbb{N}$,

$$\Gamma(n+1) = n!$$

Finally, one can compute (see exercise 3.5)

$$\Gamma\left(\tfrac{1}{2}\right) = \sqrt{\pi}.$$

Now let $x \in \mathbb{C}$. We put $x = re^{i\theta}$, with $-\pi < \theta \le \pi$, and $x^v = r^v e^{iv\theta}$. The *Bessel function* J_v is defined for every $v > -1$ and every $x \in \mathbb{C}$ by

$$J_v(x) = \left(\frac{x}{2}\right)^v \sum_{k=0}^{+\infty} \frac{(-1)^k}{k!\,\Gamma(v+k+1)} \left(\frac{x}{2}\right)^{2k}. \tag{3.21}$$

Especially, we have (exercise 3.6)

$$\begin{cases} J_{\frac{1}{2}}(x) = \sqrt{\dfrac{2}{\pi}} x^{-\frac{1}{2}} \sin x \\[2mm] J_{-\frac{1}{2}}(x) = \sqrt{\dfrac{2}{\pi}} x^{-\frac{1}{2}} \cos x \end{cases} \tag{3.22}$$

The Bessel function satisfies (exercise 3.7) the differential equation

$$x^2 y'' + xy' + (x^2 - v^2) y = 0. \tag{3.23}$$

Moreover, three contiguous Bessel functions (that is Bessel functions whose parameters differ from ± 1) satisfy the relation

$$J_{\nu-1}(x) + J_{\nu+1}(x) = \frac{2\nu}{x} J_\nu(x) \quad \text{for } x \neq 0 \text{ and } \nu > 0. \quad (3.24)$$

For the proof of (3.24), see exercise 3.8. Relation (3.24) allows to get an expansion in continued fraction of $J_{\nu-1}(x)/J_\nu(x)$, as follows.

First we divide both sides of (3.24) by $J_\nu(x)$, and obtain

$$\frac{J_{\nu-1}(x)}{J_\nu(x)} = \frac{2\nu}{x} - \frac{J_{\nu+1}(x)}{J_\nu(x)}.$$

This result is always true in Σ, because $J_\nu(x)$ and $J_{\nu-1}(x)$ are not both zero when $x \neq 0$ (exercise 3.9). Hence

$$\frac{J_\nu(x)}{J_{\nu-1}(x)} = \frac{1}{\dfrac{2\nu}{x} - \dfrac{J_{\nu+1}(x)}{J_\nu(x)}}. \quad (3.25)$$

Put $\alpha_n(x) = i J_{\nu+n-1}(x)/J_{\nu+n-2}(x)$. Then (3.25) yields

$$\alpha_n(x) = \frac{1}{\dfrac{2(\nu+n-1)}{ix} + \alpha_{n+1}(x)}. \quad (3.26)$$

But we have $J_{\nu+n}(x) \underset{n \to +\infty}{\sim} \left(\dfrac{x}{2}\right)^{\nu+n} \dfrac{1}{\Gamma(\nu+n+1)}$ by exercise 3.9. Therefore $J_{\nu+n-1}(x) \neq 0$ for n sufficiently large, and consequently $\alpha_n(x) \in \mathbb{C}$ for n sufficiently large. Applying theorem 3.7 with $c_n = 2(\nu+n-1)/ix$ yields

$$\frac{iJ_\nu(x)}{J_{\nu-1}(x)} = \frac{1}{\dfrac{2\nu}{ix}} + \frac{1}{\dfrac{2(\nu+1)}{ix}} + \frac{1}{\dfrac{2(\nu+2)}{ix}} + \frac{1}{\dfrac{2(\nu+3)}{ix}} + \cdots \quad (3.27)$$

We can simplify this expression by arguing as in theorem 3.5. First we multiply numerator and denominator of the continued fraction by ix.

$$\frac{iJ_\nu(x)}{J_{\nu-1}(x)} = \frac{ix}{2\nu} + \frac{ix}{2(\nu+1)} + \frac{1}{\dfrac{2(\nu+2)}{ix}} + \frac{1}{\dfrac{2(\nu+3)}{ix}} + \cdots$$

Then we do the same to the second term (the continued fraction in the denominator after 2ν) and obtain

$$\frac{iJ_\nu(x)}{J_{\nu-1}(x)} = \frac{ix}{2\nu} - \frac{x^2}{2(\nu+1)} + \frac{ix}{2(\nu+2)} + \frac{1}{\dfrac{2(\nu+3)}{ix}} + \cdots$$

And so on... Finally,

$$\frac{iJ_\nu(x)}{J_{\nu-1}(x)} = \frac{ix}{2\nu} - \frac{x^2}{2(\nu+1)} - \frac{x^2}{2(\nu+2)} - \cdots \frac{x^2}{2(\nu+n)} - \cdots \qquad (3.28)$$

This result remains true if $x=0$, because in this case the both members of (3.28) are equal to zero. If we divide both sides of (3.28) by i, we obtain an expansion of $J_{\nu-1}(x)/J_\nu(x)$ into a continued fraction:

$$\frac{J_\nu(x)}{J_{\nu-1}(x)} = \frac{x}{2\nu} - \frac{x^2}{2(\nu+1)} - \frac{x^2}{2(\nu+2)} - \cdots \frac{x^2}{2(\nu+n)} - \cdots \qquad (3.29)$$

This result holds in Σ, for every $x \in \mathbb{C}$ and $\nu > 0$.

For $\nu = 1/2$, by taking (3.22) into account, we obtain *Lambert continued fraction*. For every $x \in \mathbb{C}$,

$$\tan x = \frac{x}{1} - \frac{x^2}{3} - \frac{x^2}{5} - \cdots \frac{x^2}{2n+1} - \cdots \qquad (3.30)$$

Remark 3.2   Taking $\nu = 1$ and $x = -2i$ in (3.27) yields

$$\frac{1}{1} + \frac{1}{2} + \cdots + \frac{1}{n} + \cdots = \frac{iJ_1(-2i)}{J_0(-2i)} = \frac{\displaystyle\sum_{k=0}^{+\infty} \frac{1}{k!(k+1)!}}{\displaystyle\sum_{k=0}^{+\infty} \frac{1}{(k!)^2}}.$$

3.6   Continued fractions and irrationality

The following theorem gives a sufficient condition for the value of a continued fraction with integer coefficients to be irrational.

Theorem 3.8   *Let* $\alpha = \dfrac{a_1}{b_1} + \dfrac{a_2}{b_2} + \cdots \dfrac{a_n}{b_n} + \cdots$ *be a convergent continued fraction, with* $a_n \in \mathbb{Z}$, $b_n \in \mathbb{Z}$ *for every* $n \in \mathbb{N}^*$. *Assume there exists an integer N such that* $0 < |a_n| < |b_n|$ *for every* $n \geq N$, *and that* $|a_n| \neq |b_n| - 1$ *for infinitely many* n. *Then* α *is irrational.*

Proof   Define $\alpha_k = \dfrac{a_k}{b_k} + \dfrac{a_{k+1}}{b_{k+1}} + \dfrac{a_{k+2}}{b_{k+2}} + \cdots$.

First we prove that $|\alpha_k| \leq 1$ for every $k \geq N$. Observe that

$$|a_{k+1}| < |b_{k+1}| \Rightarrow \left|\frac{a_{k+1}}{b_{k+1}}\right| < 1 \Rightarrow b_k - 1 < b_k + \frac{a_{k+1}}{b_{k+1}} < b_k + 1.$$

Hence, if $b_k > 0$, then $b_k + a_{k+1}/b_{k+1} > |b_k| - 1 \geq |a_k|$ (since a_k and b_k are integers). But if $b_k < 0$, then $b_k + a_{k+1}/b_{k+1} < -|b_k| + 1 \leq -|a_k|$. In both cases, $|b_k + a_{k+1}/b_{k+1}| > |a_k|$. Therefore, for every $k \geq N$, $\left|\dfrac{a_k}{b_k} \dfrac{a_{k+1}}{+ b_{k+1}}\right| < 1$.

If we replace $\dfrac{a_{k+1}}{b_{k+1}}$ by $\dfrac{a_{k+1}}{b_{k+1}} \dfrac{a_{k+2}}{+ b_{k+2}}$ and argue the same way, we see that,

for every $k \geq N$, $\left|\dfrac{a_k}{b_k} \dfrac{a_{k+1}}{+ b_{k+1}} \dfrac{a_{k+2}}{+ b_{k+2}}\right| < 1$.

More generally, we get by induction that

$$\forall k \geq N, \ \forall n \in \mathbb{N}, \ \left|\frac{a_k}{b_k} \frac{a_{k+1}}{+ b_{k+1}} \frac{a_{k+n}}{+ \cdots b_{k+n}}\right| < 1.$$

Hence $|\alpha_k| \leq 1$ by letting n tend to infinity.

We now prove that there exist infinitely many $k \geq N$ such that $|\alpha_k| < 1$. Assume on the contrary that, for k sufficiently large ($k \geq N_1 \geq N$, say) we have $|\alpha_k| = 1$. Then $|\alpha_{k+1}| = 1$ and the equality $\alpha_k = a_k/(b_k + \alpha_{k+1})$ implies that $|a_k| = |b_k + \alpha_{k+1}| \geq ||b_k| - 1| = |b_k| - 1$ because $|b_k| \geq 2$. But $|a_k| \leq |b_k| - 1$. Therefore $|a_k| = |b_k| - 1$ for every $k \geq N_1$, contradiction.

Finally, assume that α is rational. We can write

$$\alpha = \frac{a_1}{b_1} \frac{a_{k-1}}{+ \cdots b_{k-1}} \frac{\alpha_k}{+ 1} = \frac{P_{k-1} + \alpha_k P_{k-2}}{Q_{k-1} + \alpha_k Q_{k-2}}.$$

Therefore $\alpha_k(\alpha Q_{k-2} - P_{k-2}) = P_{k-1} - \alpha Q_{k-1}$. Now $\alpha Q_{k-2} - P_{k-2} \neq 0$. Indeed, otherwise $\alpha = P_{k-2}/Q_{k-2} = P_{k-1}/Q_{k-1}$, impossible by (3.5).

Hence $\alpha_k = \dfrac{P_{k-1} - \alpha Q_{k-1}}{\alpha Q_{k-2} - P_{k-2}} \in \mathbb{Q}$ for every k.

In particular α_N is rational. Put $\alpha_N = A_2/A_1$, with $A_1, A_2 \in \mathbb{Z}$. Then

$$\alpha_N = \frac{A_2}{A_1} = \frac{a_N}{b_N + \alpha_{N+1}}, \text{ whence } \alpha_{N+1} = \frac{A_1 a_N - b_N A_2}{A_2} = \frac{A_3}{A_2}, \text{ with } A_3 \in \mathbb{Z}.$$

Similarly $\alpha_{N+2} = A_4/A_3$, $\alpha_{N+3} = A_5/A_4$,.... But we know that $|\alpha_k| \leq 1$ for every $k \geq N$, and that for infinitely many k, $|\alpha_k| < 1$. We deduce that $|A_{n+1}| \leq |A_n|$ for every n, and that for infinitely many n, $|A_{n+1}| < |A_n|$. The sequence of *integers* $|A_n|$ tends therefore to $-\infty$. Contradiction because $|A_n| \geq 0$. Theorem 3.8 is proved.

Theorem 3.9 *Let* $v \in \mathbb{Q}$, $v > 0$. *Let* $x \in \mathbb{Q}^*$. *Assume* $J_{v-1}(x) \neq 0$, *Then* $J_v(x)/J_{v-1}(x)$ *is irrational.*

Proof Put $x = a/b$ and $v = c/d$. Formula (3.29) provides an expansion of $J_v(x)/J_{v-1}(x)$ as a continued fraction with rational coefficients. We transform it into a continued fraction with integer coefficients:

$$\frac{J_v(x)}{J_{v-1}(x)} = \cfrac{\dfrac{a}{b}}{2\dfrac{c}{d} - \cfrac{\dfrac{a^2}{b^2}}{2\dfrac{c+d}{d} - \cfrac{\dfrac{a^2}{b^2}}{2\dfrac{c+2d}{d} - \cdots 2\dfrac{c+nd}{d} - \cdots}}}$$

$$= \cfrac{ad}{2bc - \cfrac{\dfrac{a^2 d}{b}}{2\dfrac{c+d}{d} - \cfrac{\dfrac{a^2}{b^2}}{2\dfrac{c+2d}{d} - \cdots 2\dfrac{c+nd}{d} - \cdots}}}$$

$$= \cfrac{ad}{2bc - \cfrac{a^2 d^2}{2b(c+d) - \cfrac{\dfrac{a^2 d}{b}}{2\dfrac{c+2d}{d} - \cdots 2\dfrac{c+nd}{d} - \cdots}}}$$

$$= \cfrac{ad}{2bc - \cfrac{a^2 d^2}{2b(c+d) - \cfrac{a^2 d^2}{2b(c+2d) - \cdots 2b(c+nd) - \cdots}}}$$

Theorem 3.8 applies, and $J_v(x)/J_{v-1}(x)$ is irrational.

Corollary 3.1 *Let* $x \in \mathbb{Q}^*$, $x \neq \dfrac{\pi}{2} + k\pi$. *Then* $\tan x \notin \mathbb{Q}$.

Proof Use (3.22).

Corollary 3.2 π *is irrational.*

Proof If π was rational, $\pi/4$ would also be rational. By corollary 3.1, $\tan(\pi/4)$ would be irrational. But $\tan(\pi/4)=1$.

This proof of the irrationality of π was the first one in time. It is due to Lambert (1761).

Exercises

3.1 Prove theorem 3.2.

3.2 Prove theorem 3.3.

3.3 Prove theorem 3.4.

3.4 Let $(u_n)_{n\geq 0}$ be a sequence of complex numbers. The infinite product $\prod_{n=0}^{+\infty}(1+u_n)$ is said to be *convergent* if the sequence $v_n = \prod_{k=0}^{n}(1+u_k)$ has a finite limit when $n \to +\infty$.

1) Assume that the series $\sum_{n=0}^{+\infty} u_n$ is convergent. Prove that the infinite product $\prod_{n=0}^{+\infty}(1+u_n)$ is convergent.

2) Under the same assumption, prove that $\prod_{n=0}^{+\infty}(1+u_n)=0$ if and only if there exist an integer m such that $1+u_m=0$.

3.5 Prove that $\Gamma(1/2)=\sqrt{\pi}$.

3.6 Prove that $J_{\frac{1}{2}}(x)=\sqrt{\dfrac{2}{\pi}}x^{-\frac{1}{2}}\sin x$ et $J_{-\frac{1}{2}}(x)=\sqrt{\dfrac{2}{\pi}}x^{-\frac{1}{2}}\cos x$.

3.7 Prove that J_v satisfies the differential equation
$$x^2 y'' + xy' + (x^2 - v^2)y = 0.$$

3.8   Prove that, for $x \neq 0$ and $\nu > 0$,

$$J_{\nu-1}(x) + J_{\nu+1}(x) = \frac{2\nu}{x} J_\nu(x).$$

3.9   Prove that $J_{\nu+n}(x) \underset{n \to +\infty}{\sim} \left(\frac{x}{2}\right)^{\nu+n} \frac{1}{\Gamma(\nu+n+1)}$.

Deduce that, for $x \neq 0$, $J_\nu(x)$ and $J_{\nu-1}(x)$ cannot both be zero.

3.10   Prove that $\dfrac{1}{\text{Log}\,2} = 1 + \dfrac{1^2}{1} \dfrac{2^2}{+1} \dfrac{3^2}{+1} \dfrac{n^2}{+\cdots 1} +\cdots$.

3.11   Let $(F_n)_{n \geq 0}$ be the *Fibonacci sequence*, defined by $F_0 = 0$, $F_1 = 1$, and the recurrence relation $F_{n+1} = F_n + F_{n-1}$ $(n \geq 1)$.

1) Compute F_n by using the *golden number* $\Phi = (1 + \sqrt{5})/2$.

2) Prove that the continued fraction $\alpha = 1 + \dfrac{1}{1} \dfrac{1}{+1} \dfrac{1}{+\cdots 1} +\cdots$ is convergent and compute its value.

3) Compute the sum of the series $\sum_{n=1}^{+\infty} (-1)^{n-1}/F_n F_{n+1}$.

3.12   Continued fractions with positive coefficients

1) Consider the continued fraction $\alpha = \dfrac{1}{c_1} \dfrac{1}{+c_2} \dfrac{1}{+\cdots c_n} +\cdots$, where $c_n > 0$.

    a. Prove that the sequence $Q_n Q_{n+1}$ is not decreasing, and that $Q_n \leq C \prod_{k=1}^{n} (1+c_k)$, where C is a constant.

    b. Prove that $Q_{2p} \geq 1 + c_1(c_2 + c_4 + \cdots + c_{2p})$, and also that $Q_{2p+1} \geq c_1 + c_3 + \cdots + c_{2p+1}$.

    c. Assume that the series $\beta = \sum_{n=1}^{+\infty} c_n$ is convergent. Prove that the continued fraction α is divergent.

    d. Prove *Seidel's criterion*: if $c_n > 0$ for every $n \geq 1$, the continued fraction α is convergent if and only if the series β is divergent.

2) By using the inequality $\sqrt{ab} \leq \dfrac{a+b}{2}$, which is valid for every

$(a,b) \in \mathbb{R}_+ \times \mathbb{R}_+$, prove *Pringsheim's criterion*: if $a_n > 0$ and $b_n > 0$ for every $n \geq 1$, and if the series $\sum_{n=1}^{+\infty} \sqrt{(b_n b_{n+1})/a_{n+1}}$ diverges, then the continued fraction $\dfrac{a_1}{b_1 +} \dfrac{a_2}{b_2 +} \cdots \dfrac{a_n}{b_n +} \cdots$ converges.

3.13 Let $(a_n)_{n \geq 1}$ and $(b_n)_{n \geq 1}$ be two sequences of integers satisfying $0 < a_n \leq b_n$ for every n sufficiently large. Prove that the continued fraction $\dfrac{a_1}{b_1 +} \dfrac{a_2}{b_2 +} \cdots \dfrac{a_n}{b_n +} \cdots$ converges, and that its value is an irrational number. Deduce the irrationality of e.

3.14 Prove that $\tanh(x)$ is irrational for every $x \in \mathbb{Q}^*$. Deduce that e^x is irrational for every $x \in \mathbb{Q}^*$.

3.15 Let $v \in \mathbb{Q}$, $v > -1$. Let $x \in \mathbb{R}^*$ such that $J_v(x) = 0$. Prove that $x^2 \notin \mathbb{Q}$. Deduce that π^2 is irrational (Legendre 1794).

3.16 Rogers-Ramanujan continued fraction

Let $q \in \mathbb{C}$, such that $|q| < 1$. For every complex number a and every natural integer n, we define

$$\begin{cases} (a;q)_n = (1-a)(1-aq)\cdots(1-aq^{n-1}) & (n \geq 1) \\ (a;q)_0 = 1 \end{cases}$$

Let F_q be the entire function defined by

$$F_q(x) = \sum_{n=0}^{+\infty} \frac{q^{n^2}}{(q;q)_n} x^n = \sum_{n=0}^{+\infty} \frac{q^{n^2}}{(1-q)(1-q^2)\cdots(1-q^n)} x^n.$$

1) Prove that F_q satisfies the functional equation

$$F_q(x) = F_q(qx) + qx F_q(q^2 x).$$

2) For every non zero complex number x, define

$$\alpha_{2n} = q^n x \frac{F_q(q^{2n} x)}{F_q(q^{2n-1} x)} \quad (n \geq 1), \quad \alpha_{2n+1} = q^n \frac{F_q(q^{2n+1} x)}{F_q(q^{2n} x)} \quad (n \geq 0).$$

Compute c_n such that, for every $n \geq 1$,

$$\alpha_n = \frac{1}{c_n + \alpha_{n+1}}.$$

3) Prove that, for every complex number x,

$$\frac{\displaystyle\sum_{n=0}^{+\infty} \frac{q^{n^2}}{(q;q)_n} x^n}{\displaystyle\sum_{n=0}^{+\infty} \frac{q^{n^2+n}}{(q;q)_n} x^n} = \frac{1}{1+} \frac{qx}{1+} \frac{q^2 x}{1+} \frac{q^3 x}{1+} + \cdots .$$

Chapter 4

Regular continued fractions

We define in section 4.1 the regular continued fraction expansion of a positive real number and study its first properties, which result directly from the general properties of continued fractions (chapter 3). Then, we give two examples, by computing the regular continued fraction expansion of e (section 4.2) and explaining how to solve the diophantine equation of the first degree in two unknowns $ax + by = c$ (section 4.3). In section 4.4 we show that the convergents of their expansion in regular continued fraction provide "good" diophantine approximations of irrational numbers. Section 4.5 is devoted to the study of the regular continued fraction expansion of quadratic irrational numbers, with an application to the resolution of the Pell's equation $x^2 - Dy^2 = 1$.

4.1 Regular continued fraction expansion of a positive real number

4.1.1 Let $\alpha_0 > 0$. Put $\alpha_0 = c_0 + b_0$, where $c_0 = [\alpha_0]$ (integer part of α_0) and $b_0 \in [0,1[$. If $b_0 \neq 0$, that is if α_0 is not an integer, we can write $b_0 = 1/\alpha_1$, with $\alpha_1 > 1$ since $b_0 < 1$. Hence

$$\alpha_0 = c_0 + \frac{1}{\alpha_1}. \tag{4.1}$$

Again with α_1, if α_1 is not an integer, we can write $\alpha_1 = c_1 + 1/\alpha_2$, with $c_1 = [\alpha_1] \geq 1$ (because $\alpha_1 > 1$) and $\alpha_2 > 1$. Therefore

$$\alpha_0 = c_0 + \cfrac{1}{c_1 + \cfrac{1}{\alpha_2}}. \tag{4.2}$$

We can go on the same way as long as α_n is not an integer, and we get

$$\alpha_0 = c_0 + \frac{1}{c_1} \frac{1}{+c_2} + \cdots c_{n-1} + \frac{1}{c_n + \cfrac{1}{\alpha_{n+1}}} = c_0 + \frac{1}{c_1} + \cdots c_n \frac{1}{+\alpha_{n+1}} \qquad (4.3)$$

by using the following algorithm:

$$\alpha_n = c_n + \frac{1}{\alpha_{n+1}}, \qquad c_n = [\alpha_n]. \qquad (4.4)$$

Two cases can occur:

- One of the α_n's is an integer, in which case (4.4) becomes

  $\alpha_n = c_n = [\alpha_n]$, and (4.3) yields $\alpha_0 = c_0 + \dfrac{1}{c_1} \cdots \dfrac{1}{c_n}$. Then α_0 is

  clearly a *rational number*, since all the c_i's are integers.

- None of the α_n's is an integer, in which case we can write (4.3)
  for every n. We will see that, in this case, α_0 is irrational and the

  continued fraction $c_0 + \dfrac{1}{c_1} \dfrac{1}{+c_2} + \cdots \dfrac{1}{c_n} + \cdots$ converges to α_0.

4.1.2 In order to prove that α_0 is irrational if algorithm (4.4) can be used indefinitely, we will show that it stops for some n if α_0 is a rational number. Suppose therefore that $\alpha_0 = A_0/B_0 \in \mathbb{Q}$. We use the euclidean division algorithm. First we divide A_0 by B_0 and obtain

$$A_0 = c_0 B_0 + B_1, \qquad \text{with} \quad 0 \le B_1 < B_0.$$

Therefore $\alpha_0 = \dfrac{A_0}{B_0} = c_0 + \dfrac{B_1}{B_0}$.

If $B_1 = 0$, α_0 is an integer and the process stops. Otherwise, we can write $\alpha_0 = c_0 + 1/\alpha_1$, with $\alpha_1 = B_0/B_1$. We divide B_0 by B_1:

$$B_0 = c_1 B_1 + B_2, \qquad \text{with} \quad 0 \le B_2 < B_1.$$

Then $\alpha_1 = B_0/B_1 = c_1 + B_2/B_1$. If $B_2 = 0$, α_1 is an integer, the process

stops, and $\alpha_0 = c_0 + \dfrac{1}{c_1}$. Otherwise

$$\alpha_0 = c_0 + \cfrac{1}{c_1 + \cfrac{1}{\alpha_2}}, \qquad \text{with} \quad \alpha_2 = \frac{B_1}{B_2}.$$

And so on... Observe that the sequence B_0, B_1, B_2, ... is decreasing. Therefore there exists an integer n such that $B_n = 0$. Then α_{n-1} is an integer, and the *expansion in regular continued fraction* of α_0 is finite.

Example 4.1 Compute the regular continued fraction expansion of $\alpha_0 = 37/13$. We write the following euclidian divisions. $37 = 13 \times 2 + 11$, whence $\alpha_0 = 2 + 11/13 = 2 + 1/\alpha_1$, with $\alpha_1 = 13/11 = 1 + 2/11 = 1 + 1/\alpha_2$. Here $\alpha_2 = 11/2 = 5 + 1/2 = 5 + 1/\alpha_3$, with $\alpha_3 = 2$. Since α_3 is an integer, the algorithm stops and

$$\frac{37}{13} = 2 + \frac{1}{\alpha_1} = 2 + \frac{1}{1} \frac{1}{+\alpha_2} = 2 + \frac{1}{1} \frac{1}{+5} \frac{1}{+2}.$$

4.1.3 Assume now that α_0 is irrational.

We will consider the regular continued fraction $c_0 + \dfrac{1}{c_1} \dfrac{1}{+c_2} + \cdots \dfrac{1}{c_n}$, defined by algorithm (4.4), and prove that it converges to α_0. We know by theorem 3.1 that its convergents $\dfrac{P_n}{Q_n} = c_0 + \dfrac{1}{c_1} + \cdots \dfrac{1}{c_n}$ satisfy

$$\begin{cases} P_{-1} = 1, \ Q_{-1} = 0, \ P_0 = c_0, \ Q_0 = 1, \\ P_n = c_n P_{n-1} + P_{n-2}, \ Q_n = c_n Q_{n-1} + Q_{n-2}. \end{cases} \tag{4.5}$$

Since $c_n \in \mathbb{N} - \{0\}$ for every $n \geq 1$, we have $P_n \in \mathbb{N} - \{0\}$, $Q_n \in \mathbb{N} - \{0\}$ for every $n \geq 1$. Moreover, $Q_n = c_n Q_{n-1} + Q_{n-2}$ yields $Q_n \geq Q_{n-1}$, whence $Q_n \geq c_n Q_{n-2} + Q_{n-2} \geq 2Q_{n-2}$. Therefore $Q_n \geq 2^{[n/2]}$. If we apply theorem 3.1 to the finite continued fraction in (4.3), we obtain

$$\alpha_0 = \frac{P'_n}{Q'_n} = \frac{\alpha_{n+1} P_n + P_{n-1}}{\alpha_{n+1} Q_n + Q_{n-1}}.$$

However, by theorem 3.2 we have

$$P_n Q_{n-1} - P_{n-1} Q_n = (-1)^{n-1}, \tag{4.6}$$

which proves in passing that P_n and Q_n are coprime by Bézout's theorem. Hence

$$\alpha_0 - \frac{P_n}{Q_n} = \frac{Q_n P_{n-1} - P_n Q_{n-1}}{Q_n (\alpha_{n+1} Q_n + Q_{n-1})} = \frac{(-1)^n}{Q_n (\alpha_{n+1} Q_n + Q_{n-1})}. \tag{4.7}$$

Now (4.7) implies that $\alpha_0 > P_n/Q_n$ if n is even, and that $\alpha_0 < P_n/Q_n$ if n is odd. Hence α_0 *always lies between two successive convergents.* Moreover, since $Q_{n-1} > 0$ and $\alpha_{n+1} > 1$, (4.7) implies

$$\left| \alpha_0 - \frac{P_n}{Q_n} \right| < \frac{1}{Q_n^2} \tag{4.8}$$

Since $Q_n \geq 2^{\left[\frac{n}{2}\right]}$, we obtain $\lim_{n \to +\infty} P_n/Q_n = \alpha_0$, as claimed.

The expansion of α_0 obtained from algorithm (4.4) is called its *regular continued fraction expansion.* We denote it by

$$\alpha_0 = c_0 + \frac{1}{c_1 +} \frac{1}{c_2 +} \cdots \frac{1}{c_n +} \cdots = [c_0, c_1, c_2, \ldots, c_n, \ldots].$$

Example 4.2   Let $\alpha_0 = \sqrt{2}$. Here $c_0 = [\alpha_0] = 1$. Therefore

$$\alpha_0 = 1 + (\sqrt{2} - 1) = 1 + \frac{1}{\sqrt{2} + 1}.$$

Hence $\alpha_1 = \sqrt{2} + 1$, $[\alpha_1] = 2$, $\alpha_1 = 2 + (\sqrt{2} - 1) = 2 + 1/(\sqrt{2} + 1)$.

Therefore $\alpha_2 = \alpha_1$, $[\alpha_2] = 2$, $\alpha_2 = 2 + 1/(\sqrt{2} + 1)$.

We see that the expansion of $\sqrt{2}$ in regular continued fraction is periodic from rank 1, more precisely $c_n = 2$ for every $n \geq 1$. Hence

$$\sqrt{2} = 1 + \frac{1}{2 +} \frac{1}{2 +} \cdots \frac{1}{2 +} \cdots = [1, 2, 2, 2, \ldots, 2, \ldots].$$

We will see in section 4.5 that periodic continued fraction expansions are characteristic of irrational quadratic numbers.

4.1.4   The following theorem shows that the expansion in regular continued fraction of *irrational numbers* is unique.

Theorem 4.1   *If* $\alpha_0 = c_0' + \dfrac{1}{c_1' +} \dfrac{1}{c_2' +} \cdots \dfrac{1}{c_n' +} \cdots$, *with* $c_n' \in \mathbb{N} - \{0\}$ *for every* $n \geq 1$, *then* $c_n' = c_n$ *for every* $n \in \mathbb{N}$.

Proof   See exercise 4.1.

Remark 4.1 It shoud be noted that there is no uniqueness for expansions of the form $c_0 + \dfrac{1}{c_1} \dfrac{1}{+c_2} \dfrac{1}{+\cdots c_n}$, with $c_n \in \mathbb{N} - \{0\}$ for $n \geq 1$

in the case of rational numbers. For example, $\dfrac{37}{13} = 2 + \dfrac{1}{1} \dfrac{1}{+5} \dfrac{1}{+2}$, as seen

in example 4.1, but we have also $\dfrac{37}{13} = 2 + \dfrac{1}{1} \dfrac{1}{+5} \dfrac{1}{+1} \dfrac{1}{+1}$.

Example 4.3 By exercise 3.14 we know that

$$\tanh\left(\frac{a}{b}\right) = \frac{a}{b} \frac{a^2}{+3b} \frac{a^2}{+5b} \frac{a^2}{+\cdots (2n+1)b} {+\cdots} \tag{4.10}$$

Substituting $a = 1$ and $b = 2$ yields

$$\tanh\left(\frac{1}{2}\right) = \frac{1}{2} \frac{1}{+6} \frac{1}{+10} \frac{1}{+\cdots 2(2n+1)} {+\cdots}, \quad \text{whence}$$

$$\frac{1}{\tanh\left(\dfrac{1}{2}\right)} = \frac{e+1}{e-1} = [2,6,10,\ldots,2(2n+1),\ldots]. \tag{4.11}$$

4.2 The regular continued fraction expansion of e

The regular continued fraction expansion of e has been discovered by Euler (1707-1783). He obtained it by computing its first terms from the approximate value $e = 2,718281828\ldots$, then conjecturing that

$$e = [2,1,2,1,1,4,1,1,6,1,\ldots,1,2n,1,\ldots]. \tag{4.12}$$

In others words, $c_0 = 2$, $c_{3k-2} = 1$, $c_{3k-1} = 2k$, $c_{3k} = 1$ for every $k \geq 1$.

Euler proved (4.12) by comparing the convergents $R_n = P_n/Q_n$ of the continued fraction $[2,\overline{1,2n,1}]_{n=1,2,3,\ldots}$ with the convergents $R'_n = P'_n/Q'_n$ of the continued fraction (4.11). First, by using (4.5), we obtain

$$\frac{P_0}{Q_0} = \frac{2}{1}, \ \frac{P_1}{Q_1} = \frac{3}{1}, \ \frac{P_2}{Q_2} = \frac{8}{3}, \ \frac{P_3}{Q_3} = \frac{11}{4}, \ \frac{P_4}{Q_4} = \frac{19}{7} \cdots, \ \frac{P'_0}{Q'_0} = \frac{2}{1}, \ \frac{P'_1}{Q'_1} = \frac{13}{6}.$$

We observe that $P_1 = P'_0 + Q'_0$, $Q_1 = P'_0 - Q'_0$, $P_4 = P'_1 + Q'_1$, $Q_4 = P'_1 - Q'_1$, which leads us to conjecture that $P_{3k+1} = P'_k + Q'_k$ and $Q_{3k+1} = P'_k - Q'_k$ for every $k \in \mathbb{N}$. In order to prove these formulas, we prove that the sequences $T_k = P'_k + Q'_k$ and $S_k = P_{3k+1}$ satisfy the *same recurrence relation of the second order*. Since $T_0 = S_0$ and $T_1 = S_1$, we therefore have $T_k = S_k$ for every k. And the same holds for $Q_{3k+1} = P'_k - Q'_k$.

By (4.5) and (4.12), we have for $k \geq 2$
$$P'_k = 2(2k+1)P'_{k-1} + P'_{k-2}, \quad Q'_k = 2(2k+1)Q'_{k-1} + Q'_{k-2}.$$

Therefore $T_k = 2(2k+1)T_{k-1} + T_{k-2}$.

Similarly, by using (4.5), we see that, for every $k \geq 2$,
$$P_{3k-3} = P_{3k-4} + P_{3k-5}, \ P_{3k-2} = P_{3k-3} + P_{3k-4}, \ P_{3k-1} = 2kP_{3k-2} + P_{3k-3},$$
$$P_{3k} = P_{3k-1} + P_{3k-2}, \ P_{3k+1} = P_{3k} + P_{3k-1}.$$

We multiply the first equality by 1, the second by -1, the third by 2, the forth by 1, the fifth by 1, and we add all of them. We get at once
$$S_k = 2(2k+1)S_{k-1} + S_{k-2}.$$

Hence we have $P_{3k+1} = P'_k + Q'_k$, as claimed. The equality $Q_{3k+1} = P'_k - Q'_k$ can be proved by arguing the same way. Therefore, by (4.11),

$$\lim_{k \to +\infty} \frac{P_k}{Q_k} = \lim_{k \to +\infty} \frac{P_{3k+1}}{Q_{3k+1}} = \lim_{k \to +\infty} \frac{\dfrac{P'_k}{Q'_k} + 1}{\dfrac{P'_k}{Q'_k} - 1} = \frac{\dfrac{e+1}{e-1} + 1}{\dfrac{e+1}{e-1} - 1} = e,$$

which proves (4.12).

4.3   The diophantine equation $ax + by = c$

A *diophantine equation* is an equation whose unknowns are integers. The most simple is the equation of the first degree
$$ax + by = c, \qquad\qquad (4.13)$$
where $x, y \in \mathbb{Z}$ are unknown and $a, b, c \in \mathbb{Z}$ are given.

If a and b are not coprime, we can put $a = da'$, $b = db'$, with a' and b' coprime. If d does not divide c, (4.13) has no solution. If d divides c, we are brought back to (4.13), with a, b, c replaced by a', b', c' respectively.

Therefore, in (4.13) we can assume a and b to be coprime. Now expand the rational a/b in regular continued fraction. The expansion ends with a convergent $R_n = P_n/Q_n$ (see 4.1.2) and $a/b = P_n/Q_n$. As the fractions a/b and P_n/Q_n are *irreducible* by (4.6), $a = P_n$ and $b = Q_n$. Then (4.6) shows that $aQ_{n-1} - bP_{n-1} = (-1)^{n-1}$.

We multiply by $\varepsilon = \pm 1$ according to the parity of n and we get

$$a(\varepsilon Q_{n-1}) + b(-\varepsilon P_{n-1}) = 1. \tag{4.14}$$

Now we multiply both sides of (4.14) by c and substract it to (4.13). We obtain $a(x - c\varepsilon Q_{n-1}) = -b(c\varepsilon P_{n-1} + y)$, which yields

$$\frac{c\varepsilon P_{n-1} + y}{x - c\varepsilon Q_{n-1}} = -\frac{a}{b}. \tag{4.15}$$

Since a/b is irreducible, there exists $t \in \mathbb{Z}$ such that

$$x = c\varepsilon Q_{n-1} + bt, \quad y = -c\varepsilon P_{n-1} - at. \tag{4.16}$$

Conversely, (x, y) given by (4.16) is solution of (4.13) for every $t \in \mathbb{Z}$ because of (4.14). Therefore, *if a and b are coprime, the equation* (4.13) *has infinitely many solutions, given by* (4.16).

Example 4.4 Solve the diophantine equation

$$37x + 13y = 5. \tag{4.17}$$

We know (example 4.1) that $37/13 = [2,1,5,2]$. Hence we get $P_{-1} = 1$, $P_0 = 2$, $P_1 = P_0 + P_{-1} = 3$, $P_2 = 5P_1 + P_0 = 17$, $P_3 = 2P_2 + P_1 = 37$, $Q_{-1} = 0$, $Q_0 = 1$, $Q_1 = Q_0 + Q_{-1} = 1$, $Q_2 = 5Q_1 + Q_0 = 6$, $Q_3 = 2Q_2 + Q_1 = 13$.
Relation (4.6) becomes here $P_3 Q_2 - P_2 Q_3 = 1$, whence

$$37 \cdot 5Q_2 - 13 \cdot 5P_2 = 5. \tag{4.18}$$

If we substract (4.18) to (4.17), we get $37(x - 5Q_2) + 13(y + 5P_2) = 0$, and finally

$$\begin{cases} x = 5Q_2 + 13t = 30 + 13t \\ y = -5P_2 - 37t = -85 - 37t \end{cases} \quad (t \in \mathbb{Z}).$$

4.4 Regular continued fractions and diophantine approximation

We have seen in section 4.1, formula (4.8), that the convergents P_n/Q_n to any irrational number α_0 provide good diophantine approximations of α_0. Indeed we have

$$\left| \alpha_0 - \frac{P_n}{Q_n} \right| < \frac{1}{Q_n^2}. \tag{4.19}$$

It should be noted that this inequality gives an *effective* statement of theorem 1.6. Indeed, this theorem asserts the existence of infinitely many fractions p/q satisfying $|\alpha_0 - p/q| < 1/q^2$ for every irrational α_0, but *it gives no way to compute them*: it is *ineffective*. On the contrary, formula (4.8) allows to compute explicitly infinitely many of these rationals p/q, as soon as the regular continued fraction expansion of α_0 is known. This is the case for e, for example (section 4.2), or for any irrational quadratic number, as we will see in section 4.5.

The following theorem shows that regular continued fraction expansions provide an even better diophantine approximation.

Theorem 4.2 *Of two successive convergents p/q of the irrational number α_0, at least one satisfies $|\alpha_0 - p/q| < 1/2q^2$.*

Proof The proof relies on the fact that, for every $n \in \mathbb{N}$, α_0 lies between P_n/Q_n and P_{n+1}/Q_{n+1} (see subsection 4.1.3, consequences of formula (4.7), page 42). It is left to the reader (exercise 4.2).

Conversely, we have the following result, which will prove useful in the resolution of Pell's equation (section 4.6).

Theorem 4.3 *Let α_0 be an irrational number, and let p/q be any rational number satisfying $|\alpha_0 - p/q| < 1/2q^2$. Then p/q is one of the convergents of α_0.*

Proof Put $\alpha_0 - p/q = \varepsilon\theta/q^2$, with $\varepsilon = \pm 1$, $0 < \theta < 1/2$, and expand the rational p/q in regular continued fraction:

$$\frac{p}{q} = [c_0, c_1, \ldots, c_n] = \frac{P_n}{Q_n}. \tag{4.20}$$

As we have seen before in a special case (remark 4.1), we can choose the parity of n. Indeed, if $c_n > 1$, we can write $p/q = [c_0, c_1, \ldots, c_n - 1, 1]$, and if $c_n = 1$, $p/q = [c_0, c_1, \ldots, c_{n-1} + 1]$. Assume therefore that, in (4.20), we have chosen n in such a way that $\varepsilon = (-1)^n$.

Then, the equality $\alpha_0 - p/q! = \varepsilon\theta/q^2$ can be written

$$\alpha_0 - \frac{P_n}{Q_n} = \frac{(-1)^n \theta}{Q_n^2}. \tag{4.21}$$

Now define w by

$$\alpha_0 = \frac{P_n w + P_{n-1}}{Q_n w + Q_{n-1}}. \tag{4.22}$$

From (4.21), (4.22) and (4.6) we deduce that

$$\frac{(-1)^n \theta}{Q_n^2} = \frac{P_{n-1} Q_n - P_n Q_{n-1}}{Q_n (Q_n w + Q_{n-1})} = \frac{(-1)^n}{Q_n (Q_n w + Q_{n-1})}.$$

Hence $\theta = \dfrac{Q_n}{Q_n w + Q_{n-1}}$, which yields $w = \dfrac{Q_n - \theta Q_{n-1}}{\theta Q_n}$.

But $0 < \theta < 1/2$ and $Q_{n-1} < Q_n$. Therefore $w > 1$, and we can expand w in a regular continued fraction:

$$w = [c_{n+1}, c_{n+2}, \dots], \qquad \text{with} \quad c_{n+1} \geq 1. \tag{4.23}$$

Now from (4.22) we see that

$$c_0 + \frac{1}{c_1 +} \frac{1}{c_2 + \cdots} \frac{1}{c_n +} \frac{1}{w} = \frac{P_n'}{Q_n'} = \alpha_0, \tag{4.24}$$

because $P_n' = w P_n + P_{n-1}$ and $Q_n' = w Q_n + Q_{n-1}$. By substituting (4.23) into (4.24), we deduce that the regular continued fraction expansion of α_0 is

$$\alpha_0 = [c_0, c_1, \dots, c_n, c_{n+1}, c_{n+2}, \dots],$$

and that $p/q = P_n/Q_n$ is indeed one of its convergents, Q.E.D.

4.5   Quadratic irrational numbers and continued fractions

4.5.1   We say that a complex number α is *quadratic irrational* if it satisfies a second degree equation with integer coefficients

$$a\alpha^2 + b\alpha + c = 0, \tag{4.25}$$

with $a, b, c \in \mathbb{Z}$, $a \neq 0$, $\Delta = b^2 - 4ac \neq 0$.

In this case there exists a square root ω of Δ in $\mathbb{C}$, such that $\alpha = (-b + \omega)/2a$. The other solution of (4.25) is called the *conjugate* of α. We will denote it by α^*, whence $\alpha^* = (-b - \omega)/2a$.

In other words, we obtain α^* by changing the sign of the "irrational" part of α.

For example, the *golden number* $\Phi = (1+\sqrt{5})/2$ is quadratic irrational. It satisfies $\Phi^2 - \Phi - 1 = 0$. Its conjugate is $\Psi = (1-\sqrt{5})/2 = -1/\Phi$.

The reader will check easily that, if α is quadratic irrational, so is $1/\alpha$, and $(1/\alpha)^* = 1/\alpha^*$. Moreover, if α is quadratic irrational and $r \in \mathbb{Q}$, so is $r + \alpha$, and $(r+\alpha)^* = r + \alpha^*$. This will be generalized in chapter 5.

Theorem 4.4 *The irrational number $\alpha_0 > 0$ is quadratic if, and only if, its regular continued fraction expansion $\alpha_0 = [c_0, c_1, ..., c_n, ...]$ is ultimately periodic, that is if there exist $N \in \mathbb{N}$ and $T \in \mathbb{N} - \{0\}$ such that $c_{n+T} = c_n$ for every $n \geq N$. Then we will denote $\alpha_0 = [c_0, c_1, ..., c_{N-1}, \overline{c_N, ..., c_{N+T-1}}]$.*

Proof The proof consists in three steps. The first two steps are left as exercices.

Step one: We prove that, if $\alpha_0 = [c_0, c_1, ..., c_n, ...]$ satisfies $c_{n+T} = c_n$ for every $n \geq N$, then α_0 is quadratic (exercise 4.3).

Step two: We prove that, if α_0 is a quadratic number, with $\alpha_0 > 1$ and $\alpha_0^* < 0$, then its regular continued fraction expansion is ultimately periodic (exercise 4.4).

Step three: We reduce the general case to the second step. Let α_0 be a quadratic number, $\alpha_0 > 0$. If $\alpha_0 < 1$, we replace it by $\alpha_0' = 1/\alpha_0$. Therefore we can assume $\alpha_0 > 1$. We get the regular continued fraction expansion of α_0 by writing successively

$$\alpha_0 = c_0 + \frac{1}{\alpha_1} = c_0 + \cfrac{1}{c_1 + \alpha_2} = c_0 + \cfrac{1}{c_1 + \cdots c_{n-1} + \alpha_n}.$$

Taking the conjugates yields

$$\alpha_0^* = c_0 + \frac{1}{\alpha_1^*} = \cdots = c_0 + \cfrac{1}{c_1 + \cdots c_{n-1} + \alpha_n^*}.$$

If $\alpha_n^* > 1$ for every $n \geq 1$, we obtain here the regular continued fraction expansion of α_0^*, which would be the same as α_0's one. If so, $\alpha_0^* = \alpha_0$, which is impossible. Therefore there exists an integer $p \geq 1$ such that

$\alpha_p^* < 1$. If $\alpha_p^* > 0$, we write $\alpha_p = c_p + 1/\alpha_{p+1}$. Taking the conjugates yields $\alpha_p^* = c_p + 1/\alpha_{p+1}^*$. But $\alpha_p^* < 1 \leq c_p$ because $c_p = [\alpha_p]$ and $\alpha_p > 1$. Therefore $c_p + 1/\alpha_{p+1}^* < c_p$ and $\alpha_{p+1}^* < 0$. Hence there exists an integer $h \geq 1$ such that $\alpha_h > 1$ et $\alpha_h^* < 0$. From the second step we know that the continued fraction expansion of α_h is ultimately periodic, and so is the expansion of α_0, since $\alpha_0 = c_0 + \dfrac{1}{c_1 +} \dfrac{1}{c_2 +} \cdots \dfrac{1}{c_{h-1} +} \dfrac{1}{\alpha_h}$.

Corollary 4.1   *e is not a quadratic number.*

Indeed, (4.12) shows that its regular continued fraction expansion is not ultimately periodic. For another proof, see exercise 1.8.

4.5.2   We say that the quadratic irrational number α_0 is *reduced* if it satisfies $\alpha_0 > 1$ and $-1 < \alpha_0^* < 0$. For example, the golden number $\Phi = (1 + \sqrt{5})/2$ is a reduced quadratic irrational number.

Theorem 4.5   *If α_0 is a reduced quadratic irrational number, then its regular continued fraction expansion $\alpha_0 = [c_0, c_1, ..., c_n, ...]$ is purely periodic, which means that there exists $T \in \mathbb{N} - \{0\}$ such that $c_{n+T} = c_n$ for every $n \in \mathbb{N}$.*

Proof   We have $\alpha_0 = c_0 + 1/\alpha_1$, with $\alpha_1 > 1$ and $\alpha_0^* = c_0 + 1/\alpha_1^* \in]-1, 0[$. But $\alpha_0 > 1$ implies that $c_0 \geq 1$. Therefore $1/\alpha_1^* = \alpha_0^* - c_0 < -1$ and $-1 < \alpha_1^* < 0$. Hence α_1 is reduced quadratic, and by induction the same holds for α_n for every $n \in \mathbb{N}$. Now $\alpha_n = c_n + 1/\alpha_{n+1}$ implies that $\alpha_n^* = c_n + 1/\alpha_{n+1}^*$, whence $-1/\alpha_{n+1}^* = c_n - \alpha_n^*$. Since $0 < -\alpha_n^* < 1$, we get

$$ c_n = \left[\frac{-1}{\alpha_{n+1}^*} \right], \qquad \forall n \in \mathbb{N}. \qquad (4.26) $$

Now assume that the regular continued fraction which represents α_0 is not periodic from rank 0, that is

$$ \alpha_0 = [c_0, c_1, ..., c_{N-1}, \overline{c_N, c_{N+1}, ..., c_{N+T-1}}] $$

where $N \geq 1$ is minimum. Then $c_{N-1} \neq c_{N+T-1}$. But the periodicity implies that $\alpha_N = \alpha_{N+T}$, whence $-1/\alpha_N^* = -1/\alpha_{N+T}^*$. From (4.26) we deduce that $c_{N-1} = c_{N+T-1}$. This contradiction proves theorem 4.5.

Corollary 4.2 *Let* $D \in \mathbb{N}$. *Assume that* D *is not a perfect square. Then the regular continued fraction expansion of* $\sqrt{D}$ *is periodic from rank* $N = 1$. *More precisely, it is of the form* $\sqrt{D} = [c_0, \overline{c_1, ..., c_{T-1}, 2c_0}]$.

Proof See exercise 4.5.

Example 4.6 Let $\alpha_0 = \sqrt{19}$. Then $\alpha_0 = 4 + (\sqrt{19} - 4)$ and

$$\alpha_1 = \frac{4 + \sqrt{19}}{3} = 2 + \frac{\sqrt{19} - 2}{3}, \quad \alpha_2 = \frac{2 + \sqrt{19}}{5} = 1 + \frac{\sqrt{19} - 3}{5},$$

$$\alpha_3 = \frac{\sqrt{19} + 3}{2} = 3 + \frac{\sqrt{19} - 3}{2}, \quad \alpha_4 = \frac{3 + \sqrt{19}}{5} = 1 + \frac{\sqrt{19} - 2}{5},$$

$$\alpha_5 = \frac{2 + \sqrt{19}}{3} = 2 + \frac{\sqrt{19} - 4}{3}, \quad \alpha_6 = \sqrt{19} + 4 = 8 + (\sqrt{19} - 4).$$

Finally $\alpha_7 = \alpha_1$, whence $\sqrt{19} = [4, \overline{2, 1, 3, 1, 2, 8}]$.

4.6   Pell's equation

Pell's equation is the diophantine equation with unknowns $(x, y) \in \mathbb{Z}^2$

$$x^2 - Dy^2 = 1, \tag{4.27}$$

where D is a positive integer which is not a perfect square.

This equation is connected with regular continued fractions as well as with units of real quadratic fields. Here we will be interested only in the first aspect. The second one will be studied in chapter 5.

As a matter of fact, regular continued fractions allow to solve, for every $1 \leq |L| < \sqrt{D}$, the most general equation

$$x^2 - Dy^2 = L. \tag{4.28}$$

In (4.28), we can assume that $(x, y) \in \mathbb{N}^2$, and also that x and y are coprime. If not, let d be their greatest common divisor. Then d^2 divides

L, and we have to solve another equation (4.28), where L is replaced by L/d^2. The reader will check (exercise 4.6) that:

If $(x, y) \in \mathbb{N}^2$, x and y coprime, is a solution of (4.28), then $x = P_n$ and $y = Q_n$, where P_n/Q_n is a convergent of the expansion of $\sqrt{D}$ in regular continued fraction.

Hence we have to look for the convergents P_n/Q_n of the expansion of $\sqrt{D}$ which satisfy $P_n^2 - DQ_n^2 = L$. Put $\alpha_0 = \sqrt{D} = [c_0, c_1, c_2, \ldots]$, and $\alpha_n = c_n + 1/\alpha_{n+1}$. By proceeding as in exercise 4.4, we see that, for every $n \in \mathbb{N}$, $\alpha_n = (A_n + \sqrt{D})/B_n$, with $A_n \in \mathbb{N}$, $B_n \in \mathbb{N} - \{0\}$. Since sequence α_n is periodic from rank 1 (corollary 4.2), the same holds for sequences A_n and B_n. In particular, B_n takes only *finitely many* values.

For example, if $D = 19$, the only possible values for B_n are 1, 2, 3 and 5 (see example 4.6).

However, $P_n^2 - DQ_n^2$ can be computed in terms of B_n for all n. Indeed, we have $\alpha_0 = c_0 + \cfrac{1}{c_1 + \cdots \cfrac{1}{c_{n-1} + \alpha_n}} = \dfrac{P_n'}{Q_n'}$. Hence, by theorem 3.1,

$$\sqrt{D} = \alpha_0 = \frac{\alpha_n P_{n-1} + P_{n-2}}{\alpha_n Q_{n-1} + Q_{n-2}} = \frac{(A_n + \sqrt{D})P_{n-1} + B_n P_{n-2}}{(A_n + \sqrt{D})Q_{n-1} + B_n Q_{n-2}}.$$

Therefore $Q_{n-1}D + (A_n Q_{n-1} + B_n Q_{n-2})\sqrt{D} = A_n P_{n-1} + B_n P_{n-2} + P_{n-1}\sqrt{D}$.

Since $\sqrt{D}$ is *irrational*, we get by corollary 1.1

$$P_{n-1} = A_n Q_{n-1} + B_n Q_{n-2}, \quad Q_{n-1}D = A_n P_{n-1} + B_n P_{n-2}.$$

Now multiply the first equality by P_{n-1}, the second by Q_{n-1}, and substract them. We obtain $P_{n-1}^2 - DQ_{n-1}^2 = B_n(P_{n-1}Q_{n-2} - P_{n-2}Q_{n-1})$, which yields, by using (4.6), $P_{n-1}^2 - DQ_{n-1}^2 = (-1)^n B_n$. Hence we have proved

Theorem 4.6 *The equation (4.28) has a solution if, and only if, there exists an integer $n_0 \geq 1$ such that $L = (-1)^{n_0} B_{n_0}$. In this case, a solution of (4.28) is $(x = P_{n_0 - 1}, y = Q_{n_0 - 1})$, where $P_{n_0 - 1}/Q_{n_0 - 1}$ is the convergent of order $n_0 - 1$ of the expansion of $\sqrt{D}$ in regular continued fraction. Moreover, let T be the period of this expansion. Then, for every $k \in \mathbb{N}$,*

$(x = P_{n_0+2kT-1}, y = Q_{n_0+2kT-1})$ *is a solution of* (4.28). *Hence this equation has infinitely many solutions.*

Example 4.7 Consider the equation

$$x^2 - 19y^2 = L, \qquad 1 \leq |L| \leq 4. \tag{4.29}$$

By using the results of example 4.6, we see that B_n is periodic of period $T = 6$. Hence $(-1)^n B_n$ is also of period 6. Its values are $(-1)^1 B_1 = -3$, $(-1)^2 B_2 = 5$, $(-1)^3 B_3 = -2$, $(-1)^4 B_4 = 5$, $(-1)^5 B_5 = -3$, $(-1)^6 B_6 = 1$. Therefore (4.29) admits solutions (x, y), with x and y coprime, if and only if $L = -3$, -2 or 1.

For example, the smallest solution of the equation $x^2 - 19y^2 = -2$ is given by the convergent $\dfrac{P_2}{Q_2} = 4 + \dfrac{1}{2} + \dfrac{1}{1} = \dfrac{13}{3}$, that is $x = 13$ and $y = 3$.

The next solutions are given by the convergents P_8/Q_8, P_{14}/Q_{14}, P_{20}/Q_{20}, and so on...

If we do not assume that x and y are coprime, denote d their GCD. Then $d^2 \mid L$, whence $L = 4$ or -4, and $d = 2$. We are taken back to $x'^2 - 19y'^2 = \pm 1$, $x = 2x'$, $y = 2y'$. Therefore $x^2 - 19y^2 = -4$ has no solution and $x^2 - 19y^2 = 4$ has infinitely many solutions.

Corollary 4.3 *Pell's equation* (4.27) *has infinitely many solutions.*

Proof Indeed, corollary 4.2 implies that $\sqrt{D} - [\sqrt{D}] = \alpha_0 - c_0$ is a reduced quadratic irrational number. Therefore $B_{kT} = B_0 = 1$ for every $k \in \mathbb{N}$. Consequently $(-1)^{2kT} B_{2kT} = 1$ for every $k \in \mathbb{N}$ and $x = P_{2kT-1}$, $y = Q_{2kT-1}$ is a solution of (4.26).

Example 4.8 By using example 4.7, we see that the solutions of the Pell equation $x^2 - 19y^2 = 1$ are given by $x = P_{6k-1}$, $y = Q_{6k-1}$. For $k = 0$, we get $x = P_{-1} = 1$, $y = Q_{-1} = 0$ (*trivial solution*).

The smallest non trivial solution (or *fundamental solution*) is $x = P_5 = 170$, $y = Q_5 = 39$. Indeed we have, by example 4.6, $P_1 = 9$,

$Q_1 = 2$, $P_2 = 13$, $Q_2 = 3$, $P_3 = 48$, $Q_3 = 11$, $P_4 = 61$, $Q_4 = 14$, $P_5 = 170$, $Q_5 = 39$.

Exercises

4.1   Prove theorem 4.1.

4.2   Prove theorem 4.2.

4.3   Prove the first step in the proof of theorem 4.4.

4.4   Assume α_0 to be a quadratic number. Prove that α_1 defined by $\alpha_0 = c_0 + \dfrac{1}{\alpha_1}$ is also quadratic, and that the second degree equations satisfied by α_1 and α_0 have the same discriminant. Then prove the second step in the proof of theorem 4.4.

4.5   Prove corollary 4.2.

4.6   Assume that $(x, y) \in \mathbb{N}^2$ satisfies (4.28). Prove that x/y is a convergent of the regular continued fraction expansion of $\sqrt{D}$.

4.7   Solve the diophantine equation $11x + 43y = 10$.

4.8   Find the regular continued fraction expansion of $\sqrt{34}$. Then determine all values of L for which the diophantine equation $x^2 - 34y^2 = L$ ($1 \leq L \leq 5$) has solutions. In any case, find the fundamental solution.

4.9   **Galois theorem**

Let $\alpha_0 = \overline{[c_0, c_1, ..., c_{T-1}]}$ be a reduced irrational quadratic number. Prove that $-1/\alpha_0^* = \overline{[c_{T-1}, c_{T-2}, ..., c_1, c_0]}$. Deduce that, if $D \in \mathbb{N}$ is not a perfect square, then $\sqrt{D} = [c_0, \overline{c_1, c_2, ..., c_2, c_1, 2c_0}]$, which means that $c_k = c_{T-k}$ for $k = 1, 2, ..., T-1$.

4.10   Sums of two squares

1) Let $\alpha_0 = \sqrt{D}$. As in section 4.6, put $\alpha_n = \left(A_n + \sqrt{D}\right)/B_n$.

Prove that, for every $n \geq 0$, $A_n + A_{n+1} = c_n B_n$ and $D = A_{n+1}^2 + B_n B_{n+1}$.

2) Check that $B_0 = B_T$ and $B_1 = B_{T-1}$. Then prove that $B_k = B_{T-k}$ for every $k \in \{0, 1, 2, ..., T\}$.

3) Deduce that, if the expansion of $\sqrt{D}$ in regular continued fraction has an *odd* period, then D can be written as a sum of two squares.

4) Application: write $D = 697$ as a sum of two squares.

5) Prove that $B_k \geq 2$ for $k = 1, 2, ..., T-1$. Deduce that, if the equation $x^2 - Dy^2 = -1$ has solutions, then D is a sum of two squares. Is the converse true?

4.11   Primes which are sums of two squares

1) Let p be a prime number such that $p \equiv 1 \pmod 4$. Let (x_1, y_1) be the fundamental solution of the Pell equation $x^2 - py^2 = 1$. Prove first that x_1 is odd. Then, by considering $(x_1 + 1)$ and $(x_1 - 1)$, prove that the equation $x^2 - py^2 = -1$ has solutions.

2) Let p be an odd prime. Prove that p is a *sum of two squares* if, and only if, $p \equiv 1 \pmod 4$.

4.12   Expansion of e^2 in regular continued fraction

Prove that $e^2 = [7, \overline{3n+2, 1, 1, 3n+3, 12n+18}]_{n=0}^\infty$.

Deduce that e is not biquadratic, which means that $ae^4 + be^2 + c \neq 0$ whenever a, b, c are not all zero integers.

4.13   Assume $D \in \mathbb{N}$ is not a perfect square. Use Stratemeyer's formula (exercise 2.8) to express $\sqrt{D}$ by means of an Engel series:

$$\sqrt{D} = \frac{1}{a}\left(b - \sum_{n=0}^{+\infty} \frac{1}{u_0 u_1 ... u_n}\right), \qquad a, b \in \mathbb{N}.$$

4.14   Assume $D \in \mathbb{N}$ is not a perfect square. Find an expression of $\sqrt{D}$ by means of a Cantor infinite product (see section 2.3, page 15):

$$\sqrt{D} = \frac{a}{b} \prod_{n=0}^{+\infty} \left(1 + \frac{1}{q_n} \right), \qquad a, b \in \mathbb{N}.$$

4.15   A striking number (following exercise 2.10)

Consider the number α defined in exercise 2.10 by $\alpha = 3 \prod_{n=0}^{+\infty} \left(1 + 10^{-4^n} \right)$.

Our aim is to determine the regular continued fraction expansion of α.

1) For every integer $n \geq 1$, define $r_n = 3 \prod_{k=0}^{n-1} \left(10^{4^k} + 1 \right)$ and $s_n = 10^{\left(4^n - 1 \right)/3}$.

Prove that, for every $n \geq 1$, $\left| \alpha - \dfrac{r_n}{s_n} \right| < \dfrac{1}{2s_n^3}$. Conclusion regarding $\dfrac{r_n}{s_n}$?

2) We put $c_0 = 3$, $c_1 = 3$, $c_2 = 3$, $c_3 = 30$, and for every $n \geq 2$

$$c_{2n} = 3.10^{\left(4^{n-1} - 1 \right)/3} \prod_{k=0}^{n-2} \left(10^{4^k} + 1 \right), \quad c_{2n+1} = 3.10^{\left(2.4^{n-1} + 1 \right)/3} \prod_{k=0}^{n-2} \left(10^{2.4^k} + 1 \right).$$

Let $\beta = [c_1, c_2, \cdots, c_n, \cdots]$. We denote by p_n / q_n the convergents of β. Prove that, for every $n \geq 1$,

$$\begin{cases} p_{2n} = 3 \displaystyle\prod_{k=0}^{n-1} \left(10^{4^k} + 1 \right), \quad q_{2n} = 10^{\left(4^n - 1 \right)/3}, \\[4mm] p_{2n+1} = 10^{\left(2.4^n + 1 \right)/3}, \quad q_{2n+1} = \dfrac{10^{4^n - 1}}{3 \displaystyle\prod_{k=0}^{n-1} \left(10^{4^k} + 1 \right)}. \end{cases}$$

3) Deduce the regular continued fraction expansion of α.

4.16   Continued fractions and series

1) Let $[0, c_1, c_2, \cdots, c_n, \cdots]$ be an infinite regular contiued fraction. Denote by P_n / Q_n its convergents. Prove that

$$[0, c_1, c_2, \cdots, c_n, \cdots] = \sum_{n=0}^{+\infty} \frac{(-1)^n}{Q_n Q_{n+1}}.$$

2) Let $c_1, c_2, \cdots, c_n, \cdots$ be a sequence of natural integers such that $c_n \geq 1$

for every $n \geq 1$. Let u_n satisfy $u_0 = 1$, $u_1 = c_1$, and $u_{n+1} = c_{n+1}u_n + u_{n-1}$ for every $n \geq 1$. Prove that

$$\sum_{n=0}^{+\infty} \frac{(-1)^n}{u_n u_{n+1}} = [0, c_1, c_2, \cdots, c_n, \cdots].$$

3) Let $c_1, c_2, \cdots, c_n, \cdots$ be a sequence of natural integers such that $c_n \geq 3$ for every $n \geq 1$. Let u_n satisfy $u_0 = 1$, $u_1 = c_1 - 1$, and $u_{n+1} = c_{n+1}u_n - u_{n-1}$ for every $n \geq 1$. Prove that

$$\sum_{n=0}^{+\infty} \frac{1}{u_n u_{n+1}} = [0, c_1 - 2, 1, c_2 - 2, 1, c_3 - 2, 1, \cdots].$$

4) Define the *Fibonacci sequence* F_n by $F_0 = 0$, $F_1 = 1$, $F_{n+1} = F_n + F_{n-1}$ for every $n \geq 1$. Compute the sums of the following series.

$$S = \sum_{n=1}^{+\infty} \frac{(-1)^{n-1}}{F_n F_{n+1}}, \quad T = \sum_{n=1}^{+\infty} \frac{1}{F_{2n} F_{2n+2}}, \quad U = \sum_{n=0}^{+\infty} \frac{1}{F_{2n+1} F_{2n+3}}, \quad V = \sum_{n=1}^{+\infty} \frac{1}{F_n F_{n+2}}.$$

5) Compute V by another method, more elementary.

4.17   Lagrange's method

1) Prove that the equation $x^3 - 2x - 5 = 0$ (E) has a unique root α in the interval $[2, 3]$. Prove that α is irrational.

2) Compute the first five convergents of the continued fraction expansion of α. Deduce a lower and an upper bound for α.

Chapter 5

Quadratic fields
and diophantine equations

We first define what a quadratic field $\mathbb{K} = \mathbb{Q}\left(\sqrt{d}\right)$ is, and study the properties of its ring of integers $\mathbb{A}_{\mathbb{K}}$ (sections 5.1 and 5.2). Section 5.3 is devoted to the study of the units of $\mathbb{A}_{\mathbb{K}}$, which are connected to certain diophantine equations. For example, they allow to find the complete solution of the Pell equation (theorem 5.8). In section 5.4, we solve the Fermat equations $x^2 + y^2 = z^2$ and $x^4 + y^4 = z^4$ by using the unique factorization theorem in $\mathbb{Z}$, which leads us to define prime and irreducible elements in $\mathbb{A}_{\mathbb{K}}$ and to ask if $\mathbb{A}_{\mathbb{K}}$ is, like $\mathbb{Z}$, a unique factorization domain. The answer is affirmative when $\mathbb{A}_{\mathbb{K}}$ is an euclidean domain (section 5.6), which allows us to solve, in some cases, Mordell's equation $y^2 = x^3 + k$, and Fermat's equation $x^3 + y^3 = z^3$.

5.1   Quadratic fields

We know from section 4.5 that quadratic irrational numbers can be written as $\left(b + \sqrt{D}\right)/2a$ with a, b, $D \in \mathbb{Z}$. Write $\sqrt{D} = e\sqrt{d}$, where no squared prime divides d. We say that d is a *squarefree integer*. In particular, $\sqrt{d}$ is irrational by theorem 1.1. For example, we have $\sqrt{12} = 2\sqrt{3}$, $\sqrt{-9} = 3\sqrt{-1}$, and evidently $\sqrt{-1} = i$. Thus a quadratic irrational number can be written as $\alpha + \beta\sqrt{d}$, where α, $\beta \in \mathbb{Q}$, $d \in \mathbb{Z}$ is squarefree. Hence, for any squarefreee $d \in \mathbb{Z}$, it is natural to define

$$\mathbb{K} = \mathbb{Q}\left(\sqrt{d}\right) = \left\{ x \in \mathbb{C} \,/\, x = \alpha + \beta\sqrt{d} \,/\, \alpha, \beta \in \mathbb{Q} \right\}. \qquad (5.1)$$

Theorem 5.1 $\mathbb{K} = \mathbb{Q}\left(\sqrt{d}\right)$ *is a subfield of* $\mathbb{C}$ *and contains* $\mathbb{Q}$. *It is also a vector space of dimension 2 over* $\mathbb{Q}$, *with base* $\left(1, \sqrt{d}\right)$.

Proof   See exercise 5.1.

We say that the field $\mathbb{K} = \mathbb{Q}\left(\sqrt{d}\right)$ is a *quadratic field*. It is easy to check (exercise 5.2) that every element of $\mathbb{K} = \mathbb{Q}\left(\sqrt{d}\right)$ is either rational or quadratic irrational. If $d > 0$, we say that $\mathbb{K} = \mathbb{Q}\left(\sqrt{d}\right)$ is a *real quadratic field*. If $d < 0$, that it is an *imaginary quadratic field*.

Now let $x = \alpha + \beta\sqrt{d} \in \mathbb{Q}\left(\sqrt{d}\right)$. By definition, the *conjugate* of x is $x^* = \alpha - \beta\sqrt{d}$. Note that, if $\mathbb{K} = \mathbb{Q}\left(\sqrt{d}\right)$ is an imaginary quadratic field, then for every $x \in \mathbb{K}$, $x^* = \bar{x}$, where $\bar{x}$ is the complex conjugate of x.

The *norm* of $x = \alpha + \beta\sqrt{d} \in \mathbb{Q}\left(\sqrt{d}\right)$ is defined by

$$N(x) = xx^* = \alpha^2 - d\beta^2. \tag{5.2}$$

The norm of x is clearly always a rational number, and if $x \in \mathbb{Q}$, then $N(x) = x^2$. If $d < 0$, that is if $\mathbb{K} = \mathbb{Q}\left(\sqrt{d}\right)$ is an imaginary quadratic field, then $N(x) = |x|^2$ for every $x \in \mathbb{K}$.

Theorem 5.2   *Let* $\mathbb{K}$ *be a quadratic field. Then, for every* $x, y \in \mathbb{K}$,
$$(x + y)^* = x^* + y^*, \quad (xy)^* = x^* y^*, \quad N(xy) = N(x)N(y).$$

Proof   See exercice 5.3.

5.2   Ring of integers of a quadratic field

Let $x = \alpha + \beta\sqrt{d} \in \mathbb{Q}\left(\sqrt{d}\right)$. We say that x is an integer of $\mathbb{Q}\left(\sqrt{d}\right)$ (or *quadratic integer*) if x satisfies an equation of the form

$$x^2 + ax + b = 0, \quad \text{with } a, b \in \mathbb{Z}.$$

For example, the *golden number* $\Phi = (1+\sqrt{5})/2$ is an integer of the real quadratic field $\mathbb{Q}(\sqrt{5})$, since $\Phi^2 - \Phi - 1 = 0$. The number $j = e^{2i\pi/3}$ $= (-1+i\sqrt{3})/2$ is an integer of the imaginary quadratic field $\mathbb{Q}(i\sqrt{3})$, since $j^2 + j + 1 = 0$. And the number i is an integer of $\mathbb{Q}(i)$ because $i^2 + 1 = 0$. Evidently, every number $m \in \mathbb{Z}$ is an integer of $\mathbb{Q}(\sqrt{d})$ (for every d), since m is a root of the equation $(X - m)^2 = 0$. Conversely, if $m \in \mathbb{Q}$ is an integer of $\mathbb{Q}(\sqrt{d})$, then $m \in \mathbb{Z}$ (exercise 1.6).

Remark 5.1   *If x is a quadratic integer, then $N(x) \in \mathbb{Z}$* (exercise 5.4, question 1). The converse is false (take for example one of the roots of $2x^2 + 3x + 2 = 0$).

Theorem 5.3   *Let $\mathbb{K} = \mathbb{Q}(\sqrt{d})$ be a quadratic field. The set $\mathbb{A}_\mathbb{K}$ of all integers of $\mathbb{K}$ is a subring of $\mathbb{K}$, called the ring of integers of $\mathbb{K}$. Moreover,*

a. *If $d \equiv 2$ or $3 \pmod 4$,*
$$\mathbb{A}_\mathbb{K} = \left\{ x \in \mathbb{K} \,/\, x = \alpha + \beta\sqrt{d} \,/\, \alpha, \beta \in \mathbb{Z} \right\} = \mathbb{Z}(\sqrt{d}).$$

b. *If $d \equiv 1 \pmod 4$,*
$$\mathbb{A}_\mathbb{K} = \left\{ x \in \mathbb{K} \,/\, x = \frac{\alpha + \beta\sqrt{d}}{2} \,/\, \alpha, \beta \in \mathbb{Z}, \ \alpha \text{ and } \beta \text{ of the same parity} \right\}$$
$$= \mathbb{Z}\left(\frac{1+\sqrt{d}}{2} \right).$$

Proof   See exercise 5.4.

It should be noted that, if $d \equiv 1 \pmod 4$, $\mathbb{A}_\mathbb{K}$ contains $\mathbb{Z}(\sqrt{d})$ (take $\alpha = 2m$ and $\beta = 2p$ even), but there are other integers in $\mathbb{A}_\mathbb{K}$, of the form $\left[2m + 1 + (2p+1)\sqrt{d} \right]/2$, $m, p \in \mathbb{Z}$.

In the case where $d < 0$, the ring $\mathbb{A}_\mathbb{K}$ of integers of $\mathbb{K} = \mathbb{Q}(\sqrt{d})$ is easy to represent in the complex plane. Figure 5.1 below shows the ring $\mathbb{Z}(i)$

of $\mathbb{Q}(i)$, or *ring of gaussian integers*. Figure 5.2 shows the ring $\mathbb{Z}(j)$ of integers of $\mathbb{Q}(i\sqrt{3})$.

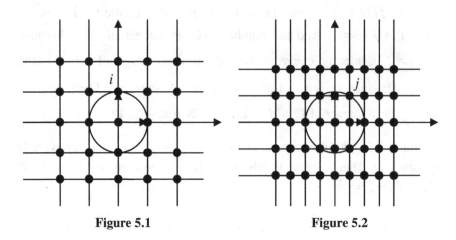

Figure 5.1   **Figure 5.2**

5.3   Units of the ring of integers of a quadratic field

5.3.1   Introduction

Let $\mathbb{A}$ be an integral domain, that is a commutative ring without divisors of zero. We say that $\varepsilon \in \mathbb{A}$ is a *unit* of $\mathbb{A}$ if ε is *invertible* in $\mathbb{A}$, which means that there exists $\eta \in \mathbb{A}$ such that $\varepsilon\eta = 1$. For example, the only units of $\mathbb{Z}$ are $+1$ and -1. It is easy to check (see exercise 5.5) that the set $\mathbb{A}^{\times}$ of all units of $\mathbb{A}$ is a *multiplicative group*. In the case where $\mathbb{A} = \mathbb{A}_{\mathbb{K}}$, ring of integers of a quadratic number field $\mathbb{K}$, we have

Theorem 5.4   *Let* $\varepsilon \in \mathbb{A}_{\mathbb{K}}$. *Then* $\varepsilon \in \mathbb{A}_{\mathbb{K}}^{\times} \Leftrightarrow |N(\varepsilon)| = 1$.

Proof   See exercice 5.6.

5.3.2   Units of the ring of integers of an imaginary quadratic field

Theorem 5.5   *Let* $\mathbb{K} = \mathbb{Q}(i\sqrt{d})$ *be an imaginary quadratic field. Then:*
a. *If* $d = 1$, *that is if* $\mathbb{A}_{\mathbb{K}}$ *is the ring of the gaussian integers* $\mathbb{Z}(i)$,

$$\mathbb{A}_{\mathbb{K}}^{\times} = \{-1, 1, i, -i\}.$$

b. *If* $d = 3$, *that is if* $\mathbb{A}_{\mathbb{K}} = \mathbb{Z}(j)$, $j = e^{2i\pi/3}$,

$$\mathbb{A}_{\mathbb{K}}^{\times} = \left\{ -1, 1, \frac{1+i\sqrt{3}}{2}, \frac{1-i\sqrt{3}}{2}, \frac{-1+i\sqrt{3}}{2}, \frac{-1-i\sqrt{3}}{2} \right\}.$$

Hence $\mathbb{A}_{\mathbb{K}}$ *is the group of the sixth roots of unity in* $\mathbb{C}$.

c. *If* $d \neq 1$ *and* $d \neq 3$, $\mathbb{A}_{\mathbb{K}}^{\times} = \{-1, 1\}$.

Proof See exercise 5.7.

Remark 5.2   Let $\mathbb{A}$ be any domain. We say that two elements x and y of $\mathbb{A}$ are *associated* if there exists $\varepsilon \in \mathbb{A}^{\times}$ such that $x = \varepsilon y$. Theorem 5.5 shows that the ring $\mathbb{A}_{\mathbb{K}} = \mathbb{Z}(j)$ of integers of $\mathbb{Q}(i\sqrt{3})$, *which is not equal to* $\mathbb{Z}(i\sqrt{3})$ since $-3 \equiv 1 \pmod 4$, counts "a lot of units". This fact has an interesting consequence, which will be useful later:

Theorem 5.6   *Every element of the ring* $\mathbb{Z}(j)$ *of integers of* $\mathbb{Q}(i\sqrt{3})$ *can be associated with an element of* $\mathbb{Z}(i\sqrt{3})$. *In other words,*

$$\forall x \in \mathbb{Z}(j), \exists \varepsilon \in \mathbb{Z}(j)^{\times}, \exists y \in \mathbb{Z}(i\sqrt{3}) \text{ such that } x = \varepsilon y.$$

Proof See exercise 5.8.

5.3.3   Units of the ring of integers of a real quadratic field

Lemma 5.1   *Let* G *be a subgroup of the multiplicative group* $(\mathbb{R}_{+}^{*}, \cdot)$. *Assume that* $G_1 = \{x \in G \, / \, x > 1\}$ *has a least element, denoted by* ω. *Then* $G = \{\omega^n \, / \, n \in \mathbb{Z}\}$. *In other words,* G *is cyclic with generator* ω.

Proof Let $x \in G$. Since $\lim_{n \to +\infty} \omega^n = +\infty$ and $\lim_{n \to -\infty} \omega^n = 0$, there exists $n \in \mathbb{Z}$ such that $\omega^n \leq x < \omega^{n+1}$. Then $1 \leq x/\omega^n < \omega$. Since $x/\omega^n \in G$ and ω is the smallest element of G_1, we have $x = \omega^n$. Hence $G \subset \{\omega^n \, / \, n \in \mathbb{Z}\}$. As evidently $G \supset \{\omega^n \, / \, n \in \mathbb{Z}\}$, lemma 5.1 is proved.

Lemma 5.2   *Let* $d > 0$, *and let* $\mathbb{A}_{\mathbb{K}}$ *be the ring of integers of*

$\mathbb{K} = \mathbb{Q}\left(\sqrt{d}\right)$. *Then there exists* $\varepsilon \in A_{\mathbb{K}}^{\times}$ *such that* $\varepsilon > 1$, *and every* $\varepsilon \in A_{\mathbb{K}}^{\times}$ *such that* $\varepsilon > 1$ *satisfies* $\varepsilon \geq (1 + \sqrt{d})/2$.

Proof We know by corollary 4.3 (or by exercise 1.11), that Pell's equation $x^2 - dy^2 = 1$ has at least one solution $x > 0$, $y > 0$. Then $\varepsilon = x + y\sqrt{d} \in A_{\mathbb{K}}^{\times}$ by theorem 5.4, and $\varepsilon > 1$. Consider now $\varepsilon \in A_{\mathbb{K}}^{\times}$ such that $\varepsilon > 1$. Then $\varepsilon = x + y\sqrt{d}$ with $2x \in \mathbb{Z}$ and $2y \in \mathbb{Z}$ by theorem 5.3, and $|N(\varepsilon)| = |x^2 - dy^2| = 1$ by theorem 5.4. Now we prove that $x > 0$ and $y > 0$. First, $x \neq 0$ and $y \neq 0$ since $|x^2 - dy^2| = 1$ and $\varepsilon > 1$. Moreover, x and y are not both negative since $\varepsilon > 1$. Finally, if $xy < 0$, then $|x + y\sqrt{d}| < |x - y\sqrt{d}|$, whence $1 = |x^2 - dy^2| > (x + y\sqrt{d})^2$, which is impossible since $\varepsilon > 1$. Hence, if $\varepsilon = x + y\sqrt{d} \in A_{\mathbb{K}}^{\times}$ satisfies $\varepsilon > 1$, we have $2x \in \mathbb{N}^*$ and $2y \in \mathbb{N}^*$. Thus $x \geq 1/2$, $y \geq 1/2$, $\varepsilon \geq (1 + \sqrt{d})/2$.

Theorem 5.7 *Let* $d > 0$, *and let* $A_{\mathbb{K}}$ *be the ring of integers of* $\mathbb{K} = \mathbb{Q}\left(\sqrt{d}\right)$. *Then there exists a unit* $\omega > 1$, *called fundamental unit, such that* $A_{\mathbb{K}}^{\times} = \left\{\pm\omega^n \mid n \in \mathbb{Z}\right\}$.

Proof The set of positive units is clearly a subgroup of the multiplicative group $\mathbb{R}_+^*$ and theorem 5.7 is an easy consequence of lemma 5.1. To apply this lemma, we only need to prove the existence of the smallest unit $\omega > 1$. Assume there is no such unit. Then we can find a sequence ε_k of units so that $\varepsilon_1 > \varepsilon_2 > \varepsilon_3 > \cdots > \varepsilon_k > \cdots > 1$. The sequence ε_k is decreasing and has a lower bound, therefore it converges to $L \geq 1$. Hence $\lim_{k \to +\infty} \varepsilon_k / \varepsilon_{k+1} = 1$. But $\varepsilon_k / \varepsilon_{k+1} > 1$ and it is a unit. Lemma 5.2 applies and we have $\varepsilon_k / \varepsilon_{k+1} \geq (1 + \sqrt{d})/2 > 1$, contradiction.

Example 5.1 Let $d = 5$. Then $\mathbb{K} = \mathbb{Q}(\sqrt{5})$ and $A_{\mathbb{K}} = \mathbb{Z}\left((1 + \sqrt{5})/2\right)$ $= \mathbb{Z}(\Phi)$, where Φ is the golden number. In this case, the fundamental unit ω satisfies $\omega \geq (1+\sqrt{5})/2$ (lemma 5.2). But $N((1+\sqrt{5})/2) = -1$. Hence Φ is a unit by theorem 5.4. It is the fundamental unit and

$$\mathbb{A}_{\mathbb{K}}^{\times} = \{\pm\Phi^n \mid n \in \mathbb{Z}\}.$$

By using lemmas 5.1 and 5.2 we also obtain a complete solution of Pell's equation.

Theorem 5.8 *Let* (x_1, y_1) *be the fundamental solution of Pell's equation* $x^2 - dy^2 = 1$. *Let* (x_n, y_n) *be the sequence of the positive solutions* $x_n \geq 0$, $y_n \geq 0$, *in increasing order. Then*

    a. $x_n + y_n\sqrt{d} = (x_1 + y_1\sqrt{d})^n$ *for every* $n \in \mathbb{N}$.

    b. $x_{n+2} = 2x_1 x_{n+1} - x_n$ *and* $y_{n+2} = 2x_1 y_{n+1} - y_n$ *for every* $n \in \mathbb{N}$.

Proof See exercise 5.9.

Exemple 5.2 Consider the Pell's equation $x^2 - 19y^2 = 1$. The trivial solution is $x_0 = 1$, $y_0 = 0$. The fundamental solution is $x_1 = 170$, $y_1 = 39$ by example 4.8. Sequences (x_n) and (y_n) satisfy the recurrence relations $x_{n+2} = 340x_{n+1} - x_n$, $y_{n+2} = 340y_{n+1} - y_n$. Therefore the next solutions are $x_2 = 57799$, $y_2 = 13260$, $x_3 = 19651490$, $y_3 = 4508361$, $x_4 = 6681448801$, $y_4 = 1532829480$...

5.4 Unique factorization theorem in $\mathbb{Z}$

One of the most important properties of $\mathbb{Z}$ is the **unique factorization theorem**: *every* $n \in \mathbb{Z}$ *can be factorized as a product of prime numbers, that is* $n = p_1 p_2 \ldots p_k$, *and this factorization is essentially unique.* The two characteristic properties of prime numbers p in $\mathbb{Z}$ are

$$p = ab \Rightarrow a = \pm 1 \text{ or } b = \pm 1. \tag{5.3}$$

$$p \mid ab \Rightarrow p \mid a \text{ or } p \mid b. \tag{5.4}$$

What does "the factorization is essentially unique" mean ?

If $n = p_1 p_2 \ldots p_k$, then evidently also $n = (-p_1)(-p_2) \ldots p_k$, for example, and the primes p_1 and $-p_1$ are *associated*, since -1 is invertible in $\mathbb{Z}$.

"*Unique* factorization" therefore means: if $n = p_1 p_2 \cdots p_k = p_1' p_2' \cdots p_m'$, then $m = k$ and the p_i' can be ordered in such a way that p_i' is *associated* to p_i for every $i = 1, 2, \ldots, k$.

Now we give a first example of the use of the unique factorization theorem in $\mathbb{Z}$ by solving the diophantine equation

$$x^2 + y^2 = z^2. \tag{5.5}$$

We can assume that x, y and z are *coprime*. Indeed, if x, y and z have a common divisor δ, we write $x = \delta x'$, $y = \delta y'$, $z = \delta z'$, and simplify equation (5.5). This equation is a special case of *Fermat's equation*

$$x^n + y^n = z^n. \tag{5.6}$$

Equation (5.5) has infinitely many solutions, the *pythagorean triples* (for example $x = 3$, $y = 4$, $z = 5$), as shown by the following theorem.

Theorem 5.9 *Up to a permutation of x and y, the solutions of equation (5.5), with x, y and z coprime, are given by*

$$x = u^2 - v^2, \quad y = 2uv, \quad z = u^2 + v^2,$$

where $u, v \in \mathbb{Z}$ are coprime and have different parities.

Proof If $x^2 + y^2 = z^2$, one of the two numbers x or y is even. Indeed, if $x = 2k + 1$ and $y = 2m + 1$, then $x^2 \equiv 1 \pmod 4$ and $y^2 \equiv 1 \pmod 4$, whence $z^2 \equiv 2 \pmod 4$, which is impossible because the squares modulo 4 are 0 and 1. Hence we can assume, for example, that y is even. Then x is odd. Otherwise, z would also be even, which is impossible because x, y and z are coprime. We now write (5.5) as $(z + x)(z - x) = y^2$, and *look for the common prime factors of $z + x$ and $z - x$*. Observe first that 2 divises $z + x$ and $z - x$ since x and z are odd. Now let $p \neq 2$ divide $z + x$ and $z - x$. Then $p \mid ((z + x) + (z - x)) = 2z$, $p \mid ((z + x) - (z - x)) = 2x$ As p is an odd prime, $p \mid z$ and $p \mid x$ by (5.4). This implies that $p \mid y$ by equation (5.5), which is impossible.

Hence $\mathrm{GCD}(z + x, z - x) = 2$ and we can write $z + x = 2a$, $z - x = 2b$, $4ab = y^2$, with a and b coprime. But the prime number factorization of $(y/2)^2$ has only even exponents, and the primes factors of $(y/2)^2$ must divide *either a or b*, since a and b are coprime. Therefore the factorizations of a and b have only even exponents, whence $a = u^2$ and $b = v^2$, with $u, v \in \mathbb{Z}$, u and v coprime. Thus $z + x = 2u^2$ and $z - x = 2v^2$, and $z = u^2 + v^2$, $x = u^2 - v^2$, $y = 2uv$. Now u and v must have different parities since z is odd. Conversely, it is easy to check that, if $x = u^2 - v^2$, $y = 2uv$, $z = u^2 + v^2$, where u, v are coprime and have different parities, then (x, y, z) is a pythagorean triple.

Fermat's equation for $n = 4$ provides a second example of the use of the unique factorization theorem.

Theorem 5.10 *The diophantine equation $x^4 + y^4 = z^4$ has no solution if $x \neq 0$, $y \neq 0$, $z \neq 0$.*

Proof We follow Fermat's proof (1640). Its starting point consists in observing that every solution of the equation $x^4 + y^4 = z^4$, with x, y, z coprime, also provides a solution of the equation

$$x^4 + y^4 = z^2, \tag{5.7}$$

with x, y, z coprime. Hence it suffices to prove that (5.7) has no trivial solution. Assume the contrary. Let (x, y, z) be a solution of (5.7), with $z > 0$ and z *minimum*. As for (5.5), we can assume that x and z are odd and y is even. Then, by theorem 5.9, there exist u and v coprime, with different parities, such that $x^2 = u^2 - v^2$, $y^2 = 2uv$, $z = u^2 + v^2$. Thus $x^2 + v^2 = u^2$. As x and v cannot both be odd (see the proof of theorem 5.9), we have x odd, v even, u odd. By applying again theorem 5.9, we obtain $x = a^2 - b^2$, $v = 2ab$, $u = a^2 + b^2$, where a and b are coprime and have different parities. Therefore $y^2 = 4ab(a^2 + b^2)$. Now the numbers a, b, $a^2 + b^2$ are pairwise coprime. By the unique factorization theorem, $a = \alpha^2$, $b = \beta^2$, $a^2 + b^2 = \gamma^2$, $\alpha, \beta, \gamma \in \mathbb{N}$. Hence $\gamma^2 = a^2 + b^2 = \alpha^4 + \beta^4$. The triple (α, β, γ) satisfies (5.7), and $0 < \gamma \leq u < z$, which contradicts the minimality of z. Thus (5.7) has no non trivial solution, and so does $x^4 + y^4 = z^4$.

In this proof, we have used the *descent method* (see section 1.6) . Here it consists in starting from a solution of (5.7) and constructing another solution, "smaller" than the first one.

5.5   Primes and irreducibles

In $\mathbb{Z}$, properties (5.3) and (5.4) are equivalent. Unfortunately, this is not always the case in the ring of integers $\mathbb{A}_{\mathbb{K}}$ of any quadratic field $\mathbb{K}$. This leads us to the following definitions.

Definition 5.1 *Let $\mathbb{A}$ be an integral domain. We say that $p \in \mathbb{A}$ is irreducible if p is not invertible and if $p = ab$, with $a, b \in \mathbb{A}$, implies that a or b is invertible in $\mathbb{A}$.*

Definition 5.2 *Let $\mathbb{A}$ be an integral domain. We say that $p \in \mathbb{A}$ is prime if p is not invertible and if $p \mid ab$, with $a, b \in \mathbb{A}$, implies that $p \mid a$ or $p \mid b$.*

Example 5.3   Let $\mathbb{A}_{\mathbb{K}}$ be the ring of integers of $\mathbb{Q}(i\sqrt{5})$. By theorem 5.5, $\mathbb{A}_{\mathbb{K}} = \mathbb{Z}(i\sqrt{5}) = \left\{ x + iy\sqrt{5} \mid (x, y) \in \mathbb{Z}^2 \right\}$, since $-5 \equiv 3 \pmod{4}$. Let $p = 2$. First we prove that p is *irreducible* in $\mathbb{A}_{\mathbb{K}}$. Assume that $2 = ab$. Then $N(2) = N(a)N(b) \Rightarrow 4 = N(a)N(b)$. But the norm of $\alpha + i\beta\sqrt{5}$ is $\alpha^2 + 5\beta^2$, whence $N(a) = \pm 2$ is impossible. Thus the equation $4 = N(a)N(b)$ implies $N(a) = 1$ or $N(b) = 1$, since $N(a)$ and $N(b)$ are natural integers. Therefore a, or b, is invertible in $\mathbb{Z}(i\sqrt{5})$ by theorem 5.4, which proves that 2 is irreducible in $\mathbb{Z}(i\sqrt{5})$. Now we prove that 2 *is not prime* in $\mathbb{Z}(i\sqrt{5})$. Evidently we have $2 \mid 6$, in other words $2 \mid (1 + i\sqrt{5})(1 - i\sqrt{5})$. However, 2 does not divide $1 + i\sqrt{5}$ nor $1 - i\sqrt{5}$. Indeed, if we had $1 + i\sqrt{5} = 2\alpha$ with $\alpha \in \mathbb{Z}(i\sqrt{5})$, we could take the norms and write $6 = 4N(\alpha)$, whence $4 \mid 6$ in $\mathbb{Z}$, which is false. Thus, in $\mathbb{Z}(i\sqrt{5})$, *there exist irreducibles which are not primes*. On the contrary:

Theorem 5.11 *Let $\mathbb{A}$ be an integral domain. If p is prime in $\mathbb{A}$, then p is irreducible in $\mathbb{A}$.*

Proof   Assume that p is prime in $\mathbb{A}$, and that $p = ab$. Then $1 \cdot p = ab$. Therefore $p \mid ab$, whence $p \mid a$, for example. Thus $a = a'p$, $a' \in \mathbb{A}$, and

$p = ab = a'pb$. Since $\mathbb{A}$ has no divisor of zero, we get $1 = a'b$ and b is invertible in $\mathbb{A}$

Theorem 5.12 *Let $x \in \mathbb{A}_K$. If $N(x)$ is prime in $\mathbb{Z}$, then x is irreducible in $\mathbb{A}_K$.*

Proof See exercise 5.10. Even when theorem 5.12 does not apply, we have seen in example 5.3 how the norm can be used to prove a number to be irreducible.

Theorem 5.13 *Let $\mathbb{A}_K$ be the ring of integers of a quadratic field $\mathbb{K} = \mathbb{Q}(\sqrt{d})$. Then every element of $\mathbb{A}_K$ can be written as a product of irreducible elements.*

Proof See exercise 5.11. Unfortunately, in the general case the expansion in product of irreducibles is not unique. For example, if $\mathbb{A}_K = \mathbb{Z}(i\sqrt{5})$, we have, by example 5.2, $6 = 2 \cdot 3 = (1 + i\sqrt{5})(1 - i\sqrt{5})$. But in $\mathbb{Z}(i\sqrt{5})$ the numbers 2, 3, $1 + i\sqrt{5}$ are $1 - i\sqrt{5}$ are irreducible and not associated (see exercise 5.12). Hence the expansion into a product of irreducibles is not unique in $\mathbb{Z}(i\sqrt{5})$.

Definition 5.3 *Let $\mathbb{A}$ be an integral domain. We say that $\mathbb{A}$ is a unique factorization domain if every non invertible element of $\mathbb{A}$ can be written, in a unique way, as a product of irreducible elements of $\mathbb{A}$.*

"Unique way" means that, if $x = p_1 p_2 \ldots p_k = p_1' p_2' \ldots p_m'$, then $k = m$ and the p_i' can be ordered in such a way that p_i' is *associated* to p_i for every $i = 1, 2, \ldots, k$ (which means that there exist ε_i invertible in $\mathbb{A}$ such that $p_i' = \varepsilon_i p_i$ for every $i = 1, 2, \ldots, k$).

Theorem 5.14 *The ring of integers $\mathbb{A}_K$ of a quadratic number field $\mathbb{K} = \mathbb{Q}(\sqrt{d})$ is a unique factorization domain if, and only if, every irreducible in $\mathbb{A}_K$ is prime.*

Proof We already know by theorem 5.13 that $x \in \mathbb{A}_K$ can be factorized as a product of irreducibles. Assume first that every irreducible is prime, and that $x = p_1 p_2 \ldots p_k = p_1' p_2' \ldots p_m'$. Since $p_1 \mid p_1' p_2' \ldots p_m'$ and p_1 is prime, p_1 divides one of the p_i'. We can renumber and assume that $p_1 \mid p_1'$. Since p_1' is irreducible, we have $p_1' = \varepsilon_1 p_1$, where ε_1 is a unit of $\mathbb{A}_\mathbb{K}$, which means that p_1 and p_1' are associated. After simplification by p_1, we get $p_2 \ldots p_k = \varepsilon_1 p_2' \ldots p_m'$. We can argue the same way as before and prove that p_2 and p_2' are associated, and so on. Hence $\mathbb{A}_\mathbb{K}$ is a unique factorization domain. Conversely, suppose that $\mathbb{A}_\mathbb{K}$ is a unique factorization domain. If p, irreducible, divides ab, then p is one of the irreducible factors of a or b. Hence p divides a or b, and p is prime.

Thus, *in a unique factorization domain*, there is no difference between primes and irreducibles. Like in $\mathbb{Z}$, we will have a *unique factorization theorem*. An important case of unique factorization domains is the case of euclidean domains, which we will study now.

5.6   Euclidean domains

Definition 5.4 *We say that the ring $\mathbb{A}_\mathbb{K}$ of integers of a quadratic field $\mathbb{K} = \mathbb{Q}(\sqrt{d})$ is an euclidean domain if, for every $(a,b) \in \mathbb{A}_\mathbb{K} \times \mathbb{A}_\mathbb{K}^*$, there exist $(q,r) \in \mathbb{A}_\mathbb{K}^2$ satisfying $a = bq + r$ and $|N(r)| < |N(b)|$.*

In other words, there is in $\mathbb{A}_\mathbb{K}$ an euclidean division algorithm like in $\mathbb{Z}$ (in this last case, the condition $|N(r)| < |N(b)|$ simply means $|r| < |b|$).

Assume now that a and b are *coprime* in the euclidean domain $\mathbb{A}_\mathbb{K}$, which means that the only common divisors to a and b are the invertible elements of $\mathbb{A}_\mathbb{K}$. Then *Bézout's identity* holds: there exist two elements u and v in $(u,v) \in \mathbb{A}_\mathbb{K}$ such that

$$au + bv = 1. \qquad (5.8)$$

The proof of (5.8) relies on the euclidean algorithm (exercise 5.13).

Theorem 5.15 *If $\mathbb{A}_\mathbb{K}$ is euclidean, it is a unique factorization domain.*

Proof By theorem 5.14, it suffices to prove that every irreducible of $\mathbb{A}_\mathbb{K}$ is a prime. Let p be irreducible. Assume that $p \mid ab$. If p does not divide a, then p and a are coprime. Indeed, if $d \mid p$ and $d \mid a$, then $d = \varepsilon$ or $d = \varepsilon p$ with $\varepsilon \in A_{\mathbb{K}}^{\times}$ since p is irreducible, and $d = \varepsilon p$ is impossible because p does not divide a. Therefore Bézout's identity applies. There exist $(u, v) \in A_{\mathbb{K}}^{2}$ such that $au + pv = 1$, whence $abu + bpv = b$. Since $p \mid ab$, p divides b, Q.E.D.

Theorem 5.16 *The ring of integers* $A_{\mathbb{K}}$ *of* $\mathbb{K} = \mathbb{Q}(\sqrt{d})$ *is an euclidean domain for* $d = -11, \ -7, \ -3, \ -2, \ -1, \ 2, \ 3, \ 5$ *and* 13.

Proof a. Assume first that $d = -2, \ -1, \ 2$ or 3. Then $d \equiv 2$ or 3 (mod 4), and $A_{\mathbb{K}} = \mathbb{Z}(\sqrt{d})$ by theorem 5.3.

Let $a, b \in A_{\mathbb{K}}$, $b \neq 0$. Then $a/b = x + y\sqrt{d}$, with $x, y \in \mathbb{Q}$. Let α and β be the closest rational integers to x an y respectively. Thus $|x - \alpha| \leq 1/2$ and $|y - \beta| \leq 1/2$. Define $q = \alpha + \beta\sqrt{d}$ and $r = a - bq$. Then $a = b(x + y\sqrt{d})$ implies $r = b((x - \alpha) + (y - \beta)\sqrt{d})$, and therefore $N(r) = N(b)((x - \alpha)^2 - d(y - \beta)^2)$. We see that:

If $|d| \leq 2$, $\left|(x - \alpha)^2 - d(y - \beta)^2\right| \leq (x - \alpha)^2 + 2(y - \beta)^2) \leq 3/4$.

If $d = 3$, $\left|(x - \alpha)^2 - 3(y - \beta)^2\right| \leq \max((x - \alpha)^2, 3(y - \beta)^2) \leq 3/4$.

In both cases, $|N(r)| < |N(b)|$ and $A_{\mathbb{K}}$ is euclidean.

b. Assume now that $d = -11, \ -7, \ -3, \ 5$ or 13. In these cases $d \equiv 1$ (mod 4). Again, let $a, b \in A_{\mathbb{K}}$, with $a/b = x + y\sqrt{d}$, $x, y \in \mathbb{Q}$. Let β be the closest rational integer to $2y$. Then $|\beta/2 - y| \leq 1/4$. Let α be the integer *of same parity* than β, closest to $2x$. Then $|\alpha - 2x| \leq 1$, and $|\alpha/2 - x| \leq 1/2$. Define $q = (\alpha + \beta\sqrt{d})/2$, $r = a - bq$. By theorem 5.3, $q, r \in A_{\mathbb{K}}$. Since $a = b(x + y\sqrt{d})$, $r = b\left((x - \alpha/2) + (y - \beta/2)\sqrt{d}\right)$ and $N(r) = N(b)\left((x - \alpha/2)^2 - d(y - \beta/2)^2\right)$.

If $|d| \leq 11$, we have $\left|\left(x - \dfrac{\alpha}{2}\right)^2 - d\left(y - \dfrac{\beta}{2}\right)^2\right| \leq \dfrac{1}{4} + \dfrac{11}{16} < 1$.

If $d = 13$, $\left| \left(x - \dfrac{\alpha}{2} \right)^2 - 13 \left(y - \dfrac{\beta}{2} \right)^2 \right| \leq \max\left(\dfrac{1}{4}, \dfrac{13}{16} \right) < 1$.

In both cases, $|N(r)| < |N(b)|$, and theorem 5.16 is proved.

5.7   Diophantine equations

5.7.1   Mordell's equation $y^2 = x^3 + k$

Let $k \in \mathbb{Z}$ be given. The equation $y^2 = x^3 + k$ has been studied by Mordell. In some cases it can be solved by using elementary means. The general case is much more difficult, and we will give some complements in chapter 11. As an elementary example, consider the equation

$$y^2 = x^3 - 1. \qquad (5.9)$$

First we observe that x is odd (otherwise $y^2 \equiv -1$ (mod 4), which is impossible). We factorize in the ring of *gaussian integers* $\mathbb{Z}(i)$, which is euclidean by theorem 5.16. We have $(y+i)(y-i) = x^3$. We begin by showing that $y+i$ and $y-i$ are coprime. For this, suppose that p prime in $\mathbb{Z}(i)$ divides $y+i$ and $y-i$. Then p divides their difference, therefore $p \mid 2i$. Taking the norm yields $N(p) \mid 4$, whence $N(p) = 2$ or $N(p) = 4$, since p is not invertible. But $N(p) \mid N(y+i)$ since $p \mid y+i$. This yields $N(p) \mid y^2 + 1 = x^3$ in $\mathbb{Z}$, which is impossible because x is odd. Thus $y+i$ and $y-i$ have no common prime factor. Since all prime factor of x^3 are cubed, the unique factorization in $\mathbb{Z}(i)$ (*up to an invertible factor*) implies that $y+i = \varepsilon(p_1 p_2 \ldots p_k)^3$, $\varepsilon \in \mathbb{Z}(i)^\times$. But every invertible element ε is a cube in $\mathbb{Z}(i)$ by theorem 5.5, since $1 = 1^3$, $-1 = (-1)^3$, $i = (-i)^3$, $-i = i^3$. Therefore $y+i = (a+ib)^3$, with a, $b \in \mathbb{Z}$. Developing yields $y = a^3 - 3ab^2$ and $1 = 3a^2 b - b^3$. By the second equality we see that $b \mid 1$, whence $b = 1$ or -1. Therefore $\pm 1 = 3a^2 - 1$, which implies $b = -1$ and $a = 0$. Then by the first equality $y = 0$. Replacing in (5.9) finally yields $x^3 = 1$, whence $x = 1$.

Therefore Mordell's equation (5.9) has only one solution, $x = 1, y = 0$.

5.7.2   Fermat's equation for $n = 3$

Theorem 5.17 *The equation $x^3 + y^3 = z^3$ has no solution when x, y, z are non zero rational integers.*

Proof Assume that $x^3 + y^3 + z^3 = 0$, with x, y, $z \in \mathbb{Z}$, $x \neq 0$, $y \neq 0$, $z \neq 0$. We can assume that x, y, z are coprime, and that x and z are odd, while y is even. Indeed, the combinations even-even-even, even-even-odd and odd-odd-odd are impossible. We use the descent method and choose such a solution with y *minimum*. Following Euler, put $x = a + b$ and $z = a - b$. Then a and b are coprime (otherwise x and z, and thus x, y and z would have a common divisor) and have different parities (since x and z are odd). Fermat's equation $x^3 + z^3 = -y^3$ becomes

$$2a(a^2 + 3b^2) = -y^3. \tag{5.10}$$

Since a and b have different parities, $a^2 + 3b^2$ is odd. Therefore 8 divides $2a$ since y is even. Moreover, every common divisor d to $2a$ and $a^2 + 3b^2$ is odd. Hence d divides a, and therefore d divides $3b^2$. Since a and b are coprime, this yields GCD $(2a, a^2 + 3b^2) = 1$ or 3.

First case: GCD $(2a, a^2 + 3b^2) = 1$

The unique factorization theorem in $\mathbb{Z}$ and (5.10) imply $2a = r^3$ and $a^2 + 3b^2 = s^3$, r, $s \in \mathbb{Z}$, s odd. Factorizing in $\mathbb{Z}(j)$ yields

$$(a + ib\sqrt{3})(a - ib\sqrt{3}) = s^3.$$

But $a + ib\sqrt{3}$ and $a - ib\sqrt{3}$ are coprime in $\mathbb{Z}(j)$, which is euclidean by theorem 5.16, case $d = -3$. Indeed, if p prime in $\mathbb{Z}(j)$ divides $a + ib\sqrt{3}$ and $a - ib\sqrt{3}$, it also divides their sum $2a$. Taking the norms yields $N(p) | a^2 + 3b^2$ and $N(p) | 4a^2$ in $\mathbb{Z}$. Now $N(p)$ is odd since $N(p) | a^2 + 3b^2$, whence $N(p) | a^2$, $N(p) | a^2 + 3b^2$. This is impossible because a and $a^2 + 3b^2$ are coprime.

Now, the unique factorization theorem in $\mathbb{Z}(j)$ shows that there exist $t \in \mathbb{Z}(j)$ and $\eta \in \mathbb{Z}(j)^\times$ such that $a + ib\sqrt{3} = \eta t^3$. Since $-1 = (-1)^3$, we can assume by theorem 5.5.b that $\eta = 1$ or $\frac{1}{2}(1 + i\sqrt{3})$ or $\frac{1}{2}(1 - i\sqrt{3})$.

Let $\varepsilon \in \mathbb{Z}(j)^{\times}$ satisfying $\varepsilon t \in \mathbb{Z}(i\sqrt{3})$ (theorem 5.6). Since ε is a sixth root of unity in $\mathbb{C}$ by theorem 5.5, we have $\varepsilon^{-3} = \pm 1$. This yields $a + ib\sqrt{3} = \eta\varepsilon^{-3}(\varepsilon t)^{3} = \eta(\pm\varepsilon t)^{3} = \eta(u + iv\sqrt{3})^{3}$, with $u \in \mathbb{Z}$ and $v \in \mathbb{Z}$. Developing yields

$$a + ib\sqrt{3} = \eta\left[u(u+3v)(u-3v) + i\sqrt{3}.3v(u-v)(u+v) \right]. \qquad (5.11)$$

We prove now that $\eta \neq (1 + i\sqrt{3})/2$. Assume the contrary, and develop the right-hand member of (5.11). We obtain

$$\begin{cases} a = \dfrac{1}{2}\left[u(u+3v)(u-3v) - 9v(u-v)(u+v) \right] \\[2mm] b = \dfrac{1}{2}\left[3v(u-v)(u+v) + u(u+3v)(u-3v) \right] \end{cases}$$

But this is impossible. If u and v have the same parity, a and b are even, hence not coprime. If u and v have different parities, one of the numbers a or b is not an integer. Therefore $\eta \neq (1 + i\sqrt{3})/2$ and similarly we see that $\eta \neq (1 - i\sqrt{3})/2$.

Thus $\eta = 1$ and (5.11) yields

$$\begin{cases} a = u(u+3v)(u-3v) \\ b = 3v(u-v)(u+v) \end{cases}$$

But a is even by (5.10), b is odd, a and b are coprime. Consequently v is odd, u is even, u and $3v$ are coprime. Since $2a = r^{3}$, we have $r^{3} = 2u(u+3v)(u-3v)$. It is easy to check that $2u$, $u+3v$, $u-3v$ are pairwise coprime. By the unique factorization theorem, there exist q, m, $n \in \mathbb{Z}$ such that $2u = q^{3}$, $u + 3v = m^{3}$, $u - 3v = n^{3}$. By addition, we obtain $m^{3} + n^{3} = q^{3}$, with q even, q, m, n coprime. But we have

$$\left| y^{3} \right| = \left| 2a(a^{2} + 3b^{2}) \right| = \left| q^{3}(u^{2} - 9v^{2})(a^{2} + 3b^{2}) \right| \geq 3\left| q^{3} \right| > \left| q \right|^{3}.$$

This contradicts the minimality of y.

Second case: GCD $(2a, a^{2} + 3b^{2}) = 3$

This part of the proof is left to the reader. See exercise 5.14.

Exercises

5.1   Prove theorem 5.1.

5.2   Prove that every $x \in \mathbb{Q}(\sqrt{d})$ is rational or quadratic irrational.

5.3   Prove theorem 5.2.

5.4   Let $x = \alpha + \beta\sqrt{d} \in \mathbb{Q}(\sqrt{d})$. Define the *trace* of x by
$$\mathrm{Tr}(x) = x + x^* = 2\alpha.$$
1) Assume that x is an integer in $\mathbb{Q}(\sqrt{d})$. Prove that $\mathrm{Tr}(x) \in \mathbb{Z}$ and $N(x) \in \mathbb{Z}$.

2) Assume that $x = \dfrac{\alpha}{\beta} + \dfrac{\gamma}{\delta}\sqrt{d}$ is an integer of $\mathbb{Q}(\sqrt{d})$, with α, $\gamma \in \mathbb{Z}$, β, $\delta \in \mathbb{N}$, α and β coprime, γ and δ coprime. Prove that $\beta = 1$ or 2. If $\beta = 1$, prove that $x = \alpha + \gamma\sqrt{d}$, and if $\beta = 2$, prove that α and γ are odd, $\delta = 2$, and $d \equiv 1 \pmod 4$.
3) Prove parts a) and b) of theorem 5.3.
4) Prove that $\mathbb{A}_{\mathbb{K}}$ is a ring.

5.5   Prove that the set $\mathbb{A}^\times$ of all units of a domain $\mathbb{A}$ is a multiplicative group.

5.6   Prove theorem 5.4.

5.7   Prove theorem 5.5.

5.8   Assume $a \equiv 1 \pmod 4$ and $b \equiv -1 \pmod 4$. Check that
$$\frac{a + ib\sqrt{3}}{2} \times \frac{1 + i\sqrt{3}}{2} \in \mathbb{Z}(i\sqrt{3}).$$
Then prove theorem 5.6.

5.9   Proof of theorem 5.8
1) Prove that the set

$$G = \left\{ \varepsilon = x + y\sqrt{d} \in \mathbb{Z}(\sqrt{d}) \, / \, \varepsilon > 0 \text{ and } N(\varepsilon) = 1 \right\}$$

is a multiplicative subgroup of $\mathbb{R}_+^*$.

2) Prove part a) of theorem 5.8.
3) Prove part b) of theorem 5.8.

5.10   Prove theorem 5.12

5.11   Let $x \in \mathbb{A}_{\mathbb{K}}$. Assume that x is not invertible. Prove that x admits an irreducible divisor. Deduce the proof of theorem 5.13.

5.12   Prove that the numbers 2, 3, $1+i\sqrt{5}$ et $1-i\sqrt{5}$ are irreducible in $\mathbb{Z}(i\sqrt{5})$. Then prove that $1+i\sqrt{5}$ is not associated with 2 nor with 3.

5.13   Bézout's identity in an euclidean domain $\mathbb{A}_{\mathbb{K}}$
Assume $\mathbb{A}_{\mathbb{K}}$ to be euclidean, and consider $a, b \in \mathbb{A}_{\mathbb{K}}$, a and b coprime. We perform the sequence of euclidean divisions:

$$a = bq_1 + r_1, \quad b = r_1 q_2 + r_2, \quad r_1 = r_2 q_3 + r_3 \dots \text{ (Euclidean algorithm).}$$

We put $b = r_0$.

1) Prove that there exists k such that $r_k = 0$. Deduce that r_{k-1} is a common divisor to a and b, and consequently that $r_{k-1} \in \mathbb{A}_{\mathbb{K}}^{\times}$.
2) Prove Bézout's identity.

5.14   Complete the proof of theorem 5.17.

5.15   Prove that $1+i\sqrt{13}$ is irreducible in $\mathbb{Z}(i\sqrt{13})$. Deduce that $\mathbb{Z}(i\sqrt{13})$ is not a unique factorization domain.

5.16   By using continued fractions (theorem 4.6), prove that the equation $a^2 - 87b^2 = \pm 2$ has no solution in $\mathbb{Z}^2$. Deduce that $\mathbb{Z}(\sqrt{87})$ is not a unique factorization domain.

5.17 By using properties of the euclidean domain $\mathbb{Z}(i\sqrt{2})$, solve the Mordell equation $y^2 = x^3 - 2$.

5.18 Solve the Mordell equation $y^2 = x^3 - 11$.

5.19 Solve the diophantine equation $x^5 - y^2 = 1$.

5.20 Use the descent method to prove that the diophantine equation $x^4 - y^4 = z^2$ has no solution $x > 0$, $y > 0$, $z > 0$.

5.21 Find the group of unities of $\mathbb{Z}(\sqrt{2})$, ring of integers of $\mathbb{Q}(\sqrt{2})$. Then find the group of unities of the ring $\mathbb{Z}\left(\dfrac{1+\sqrt{13}}{2}\right)$, which is the ring of integers of $\mathbb{Q}(\sqrt{13})$.

5.22 An irrationality result
Let (F_n) be the *Fibonacci sequence*, defined by $F_0 = 0$, $F_1 = 1$, and the recurrence relation $F_{n+1} = F_n + F_{n-1}$ $(n \geq 1)$ (see also exercise 3.11). We define $\theta = \displaystyle\sum_{n=0}^{+\infty} \frac{1}{F_n}$. The aim of this exercise is to prove that $\theta \notin \mathbb{Q}(\sqrt{5})$. In particular, θ is irrational (André-Jeannin, 1989).
1) Define, for $|q| > 1$,

$$f_q(x) = 1 + \sum_{n=1}^{+\infty} \frac{x^n}{(q-1)(q^2-1)\ldots(q^n-1)}, \qquad g_q(x) = \sum_{n=1}^{+\infty} \frac{x^n}{q^n-1}.$$

   a) Check that $f_q(x)$ exists for every $x \in \mathbb{C}$, and that $g_q(x)$ exists for every x such that $|x| < |q|$.

   b) Express $f_q(x)$ as a function of $f_q\left(\dfrac{x}{q}\right)$. Deduce that

$$f_q(x) = \prod_{n=1}^{+\infty}\left(1 + \frac{x}{q^n}\right) \quad \text{for every } x \in \mathbb{C}.$$

c) Prove that $g_q(x) = \sum_{n=1}^{+\infty} \dfrac{x}{q^n - x}$, $|x| < |q|$.

d) Prove that, for all reals x and q satisfying $|q| > 1$ and $|x| < |q|$,

$$g_q(x) = x \dfrac{f'_q(-x)}{f_q(-x)}.$$

2) Let $\Phi = (1 + \sqrt{5})/2$ be the golden number, and let $\Psi = -1/\Phi = \Phi^*$.

a) Prove that $\theta = (\Phi - \Psi) g_{-\Phi^2}(-\Phi)$.

b) Assume that $\theta \in \mathbb{Q}(\sqrt{5})$, $\theta = A/B$, with $A \in \mathbb{Z}(\sqrt{5})$, $B \in \mathbb{N}$.
Prove that

$$A + \sum_{n=1}^{+\infty} \dfrac{A + Bn(\Phi - \Psi)}{(1 + \Phi^2)(1 - \Phi^4)\dots(1 - (-\Phi^2)^n)} (-\Phi)^n = 0.$$

c) For every $k \in \mathbb{N} - \{0\}$, we put

$$X_k = \left(A + \sum_{n=1}^{k} \dfrac{A + Bn(\Phi - \Psi)}{(1 + \Phi^2)(1 - \Phi^4)\dots(1 - (-\Phi^2)^n)} (-\Phi)^n \right)$$

$$\times \left((1 + \Phi^2)(1 - \Phi^4)\dots(1 - (-\Phi^2)^k) \right).$$

Prove that $X_k \in \mathbb{Z}(\Phi)$ and that $|X_k| \le Ck^2 \left(\dfrac{2\Phi}{2\Phi + 1} \right)^k$ for k
sufficiently large, where C is a constant.

d) Prove that there exists a constant C' such that $|X_k^*| \le C'k^2$ for k
sufficiently large.

e) Prove that $N(X_k) = 0$ for k sufficiently large and conclude.

Chapter 6

Squares and sums of squares

This chapter is mainly devoted to the problem of representing natural numbers as sums of squares. The representation by a sum of two squares $n = x^2 + y^2$ is the simplest case. It is studied in section 6.1. Section 6.2 is devoted to complements about finite groups, rings and fields. Its content is used in sections 6.3 and 6.4, where we define and study Legendre symbol, which allows to recognize if a given natural number n is a square modulo a prime p, that is if the equation $x^2 \equiv n \pmod{p}$ has any solution. The representation of integers by general binary quadratic forms $ax^2 + bxy + cy^2$ is studied in section 6.5. Finally, in section 6.6 we prove the celebrated Bachet's theorem, which asserts that every natural integer n can be written as a sum of four squares.

6.1 Sums of two squares

Our aim is to find on which conditions a natural integer n can be written as a sum of two squares, that is $n = a^2 + b^2$, $a, b \in \mathbb{N}$.

We know the answer in the case where n is prime (exercise 4.11):

Theorem 6.1 *Let $p \in \mathbb{N}$, p prime. Then p is a sum of two squares if, and only if, $p = 2$ or $p \equiv 1 \pmod 4$.*

Moreover, if the integers n and m are sums of two squares, so is their product mn. This results directly from *Lagrange's identity*:

$$(a^2 + b^2)(c^2 + d^2) = (ad + bc)^2 + (ac - bd)^2. \qquad (6.1)$$

One can check (6.1) by a direct computation. Alternatively, on can also observe that, by putting $x = a + ib$, $y = c + id$, Lagrange's identity is equivalent to $N(x)N(y) = N(xy)$ in the ring of gaussian integers $\mathbb{Z}(i)$.

Theorem 6.2 *Let $n \in \mathbb{N}$. Then n is a sum of two squares if, and only if, the exponents of its prime factors congruent to -1 (mod 4) are even.*

Proof Assume first that n is a sum of two squares, $n = a^2 + b^2$, with $a, b \in \mathbb{N}$. Let $\delta = \text{GCD}(a,b)$. We can write $n = \delta^2(c^2 + d^2)$, with c and d coprime. Let p be an odd prime dividing $c^2 + d^2$. Then in $\mathbb{Z}(i)$ we have $p \mid (c + id)(c - id)$. If p is prime in $\mathbb{Z}(i)$, it divides $c + id$ in $\mathbb{Z}(i)$, and also $c - id$ by taking the conjugates. Hence it divides their sum and difference, whence $p \mid 2c$ and $p \mid 2id$ in $\mathbb{Z}(i)$. Taking the norms yields $p^2 \mid 4c^2$ and $p^2 \mid 4d^2$ in $\mathbb{Z}$. Since p is an odd prime, $p \mid c$ and $p \mid d$, contradiction. Therefore p is not prime in $\mathbb{Z}(i)$, and $p = xy$, x and y non invertible in $\mathbb{Z}(i)$. Taking the norms yields $p^2 = N(x)N(y)$, and $N(x) = p$ since $N(x) \neq 1$ and $N(y) \neq 1$. If we put $x = \alpha + i\beta$, we see that $p = \alpha^2 + \beta^2$. Hence p is a sum of two squares and $p \equiv 1 \pmod 4$ by theorem 6.1. The only prime divisors of $c^2 + d^2$ are therefore 2 or the primes congruent to 1 (mod 4). Hence the primes congruent to -1 (mod 4) which divide n must divide δ; and their exponant in the prime number factorization of n is even, since $n = \delta^2(c^2 + d^2)$ divides n.

Conversely, if the prime factors p_i congruent to -1 (mod 4) which divide n have an even exponent, then $n = \left(p_1^{\alpha_1} p_2^{\alpha_2} ... p_k^{\alpha_k} \right)^2 \cdot q_1^{\beta_1} q_2^{\beta_2} ... q_m^{\beta_m}$.

Now, the q_i's are sums of two squares by theorem 6.1. Hence, by Lagrange's identity (6.1), $q_1^{\beta_1} q_2^{\beta_2} ... q_m^{\beta_m} = a^2 + b^2$.

Therefore $n = (\gamma a)^2 + (\gamma b)^2$ is a sum of two squares, with $\gamma = p_1^{\alpha_1} ... p_k^{\alpha_k}$.

Example 6.1 $585 = 3^2 \cdot 5 \cdot 13$ is a sum of two squares. Its only prime divisor congruent to -1 (mod 4) is 3 and its exponent is even. We have $5 = 2^2 + 1^2$ and $13 = 3^2 + 2^2$. By Lagrange's identity (6.1) with $a = 2$, $b = 1$, $c = 3$, $d = 2$, we get $5 \cdot 13 = 7^2 + 4^2$. And for $a = 1$, $b = 2$, $c = 3$, $d = 2$, we get $5 \cdot 13 = 8^2 + 1^2$. Therefore $585 = 21^2 + 12^2 = 24^2 + 3^2$.

6.2   Finite algebraic structures

The basic groups, rings and fields we encounter in number theory are infinite. It will be useful to classify their elements in order to operate in finite sets. This will be done by means of equivalence relations. A binary relation $\mathcal{R}$ on a set E is an *equivalence relation* if it is

- Reflexive: $\forall x \in E, \; x\mathcal{R}x$.
- Symmetric: $\forall (x, y) \in E^2, \; x\mathcal{R}y \Rightarrow y\mathcal{R}x$.
- Transitive: $\forall (x, y, z) \in E^3, \; x\mathcal{R}y$ and $y\mathcal{R}z \Rightarrow x\mathcal{R}z$.

An equivalence relation on E defines a partition of E in *equivalence classes*. All elements of the same class are equivalent, and two elements x and y belonging to different classes are not equivalent.

Now let $(G, +)$ [respectively $(G, \cdot)$] be a commutative group, and H a subgroup of G. Define relation $\mathcal{R}$ by $x\mathcal{R}y \Leftrightarrow x - y \in H$ [respectively $xy^{-1} \in H$]. It is easy to check (see exercise 6.1) that $\mathcal{R}$ is an equivalence relation. The set of equivalence classes is called the *quotient* of G by H and denoted by G/H.

Example 6.2   Take $(G, +) = (\mathbb{Z}, +)$. Let $n \in \mathbb{Z} - \{0\}$ be given. Consider the subgroup H of all multiples of n in $\mathbb{Z}$, $H = n\mathbb{Z} = \{nm \, / \, m \in \mathbb{Z}\}$. In this case $x\mathcal{R}y \Leftrightarrow x - y \in H \Leftrightarrow x - y$ is a multiple of $n \Leftrightarrow x \equiv y \pmod{n}$. The quotient is $\mathbb{Z}/n\mathbb{Z}$. It is a finite set with n elements, which are the rests modulo n, namely 0, 1, 2, ..., $n-1$.

Now return to the general case. For every $x \in G$, denote by $\hat{x}$ the class of x modulo $\mathcal{R}$. We observe that the addition in G [respectively the multiplication] is *compatible with* $\mathcal{R}$, which means that $x\mathcal{R}y$ and $z\mathcal{R}t \Rightarrow (x+z)\mathcal{R}(y+t)$ [respectively $(xz)\mathcal{R}(yt)$]. This allows to define an addition in the set G/H [respectively a multiplication] by putting $\hat{x} + \hat{y} = \widehat{x+y}$ [respectively $\hat{x}.\hat{y} = \widehat{x.y}$], which provides G/H with a group structure (exercice 6.2). The group G/H is called the *quotient group* of G by H.

Now, we assume that $(\mathbb{A}, +, \cdot)$ is a domain. We will call an additive subgroup I of $\mathbb{A}$ an *ideal* if it satisfies the property:

$$\forall x \in I, \forall y \in \mathbb{A}, \qquad xy \in I. \tag{6.2}$$

Since I is an additive subgroup of $\mathbb{A}$, the quotient $\mathbb{A}/I$ is an additive group. Moreover, it is easy to check that the binary relation $x \mathcal{R} y$ $\Leftrightarrow x - y \in I$ is compatible with the multiplication. Therefore, *if I is an ideal of the domain $\mathbb{A}$, then $\mathbb{A}/I$ is also a domain* (see exercise 6.3). This is the case, in particular, of $\mathbb{Z}/n\mathbb{Z}$. The situation is similar if we replace $\mathbb{Z}$ by the ring $\mathbb{A}_{\mathbb{K}}$ of integers of a quadratic field $\mathbb{K}$.

Theorem 6.3   *Let $I = a\mathbb{A}_{\mathbb{K}}$ be the set of all multiples, in $\mathbb{A}_{\mathbb{K}}$, of the quadratic integer $a \in \mathbb{A}_{\mathbb{K}}$. Then I is an ideal of $\mathbb{A}_{\mathbb{K}}$, and the quotient ring $\mathbb{A}_{\mathbb{K}}/a\mathbb{A}_{\mathbb{K}}$ is finite.*

Proof   See exercise 6.4.

We now consider a finite group G, and denote by $|G|$ its cardinality. The following result is known as *Lagrange's theorem*.

Theorem 6.4   *Let G be a finite commutative group, and H a subgroup of G. Then the groups H and G/H are finite, and $|G| = |H| \times |G/H|$. In particular, $|H|$ divides $|G|$.*

Proof   Assume for example that G is an additive group. Then G/H is the set of equivalence classes for the relation $x \mathcal{R} y \Leftrightarrow (x - y \in H)$. Hence H and G/H are both finite. Let C be any equivalence class. Consider any fixed $x_0 \in C$. Any element of C can be written as $x_0 + x$, with $x \in H$, whence $|C| = |H|$. Thus all equivalence classes have the same cardinality. Since they define a partition of G, $|G| = |G/H| \times |H|$.

Example 6.3   Let G be a finite multiplicative group. Let $a \in G$. Denote by $G_a = \{a^n \mid n \in \mathbb{Z}\}$ the cyclic group of generator a. Since G is finite, we can find two integers m and n, with $m < n$, such that $a^n = a^m$ (pigeon-hole principle), whence $a^{n-m} = 1$. Therefore there exists $k \in \mathbb{N}^*$ such that $a^k = 1$. The least integer k satisfying this property is called the *order* of a in G, and it is clear that $G_a = \{1, a, a^2, ..., a^{k-1}\}$. Hence the group G_a is cyclic and $|G_a| = k =$ order of a in G. By theorem 6.4, we deduce that the order of a in G is a divisor of $|G|$. For example, let $\omega = e^{2i\pi/15}$ and $a = \omega^6 = e^{4i\pi/5}$ in the multiplicative group G of the 15th roots of unity in $\mathbb{C}$. Then the order of a in G is equal to 5.

Theorem 6.5 *Every finite domain $\mathbb{A}$ is a field.*

Proof It suffices to prove that every non-zero element x of $\mathbb{A}$ is invertible for the multiplication. Consider the mapping $\varphi : \mathbb{A} \to \mathbb{A}$ defined by $\varphi(y) = xy$ for every $y \in \mathbb{A}$. It is clear that φ is injective. Indeed $[\varphi(y) = \varphi(y')] \Rightarrow [xy = xy'] \Rightarrow [x(y - y') = 0] \Rightarrow [y - y' = 0]$ since $x \neq 0$ and $\mathbb{A}$ is an integral domain. As $\mathbb{A}$ is finite and φ is injective, φ is bijective. Hence there exists a unique $y \in \mathbb{A}$ such that $xy = 1$, Q.E.D.

Theorem 6.6 *Let $\mathbb{A}$ be a domain, and let $p \in \mathbb{A}$, p non invertible. Then p is prime if, and only if, $\mathbb{A}/p\mathbb{A}$ is an integral domain.*

Proof See exercise 6.5.

Example 6.4 By using theorems 6.3, 6.5 and 6.6, we see that, if p is prime in the ring $\mathbb{A}_{\mathbb{K}}$ of integers of the quadratic field $\mathbb{K}$, then $\mathbb{A}_{\mathbb{K}}/p\mathbb{A}_{\mathbb{K}}$ is a field. In particular, if $p > 0$ is prime in $\mathbb{Z}$, then $\mathbb{Z}/p\mathbb{Z}$ is a finite field of cardinal p, denoted $\mathbb{F}_p$. By using the notion of order in the multiplicative group $\mathbb{F}_p^*$, we obtain *Fermat's theorem*:

Theorem 6.7 *Let $p \in \mathbb{N}$ be a prime. Then $a^p \equiv a$ (mod p).*

Proof See exercise 6.6.

6.3 Legendre symbol

6.3.1 Introduction

Let $p \in \mathbb{N}$ be a prime. Solving the equation $ax^2 + bx + c \equiv 0$ (mod p) in the field $\mathbb{F}_p$ leads to the following question: does it exist $\omega \in \mathbb{F}_p$ such that $\omega^2 = \Delta = b^2 - 4ac$? In other words, how can we identify the rational integers which are squares modulo p? To address this problem, we define, for every $n \in \mathbb{Z}$ and p prime, *Legendre symbol* by

$$
\begin{cases}
\left(\dfrac{n}{p}\right) = 0 & \text{if } n \equiv 0 \ (\text{mod } p) \\[2ex]
\left(\dfrac{n}{p}\right) = 1 & \text{if } n \not\equiv 0 \ (\text{mod } p) \text{ and } n \text{ is a square } (\text{mod } p) \qquad (6.3) \\[2ex]
\left(\dfrac{n}{p}\right) = -1 & \text{if } n \not\equiv 0 \ (\text{mod } p) \text{ and } n \text{ is not a square } (\text{mod } p)
\end{cases}
$$

For example, $\left(\frac{2}{7}\right) = 1$ since $2 \equiv 3^2$ (mod 7), whereas $\left(\frac{3}{7}\right) = -1$, because the squares modulo 7 are 0, 1, 2, 4.

We first prove *Euler's criterion*:

$$
\left(\frac{n}{p}\right) \equiv n^{\frac{p-1}{2}} \quad (\text{mod } p). \tag{6.4}
$$

If $n \equiv 0$ (mod p), the result is evident. Hence we assume $n \not\equiv 0$ (mod p), and we work in the multiplicative group $\mathbb{F}_p^*$. The set of all squares in $\mathbb{F}_p^*$ is the subgroup $G_p = \left\{1^2, 2^2, \ldots, ((p-1)/2)^2\right\}$, the elements of which are all different. Indeed, $a^2 = b^2 \Leftrightarrow a = b$ or $a = -b$ in $\mathbb{F}_p^*$. But $a = -b \Leftrightarrow a + b \equiv 0$ (mod p), impossible since $2 \le a + b \le p - 1$). Therefore, in $\mathbb{F}_p^*$, exactly $(p-1)/2$ elements are squares, and $(p-1)/2$ are not. Now for every $x \in \mathbb{F}_p^*$, $x^{\frac{p-1}{2}} \equiv \pm 1$ (mod p) since $x^{p-1} \equiv 1$ (mod p) by Fermat's theorem 6.7. If x is a square, then $x^{\frac{p-1}{2}} = y^{p-1} = 1$. Thus the equation $X^{\frac{p-1}{2}} = 1$ has $(p-1)/2$ distinct roots in the field $\mathbb{F}_p$, the elements of G_p. If x is not a square, $x^{\frac{p-1}{2}} \equiv -1$ (mod p), since $X^{\frac{p-1}{2}} = 1$ has at most $(p-1)/2$ roots. Euler's criterion is proved.

Example 6.5 For small p, Legendre symbol can be computed by means of Euler's criterion. For example, $\left(\frac{7}{11}\right) \equiv 7^5$ (mod 11). Now $7^2 \equiv 49 \equiv 5$, $7^3 \equiv 35 \equiv 2$, $7^4 \equiv 14 \equiv 3$, $7^5 \equiv 21 \equiv -1$, hence $\left(\frac{7}{11}\right) = -1$, and 7 is not a square modulo 11.

Corollary 6.1 *Legendre symbol is multiplicative, which means that, for every prime p,*

$$\left(\frac{mn}{p}\right)=\left(\frac{m}{p}\right)\times\left(\frac{n}{p}\right).$$

Indeed $\left(\frac{mn}{p}\right)\equiv(mn)^{\frac{p-1}{2}}\equiv m^{\frac{p-1}{2}}n^{\frac{p-1}{2}}\equiv\left(\frac{m}{p}\right)\left(\frac{n}{p}\right)$ (mod p).

Therefore we will be able to compute $\left(\frac{n}{p}\right)$ for all n and all p as soon as we know $\left(\frac{-1}{p}\right)$, $\left(\frac{2}{p}\right)$ and $\left(\frac{q}{p}\right)$ for every odd prime number q.

Now the first of these symbols is easy to find, since by Euler's criterion

$$\left(\frac{-1}{p}\right)=(-1)^{\frac{p-1}{2}}. \tag{6.5}$$

This means that $\left(\frac{-1}{p}\right)=1$ if $p=4k+1$ and $\left(\frac{-1}{p}\right)=-1$ if $p=4k-1$.

6.3.2 Gauss lemma

Let $p\in\mathbb{N}$ be prime. For every $a\in\mathbb{Z}$, there exists a unique $\tilde{a}\in\mathbb{Z}$ such that $a\equiv\tilde{a}$ (mod p) and $-\frac{p-1}{2}<\tilde{a}\le\frac{p-1}{2}$. Gauss lemma is the following

Lemma 6.1 *Let h be the number of negative integers in the set $E=\left\{\tilde{a},\widetilde{2a},\widetilde{3a},...,\widetilde{\frac{p-1}{2}a}\right\}$. Then, if $a\not\equiv 0$ (mod p), $\left(\frac{a}{p}\right)=(-1)^{h}$.*

Proof Let $x=\widetilde{ka}$, $y=\widetilde{ma}\in E$, with $k\ne m$, $k,m\in\{1,2,...,(p-1)/2\}$. Then $|x|\ne|y|$. Indeed, $|x|=|y|\Rightarrow(k-m)|a|=rp$, with $r\in\mathbb{Z}$, whence $p|k-m$ since $p\nmid a$. This is impossible because $0<|k-m|\le p-1$. This proves that $F=\{|x|/x\in E\}$ has $(p-1)/2$ elements. As $|x|\le(p-1)/2$ for every $x\in E$, $F=\{1,2,...,(p-1)/2\}$. Hence, the product of all elements of E is $\tilde{a}\cdot\widetilde{2a}\cdots\widetilde{\frac{p-1}{2}a}=(-1)^{h}(\frac{p-1}{2})!$ and $(-1)^{h}(\frac{p-1}{2})!\equiv(\frac{p-1}{2})!a^{\frac{p-1}{2}}$ (mod p), which yields $a^{\frac{p-1}{2}}\equiv(-1)^{h}$ (mod p).

Thus Gauss lemma results from Euler's criterion (6.4).

Example 6.5   Compute Legendre symbol $\left(\frac{2}{p}\right)$ for odd prime p. Here $E=\left\{2,4,6,...,2[\frac{p-1}{4}],2\left([\frac{p-1}{4}]+1\right)-p,...,-3,-1\right\}$. Therefore $h=\frac{p-1}{2}-[\frac{p-1}{4}]$. Hence h is even if $p\equiv\pm1$ (mod 8). In this case $\left(\frac{2}{p}\right)=1$. And h is odd if $p\equiv\pm3$ (mod 8), in which case $\left(\frac{2}{p}\right)=-1$. Summarizing yields

$$\left(\frac{2}{p}\right)=(-1)^{\frac{p^2-1}{8}}. \tag{6.6}$$

6.3.3   Law of quadratic reciprocity

Theorem 6.8   *Let p and q be two odd prime numbers. Then*

$$\left(\frac{p}{q}\right)\left(\frac{q}{p}\right)=(-1)^{\frac{(p-1)(q-1)}{4}}.$$

Proof   It is based on a geometrical interpretation of Gauss lemma 6.1. We know that $\left(\frac{p}{q}\right)=(-1)^h$, where h is the number of negative elements of the set $E=\left\{\widetilde{xp}\,/\,0<x<\frac{1}{2}q\right\}$. Each of these elements is characterized by the existence of an integer y satisfying $-\frac{1}{2}q<px-qy<0$. Thus h is the number of lattice points $(x,y)\in\mathbb{N}^2$ in the region

$$\left\{0<x<\frac{1}{2}q\;;\;-\frac{1}{2}q<px-qy<0\right\}.$$

The second double inequality can be written $\frac{p}{q}x<y<\frac{p}{q}x+\frac{1}{2}$, which implies that $y<\frac{p+1}{2}<\frac{p}{2}$ since $x<\frac{1}{2}q$ and p is odd. Therefore h is the number of lattice points in the rectangle

$$R=\left\{(x,y)\in\mathbb{R}^2/\,0<x<q/2\,,\,0<y<p/2\right\}$$

such that $-\frac{1}{2}q<px-qy<0$. Similarly, $\left(\frac{q}{p}\right)=(-1)^m$, where m is the number of lattice points of R satisfying $-\frac{1}{2}p<qy-px<0$ (figure 6.1). Therefore it suffices to prove that $\frac{1}{4}(p-1)(q-1)-(h+m)=r$ is even. Now $\frac{1}{4}(p-1)(q-1)$ is the total number of lattice points in R. Hence r is the number of lattice points in R satisfying $px-qy\leq-\frac{1}{2}q$ or

$qy - px \leq -\frac{1}{2}p$. The regions A and B defined by these inequalities are disjoint and they contain the same number of lattice points (exercise 6.7). Hence the cardinality of their union is even, Q.E.D.

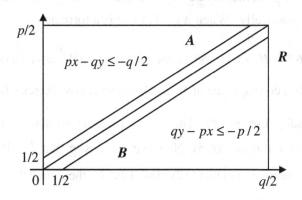

Figure 6.1

Example 6.6 Compute $\left(\frac{11}{83}\right)$. The law of quadratic reciprocity yields $\left(\frac{11}{83}\right) = \left(\frac{83}{11}\right) \times (-1)^{\frac{10 \times 82}{4}} = -\left(\frac{83}{11}\right) = -\left(\frac{6}{11}\right) = -\left(\frac{2}{11}\right)\left(\frac{3}{11}\right)$. By using formula (6.6) and again the law of quadratic reciprocity, we get $\left(\frac{11}{83}\right) = -(-1)\left(\frac{3}{11}\right)$ $= \left(\frac{11}{3}\right) \times (-1)^{\frac{2 \times 10}{4}} = -\left(\frac{11}{3}\right) = -\left(\frac{2}{3}\right) = 1$. Therefore 11 is a square modulo 83.

Example 6.7 We now compute $\left(\frac{3}{p}\right)$ for every odd prime p. We have $\left(\frac{3}{p}\right) = (-1)^{\frac{p-1}{2}}\left(\frac{p}{3}\right)$. But $p \equiv 1$, 5, -1 or -5 (mod 12). If $p \equiv 1$ (mod 12), then $p \equiv 1$ (mod 3), whence $\left(\frac{p}{3}\right) = \left(\frac{1}{3}\right) = 1$ and $\left(\frac{3}{p}\right) = 1$. The three other cases are similar, and we finally obtain

$$\begin{cases} \left(\dfrac{3}{p}\right) = 1 & \text{if } p \equiv \pm 1 \pmod{12} \\[2mm] \left(\dfrac{3}{p}\right) = -1 & \text{if } p \equiv \pm 5 \pmod{12} \end{cases} \tag{6.7}$$

6.4   Computations in $\mathbb{F}_p^*$

Let p be a prime number. Legendre symbol enables us to know if an element $x \in \mathbb{F}_p^*$ is the square of an element $y \in \mathbb{F}_p^*$. But how can we compute y practically? When $x = -1$, there is a formula:

Theorem 6.9   *If $p \equiv 1 \pmod 4$, then* $\left[((p-1)/2)! \right]^2 \equiv -1 \pmod p$.

Proof   It is a consequence of *Wilson's theorem* (see exercise 6.8).

Example 6.8   Let $p = 29$. Then $p \equiv 1 \pmod 4$, so that -1 is a square modulo 29 by formula (6.5). Now we see easily that $\left(\frac{p-1}{2} \right)! = 14! \equiv 12 \pmod{29}$. Therefore $(-1) = (\pm 12)^2 \pmod{29}$ by theorem 6.9.

In the general case, there is no formula, but we can use tables such as table 6.1, which relies on the notion of primitive root modulo p.

Theorem 6.10   *The multiplicative group $\mathbb{F}_p^*$ is cyclic, which means that there exists $g \in \mathbb{F}_p^*$ such that $\mathbb{F}_p^* = \{ g^0 = 1, g, g^2, ..., g^{p-2} \}$. Such an element g is called a primitive root modulo p.*

Proof   See exercise 6.9. For example, 3 is a primitive root modulo 7, since $\mathbb{F}_7^* = \{ 3^0 = 1, \; 3^1 = 3, \; 3^2 = 2, \; 3^3 = 6, \; 3^4 = 4, \; 3^5 = 5 \}$.

Now let $x \in \mathbb{F}_p^*$, and let g be a primitive root modulo p. We define the *index* i of x by the relation $g^i = x$ (which means that $g^i \equiv x \pmod p$). It is defined up to a multiple of $p-1$, since $x^{p-1} = 1$ in $\mathbb{F}_p^*$ by Fermat's theorem 6.7.

There is no general method for finding the primitive roots in $\mathbb{F}_p^*$ nor for computing the index of a given element x. Only by trial and error can one establish tables. The first of them has been issued by Jacobi in 1839 for the prime numbers less than 1000. Table 6.1 gives the least primitive root and the corresponding indices for all primes less than 40.

It is clear that, for every x and y in $\mathbb{F}_p^*$, of respective indices a and b relatively to g, $xy = g^a g^b = g^{a+b}$ in $\mathbb{F}_p^*$. Therefore, the properties of indices are similar to those of logarithms, and tables such that table 6.1 can be used to perform computations modulo p.

Table 6.1: Primitive roots modulo p and indices

	p	3	5	7	11	13	17	19	23	29	31	37
x	g	2	2	3	2	2	3	2	5	2	3	2
2		1	1	2	1	1	14	1	2	1	24	1
3			3	1	8	4	1	13	16	5	1	26
4			2	4	2	2	12	2	4	2	18	2
5				5	4	9	5	16	1	22	20	23
6				3	9	5	15	14	18	6	25	27
7					7	11	11	6	19	12	28	32
8					3	3	10	3	6	3	12	3
9					6	8	2	8	10	10	2	16
10					5	10	3	17	3	23	14	24
11						7	7	12	9	25	23	30
12						6	13	15	20	7	19	28
13							4	5	14	18	11	11
14							9	7	21	13	22	33
15							6	11	17	27	21	13
16							8	4	8	4	6	4
17								10	7	21	7	7
18								9	12	11	26	17
19									15	9	4	35
20									5	24	8	25
21									13	17	29	22
22									11	26	17	31
23										20	27	15
24										8	13	29
25										16	10	10
26										19	5	12
27										15	3	6
28										14	16	34
29											9	21
30											15	14
31												9
32												5
33												20
34												8
35												19
36												18

For example, for computing 15×7 in $\mathbb{F}_{23}^*$, we add the corresponding indices, that is $17 + 19 = 36$, which amounts to $14 = 36 - 22$. In table 6.1, index 14 corresponds to 13. Hence $15 \times 7 \equiv 13 \pmod{23}$. Similarly, to find a square root, we observe that $\sqrt{x} = g^{a/2}$ if $x = g^a$. For example, 5 is a square modulo 31 since $\left(\frac{5}{31}\right) = \left(\frac{31}{5}\right) = \left(\frac{1}{5}\right) = 1$. In table 6.1, we read that the index of 5 modulo 31 is 20. Therefore its square root has index 10, and table 6.1 again tells us that $5 = (\pm 25)^2 \pmod{31}$.

6.5   Binary quadratic forms with integer coefficients

6.5.1   Introduction

A binary quadratic form with integer coefficients is a function of the form $\varphi(x, y) = ax^2 + bxy + cy^2$, with $(a, b, c) \in \mathbb{Z}^3$, the variables x and y being rational integers. The form φ will be denoted by (a, b, c). The *discriminant* of the form (a, b, c) is defined by $\Delta = b^2 - 4ac$. We observe that $\Delta \equiv b^2 \pmod 4$. Therefore Δ is a multiple of 4 if b is even and $\Delta \equiv 1 \pmod 4$ if b is odd.

The form (a, b, c) is positive definite when $\Delta < 0$ and $a > 0$. In this case c is also positive, and $\varphi(x, y) > 0$ if $(x, y) \neq (0, 0)$.

The most simple example of binary quadratic form is $\varphi(x, y) = x^2 + y^2$, that is $(1, 0, 1)$. It is positive definite, with $\Delta = -4$. Theorem 6.2 gives a necessary and sufficient condition for a given integer n to be represented by the form $x^2 + y^2$. More generally, we will say that a given integer n is represented by the form (a, b, c) if there exists $(x, y) \in \mathbb{Z}^2$ such that

$$n = ax^2 + bxy + cy^2.$$

6.5.2   Equivalent forms

Let f be a linear mapping from $\mathbb{Z}^2$ to $\mathbb{Z}^2$, defined by $x = px' + qy'$, $y = rx' + sy'$, with p, q, r, $s \in \mathbb{Z}$. We say that f is *unimodular* if its determinant $ps - qr$ is equal to 1. An unimodular transformation is bijective from $\mathbb{Z}^2$ to $\mathbb{Z}^2$, because by Cramer's formulas we see that $x' = sx - qy$ and $y' = -rx + py$.

The most simple examples of unimodular transformations are

$$x = y', \quad y = -x', \tag{6.8}$$
$$x = x' + y', \quad y = y', \tag{6.9}$$
$$x = x' - y', \quad y = y'. \tag{6.10}$$

If we replace, in the quadratic form $\varphi(x,y) = ax^2 + bxy + cy^2$, x by $px' + qy'$ and y by $rx' + sy'$, we clearly obtain a new quadratic form $\varphi'(x',y') = a'x'^2 + b'x'y' + c'y'^2$, with

$$\begin{cases} a' = ap^2 + bpr + cr^2 = \varphi(p,r) \\ b' = 2apq + b(ps + qr) + 2crs \\ c' = aq^2 + bqs + cs^2 = \varphi(q,s) \end{cases} \tag{6.11}$$

Two quadratic forms are said to be *equivalent* if one can be transformed into the other by an unimodular transformation. Thus we will denote $(a,b,c) \sim (a',b',c')$ if (6.11) is satisfied with $ps - qr = 1$. It is easy to check that this defines an equivalence relation and that two equivalent forms have the same discriminant (exercise 6.10). For example, by using the unimodular transformations (6.8), (6.9) and (6.10), we obtain

$$(a,b,c) \sim (c,-b,a) \tag{6.12}$$
$$(a,b,c) \sim (a, b+2a, a+b+c) \tag{6.13}$$
$$(a,b,c) \sim (a, b-2a, a-b+c) \tag{6.14}$$

Remark 6.1 Let φ be positive definite. Assume we have ordered the values $v_0 = 0 < v_1 \leq v_2 \cdots$ taken by $\varphi(x,y)$ as a non decreasing sequence, each value v being repeated as many times as there exist couples (x,y) such that $\varphi(x,y) = v$. Then two equivalent positive definite forms have the same sequence v_n. In particular, they represent the same integers.

6.5.3   Reduction of positive definite forms

Theorem 6.11 *Every positive definite form is equivalent to a binary form (a,b,c) satisfying*

$$-a < b \leq a < c \text{ or } 0 \leq b \leq a = c. \tag{6.15}$$

A positive definite form satisfying (6.15) is said to be *reduced*.

Proof The transformation (6.12) interchanges a and c, leaving $|b|$ unchanged. Moreover, one of the transformations (6.13) or (6.14) reduces $|b|$ if $|b| > a$, leaving a unchanged. Hence we see that every positive definite form is equivalent to a form (a_0, b_0, c_0) with $-a_0 \le b_0 \le a_0 \le c_0$. If $b_0 = -a_0$, the transformation (6.13) shows that $(a_0, b_0, c_0) \sim (a_1, b_1, c_1)$, with $-a_1 < b_1 \le a_1 \le c_1$, since $c_1 = c_0$. If $a_1 = c_1$, the transformation (6.12) allows to replace b_1 by $-b_1$.

Example 6.8 Let us reduce the form $(10, 34, 29)$. We have $(10, 34, 29)$ $\sim (10, 14, 5) \sim (10, -6, 1) \sim (1, 6, 10) \sim (1, 4, 5) \sim (1, 2, 2) \sim (1, 0, 1)$ by using (6.14), (6.14), (6.12), (6.14), (6.14), (6.14). Thus $\varphi(x, y) = 10x^2 + 34xy$ $+ 29y^2$ is equivalent to $\psi(x, y) = x^2 + y^2$. Therefore n is represented by $10x^2 + 34xy + 29y^2$ if and only if n is a sum of two squares.

Theorem 6.12 *Two reduced definite positive quadratic forms are not equivalent.*

Proof Let $\varphi = (a, b, c)$ be reduced (which implies $|b| \le a \le c$).

If $|x| \ge |y| \ge 1$, then $\varphi(x, y) \ge |x|(a|x| - |by|) + c|y|^2 \ge |x|^2 (a - |b|) + c|y|^2$ $\ge a - |b| + c$. By symmetry, this remains true if $x \ne 0$ and $y \ne 0$. Hence the least values of φ (see remark 6.1) are $v_0 = 0$, $v_1 = a$, $v_2 = a$, $v_3 = c$, $v_4 = c$, $v_5 = a - |b| + c$. They are obtained for $(x, y) = (0, 0)$, $(1, 0)$, $(-1, 0)$, $(0, 1)$, $(0, -1)$, and either $(1, 1)$ or $(1, -1)$ respectively.

If the reduced form $\varphi' = (a', b', c')$ is equivalent to φ, it takes the same values, whence $a' = a$, $c' = c$, et $a' - |b'| + c' = a - |b| + c$, which yields $|b'| = |b|$. It remains to prove that $b' = b$. We distinguish three cases.

a) If $b = a$, then $|b'| = a$. But $b' = -a = -a'$ is impossible by (6.15), since φ' is reduced. Therefore $b' = a = b$.

b) If $a = c$, then $a' = c'$, whence $b \ge 0$ and $b' \ge 0$ by (6.15). Thus $|b'| = |b| \Rightarrow b' = b$.

c) As $b \ne a$ and $a \ne c$, then we have by (6.15) $-a < b < a < c$. Let $x = px' + qy'$, $y = rx' + sy'$, $ps - qr = 1$, be the unimodular transformation which transforms φ into φ'. Then the first equality in (6.11) yields $a' = a = \varphi(p,r) = v_1 = v_2 < v_3$. Therefore $p = \pm 1$, $r = 0$, whence $ps = 1$. Now the second equality in (6.11) implies $b' \equiv b \pmod{2a}$. As $|b'| = |b|$ $-a < b < a$, and $-a < b' \le a$, we see that $b' = b$. Theorem 6.12 is proved.

Theorem 6.13 *The number of equivalence classes of binary quadratic forms with discriminant* $\Delta < 0$ *is finite. This number is denoted by* $h(\Delta)$ *and is called the class number. It is equal to the number of solutions* $(a,b,c) \in \mathbb{N}^* \times \mathbb{Z} \times \mathbb{N}^*$ *of the equation* $b^2 - 4ac = \Delta$, *satisfying* (6.15) *together with the condition* $a \le \sqrt{-\Delta/3}$.

Proof We know that every form is equivalent to a reduced form (theorem 6.11) and that two reduced forms are not equivalent (theorem 6.12). Hence the number of equivalence classes of forms with discriminant Δ is equal to the number of reduced forms $\varphi = (a,b,c)$ with discriminant Δ. Then $b^2 - 4ac = \Delta$. Moreover $|b| \le a$ and $|b| \le c$ $\Rightarrow b^2 \le ac$. Therefore $\Delta \le -3ac$, whence $ac \le -\Delta/3$. Since $0 \le a \le c$ by (6.15), we see that $a \le \sqrt{-\Delta/3}$. Therefore a can take only finitely many values. The same holds for b since $|b| \le a$, and for c since $b^2 - 4ac = \Delta$.

Example 6.9 Let $\Delta = -4$. Then $a \le \sqrt{4/3} \Rightarrow a = 1$. Therefore $|b| \le a$ $\Rightarrow |b| = 0$ or 1. Now $b^2 - 4ac = \Delta$ yields $b^2 - 4ac = -4$, whence $b = 0$ and $c = 1$. There is only one reduced quadratic form with discriminant -4, which is, $\varphi = (1,0,1)$, the sum of two squares, and $h(-4) = 1$.

Example 6.10 Let $\Delta = -20$. Then $a \le \sqrt{20/3} \Rightarrow a = 1$ or $a = 2$. If $a = 1$, then $b = 0$ and $c = 5$. If $a = 2$, then $b = 0$, ± 1 or 2, which yields $-8c = -20$, or $1 - 8c = -20$, or $4 - 8c = -20$. Only the last equation has a solution in $\mathbb{N}$, namely $c = 3$, which implies $b = 2$. Therefore there are two reduced binary quadratic forms with discriminant -20: $\varphi_1 = (1,0,5)$ and $\varphi_2 = (2,2,3)$. Consequently $h(-20) = 2$.

Table 6.2, page 92, gives the class number $h(\Delta)$ for $-79 \le \Delta \le -3$.

Table 6.2: Class number $h(-\Delta)$

Δ	3	4	7	8	11	12	15	16	19	20	23	24	27
$h(-\Delta)$	1	1	1	1	1	2	2	2	1	2	3	2	2

Δ	28	31	32	35	36	39	40	43	44	47	48	51	52
$h(-\Delta)$	2	3	3	2	3	4	2	1	4	5	4	2	2

Δ	55	56	59	60	63	64	67	68	71	72	75	76	79
$h(-\Delta)$	4	4	3	4	5	4	1	4	7	3	3	4	5

6.5.4   Representation of an integer by a binary quadratic form

We say that the integer n is *properly represented* by the quadratic form $\varphi = (a,b,c)$ if $n = ax^2 + bxy + cy^2$, with x and y coprime.

Theorem 6.14 *The integer n is properly represented by one of the quadratic forms with discriminant Δ if, and only if, the equation $\Delta \equiv k^2$ (mod $4n$) is soluble.*

Proof Assume first that the equation $\Delta \equiv k^2$ (mod $4n$) is soluble. Define m by $\Delta = 4nm + k^2$. Let $\varphi(x,y) = nx^2 + kxy - my^2$; then φ has discriminant Δ, and it represents properly n since $n = \varphi(1,0)$. Conversely, assume that $n = \varphi(p,r) = ap^2 + bpr + cr^2$, where p and r are coprime, and $b^2 - 4ac = \Delta$. By Bézout's theorem, there exist q and s satisfying $pq - rs = 1$. By using (6.11), we see that $\varphi \sim \varphi' = (a',b',c')$, with $a' = \varphi(p,r) = n$. Since φ and φ' have the same discriminant $\Delta = b'^2 - 4a'c'$, the equation $\Delta \equiv k^2$ (mod $4n$) is soluble.

Theorem 6.4 gives the best results when $h(\Delta) = 1$. As an example, we give a new proof of theorem 6.1. Let n be an odd prime. Then n is represented (which is equivalent here to be properly represented since n is prime) by one of the quadratic forms with discriminant -4 if, and only if, the equation $-4 \equiv k^2$ (mod $4n$) is soluble, that is if and only if the equation $-1 \equiv x^2$ (mod n) is soluble. This is equivalent to $n \equiv 1$ (mod 4) by (6.5). Since $h(-4) = 1$, all quadratic forms with discriminant -4 are equivalent to $\varphi = (1,0,1)$ (example 6.9). Hence n is represented by $\varphi(x,y) = x^2 + y^2$ if, and only if, $n \equiv 1 \pmod 4$.

By arguing the same way, we see that an odd prime n is represented by one of the two reduced forms with discriminant -20, namely $\varphi_1 = (1,0,5)$ and $\varphi_2 = (2,2,3)$ (example 6.10) if and only if $-20 \equiv k^2 \pmod{4n}$ is soluble, that is if $-5 \equiv x^2 \pmod n$. This means that $\left(\frac{-5}{n}\right) = 1$ or $n = 5$. But $\left(\frac{-5}{n}\right) = \left(\frac{-1}{n}\right)\left(\frac{5}{n}\right) = (-1)^{\frac{n-1}{2}}\left(\frac{n}{5}\right)$ by (6.5) and theorem 6.8. Hence $\left(\frac{-5}{n}\right) = 1$ if and only if $n \equiv 1,\ 3,\ 7,\ 9 \pmod{20}$. Therefore the odd prime n can be written as $x^2 + 5y^2$ or $2x^2 + 2xy + 3y^2$ if, and only if, $n \equiv 1,\ 3,\ 7$ or 9 $\pmod{20}$, or if $n = 5$. Theorem 6.14 does not allow to choose between these two representations. This can be done by observing that $n = x^2 + 5y^2$ is only possible if n is a square mod 5, that is if $\left(\frac{n}{5}\right) = 1$ or $n = 5$, whereas $n = 2x^2 + 2xy + 3y^2$ implies $2n = (2x + y)^2 + 5y^2$, which means $\left(\frac{2n}{5}\right) = 1$, and finally $\left(\frac{n}{5}\right) = -1$. In conclusion, if n is an odd prime, then $n = 5$ or $n \equiv 1$ or $9 \pmod{20} \Rightarrow n = x^2 + 5y^2$. And $n \equiv 3$ ou 7 (mod $20) \Rightarrow n = 2x^2 + 2xy + 3y^2$. Otherwise, n is not represented by any form with discriminant -20. Thus, for example, we have $61 = 4^2 + 5 \cdot 3^2$, whereas $67 = 2(-4)^2 + 2(-4)5 + 3 \cdot 5^2$.

6.6   Sums of four squares

This is one of the most celebrated results in number theory. It was conjectured by Bachet in 1621, and proved by Lagrange in 1770.

Theorem 6.15 *Every $n \in \mathbb{N}$ can be written as a sum of four squares.*

Evidently, some of these squares can be zero. Thus, for example, $2 = 1^2 + 1^2 + 0^2 + 0^2$ and $12 = 2^2 + 2^2 + 2^2$. However, this result is the best possible, because no number of the form $8n + 7$ can be written as a sum of only 3 squares (exercise 6.11). For example $7 = 2^2 + 1^2 + 1^2 + 1^2$ and $15 = 3^2 + 2^2 + 1^2 + 1^2$.

The proof of theorem 6.15 relies on an identity connected with the theory of quaternions, but which can also be checked by a direct computation:

$$(x^2 + y^2 + z^2 + w^2)(x'^2 + y'^2 + z'^2 + w'^2)$$
$$= (xx' + yy' + zz' + ww')^2 + (xy' - yx' + wz' - zw')^2 \qquad (6.16)$$
$$+ (xz' - zx' + yw' - wy')^2 + (xw' - wx' + zy' - yz')^2.$$

Hence the product of two sums of 4 squares is also a sum of 4 squares. Since 2 is a sum of 4 squares, theorem 6.15 will be proved if we succeed in proving that every odd prime is a sum of 4 squares. We will use the descent method and the following lemma.

Lemma 6.2   *Let p be an odd prime. Then there exists $(x, y) \in \mathbb{Z}^2$ such that $1 + x^2 + y^2 \equiv 0$ (mod p).*

Proof   See exercise 6.12.

Now we prove theorem 6.15. In lemma 2, we choose x and y such that $|x| < p/2$ and $|y| < p/2$. Hence $1 + x^2 + y^2 < 1 + p^2/2 < p^2$, and we can find k such that $1 + x^2 + y^2 = kp$, with $1 \le k < p$. Let m be the least positive integer satisfying $a^2 + b^2 + c^2 + d^2 = mp$, with $m < p$.

Then $m \le k$ and m is odd. Indeed, if m was even, then 0, 2 or 4 among the numbers a, b, c, d would be odd. Therefore a, b, c, d could be be associated by pairs odd-odd or even-even, for example (a,b), (c,d). In which case $(a+b)$, $(a-b)$, $(c+d)$, $(c-d)$ would be even, and

$$\left(\frac{a+b}{2}\right)^2 + \left(\frac{a-b}{2}\right)^2 + \left(\frac{c+d}{2}\right)^2 + \left(\frac{c-d}{2}\right)^2 = \frac{m}{2}p.$$

This is impossible because m is minimal.

Thus m is odd. Assume that $m \ge 3$, and let a', b', c', d' satisfying $|a'| < m/2$, $|b'| < m/2$, $|c'| < m/2$, $|d'| < m/2$, $a' \equiv a$, $b' \equiv b$, $c' \equiv c$, $d' \equiv d$ (mod m). Put $n = a'^2 + b'^2 + c'^2 + d'^2$. Then $n \equiv a^2 + b^2 + c^2 + d^2 \equiv 0$ (mod m) and $n > 0$. Indeed $n = 0 \Rightarrow a' = b' = c' = d' = 0 \Rightarrow mp \equiv 0$ (mod m^2) $\Rightarrow m \mid p$, which is impossible. Moreover, $n < 4(m^2/4) = m^2$, whence $n = um$ with $0 < u < m$. Since $um = a'^2 + b'^2 + c'^2 + d'^2$ and $mp = a^2 + b^2 + c^2 + d^2$, (6.16) yields $(um)(mp) = A^2 + B^2 + C^2 + D^2$. Even better, (6.16) shows that $A \equiv B \equiv C \equiv D \equiv 0$ (mod m) by definition of a', b', c', d'. Simplifying by m^2 yields $up = A'^2 + B'^2 + C'^2 + D'^2$

with $0 < u < m$. This contradicts the minimality of m. Therefore $m = 1$, and Bachet's theorem is proved.

Exercises

6.1   Let $(G, +)$ be a commutative group. Let H be a subgroup of G. Prove that the relation $x \mathcal{R} y \Leftrightarrow (x - y \in H)$ is an equivalence relation.

6.2   With the notations of exercise 6.1, prove that the addition is compatible with $\mathcal{R}$. Then prove that G/H is a commutative group.

6.3   Let $\mathbb{A}$ be a domain, I an ideal of $\mathbb{A}$. Prove $\mathbb{A}/I$ is a domain.

6.4   Let $a \in \mathbb{A}_\mathbb{K}$, where $\mathbb{A}_\mathbb{K}$ is the ring of integers of the quadratic field $\mathbb{K} = \mathbb{Q}(\sqrt{d})$. Draw inspiration from the proof of theorem 5.16 to prove that, for every $x \in \mathbb{A}_\mathbb{K}$, there are $q, r \in \mathbb{A}_\mathbb{K}$ satisfying $x = aq + r$, with $r = \alpha + \beta\sqrt{d}$, $|\alpha| \leq |a|$, $|\beta| \leq |a|$. Then prove theorem 6.3.

6.5   Prove theorem 6.6.

6.6   1) Prove Fermat's theorem by using the group $(\mathbb{F}_p^*, \cdot)$.

2) Let $p \in \mathbb{N}$ be prime. Prove that p divides the binomial coefficient $\binom{p}{k}$ for every $1 \leq k \leq p - 1$. Deduce an alternative proof (by induction) of Fermat's theorem.

3) We say that p is a *Carmichael number* if p is not prime and if $a^p \equiv a$ (mod p) for every $a \in \mathbb{Z}$. Check that 561 is a Carmichael number, which proves that the converse of Fermat's theorem is false.

6.7   Prove that $f : (x, y) \mapsto (x', y')$ defined by $x = \frac{1}{2}(q + 1) - x'$ and $y = \frac{1}{2}(p + 1) - y'$ is bijective from $A = \{(x, y) \in R \,/\, px - qy \leq -\frac{1}{2}q\}$ to $B = \{(x, y) \in R \,/\, qy - px \leq -\frac{1}{2}p\}$.

6.8   Wilson's theorem

1) Prove that, for every prime $p \in \mathbb{N}$, $(p-1)! \equiv -1 \pmod{p}$ by using the fact that every $x \in \mathbb{F}_p^*$ has an inverse.

2) Deduce a proof of theorem 6.9.

6.9   Our aim is to prove theorem 6.10 by using a *constructive* method, that is a method allowing to compute explicitly a primitive root g of $\mathbb{F}_p^*$ when p is given. We denote by $p-1 = p_1^{a_1} p_2^{a_2} \ldots p_j^{a_j}$ the prime number factorization of $p-1$.

1) Assume that a has order α and b has order β, with α and β coprime. Prove that ab has order $\alpha\beta$.

2) Prove that the congruence $x^d - 1 \equiv 0 \pmod{p}$ has exactly d solutions in $\mathbb{F}_p^*$ if $d \mid p-1$. Deduce that, for every prime divisor p_k of $p-1$, there exists $x \in \mathbb{F}_p^*$ such that x is of order $p_k^{a_k}$.

3) Prove the existence of primitive roots modulo p.

4) Application : find all primitive roots modulo 41.

6.10   1) Write the quadratic form $\varphi(x, y) = ax^2 + bxy + cy^2$ in matricial form. Let M be the matrix of φ in the canonical basis of $\mathbb{R}^2$. What is the relation between Δ and $\det(M)$?

2) A matrix U is said to be unimodular if $U = \begin{pmatrix} p & q \\ r & s \end{pmatrix}$, with $p, q, r, s \in \mathbb{Z}$ and $\det U = 1$. Translate the relation $\sim$ defined in subsection 6.5.2 into matricial form. Deduce that $\sim$ is an equivalence relation.

3) Prove that two equivalent forms have the same discriminant.

6.11   Let $m = 8n + 7$.

Prove that m cannot be written as $m = a^2 + b^2 + c^2$, with $a, b, c \in \mathbb{Z}$.

6.12   Prove that the numbers $1 + x^2$, with $x \in \{0, 1, \ldots, (p-1)/2\}$, are all different modulo p.

Then prove lemma 6.2 by using the pigeon-hole principle.

6.13 Compute the Legendre symbols $\left(\frac{26}{31}\right)$ and $\left(\frac{33}{37}\right)$. Then solve the equations $x^2 + 7x - 2 \equiv 0 \pmod{31}$ and $2x^2 + 5x - 1 \equiv 0 \pmod{37}$.

6.14 Reduce the quadratic form $\varphi = (12, 31, 21)$.

6.15 Determine the reduced forms and the class number for $\Delta = -3$, $\Delta = -12$, $\Delta = -31$.

6.16 Prove that $h(-7) = 1$, then find the prime numbers which can be represented by $\varphi(x, y) = x^2 + xy + 2y^2$. Numerical application: $p = 211$.

6.17 1) Prove that $h(-8) = 1$, then find the prime numbers which can be represented by $x^2 + 2y^2$.
2) By drawing inspiration from the proof of theorem 6.2, find a necessary and sufficient condition for an integer n to be written as $x^2 + 2y^2$.

6.18 Fermat numbers
The Fermat numbers are defined by $F_n = 2^{2^n} + 1$, $n \in \mathbb{N}$.
1) Check that F_0, F_1, F_2 and F_3 are prime.
2) We assume that $n \geq 2$, and consider a prime divisor p of F_n.
   a. Compute the order of 2 in $\mathbb{F}_p$. Deduce that 2^{n+1} divides $p - 1$.
   b. Compute Legendre's symbol $\left(\frac{2}{p}\right)$. Deduce that $p = k2^{n+2} + 1$.
   c. Prove that F_4 is prime and that F_5 and F_6 are not prime.

6.19 Sophie Germain theorem, 1823
The aim of this exercise is to prove the celebrated Sophie Germain theorem on Fermat's equation $x^p + y^p = z^p$:
If p is an odd prime such that $2p + 1$ or $4p + 1$ is also prime, and if there exist pairwise coprime x, y, z such that $x^p + y^p = z^p$, then p divides one of the three numbers x, y or z.
The theorem applies for $p = 3, 5, 7, 11, 13, 23, 29, 37, \ldots$
To prove the theorem, we argue by *reductio ad absurdum*, and assume that $x^p + y^p + z^p = 0$, where p is an odd prime and x, y, z are pairwise

coprime satisfying $x \not\equiv 0$, $y \not\equiv 0$, $z \not\equiv 0$ (mod p). Moreover, we assume that there exists an odd prime q satisfying the property

$$x^p + y^p + z^p \equiv 0 \ (\text{mod } q) \implies q \,|\, x \text{ or } q \,|\, y \text{ or } q \,|\, z. \qquad (P)$$

1) Prove the *Abel-Barlow relations* (1810):

$$\begin{cases} x+y=t^p, & \dfrac{x^p+y^p}{x+y}=t_1^p, & z=-tt_1 \\[2ex] y+z=r^p, & \dfrac{y^p+z^p}{y+z}=r_1^p, & x=-rr_1 \\[2ex] z+x=s^p, & \dfrac{z^p+x^p}{z+x}=s_1^p, & y=-ss_1 \end{cases}$$

where $GCD(t,t_1) = GCD(r,r_1) = GCD(s,s_1) = 1$.

2) Prove that there exists an integer k such that $p \equiv k^p$ (mod q).

3) Prove Sophie Germain theorem when $2p+1=q$ is prime.

4) Prove Sophie Germain theorem when $4p+1=q$ is prime.

6.20   Principal and euclidean domains

An ideal I of a ring $\mathbb{A}$ is said to be *principal* if there exists $a \in \mathbb{A}$ such that $I = a\mathbb{A}$ (see theorem 6.3). A domain $\mathbb{A}$ is *principal* if it is not a field and if every ideal I of $\mathbb{A}$ is principal. A domain $\mathbb{A}$ is *euclidean* if there exists a mapping $\varphi : \mathbb{A}^* \longrightarrow \mathbb{N}$ such that, for every $(a,b) \in \mathbb{A} \times \mathbb{A}^*$, there exists $(q,r) \in \mathbb{A}^2$ such that $a = bq + r$ and $\varphi(r) < \varphi(b)$ or $r=0$.

1) Give examples of euclidean domains.

2) Prove that every euclidean domain is principal.

3) Let $\mathbb{A}$ be a domain. Prove that, for every $(a,b) \in \mathbb{A}^2$, the set

$$I = \left\{ ax+by / (x,y) \in \mathbb{A}^2 \right\}$$

is an ideal of $\mathbb{A}$. Deduce that Bezout's identity (5.8) remains true in a principal domain.

4) Prove that every principal domain is a unique factorization domain.

Chapter 7

Arithmetical functions

This chapter is devoted to the study of elementary arithmetical functions, such as $d(n)$, the number of divisors of n, or $r_2(n)$ and $r_4(n)$, the numbers of representations of n as a sum of two squares and four squares respectively. The main tools we use here are analytical: generating functions (section 7.1), Lambert series (section 7.2), Jacobi's triple product identity (section 7.3). These tools allow us to prove in section 7.5 precise formulas for $r_2(n)$ and $r_4(n)$, thus obtaining a quantitative improvement of the results of chapter 6. We also study Euler function $\varphi(n)$ (section 7.6) and devote section 7.7 to the computation of the mean value of $r_2(n)$, which rests on a geometrical interpretation. Finally, we give a quick insight into elementary methods in prime number theory and the function $\pi(n)$, by proving the upper bound $\mathrm{LCM}(1,2,\ldots,n) \le 3^n$ and Chebyshev's estimates.

7.1   Ordinary generating function

Let $(a_n)_{n \in \mathbb{N}}$ be a sequence of complex numbers. The *ordinary generating function* of (a_n) is defined by

$$f(x) = \sum_{n=0}^{+\infty} a_n x^n, \qquad (7.1)$$

provided that the power series in (7.1) has a non-zero radius of convergence. This will be the case if (a_n) has *exponential order*, that is if there exists $A > 0$ such that $|a_n| \le A^n$.

Example 7.1   Let (F_n) be the *Fibonacci sequence*, defined by

$$F_0 = 0, \ F_1 = 1, \ F_{n+2} = F_{n+1} + F_n \ \text{for every } n \in \mathbb{N}. \qquad (7.2)$$

The recurrence relation $F_{n+2} = F_{n+1} + F_n$ shows that $F_n \geq 0$, and therefore $F_n \leq F_{n+1}$ for all n. Hence $F_{n+1} \leq 2F_n$ for every $n \geq 1$. From this we deduce that $F_n \leq 2^n$ for $n \geq 1$. Thus the series $f(x) = \sum_{n=0}^{+\infty} F_n x^n$ has a radius of convergence $R \geq \frac{1}{2}$. We can compute $f(x)$ the following way.

$$f(x) = x + \sum_{n=2}^{+\infty} F_n x^n = x + \sum_{n=2}^{+\infty} (F_{n-1} + F_{n-2}) x^n = x + x \sum_{n=2}^{+\infty} F_{n-1} x^{n-1}$$

$$+ x^2 \sum_{n=2}^{+\infty} F_{n-2} x^{n-2} = x + x \sum_{p=1}^{+\infty} F_p x^p + x^2 \sum_{p=0}^{+\infty} F_p x^p = x + (x + x^2) f(x).$$

Therefore the ordinary generating function of the Fibonacci sequence is

$$f(x) = \frac{-x}{x^2 + x - 1}. \tag{7.3}$$

Knowing $f(x)$ allows us here to compute explicitely F_n as a function of n. Indeed, we have $x^2 + x - 1 = (x + \Phi)(x + \Psi)$, where $\Phi = (1 + \sqrt{5})/2$ is the Golden Number, and $\Psi = -1/\Phi = (1 - \sqrt{5})/2$. Therefore the rational function $f(x)$ defined in (7.3) has the partial fraction decomposition

$$f(x) = \frac{1}{\sqrt{5}} \left(\frac{-\Phi}{x + \Phi} + \frac{\Psi}{x + \Psi} \right) = \frac{1}{\sqrt{5}} \left(\frac{1}{1 + x/\Psi} - \frac{1}{1 + x/\Phi} \right).$$

Hence, since $1/\Phi = -\Psi$ and $1/\Psi = -\Phi$,

$$f(x) = \frac{1}{\sqrt{5}} \left(\frac{1}{1 - \Phi x} - \frac{1}{1 - \Psi x} \right) = \frac{1}{\sqrt{5}} \left(\sum_{n=0}^{+\infty} \Phi^n x^n - \sum_{n=0}^{+\infty} \Psi^n x^n \right) \quad (|x| < -\Psi).$$

By comparing to $f(x) = \sum_{n=0}^{+\infty} F_n x^n$, we obtain

$$F_n = \frac{1}{\sqrt{5}} (\Phi^n - \Psi^n) \quad \text{for every } n \in \mathbb{N}. \tag{7.4}$$

See exercice 3.11 for another way of computing F_n.

Example 7.2   For every $n \in \mathbb{N}$, denote by $r_2(n)$ the number of representations of n as a sum of the squares of two rational integers, that is the number of *ordered* couples $(a, b) \in \mathbb{Z}^2$ such that $n = a^2 + b^2$.

For instance, $5 = 1^2 + 2^2 = 2^2 + 1^2 = (-1^2) + 2^2 = 2^2 + (-1)^2 = 1^2 + (-2)^2$ $= (-2)^2 + 1^2 = (-1)^2 + (-2)^2 = (-2)^2 + (-1)^2$, whence $r_2(5) = 8$. Similarly

$9 = 3^2 + 0^2 = 0^2 + 3^2 = (-3)^2 + 0^2 = 0^2 + (-3)^2$, which implies $r_2(9) = 4$. And $r_2(7) = 0$ since $7 \equiv -1 \pmod 4$ (theorem 6.1).

The ordinary generating function $f(x) = \sum_{n=0}^{+\infty} r_2(n) x^n$ can be obtained by arguing the following way. Consider, for $|x| < 1$,

$$g(x) = \sum_{n=-\infty}^{+\infty} x^{n^2} = 1 + 2 \sum_{n=1}^{+\infty} x^{n^2},$$

and compute $(g(x))^2 = \left(\sum_{n=-\infty}^{+\infty} x^{n^2} \right)^2$. We obtain the coefficient of x^n in $(g(x))^2$ by multiplying the x^{ℓ^2} terms of the first series by the x^{m^2} terms of the second one, in such a way that $x^{\ell^2} \cdot x^{m^2} = x^n$. Hence $n = \ell^2 + m^2$. Each of these products yields a coefficient equal to 1, and there are exactly as many of these products as there are solutions $(\ell, m) \in \mathbb{Z}^2$ of the equation $\ell^2 + m^2 = n$. Therefore the coefficient of x^n in $(g(x))^2$ is exactly $r_2(n)$, and we obtain

$$\sum_{n=0}^{+\infty} r_2(n) x^n = \left(\sum_{n=-\infty}^{+\infty} x^{n^2} \right)^2. \tag{7.5}$$

Remark 7.1   The function $r_2(n)$ is an example of *arithmetical function*, that is a function from $\mathbb{N}$ to $\mathbb{C}$. Another classic example is the *divisor function* $d(n)$, which counts the number of positive divisors of n. Thus $d(1) = 1$, $d(2) = 2$, $d(3) = 2$, $d(4) = 3$, ..., $d(24) = 8$, and so on. It should be noted that $d(n) = \sum_{d|n} 1$. Similarly, the function *sum of divisors* can be expressed as $\sigma(n) = \sum_{d|n} d$.

Remark 7.2   We have seen in example 7.1 that the ordinary generating function of the Fibonacci sequence (F_n), which satisfies the *linear recurrence relation* $F_{n+2} = F_{n+1} + F_n$, is a rational function. This is a special case of the following result.

Theorem 7.1   *The generating function $f(x) = \sum_{n=0}^{+\infty} u_n x^n$ is a rational function $P(x)/Q(x)$, with $\deg P < \deg Q$, if and only if (u_n) satisfies a linear recurrence relation of the form*

$$u_{n+p} = \alpha_{p-1}u_{n+p-1} + \alpha_{p-2}u_{n+p-2} + \cdots + \alpha_0 u_n,$$

where $p \in \mathbb{N}^*$ and $\alpha_0 \neq 0$, $\alpha_1, \ldots, \alpha_{p-1} \in \mathbb{C}$.

Proof See exercice 7.1.

7.2   Lambert series

We first recall the following classic result.

Lemma 7.1 *Let Ω be an open subset of $\mathbb{C}$, and let (f_n) be a sequence of functions defined and holomorphic in Ω. Assume that $f_n \to f$, uniformly on every compact subset of Ω. Then f is holomorphic in Ω.*

Proof Let (C) be any circle with center $z_0 \in \Omega$, $(C) \subset \Omega$. By *Cauchy's formula*, for every z inside (C) and every $n \in \mathbb{N}$, we have

$$f_n(z) = \frac{1}{2i\pi} \int_C \frac{f_n(t)}{t-z} dt .$$ Since $f_n \to f$ uniformly on (C), we get

$$f(z) = \frac{1}{2i\pi} \int_C \frac{f(t)}{t-z} dt \qquad (7.6)$$

Therefore we can write

$$\frac{f(z) - f(z_0)}{z - z_0} = \frac{1}{2i\pi} \int_C \frac{f(t)}{z - z_0}\left(\frac{1}{t-z} - \frac{1}{t-z_0}\right) dt = \frac{1}{2i\pi} \int_C \frac{f(t)}{(t-z)(t-z_0)} dt.$$

Thus $\lim\limits_{z \to z_0} \dfrac{f(z) - f(z_0)}{z - z_0} = \dfrac{1}{2i\pi} \int_C \dfrac{f(t)}{(t-z_0)^2} dt$.

This proves that f is holomorphic at z_0, and even that f can be derived by taking the partial derivative with respect to z in (7.6), that is

$$f'(z) = \frac{1}{2i\pi} \int_C \frac{f(t)}{(t-z)^2} dt. \qquad (7.7)$$

Now, we consider a sequence $a(n)$ of exponential order, which means that there exists $A > 0$ such that $|a(n)| \leq A^n$.

Define $\rho = \inf\left(\dfrac{1}{A}, 1\right)$, and consider the series

$$L(x) = \sum_{n=1}^{+\infty} \frac{a(n)x^n}{1-x^n}. \qquad (7.8)$$

Let $D(\rho) = \{x \in \mathbb{C} \,/\, |x| < \rho\}$ be the open disk with center 0 and radius ρ. For every compact $K \subset D(\rho)$, there exists $k > 0$ such that $x \in K \Rightarrow |x| \le k < \rho$. Therefore, if $x \in K$,

$$\left| \frac{a(n)x^n}{1-x^n} \right| \le \frac{|a(n)||x|^n}{1-|x|^n} \le \frac{(kA)^n}{1-k}.$$

Since $0 < kA < 1$, this implies that the series defined by (7.8) is uniformly convergent on every compact subset of $D(\rho)$. Hence, by lemma 7.1, $L(x)$ is holomorphic in $D(\rho)$. The series $L(x)$ defined by (7.8) is called the *Lambert series* associated with the sequence $a(n)$.

Here is the fundamental result concerning Lambert series:

$$\sum_{n=1}^{+\infty} \frac{a(n)x^n}{1-x^n} = \sum_{n=1}^{+\infty} \left(\sum_{d|n} a(d) \right) x^n, \quad |x| < \rho. \tag{7.9}$$

To prove (7.9), observe that the function $L(x) = \sum_{n=1}^{+\infty} a(n)x^n/(1-x^n)$ is holomorphic, and therefore analytic, in $D(\rho)$. Hence it can be expanded in a power series $L(x) = \sum_{n=1}^{+\infty} b(n)x^n$. To find $b(n)$, we write

$$L(x) = \sum_{d=1}^{+\infty} \frac{a(d)x^d}{1-x^d} = \sum_{d=1}^{+\infty} a(d)x^d \sum_{k=0}^{+\infty} x^{kd} = \sum_{d=1}^{+\infty} \sum_{m=1}^{+\infty} a(d)x^{md}.$$

Thus each term of degree n in $L(x)$ can be written as $n = md$. Therefore every divisor d of n adds $a(d)$ to $b(n)$.

This proves that $b(n) = \sum_{d|n} a(d)$.

Example 7.3 If $a(n) = 1$, we have $\sum_{d|n} a(d) = \sum_{d|n} 1 = d(n)$. On the other hand, if $a(n) = n$, then $\sum_{d|n} a(d) = \sum_{d|n} d = \sigma(n)$. From this we deduce the ordinary generating functions of the divisor function $d(n)$ and of the sum of divisors function $\sigma(n)$ expressed as Lambert series:

$$\sum_{n=1}^{+\infty} d(n)x^n = \sum_{n=1}^{+\infty} \frac{x^n}{1-x^n} \qquad (|x| < 1) \tag{7.10}$$

$$\sum_{n=1}^{+\infty} \sigma(n)x^n = \sum_{n=1}^{+\infty} \frac{nx^n}{1-x^n} \qquad (|x| < 1) \tag{7.11}$$

7.3   Jacobi's triple product identity

It gives the infinite product expansion of the function

$$\theta_3(q,x) = \sum_{n=-\infty}^{+\infty} q^{n^2} x^n, \quad |q| < 1, \ x \in \mathbb{C}^*. \tag{7.12}$$

We will need the following lemma.

Lemma 7.2   *Let Ω be an open subset of $\mathbb{C}$. Let (u_n) be a sequence of functions defined and holomorphic in Ω. Assume that the series $\sum_{n=0}^{+\infty} u_n(x)$ converges uniformly in every compact subset of Ω. Then the infinite product $h(x) = \prod_{n=0}^{+\infty}(1 + u_n(x))$ converges uniformly in every compact subset of Ω. Therefore h is holomorphic in Ω. Moreover, its logarithmic derivative is given by*

$$\frac{h'(x)}{h(x)} = \sum_{n=0}^{+\infty} \frac{u_n'(x)}{1 + u_n(x)}. \tag{7.13}$$

Proof   See exercise 7.2.

Theorem 7.2   *(Jacobi's triple product identity)*

$$\theta_3(q,x) = \sum_{n=-\infty}^{+\infty} q^{n^2} x^n = \prod_{n=1}^{+\infty} \left(1 - q^{2n}\right)\left(1 + xq^{2n-1}\right)\left(1 + \frac{1}{x} q^{2n-1}\right).$$

Proof   Lemma 7.2 shows immediatly that, for fixed $|q| < 1$, the function $h_0(x) = \prod_{n=1}^{+\infty}(1 + xq^{2n-1})$ is holomorphic in $\mathbb{C}$. Therefore, it is an entire function and we can write $h_0(x) = \sum_{n=0}^{+\infty} \alpha_n(q) x^n$ for every $x \in \mathbb{C}$. Hence $h(x) = \prod_{n=1}^{+\infty}(1 + xq^{2n-1})(1 + x^{-1}q^{2n-1}) = h_0(x)h_0(1/x)$ is holomorphic in $\mathbb{C}^*$. By multiplying the two series, we obtain $h(x) = \sum_{n=-\infty}^{+\infty} a_n(q) x^n$, with $a_{-n}(q) = a_n(q)$ since $h(1/x) = h(x)$. But we have also

$$h(q^2 x) = \prod_{n=1}^{+\infty}(1 + xq^{2n+1})\left(1 + x^{-1}q^{2n-3}\right)$$

$$= (1 + xq)^{-1} \prod_{n=1}^{+\infty}(1 + xq^{2n-1})(1 + 1/xq) \prod_{n=1}^{+\infty}\left(1 + x^{-1}q^{2n-1}\right).$$

Consequently $h(q^2 x) = h(x)/qx$, whence

$$\sum_{n=-\infty}^{+\infty} a_n(q) q^{2n} x^n = q^{-1} \sum_{n=-\infty}^{+\infty} a_n(q) x^{n-1} = q^{-1} \sum_{n=-\infty}^{+\infty} a_{n+1}(q) x^n.$$

From this we deduce that $a_{n+1}(q) = q^{2n+1} a_n(q)$, and an easy induction shows that $a_n(q) = q^{n^2} a_0(q)$. Replacing in $h(x)$ yields

$$\prod_{n=1}^{+\infty} (1+xq^{2n-1})\left(1+\frac{1}{x}q^{2n-1}\right) = a_0(q) \sum_{n=-\infty}^{+\infty} q^{n^2} x^n = a_0(q)\theta_3(q,x). \quad (7.14)$$

Now, it remains to compute $a_0(q)$. Following Gauss, we observe that $\theta_3(q,i) = \theta_3(q,1/i) = \theta_3(q,-i)$, whence

$$\theta_3(q,i) = \frac{1}{2}(\theta_3(q,i) + \theta_3(q,-i)) = \sum_{n=-\infty}^{+\infty} q^{4n^2} (-1)^n = \theta_3(q^4,-1).$$

However, by using (7.14), we see that

$$\begin{cases} a_0(q)\theta_3(q,i) = \prod_{n=1}^{+\infty}(1+q^{4n-2}) = \prod_{n=1}^{+\infty}(1-q^{8n-4})\Big/\prod_{n=1}^{+\infty}(1-q^{4n-2}) \\ a_0(q^4)\theta_3(q^4,-1) = \prod_{n=1}^{+\infty}(1-q^{8n-4})^2 \end{cases}$$

As $\theta_3(q,i) = \theta_3(q^4,-1)$, we obtain

$$\begin{aligned} a_0(q^4) &= a_0(q)\prod_{n=1}^{+\infty}(1-q^{8n-4})(1-q^{4n-2}) \\ &= a_0(q)\prod_{n=1}^{+\infty}(1-q^{8n-4})(1-q^{8n-2})(1-q^{8n-6}). \end{aligned}$$

In the last product, the exponents of q include all even numbers, except the multiples of 8. Therefore

$$a_0(q^4) = a_0(q)\prod_{n=1}^{+\infty}(1-q^{2n})\Big/\prod_{n=1}^{+\infty}(1-q^{8n}).$$

Consequently, for every $k \in \mathbb{N}$,

$$a_0(q)\prod_{n=1}^{+\infty}(1-q^{2n}) = a_0(q^4)\prod_{n=1}^{+\infty}(1-q^{8n}) = \cdots = a_0(q^{4^k})\prod_{n=1}^{+\infty}(1-q^{2\cdot 4^k n}).$$

Now fix $x=1$, for instance. Then (7.14) and lemma 7.2 show that $a_0(q)$ is an holomorphic function of q for $|q|<1$, and that $a_0(0)=1$. Lemma 7.2 also shows that $\prod_{n=1}^{+\infty}(1-q^n)$ is holomorphic for $|q|<1$. Letting k tend to infinity yields $a_0(q)\prod_{n=1}^{+\infty}(1-q^{2n})=1$, and the proof of Jacobi's triple product identity is complete.

Corollary 7.1 *If $|q|<1$, $\displaystyle\prod_{n=1}^{+\infty}\frac{1-q^n}{1+q^n} = \sum_{n=-\infty}^{+\infty}(-1)^n q^{n^2}$.*

Proof See exercise 7.3.

7.4   Sums of two squares

Theorem 7.3   *For every $n \in \mathbb{N}^*$, $r_2(n) = 4(d_1(n) - d_3(n))$, where $d_1(n)$ is the number of divisors d of n satisfying $d \equiv 1$ (mod 4) and $d_3(n)$ is the number of divisors d of n satisfying $d \equiv 3$ (mod 4).*

Proof   We will need the following formula, whose proof is given in exercise 7.4. For $|q| < 1$,

$$\prod_{n=1}^{+\infty} (1+q^n)^2 (1-q^n) = \prod_{n=1}^{+\infty} (1+q^{4n-3})(1+q^{4n-1})(1-q^{4n}). \quad (7.15)$$

Now, to prove theorem 7.3 we replace x by $-\sqrt{q}x^2$ and q by $\sqrt{q}$ in Jacobi's triple product identity, and obtain

$$\prod_{n=1}^{+\infty} (1-q^n)(1-x^2 q^n)(1-x^{-2}q^{n-1}) = \sum_{n=-\infty}^{n=+\infty} (-1)^n x^{2n} q^{\frac{n(n+1)}{2}}. \quad (7.16)$$

But we have $\prod_{n=1}^{+\infty} (1-x^{-2}q^{n-1}) = (1-x^{-2})\prod_{n=1}^{+\infty} (1-x^{-2}q^n)$.

Therefore, after multiplication by x, (7.16) can be written

$$(x-x^{-1})\prod_{n=1}^{+\infty} (1-q^n)(1-x^2 q^n)(1-x^{-2}q^n) = \sum_{n=-\infty}^{+\infty} (-1)^n x^{2n+1} q^{\frac{n(n+1)}{2}}. \quad (7.17)$$

In the right-hand side of (7.17), we separate the terms with an even index from those with an odd index, and use the triple product identity again.

$$(x-x^{-1})\prod_{n=1}^{+\infty} (1-q^n)(1-x^2 q^n)(1-x^{-2}q^n) = \sum_{n=-\infty}^{+\infty} x^{4n+1} q^{2n^2+n} - \sum_{n=-\infty}^{+\infty} x^{4n-1} q^{2n^2-n}$$

$$= x\theta_3(q^2, qx^4) - x^{-1}\theta_3(q^2, q^{-1}x^4)$$

$$= x\prod_{n=1}^{+\infty} (1-q^{4n})(1+x^4 q^{4n-1})(1+x^{-4}q^{4n-3})$$

$$\qquad\qquad - x^{-1}\prod_{n=1}^{+\infty} (1-q^{4n})(1+x^4 q^{4n-3})(1+x^{-4}q^{4n-1}).$$

All these infinite products define holomorphic functions of x in $\mathbb{C}^*$ by lemma 7.2. We derive with respect to x and use (7.13).

$$(1+x^{-2})\prod_{n=1}^{+\infty} (1-q^n)(1-x^2 q^n)(1-x^{-2}q^n)$$

$$+ (x-x^{-1})\left(\prod_{n=1}^{+\infty} (1-q^n)(1-x^2 q^n)(1-x^{-2}q^n)\right)'$$

$$= \prod_{n=1}^{+\infty} (1-q^{4n})(1+x^4 q^{4n-1})(1+x^{-4}q^{4n-3}) \times$$

$$\left[1 + x\left(\sum_{n=1}^{+\infty} \frac{4x^3 q^{4n-1}}{1+x^4 q^{4n-1}} + \sum_{n=1}^{+\infty} \frac{-4x^{-5}q^{4n-3}}{1+x^{-4}q^{4n-3}}\right)\right] \ldots \text{(continued on page 107)}$$

$$-\prod_{n=1}^{+\infty}(1-q^{4n})(1+x^4q^{4n-3})(1+x^{-4}q^{4n-1})\times$$

$$\left[-x^{-2}+x^{-1}\left(\sum_{n=1}^{+\infty}\frac{4x^3q^{4n-3}}{1+x^4q^{4n-3}}+\sum_{n=1}^{+\infty}\frac{-4x^{-5}q^{4n-1}}{1+x^{-4}q^{4n-1}}\right)\right].$$

For $x=1$, this becomes remarkably simple, and dividing by 2 yields

$$\prod_{n=1}^{+\infty}(1-q^n)^3=\prod_{n=1}^{+\infty}(1-q^{4n})(1+q^{4n-1})(1+q^{4n-3})$$

$$\times\left(1+4\sum_{n=1}^{+\infty}\frac{q^{4n-1}}{1+q^{4n-1}}-4\sum_{n=1}^{+\infty}\frac{q^{4n-3}}{1+q^{4n-3}}\right).$$

Now we use (7.15) and corollary 7.1. We obtain

$$\left(\sum_{n=-\infty}^{+\infty}(-1)^nq^{n^2}\right)^2=1+4\sum_{n=1}^{+\infty}\frac{q^{4n-1}}{1+q^{4n-1}}-4\sum_{n=1}^{+\infty}\frac{q^{4n-3}}{1+q^{4n-3}}. \tag{7.18}$$

Replace q by $-q$. By using (7.5), we obtain the ordinary generating function of $r_2(n)$ as a sum of Lambert series:

$$\sum_{n=0}^{+\infty}r_2(n)q^n=1+4\sum_{n=1}^{+\infty}\frac{q^{4n-3}}{1-q^{4n-3}}-4\sum_{n=1}^{+\infty}\frac{q^{4n-1}}{1-q^{4n-1}}.$$

In the first of these two Lambert series, $a(n)=1$ if $n\equiv 1\pmod 4$ and $a(n)=0$ otherwise. Therefore (7.9) yields

$$\sum_{n=1}^{+\infty}\frac{q^{4n-3}}{1-q^{4n-3}}=\sum_{n=1}^{+\infty}d_1(n)q^n.$$

Similarly, $\displaystyle\sum_{n=1}^{+\infty}\frac{q^{4n-1}}{1-q^{4n-1}}=\sum_{n=1}^{+\infty}d_3(n)q^n$, and theorem 7.3 is proved.

Remark 7.3 Theorem 7.3 is a *quantitative* version of theorem 6.1. Indeed, if p is prime and $p\equiv 3\pmod 4$, then $d_1(p)=1$, since the only divisor of p congruent to 1 mod 4 is 1. And $d_3(p)=1$ since the only divisor of p congruent to 3 mod 4 is p. Hence $r_2(p)=0$, and p cannot be expressed as sum of two squares. If p is prime and $p\equiv 1\pmod 4$, then $d_1(p)=2$ and $d_3(p)=0$, whence $r_2(p)=8$. *Up to the permutations of a and b and the changes of signs* $\pm a$, $\pm b$, this means that *the expression of any odd prime p in the form a^2+b^2 is unique if, and only if, $p\equiv 1$* (mod 4). Recall that this expression can be obtained by using regular continued fraction expansions (exercise 4.10).

7.5 Jacobi's theorem on sums of four squares

By using similar computations (see exercise 7.5 for details), one can prove the following result.

$$\left(\sum_{n=-\infty}^{+\infty} q^{n^2}\right)^4 = 1 + 8\left(\sum_{n=1}^{+\infty} \frac{nq^n}{1-q^n} - \sum_{n=1}^{+\infty} \frac{4nq^{4n}}{1-q^{4n}}\right) \quad (|q|<1). \qquad (7.19)$$

However, if we denote by $r_4(n)$ the numbers of representations of n as a sum of four squares, we have, as in (7.5),

$$\sum_{n=0}^{+\infty} r_4(n)q^n = \left(\sum_{n=-\infty}^{+\infty} q^{n^2}\right)^4 \quad (|q|<1). \qquad (7.20)$$

Now, we see that the difference of the two series in the right-hand side of (7.19) can be written as the Lambert's series $\sum_{n=1}^{+\infty} a(n)q^n/(1-q^n)$, with $a(n)=0$ if $4|n$, and $a(n)=8n$ if not. Hence, by using (7.19), (7.20) and (7.9), we obtain *Jacobi's theorem*:

Theorem 7.4 *For every $n \in \mathbb{N}^*$, $r_4(n) = 8 \sum_{d|n,\, d \neq 4k} d$.*

In particular, $r_4(n) \geq 8$ since 1 divides n. Therefore every integer n is a sum of four squares, and Bachet's theorem (theorem 6.15) appears here as a corollary of Jacobi's theorem.

7.6 Euler function $\varphi(n)$

The arithmetical function $\varphi(n)$ is defined as the number of integers $0 \leq x \leq n$ which are prime to n. Thus $\varphi(2)=1$, $\varphi(3)=2$, $\varphi(4)=2$, $\varphi(5)=4$, $\varphi(6)=2$, and so on. There is an explicit formula for $\varphi(n)$:

$$\varphi(n) = n \prod_i \left(1 - \frac{1}{p_i}\right), \qquad (7.21)$$

where the product runs over *all prime divisors* p_i of n.

For example, $1176 = 2^3 \cdot 3 \cdot 7^2$, whence

$$\varphi(1176) = 1176\left(1 - \frac{1}{2}\right)\left(1 - \frac{1}{3}\right)\left(1 - \frac{1}{7}\right) = 336.$$

We give a proof of (7.21) which rests on the *sieve formula*. Denote, as usual, by $|A|$ the cardinality of A. Then, if $A_1, A_2, \ldots, A_m$ *are finite sets,*

$$|A_1 \cup A_2 \cup \ldots \cup A_m| = \sum_{i=1}^{m} |A_i| - \sum_{i<j} |A_i \cap A_j|$$

$$+ \sum_{i<j<k} |A_i \cap A_j \cap A_k| - \cdots + (-1)^{m-1} |A_1 \cap A_2 \cap \ldots \cap A_m|. \tag{7.22}$$

See exercise 7.6 for a proof of (7.22).

Now we prove (7.21). Let $n = p_1^{\alpha_1} p_2^{\alpha_2} \ldots p_m^{\alpha_m}$. The integers $1 \le x \le n$ *which are not prime to n* are those which are multiples of p_1, or p_2, ... or p_m, in other words the elements of $A_1 \cup A_2 \cup \ldots \cup A_m$, where A_i is the set of all multiples of p_i which are less than n. Since p_i divides n, we have $|A_i| = n/p_i$. Similarly $|A_i \cap A_j| = n/p_i p_j$ if $i \ne j$, because the elements of $A_i \cap A_j$ are the multiples of p_i and p_j, which are coprime. And so on. By using the sieve formula (7.22), we obtain

$$\varphi(n) = n - |A_1 \cup A_2 \cup \ldots \cup A_m|$$

$$= n - \sum_i \frac{n}{p_i} + \sum_{i<j} \frac{n}{p_i p_j} - \sum_{i<j<k} \frac{n}{p_i p_j p_k} + \cdots + (-1)^m \frac{n}{p_1 p_2 \cdots p_m}$$

$$= n \left(1 - \sum_i \frac{1}{p_i} + \sum_{i<j} \frac{1}{p_i p_j} - \cdots + (-1)^m \frac{1}{p_1 p_2 \cdots p_m} \right), \text{ Q.E.D.}$$

Now we prove the summation formula:

$$\sum_{d|n} \varphi(d) = n. \tag{7.23}$$

To prove (7.23), consider the n fractions with denominator n, namely

$$\frac{1}{n}, \frac{2}{n}, \frac{3}{n}, \ldots, \frac{n-1}{n}, \frac{n}{n},$$

and write them as *irreducible* fractions. For every divisor d of n, there exist exactly $\varphi(d)$ irreducible fractions with denominator d. Indeed, after simplification by $d' = n/d$, the resulting fraction is irreducible if and only if the numerator is prime to d. Hence (7.23) is proved.

We deduce from (7.23), by using (7.9), the *Lambert generating function* of $\varphi(n)$:

$$\sum_{n=1}^{+\infty} \frac{\varphi(n)x^n}{1-x^n} = \sum_{n=1}^{+\infty} \left(\sum_{d\mid n} \varphi(d) \right) x^n = \sum_{n=1}^{+\infty} nx^n = \frac{x}{(1-x)^2}. \qquad (7.24)$$

7.7   Average order of $r_2(n)$

The arithmetical functions have, in general, a very irregular behavior. Consider, for example, the function $r_2(n)$. Its values can be easily computed by hand by using the formula:

$$r_2(n) = 4 \prod_{p\equiv 1(\mathrm{mod}\,4)} (r+1) \prod_{p\equiv 3(\mathrm{mod}\,4)} \frac{1+(-1)^s}{2}, \qquad (7.25)$$

where r is the exponent of $p \equiv 1$ (mod 4), and s the exponent of $p \equiv 3$ (mod 4), in the standard factorization of n. Formula (7.25) is a consequence of theorem 7.3 (see exercise 7.10 for details). For $n \le 60$, we obtain the following table, which shows how irregular $r_2(n)$ is.

Table 7.1: Values of $r_2(n)$

n	1	2	3	4	5	6	7	8	9	10	11	12
$r_2(n)$	4	4	0	4	8	0	0	4	4	8	0	0
n	13	14	15	16	17	18	19	20	21	22	23	24
$r_2(n)$	8	0	0	4	8	4	0	8	0	0	0	0
n	25	26	27	28	29	30	31	32	33	34	35	36
$r_2(n)$	12	8	0	0	8	0	0	4	0	8	0	4
n	37	38	39	40	41	42	43	44	45	46	47	48
$r_2(n)$	8	0	0	8	8	0	0	0	8	0	0	0
n	49	50	51	52	53	54	55	56	57	58	59	60
$r_2(n)$	4	12	0	8	8	0	0	0	0	8	0	0

However, if we consider the arithmetical mean of the first n terms of the sequence $r_2(n)$, we obtain the following remarkable result.

Theorem 7.5 $\displaystyle\lim_{n\to\infty} \frac{r_2(1) + r_2(2) + \ldots + r_2(n)}{n} = \pi.$

Proof We use a geometrical approach (figure 7.1). Let $k \in \mathbb{N}^*$, $k \leq n$. Then $r_2(k)$ is equal to the number of lattice points (x, y) on the circle $x^2 + y^2 = k$. Therefore $S(n) = \sum_{k=1}^{n} r_2(k)$ is equal to the number of lattice points inside, or on, the circle (C) with center O and radius $\sqrt{n}$, O excluded. To each of these points A, we associate the lattice square which lies above and on the right of A (figure 7.1). These squares, together with $IJKO$, cover all the disk with center O and radius $\sqrt{n} - \sqrt{2}$, and their union is included in the disk with center O and radius $\sqrt{n} + \sqrt{2}$. As the sum of the areas of all these squares is equal to $S(n) + 1$, we deduce that $\pi(\sqrt{n} - \sqrt{2})^2 \leq S(n) + 1 \leq \pi(\sqrt{n} + \sqrt{2})^2$, which proves theorem 7.5.

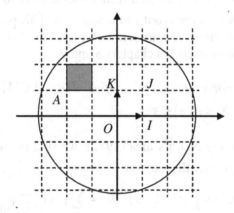

Figure 7.1

7.8   The series of primes: the function $\pi(n)$

We denote by $\pi(n)$ the number of primes less than n:

$$\pi(n) = \sum_{p \text{ prime } \leq n} 1. \tag{7.26}$$

The first effective results on the arithmetical function $\pi(n)$ have been obtained by Chebyshev in 1851. They consisted in an estimate for $\pi(n)$, similar to the following one.

Theorem 7.6   *There exists an explicitly computable integer N such that, for every $n \geq N$,*

$$\text{Log} \, 2 \frac{n}{\text{Log} \, n} \leq \pi(n) \leq \text{Log} \, 3 \frac{n}{\text{Log} \, n}.$$

A more precise result was proved by Hadamard and De la Vallée Poussin independently in 1896. It is known as the *prime number theorem*:

Theorem 7.7 $\pi(n) \sim \dfrac{n}{\text{Log} \, n} \quad (n \to +\infty).$

The prime number theorem is one of the most beautiful results in *analytic number theory*, but its proof is difficult and goes beyond the scope of this book. The interested reader should refer to [1] or [13]. We will limit ourselves to the proof of theorem 7.6. This proof relies mainly on an estimate of the least common multiple of the first n integers. This estimate will also be useful in chapters 8 and 9.

Theorem 7.8   *For every $n \in \mathbb{N}^*$, define $\delta(n) = \text{LCM}(1,2,...,n)$. Then, for every $n \geq 7$, $2^n \leq \delta(n) \leq 3^n$.*

Proof   a. Following Nairn's idea (1982), we first prove that $2 \leq \delta(n)$. For $1 \leq m \leq n$, consider the integral

$$K(m,n) = \int_0^1 x^{m-1}(1-x)^{n-m} dx = \sum_{k=0}^{n-m} (-1)^k \binom{n-m}{k} \frac{1}{m+k}.$$

First, we see that $\delta(n) K(m,n) \in \mathbb{N}$. Moreover, if we integrate $K(m,n)$ by parts $m-1$ times, we obtain

$$K(m,n) = \frac{(m-1)!}{(n-m+1)(n-m+2)\cdots(n-1)n} = \frac{1}{m\binom{n}{m}}.$$

Thus, for every m between 1 and n, $m\binom{n}{m}$ divides $\delta(n)$.

In particular, $n\binom{2n}{n} \big| \delta(2n)$. Since $\delta(2n) \big| \delta(2n+1)$, first we have

$$n\binom{2n}{n} \bigg| \delta(2n+1). \tag{7.27}$$

Since $(2n+1)\binom{2n}{n} = (n+1)\binom{2n+1}{n+1}$, we also have

$$(2n+1)\binom{2n}{n} \Big| \delta(2n+1). \tag{7.28}$$

As $\mathrm{GCD}(n, 2n+1) = 1$, (7.27) and (7.28) yield $n(2n+1)\binom{2n}{n} \Big| \delta(2n+1)$.

Therefore $\delta(2n+1) \geq n(2n+1)\binom{2n}{n}$. But for every integer k between 0

and $2n$, we have $\binom{2n}{n} \geq \binom{2n}{k}$, whence $(2n+1)\binom{2n}{n} \geq \sum_{k=0}^{2n}\binom{2n}{k} = 4^n$. Hence

$$\delta(2n+1) \geq n4^n. \tag{7.29}$$

Consequently, if $n \geq 2$ we have $\delta(2n+1) \geq 2^{2n+1}$. And if $n \geq 4$, (7.29)

yields $\delta(2n+2) \geq \delta(2n+1) \geq 2^{2n+2}$. This proves that $\delta(n) \geq 2^n$ if

$n \geq 9$, An easy computation shows that this remains true for $n = 7$ and

$n = 8$. But this lower bound is false for $n = 6$.

b. Following Hanson (1972), we now prove that $\delta(n) \leq 3^n$. This upper

bound is true, in fact, for every $n \geq 1$. A first ingredient of Hanson's

proof is the *Sylvester series* expansion of 1 (see exercise 2.9). If we

define the sequence (a_n) by $a_1 = 1$ and the recurrence relation

$a_{n+1} = a_n^2 - a_n + 1$, then (see exercise 7.7 for details)

$$\sum_{n=1}^{+\infty} \frac{1}{a_n} = 1. \tag{7.30}$$

A second ingredient of Hanson's proof is *Legendre's formula*:

Lemma 7.3   *The exponent* α_p *of the prime factor p in n! is*

$$\alpha_p = \left[\frac{n}{p}\right] + \left[\frac{n}{p^2}\right] + \left[\frac{n}{p^3}\right] + \cdots$$

Proof   See exercise 7.8.

We now proceed to the proof of the inequality $\delta(n) \leq 3^n$. Some

technical details will be left to the reader (exercice 7.9). First we have

$$\delta(n) = \prod_{p \leq n} p^{\left[\frac{\mathrm{Log}\,n}{\mathrm{Log}\,p}\right]}. \tag{7.31}$$

Indeed, $\delta(n)$ is formed from the product of prime numbers not greater than n, each of them up to the greatest exponent v_p they have in the standard factorization of any $m \leq n$. It is clear that $p^{v_p} \leq n$. Therefore $v_p \operatorname{Log} p \leq \operatorname{Log} n$, which yields (7.30) since v_p is an integer.

Now, Hanson's starting point consists in observing (see exercise 7.9 for details) that $\delta(n) \leq C(n)$, where $C(n)$ is defined by

$$C(n) = \frac{n!}{\left[\dfrac{n}{a_1}\right]! \left[\dfrac{n}{a_2}\right]! \cdots \left[\dfrac{n}{a_k}\right]!} \qquad (7.31)$$

In (7.31) the sequence (a_n) is the one defined in (7.29), and k is the greatest integer satisfying $a_k \leq n$ (whence $a_{k+1} > n$). It is clear that k is a function of n. First, we observe that $C(n)$ is an *integer*. Indeed,

$$\left[\frac{n}{a_1}\right] + \left[\frac{n}{a_2}\right] + \cdots + \left[\frac{n}{a_k}\right] \leq n\left(\frac{1}{a_1} + \frac{1}{a_2} + \cdots + \frac{1}{a_k}\right) < n.$$

Hence, if we put $b_i = [n/a_i]$ for $i \leq k$ and $b_{k+1} = n - \sum_{i=1}^{k} b_i$, we have

$$C(n) = b_{k+1}! \frac{n!}{b_1! b_2! \cdots b_{k+1}!}.$$

And since $b_1 + b_2 + \cdots + b_{k+1} = n$, we see that $n!/b_1! b_2! \cdots b_{k+1}!$ is the coefficient of $x_1^{b_1} x_2^{b_2} \cdots x_{k+1}^{b_{k+1}}$ in the expansion of $(x_1 + x_2 + \cdots + x_{k+1})^n$. Therefore it is an integer, and so is $C(n)$.

Now it can be proved (exercise 7.9) that $C(n) < (en)^{\log_2(\log_2 n)+3}(2.952)^n$, where $\log_2 x = \operatorname{Log} x/\operatorname{Log} 2$ is the logarithm to the base 2. But the function $g(x) = (ex)^{\frac{\log_2(\log_2 x)+3}{x}}$ is decreasing for $x \geq e$, which implies that $(en)^{\frac{\log_2(\log_2 n)+3}{n}} \leq (4500e)^{\frac{\log_2(\log_2 4500)+3}{4500}} \leq 1.014$ for $n \geq 4500$. Hence

$$C(n) < (1.014)^n (2.952)^n < (2.994)^n < 3^n \qquad (7.32)$$

for $n \geq 4500$. Using *Maple* or *Mathematica* allows to check that (7.32) remains true for $n < 4500$, by a direct computation of $\operatorname{Log} \delta(n) = \Psi(n)$. This proves that $\delta(n) < 3^n$ for every $n \in \mathbb{N}^*$, as claimed.

Now we proceed to the proof of theorem 7.6.

We first prove that, for every $n \geq 4$, $\pi(n) \geq \text{Log}\,2\dfrac{n}{\text{Log}\,n}$.

By using theorem 7.8 and formula (7.30), we get, for $n \geq 7$,

$$n\,\text{Log}\,2 \leq \text{Log}\left(\delta(n)\right) \leq \sum_{p \leq n} \text{Log}\,n \leq \pi(n)\text{Log}\,n, \quad \text{Q.E.D.}$$

This remains true for $n = 4$, 5 and 6, as it can be checked directly.

Finally, we prove that $\pi(n) \leq \text{Log}\,4\dfrac{n}{\text{Log}\,n}$.

By using (7.32) and (7.30), we obtain, for every $n \geq 4500$ and every m
satisfying $2 \leq m \leq n$: $\displaystyle\prod_{m < p \leq n} p \leq \prod_{p \leq n} p \leq \delta(n) \leq (2.994)^n$.

Taking the logarithms yields

$$\left(\text{Log}\,m\right)\left(\pi(n) - \pi(m)\right) \leq \sum_{m < p \leq n} \text{Log}\,p \leq n\,\text{Log}\,2.994.$$

Since evidently $\pi(m) \leq m$, we see that $\pi(n) \leq (\text{Log}\,2.994)n/\text{Log}\,m + m$.

Choose $m = \left[n/(\text{Log}\,n)^2 \right] + 1$. Then, for every $n \geq 4500$,

$$\pi(n) \leq \text{Log}\,2.994\,\frac{n}{\text{Log}\,n - 2\text{Log}\,(\text{Log}\,n)} + \frac{n}{(\text{Log}\,n)^2} + 1$$

$$\leq \frac{n}{\text{Log}\,n} \times \left(\frac{\text{Log}\,2.994}{1 - \dfrac{2\text{Log}\,(\text{Log}\,n)}{\text{Log}\,n}} + \frac{1}{\text{Log}\,n} + \frac{\text{Log}\,n}{n} \right).$$

The term between the brackets tends to $\text{Log}\,2.994$ when $n \to +\infty$. Since
$\text{Log}\,2.994 < \text{Log}\,3$, the proof of theorem 7.6 is now complete.

Exercises

7.1   Proof of theorem 6.1

1) Assume that u_n is a linear recurring sequence. Prove that its ordinary
generating function is a rational fonction $P(x)/Q(x)$, $\deg P < \deg Q$.
2) Prove the converse.

7.2   Draw inspiration from exercise 3.4 to prove lemma 7.2.

7.3   Prove corollary 7.1.

7.4   Prove formula (7.15).

7.5   Proof of formula (7.20)

1) By expanding $(1-q^n)^{-2}$ in a power series, prove that

$$\sum_{n=1}^{+\infty}\frac{q^n}{(1-q^n)^2}=\sum_{n=1}^{+\infty}\frac{nq^n}{1-q^n}\qquad(|q|<1).$$

2) Put $D_x=x\dfrac{\partial}{\partial x}$ and $D_q=q\dfrac{\partial}{\partial q}$. Check that $D_x^2\theta_3(q,x)=D_q\theta_3(q,x)$.

3) By using Jacobi's triple product identity, prove that

$$\frac{D_x^2\theta_3(x,q)}{\theta_3(q,x)}=\left[\left(\sum_{n=1}^{+\infty}\frac{xq^{2n-1}}{1+xq^{2n-1}}-\sum_{n=1}^{+\infty}\frac{x^{-1}q^{2n-1}}{1+x^{-1}q^{2n-1}}\right)^2\right.$$

$$\left.+\sum_{n=1}^{+\infty}\frac{xq^{2n-1}}{(1+xq^{2n-1})^2}+\sum_{n=1}^{+\infty}\frac{x^{-1}q^{2n-1}}{(1+x^{-1}q^{2n-1})^2}\right],$$

$$\frac{D_q\theta_3(x,q)}{\theta_3(q,x)}=\left(-\sum_{n=1}^{+\infty}\frac{2nq^{2n}}{1-q^{2n}}+\sum_{n=1}^{+\infty}\frac{(2n-1)xq^{2n-1}}{1+xq^{2n-1}}+\sum_{n=1}^{+\infty}\frac{(2n-1)x^{-1}q^{2n-1}}{1+x^{-1}q^{2n-1}}\right).$$

4) Replace q by q^2 and x by $-q$ and deduce that

$$\left(\sum_{n=1}^{+\infty}\frac{q^{4n-3}}{1-q^{4n-3}}-\frac{q^{4n-1}}{1-q^{4n-1}}\right)^2=\sum_{n=1}^{+\infty}\left(\frac{q^{4n-3}}{(1-q^{4n-3})^2}+\frac{q^{4n-1}}{(1-q^{4n-1})^2}\right)$$

$$-\sum_{n=1}^{+\infty}\frac{(2n-1)q^{4n-1}}{1-q^{4n-1}}-\sum_{n=1}^{+\infty}\frac{(2n-1)q^{4n-3}}{1-q^{4n-3}}-\sum_{n=1}^{+\infty}\frac{2nq^{4n}}{1-q^{4n}}.$$

5) Prove formula (7.20).

7.6   Prove the *sieve formula* (7.22).

7.7   Sylvester series expansion of 1

Define a_n by $a_1=2$ and $a_{n+1}=a_n^2-a_n+1$ for every $n\geq1$.

a) Prove that the series $\sum_{n=1}^{+\infty}1/a_n$ converges and compute its sum.

b) Prove that $a_n^2>a_{n+1}>(a_n-1)^2$ for every integer $n\geq1$. Deduce that the sequence $u_n = a_1^{1/a_1} \cdot a_1^{1/a_2} \cdots a_n^{1/a_n}$ is converging, and that $u_n \leq 2.952$.

c) Let $n \in \mathbb{N}^*$ and let $k = k(n)$ be the greatest integer satisfying $a_k \leq n$. Prove that $k \leq \log_2(\log_2 n) + 2$, where $\log_2 x = \mathrm{Log}\, x / \mathrm{Log}\, 2$ and $k \geq 1$.

7.8   Prove *Legendre's formula* (lemma 7.3).

7.9   Proof of theorem 7.5

Let a_n be the sequence defined in exercise 7.7.

1) Let $x \geq 1$ be a real number and $m \in \mathbb{N}^*$. Prove that $[x/m] = [[x]/m]$.

Deduce that $[x] > \sum_{i=1}^{n} [x/a_i]$ for every $n \in \mathbb{N}^*$.

2) Use (7.28), (7.29) and Legendre's formula to prove that $\delta(n) | C(n)$ and that, consequently, $\delta(n) \leq C(n)$.

3) Check that $(1 + 1/x)^x < e$ for every $x > 0$. Deduce that, if $n \geq a_i$,

$$\frac{(n/a_i)^{n/a_i}}{[n/a_i]^{[n/a_i]}} < \left(\frac{en}{a_i}\right)^{(a_i-1)/a_i}.$$

4) Let $n_1, n_2, \ldots, n_k$ be natural integers such that $n_1 + n_2 + \cdots + n_k = n$.

Prove that $n^n \geq \dfrac{n!}{n_1! n_2! \ldots n_k!} n_1^{n_1} n_2^{n_2} \ldots n_k^{n_k}$. Deduce that

$$C(n) < \frac{n^n}{[n/a_1]^{[n/a_1]} [n/a_2]^{[n/a_2]} \ldots [n/a_k]^{[n/a_k]}}.$$

Then prove that $C(n) < \dfrac{n^n (en/a_1)^{(a_1-1)/a_1} (en/a_2)^{(a_2-1)/a_2} \ldots (en/a_k)^{(a_k-1)/a_k}}{(n/a_1)^{n/a_1} (n/a_2)^{n/a_2} \ldots (n/a_k)^{n/a_k}}.$

5) Deduce that $C(n) < (en)^{\log_2(\log_2 n)+3} (2.952)^n$.

7.10   Prove formula (7.25) by using theorem 7.3.

7.11   Assume $|q| < 1$. Prove that $\displaystyle\prod_{n=1}^{+\infty}(1-q^n) = \sum_{n=-\infty}^{+\infty} (-1)^n q^{\frac{n(3n+1)}{2}}$.

7.12   Prove that $\displaystyle\left(\prod_{n=1}^{+\infty}(1-q^n)\right)^3 = \sum_{n=0}^{+\infty} (-1)^n (2n+1) q^{\frac{n(n+1)}{2}}$ $(|q| < 1)$.

7.13   Partitions

We call *partition* of the integer n every sequence $1 \le k_1 \le k_2 \le \cdots \le k_m$ of integers satisfying $k_1 + k_2 + \cdots + k_m = n$ $(m \ge 1)$. We denote by $p(n)$ the number of partitions of n. For example, the partitions of 5 are $5 = 1 + 4 = 2 + 3 = 1 + 1 + 3 = 1 + 2 + 2 = 1 + 1 + 1 + 2 = 1 + 1 + 1 + 1 + 1$, whence $p(5) = 7$. By convention, $p(0) = 1$.

1) Prove that the ordinary generating function of $p(n)$ is

$$f(q) = \sum_{n=0}^{+\infty} p(n) q^n = 1 / \prod_{k=1}^{+\infty} (1 - q^k), \quad |q| < 1.$$

2) Define $\alpha(n) = (-1)^k$ if n can be written as $k(3k+1)/2$ or $k(3k-1)/2$, $\alpha(n) = 0$ if not. By using exercise 7.11, prove that, for $n \ge 1$,

$$p(n) = -\sum_{k=1}^{n} \alpha(k) p(n-k).$$

3) Compute $p(n)$ for $1 \le n \le 12$.

7.14   Bernoulli numbers and polynomials

1) The *Bernoulli numbers* B_n are defined by the generating function

$$\frac{t}{e^t - 1} = \frac{1}{\sum_{n=1}^{+\infty} t^{n-1}/n!} = \sum_{n=0}^{+\infty} B_n \frac{t^n}{n!}. \tag{7.33}$$

Compute B_n for $0 \le n \le 6$.

2) The *Bernoulli polynomials* $B_n(x)$ are defined by

$$\frac{t e^{tx}}{e^t - 1} = \sum_{n=0}^{+\infty} B_n(x) \frac{t^n}{n!}. \tag{7.34}$$

1) Compute $B_0(x)$ and $B_1(x)$.

2) Prove that $B_n'(x) = n B_{n-1}(x)$ for every $n \in \mathbb{N} - \{0\}$.

3) Check that the function $f(t) = \dfrac{t}{e^t - 1} - B_0 - B_1 t$ is even.

   Deduce that $B_{2p+1} = 0$ if $p \ge 1$.

4) Prove that $\dfrac{t}{e^t - 1} = \sum_{n=0}^{+\infty} (-1)^n B_n(1) \dfrac{t^n}{n!}$.

   Then prove that $B_n(0) = B_n(1) = B_n$ for $n \ge 2$.

5) Prove that $B_n(x) = \sum_{k=0}^{n} \binom{n}{k} B_k x^{n-k}$ for every $n \in \mathbb{N}$.

6) Prove that $B_n(x+1) - B_n(x) = nx^{n-1}$ for every $n \in \mathbb{N}$.

Deduce the value of the sum $S_n(k) = \sum_{i=1}^{k} i^n$.

7.15 Möbius function

It is defined by $\mu(1) = 1$, $\mu(n) = 0$ if n is not squarefree, $\mu(n) = (-1)^k$ if $n = p_1 p_2 \dots p_k$ is squarefree (that is if the primes p_i are all different).

1) Compute $\mu(n)$ for $2 \le n \le 12$.

2) Check that $\varphi(n) = n \sum_{d|n} \mu(d)/d$, where $\varphi(n)$ is Euler function.

3) Prove that $\sum_{d|n} \mu(d) = 0$ if $n > 1$.

4) Prove *Möbius inversion formula*:

If $g(n) = \sum_{d|n} f(d)$, then $f(n) = \sum_{d|n} \mu(n/d) g(d)$.

7.16 Average value of $d(n)$

By using a geometrical method similar to the one used in section 7.7, prove that $d(1) + d(2) + \cdots + d(n) = n\operatorname{Log} n + O(n)$.

7.17 Riemann ζ function

It is defined by $\zeta(s) = \sum_{n=1}^{+\infty} 1/n^s$.

1) Prove that $\zeta(s)$ exists for every $s \in \mathbb{C}$ such that $\operatorname{Re}(s) > 1$ and that it is an holomorphic function in $\Omega = \{z \in \mathbb{C} \, / \, \operatorname{Re}(s) > 1\}$.

2) Expand in a Fourier series the 1-periodic function defined by $f_k(x) = B_{2k}(x)$ for $x \in [0,1]$, where the $B_{2k}(x)$ are the Bernouilli polynomials (exercise 7.14). Deduce that, for every integer $k \ge 1$,

$$\zeta(2k) = (-1)^{k-1} \frac{(2\pi)^{2k}}{(2k)!} \frac{B_{2k}}{2}.$$

3) Prove that, for $\operatorname{Re}(s) > 1$, $\zeta(s) = \prod_p \frac{1}{1 - \frac{1}{p^s}}$, where the product runs

over all prime numbers $p \in \mathbb{N}$.

7.18 Dirichlet series

Let $a(n)$ be a sequence of complex numbers of polynomial type.

That means that there exists $\alpha > 0$ such that $|a(n)| \le n^\alpha$ for every $n \in \mathbb{N}$.
We define the *generating Dirichlet series* of $a(n)$ by

$$F(s) = \sum_{n=1}^{+\infty} a(n)/n^s.$$

1) Prove that $F(s)$ exists for $\mathrm{Re}(s) > 1 + \alpha$ and that it is an holomorphic function in $\Omega = \{z \in \mathbb{C} \, / \, \mathrm{Re}(s) > 1 + \alpha\}$.

2) Assume that $F(s)$ and $G(s)$ are the generating Dirichlet series of $a(n)$ and $b(n)$ respectively. Prove that $F(s)G(s)$ is the generating Dirichlet series of $c(n) = \sum_{d|n} a(d)b(n/d)$.

3) Prove that $\sum_{n=1}^{+\infty} \mu(n)/n^s = 1/\zeta(s)$ $(\mathrm{Re}(s) > 1)$, where $\mu(n)$ is the Möbius function (ex. 7.15) and $\zeta(s)$ the Riemann function (ex. 7.17).

4) Prove that $\sum_{n=1}^{+\infty} d(n)/n^s = \zeta^2(s)$ $(\mathrm{Re}(s) > 1)$.

5) Prove that $\sum_{n=1}^{+\infty} \varphi(n)/n^s = \zeta(s-1)/\zeta(s)$ $(\mathrm{Re}(s) > 2)$.

6) Prove that $\sum_{n=1}^{+\infty} \sigma(n)/n^s = \zeta(s)\zeta(s-1)$ $(\mathrm{Re}(s) > 2)$, where $\sigma(n)$ is the sum of the divisors of n.

7.19   Bertrand's postulate (1845)
Evaluate $\delta(2n)$ by using n and $\sqrt{2n}$, and prove *Bertrand's postulate*: for every integer $n \ge 2$, there exists a prime p such that $n < p < 2n$.

7.20   Multiplicative functions
Let $f(n)$ be an arithmetical function.

We say that $f(n)$ is multiplicative if $f(pq) = f(p)f(q)$ for every pair of coprime integers (p, q).

1) Prove that Euler function $\varphi(n)$ is multiplicative.

2) Prove that the functions $d(n)$ and $\sigma(n)$ are multiplicative.

3) Deduce formulas for $d(n)$ and $\sigma(n)$.

Chapter 8

Padé approximants

In section 8.1, we define the Padé approximants of an analytic function in a neighborhood of 0, and prove lemma 8.1, which is fundamental for diophantine approximation. In section 8.2, we recall some properties of Gauss hypergeometric function, which allow us to compute the complete Padé table of the binomial function $f(x) = (1-x)^\alpha$, $\alpha \notin \mathbb{Z}$. From this we deduce the Padé approximants of the exponential function by means of confluent hypergeometric functions (section 8.3). Finally, we show in section 8.4 how Padé approximants yield good diophantine approximations and irrationality results.

8.1   Introduction

Definition 8.1   Let $f(x) = \sum_{k=0}^{+\infty} a_k x^k$ $(|x| < R)$ be analytic in a neighborhood of 0. Let $(m, n) \in \mathbb{N}^2$. We say that the triple $(Q_m, P_n, R_{m,n})$ is a $[m/n]$ Padé approximant of f if

a) Q_m and P_n are polynomials satisfying $\deg Q_m \le m$, $\deg P_n \le n$.

b) $R_{m,n}(x) = \sum_{k=0}^{+\infty} b_k x^k$ $(|x| < R)$ is analytic in a neighborhood of 0.

c) $Q_m(x) f(x) + P_n(x) = x^{m+n+1} R_{m,n}(x)$ $(|x| < R)$.

The existence of Padé approximants is an easy consequence of lemma 1.1, page 4. It should be noted that the partial sum of order n of f, $S_n(x) = \sum_{k=0}^{n} a_k x^k$, is a $[0/n]$ Padé approximant of f, since

$$1 \cdot f(x) - S_n(x) = x^{n+1} \sum_{k=0}^{+\infty} a_{n+k+1} x^k.$$

Moreover, if $(Q_m, P_n, R_{m,n})$ is a $[m/n]$ Padé approximant of f, the same holds for $(\lambda Q_m, \lambda P_n, \lambda R_{m,n})$ for every $\lambda \in \mathbb{C}$.

In number theory, we will mainly use *diagonal* Padé approximants $(n = m)$. In this case we denote $R_{n,n} = R_n$. The following result is fundamental.

Lemma 8.1   Let (Q_n, P_n, R_n) be a sequence of diagonal Padé approximants of f. Assume that $Q_n(0) \neq 0$ and $R_n(0) \neq 0$ for every $n \in \mathbb{N}$. Then, for every $n \in \mathbb{N}$,

$$\begin{vmatrix} Q_n(x) & P_n(x) \\ Q_{n+1}(x) & P_{n+1}(x) \end{vmatrix} = c_n x^{2n+1}, \text{ with } c_n \neq 0.$$

Proof   We have $Q_n(x)f(x) + P_n(x) = x^{2n+1}R_n(x)$, whence

$$\begin{vmatrix} Q_n(x) & P_n(x) \\ Q_{n+1}(x) & P_{n+1}(x) \end{vmatrix} = \begin{vmatrix} Q_n(x) & P_n(x) + Q_n(x)f(x) \\ Q_{n+1}(x) & P_{n+1}(x) + Q_{n+1}(x)f(x) \end{vmatrix}$$

$$= \begin{vmatrix} Q_n(x) & x^{2n+1}R_n(x) \\ Q_{n+1}(x) & x^{2n+3}R_{n+1}(x) \end{vmatrix} = x^{2n+1}\left(-Q_{n+1}(x)R_n(x) + x^2 Q_n(x)R_{n+1}(x) \right).$$

Therefore

$$\begin{vmatrix} Q_n(x) & P_n(x) \\ Q_{n+1}(x) & P_{n+1}(x) \end{vmatrix} = x^{2n+1}\left(-Q_{n+1}(0)R_n(0) + \sum_{k=1}^{+\infty} b_k x^k \right). \qquad (8.1)$$

Now, the left-hand side of (8.1) is a *polynomial* of degree $\leq 2n+1$, whence $b_k = 0$ for every $k \geq 1$, and $c_n = -Q_{n+1}(0)R_n(0) \neq 0$.

8.2   Gauss hypergeometric function and Padé approximants of the binomial function

For every complex number a, we denote

$$\begin{cases} (a)_0 = 1 \\ (a)_n = a(a+1)...(a+n-1) \quad \text{for } n \geq 1 \end{cases} \qquad (8.2)$$

and we define, for $|x| < 1$, the *hypergeometric function* $_2F_1$ by

$$_2F_1\left(\begin{matrix} a, b \\ c \end{matrix} \middle| x \right) = \sum_{n=0}^{+\infty} \frac{(a)_n (b)_n}{(c)_n n!} x^n. \qquad (8.3)$$

Example 8.1 For $a = b = c = 1$, we obtain the *geometric series*:

$$_2F_1\left(\left.\begin{matrix}1,1\\1\end{matrix}\right|x\right) = \sum_{n=0}^{+\infty} x^n = \frac{1}{1-x}.$$

Example 8.2 For $a = -\alpha$, $b = c$, we obtain the *binomial series*:

$$_2F_1\left(\left.\begin{matrix}-\alpha,b\\b\end{matrix}\right|x\right) = \sum_{n=0}^{+\infty} \frac{(-\alpha)(-\alpha+1)...(-\alpha+n-1)}{n!} x^n = (1-x)^\alpha.$$

Remark 8.1 It is clear that we must assume $c \neq 0, -1, -2,...$ in (8.3). However, if $a \in \mathbb{Z}_-$, then $_2F_1\left(\left.\begin{matrix}a,b\\c\end{matrix}\right|x\right)$ is a polynomial of degree $-a$. In this case, c can be a negative integer, provided that $c < a$. Indeed, in this case $(c)_n$ vanishes after $(a)_n$. Hence, *if a and c are negative integers, with $c < a$, then $_2F_1\left(\left.\begin{matrix}a,b\\c\end{matrix}\right|x\right)$ is a polynomial of degree $-a$.*

Theorem 8.1 *The hypergeometric function $_2F_1\left(\left.\begin{matrix}a,b\\c\end{matrix}\right|x\right)$ satisfies the differential equation*

$$x(1-x)y'' + (c - (a+b+1)x)y' - aby = 0. \qquad (8.4)$$

Moreover, if $c \notin \mathbb{Z}$, the general solution of (8.4) on $]0,1[$ is

$$y = A \ _2F_1\left(\left.\begin{matrix}a,b\\c\end{matrix}\right|x\right) + B \ x^{1-c} \ _2F_1\left(\left.\begin{matrix}a-c+1,b-c+1\\2-c\end{matrix}\right|x\right) \qquad (8.5)$$

Proof See exercice 8.1.

From theorem 8.1 we deduce (see exercise 8.2 for details) that

$$_2F_1\left(\left.\begin{matrix}a,b\\c\end{matrix}\right|x\right)(1-x)^{a+b-c} = \ _2F_1\left(\left.\begin{matrix}c-a,c-b\\c\end{matrix}\right|x\right) \qquad (8.6)$$

Following Padé himself (1900), we will use this relation to find the Padé approximants of the binomial function $f(x) = (1-x)^\alpha$.

Theorem 8.2 *For every $\alpha \notin \mathbb{Z}$, $(m,n) \in \mathbb{N}^2$, $|x| < 1$,*

$$_2F_1\left(\left.\begin{matrix}-m,-n+\alpha\\-m-n\end{matrix}\right|x\right)(1-x)^\alpha - \ _2F_1\left(\left.\begin{matrix}-n,-m-\alpha\\-m-n\end{matrix}\right|x\right)$$

$$= x^{m+n+1} \frac{(-1)^m (-m-\alpha)_{m+n+1}}{\binom{m+n}{m}(m+n+1)!} \ _2F_1\left(\left.\begin{matrix}n+1-\alpha,m+1\\m+n+2\end{matrix}\right|x\right).$$

Proof   Let $\varepsilon \in [-\frac{1}{2}, \frac{1}{2}]$. In relation (8.6) we take $a = -m$, $b = -(n+\varepsilon) + \alpha$ and $c = -m - (n+\varepsilon)$. We get

$$_2F_1\left(\begin{matrix} -m, -(n+\varepsilon)+\alpha \\ -m-(n+\varepsilon) \end{matrix} \middle| x\right)(1-x)^{\alpha} = {}_2F_1\left(\begin{matrix} -(n+\varepsilon), -m-\alpha \\ -m-(n+\varepsilon) \end{matrix} \middle| x\right). \qquad (8.7)$$

The hypergeometric series in the left-hand side of (8.7) is a polynomial of degree m. When $\varepsilon \to 0$, it tends to $_2F_1\left(\begin{matrix} -m, -n+\alpha \\ -m-n \end{matrix} \middle| x\right)$.

On the other hand, the hypergeometric series in the right-hand side can be written as

$$_2F_1\left(\begin{matrix} -(n+\varepsilon), -m-\alpha \\ -m-(n+\varepsilon) \end{matrix} \middle| x\right) = \sum_{k=0}^{n} \frac{(-(n+\varepsilon))_k(-m-\alpha)_k}{(-m-(n+\varepsilon))_k \cdot k!} x^k$$

$$+ \sum_{k=n+1}^{m+n} \frac{(-(n+\varepsilon))_k(-m-\alpha)_k}{(-m-(n+\varepsilon))_k \cdot k!} x^k + \sum_{k=m+n+1}^{+\infty} \frac{(-(n+\varepsilon))_k(-m-\alpha)_k}{(-m-(n+\varepsilon))_k \cdot k!} x^k.$$

When $\varepsilon \to 0$, the first sum tends to $_2F_1\left(\begin{matrix} -n, -m+\alpha \\ -m-n \end{matrix} \middle| x\right)$. Moreover the second sum tends to zero, since $(-n-\varepsilon)_k = (-n-\varepsilon)(-n-\varepsilon+1)...(-n-\varepsilon+k-1)$ contains one factor ε when $n+1 \le k \le m+n$.

Let $g(x)$ be the third sum. We first observe that

$$(-n-\varepsilon)_k = [(-n-\varepsilon)...(-\varepsilon)] \times [(-\varepsilon+1)...(-\varepsilon+m)]$$
$$\times [(-\varepsilon+m+1)...(-\varepsilon+m+1+(k-m-n-1)-1)],$$
$$(-m-n-\varepsilon)_k = [(-m-n-\varepsilon)...(-n-1-\varepsilon)] \times [(-n-\varepsilon)...(-\varepsilon)]$$
$$\times [(-\varepsilon+1)...(-\varepsilon+1+(k-m-n-1)-1)],$$
$$(-m-\alpha)_k = [(-m-\alpha)...(n-\alpha)]$$
$$\times [(n+1-\alpha)...(n+1-\alpha+(k-m-n-1)-1)],$$
$$k! = (m+n+1)![(m+n+2)...(m+n+2+(k-m-n-1)-1)].$$

But in every term of $g(x)$, we can simplify $(-n-\varepsilon)_k/(-m-n-\varepsilon))_k$ by $[(-\varepsilon)(-\varepsilon-1)...(-\varepsilon-n)]$. Factorizing and putting $h = k-m-n-1$ yields

$$g(x) = \frac{[(-\varepsilon+1)...(-\varepsilon+m)] \times [(-m-\alpha)...(n-\alpha)]}{[(-m-n-\varepsilon)...(-n-1-\varepsilon)] \times (m+n+1)!} x^{m+n+1}$$

$$\times \sum_{h=0}^{+\infty} \frac{(-\varepsilon+m+1)_h(n+1-\alpha)_h}{(-\varepsilon+1)_h(m+n+2)_h} x^h.$$

For fixed $|x| < 1$, the series converges uniformly for $\varepsilon \in [-\frac{1}{2}, \frac{1}{2}]$. Therefore we have

$$\lim_{\varepsilon \to 0} g(x) = \frac{m!(-m-\alpha)...(n-\alpha)}{(-m-n)...(-n-1)\cdot(m+n+1)!} x^{m+n+1} \sum_{h=0}^{+\infty} \frac{(m+1)_h (n+1-\alpha)_h}{h!(m+n+2)_h} x^h.$$

Thus, letting $\varepsilon \to 0$ in (8.8) proves theorem 8.2.

8.3 Confluent hypergeometric functions and Padé approximants of the exponential function

Let $x \in \mathbb{C}$, and let $b \in \mathbb{R}$, with $b > |x|$. Then

$$_2F_1\left(\begin{matrix} a,b \\ c \end{matrix}\middle| \frac{x}{b}\right) = \sum_{n=0}^{+\infty} \frac{(a)_n}{(c)_n}\left(1+\frac{1}{b}\right)\left(1+\frac{2}{b}\right)...\left(1+\frac{n-1}{b}\right)\frac{x^n}{n!}.$$

For fixed x, the series converges uniformly for $b \in [2|x|,+\infty[$, whence

$$\lim_{b \to +\infty} {}_2F_1\left(\begin{matrix} a,b \\ c \end{matrix}\middle| \frac{x}{b}\right) = \sum_{n=0}^{+\infty} \frac{(a)_n}{(c)_n}\frac{x^n}{n!}.$$

We call this entire function the *confluent hypergeometric function*, and we denote it by

$$_1F_1\left(\begin{matrix} a \\ c \end{matrix}\middle| x\right) = \sum_{n=0}^{+\infty} \frac{(a)_n}{(c)_n}\frac{x^n}{n!}. \tag{8.9}$$

Remark 8.2 More generally, the *generalized hypergeometric function* is defined for $q \geq p-1$ by

$$_pF_q\left(\begin{matrix} a_1,a_2,...,a_p \\ b_1,b_2,...,b_q \end{matrix}\middle| x\right) = \sum_{n=0}^{+\infty} \frac{(a_1)_n(a_2)_n...(a_p)_n}{(b_1)_n(b_2)_n...(b_q)_n}\frac{x^n}{n!}. \tag{8.10}$$

This explains the notations for $_1F_1$ and $_2F_1$. Now, returning to $_1F_1$ and replacing x by x/b in theorem 8.1, we obtain (see exercise 8.3)

Theorem 8.3 *The confluent hypergeometric function satisfies the differential equation*

$$xy'' + (c-x)y' - ay = 0. \tag{8.11}$$

Moreover, if $c \notin \mathbb{Z}$, the general solution of (8.11) on $]0,+\infty[$ is

$$y = A \, _1F_1\left(\begin{matrix} a \\ c \end{matrix}\middle| x\right) + B \, x^{1-c} \, _1F_1\left(\begin{matrix} a-c+1 \\ 2-c \end{matrix}\middle| x\right). \tag{8.12}$$

When a is a negative integer, $_1F_1\left(\begin{matrix} a \\ c \end{matrix}\middle| x\right)$ is a polynomial, which can be used to express the Padé approximants of the exponential function:

Theorem 8.4 *For every* $(m,n) \in \mathbb{N}^2$ *and every* $x \in \mathbb{C}$,

$$_1F_1\left(\left.{-m \atop -m-n}\right| -x\right)e^x - {}_1F_1\left(\left.{-n \atop -m-n}\right| x\right) = \frac{(-1)^m x^{m+n+1}}{\binom{m+n}{m}(m+n+1)!} {}_1F_1\left(\left.{m+1 \atop n+m+2}\right| x\right).$$

Proof In theorem 8.2, we replace x by $-x/\alpha$. First we know that $\lim_{\alpha \to +\infty} (1 + x/\alpha)^\alpha = e^x$. On the other hand,

$$\lim_{\alpha \to +\infty} {}_2F_1\left(\left.{-m,-n+\alpha \atop -m-n}\right| -\frac{x}{\alpha}\right)$$

$$= \lim_{\alpha \to +\infty} \sum_{k=0}^{+\infty} \frac{(-m)_k}{(-m-n)_k}\left(1 - \frac{n}{\alpha}\right)\cdots\left(1 - \frac{n-k+1}{\alpha}\right)\frac{(-x)^k}{k!} = {}_1F_1\left(\left.{-m \atop -m-n}\right| -x\right).$$

Similarly $\lim_{\alpha \to +\infty} {}_2F_1\left(\left.{-n,-m-\alpha \atop -m-n}\right| -\frac{x}{\alpha}\right) = {}_1F_1\left(\left.{-n \atop -m-n}\right| x\right)$.

And $\lim_{\alpha \to +\infty} {}_2F_1\left(\left.{n+1-\alpha,m+1 \atop m+n+2}\right| -\frac{x}{\alpha}\right) = {}_1F_1\left(\left.{m+1 \atop m+n+2}\right| x\right)$. Theorem 8.4 is proved.

8.4 Arithmetical applications

In this section we will prove the following result.

Theorem 8.5 *Let* $\mathbb{K} = \mathbb{Q}(i\sqrt{d})$ *be an imaginary quadratic field. Let* $\alpha \in \mathbb{K}^*$. *Then* $e^\alpha \notin \mathbb{K}$.

In particular, if $\alpha \in \mathbb{Q}^*$, then e^α is irrational (see also exercise 3.14). Another interesting consequence of theorem 8.5 is the following

Corollary 8.1 *Let* $k \in \mathbb{N}^*$. *Then* $\pi\sqrt{k}$ *is irrational.*

Indeed, let us write $\sqrt{k} = m\sqrt{d}$, where d is squarefree, and assume that $\pi\sqrt{k} \in \mathbb{Q}$. Then $\alpha = (\pi\sqrt{k})i\sqrt{k} \in \mathbb{Q}(i\sqrt{d})$, and therefore theorem 8.5 yields $e^\alpha = e^{ik\pi} = (-1)^k \notin \mathbb{Q}(i\sqrt{d})$, a contradiction. Note that corollary 8.1 also results from exercise 3.15.

To prove theorem 8.5, we will start from the diagonal Padé approximants of e^x. We get them by taking $m = n$ in theorem 8.4:

$$_1F_1\left(\begin{smallmatrix}-n\\-2n\end{smallmatrix}\Big|-x\right)e^x - {}_1F_1\left(\begin{smallmatrix}-n\\-2n\end{smallmatrix}\Big|x\right) = \frac{(-1)^n x^{2n+1}}{\binom{2n}{n}(2n+1)!}\,{}_1F_1\left(\begin{smallmatrix}n+1\\2n+2\end{smallmatrix}\Big|x\right). \qquad (8.13)$$

And we will need four lemmas:

Lemma 8.2   *Let $\mathbb{A}_{\mathbb{K}}$ be the ring of integers of the imaginary quadratic field $\mathbb{K} = \mathbb{Q}(i\sqrt{d})$. Let (a_n) be a sequence of elements of $\mathbb{A}_{\mathbb{K}}$ satisfying $\lim_{n\to+\infty} a_n = 0$. Then $a_n = 0$ for every large n.*

Proof   In $\mathbb{A}_{\mathbb{K}}$, we have $N(a_n) = |a_n|^2 \in \mathbb{Z}$ (remark 5.1). Therefore $\lim_{n\to+\infty} a_n = 0 \Rightarrow N(a_n) = 0$ and $a_n = 0$ for every large n.

Remark 8.3   Lemma 8.2 enables us to use *diophantine approximation* in $\mathbb{A}_{\mathbb{K}}$ as in $\mathbb{Z}$ (see section 1.6). It should be noted that this is not possible if $\mathbb{K}$ is a *real* quadratic field. For example, if the P_n/Q_n's are the convergents of the regular continued fraction expansion of $\sqrt{2}$, then $|P_n - Q_n\sqrt{2}| \le 1/Q_n$ by formula (4.8), whence $\lim_{n\to+\infty}(P_n - Q_n\sqrt{2}) = 0$. However $P_n - Q_n\sqrt{2}$ is never zero since $\sqrt{2}$ is irrational.

Lemma 8.3   *For $a, c \in \mathbb{R}$, $c > a > 0$,*

$$_1F_1\left(\begin{smallmatrix}a\\c\end{smallmatrix}\Big|x\right) = \frac{\Gamma(c)}{\Gamma(a)\Gamma(c-a)}\int_0^1 e^{xt} t^{a-1}(1-t)^{c-a-1}\,dt.$$

Proof   See exercise 8.4.

Lemma 8.4   *For $\alpha \in \mathbb{C}$, $a, c \in \mathbb{R}$, $c > a > 0$,*

$$\left|{}_1F_1\left(\begin{smallmatrix}a\\c\end{smallmatrix}\Big|\alpha\right)\right| \le \text{Max}(1, e^{\text{Re}\,\alpha})\frac{\Gamma(c)}{\Gamma(a)\Gamma(c-a)}.$$

Proof   See exercise 8.5.

Lemma 8.5   *For every $n \in \mathbb{Z}$ and every $k \in \mathbb{N}$, $k! \mid (n)_k$, whence $k! \mid n(n-1)...(n-k+1)$.*

Proof   See exercise 8.6.

We now proceed to the proof of theorem 8.5.

Let $\alpha \in \mathbb{Q}(i\sqrt{d})$. We write $\alpha = (a + ib\sqrt{d})/\delta = \beta/\delta$, with $(a, b, \delta) \in \mathbb{Z}^3$, $\beta \in \mathbb{A}_\mathbb{K}$. We say that δ is a *denominator* of α.

Since $\,_1F_1\left(\begin{smallmatrix} -n \\ -2n \end{smallmatrix} \middle| x\right) = \sum_{k=0}^{n} \dfrac{(-n)_k}{(-2n)_k} \dfrac{x^k}{k!}$, we know by lemma 8.5 that p_n and

q_n defined by $p_n = \,_1F_1\left(\begin{smallmatrix} -n \\ -2n \end{smallmatrix} \middle| \alpha\right) \times D_n$, $q_n = \,_1F_1\left(\begin{smallmatrix} -n \\ -2n \end{smallmatrix} \middle| -\alpha\right) \times D_n$, where

$$D_n = (2n)(2n-1)\ldots(n+1)\delta^n = \frac{(2n)!}{n!}\delta^n, \tag{8.14}$$

belong to $\mathbb{A}_\mathbb{K}$. Indeed $\dfrac{(2n)!}{n!} \times \dfrac{(-n)_k}{(-2n)_k \, k!} = \dfrac{(2n-k)!}{k!(n-k)!}$.

Replacing x by α in (8.13) and multiplying by D_n yields

$$q_n e^\alpha - p_n = r_n = \frac{(-1)^n D_n \alpha^{2n+1}}{\binom{2n}{n}(2n+1)!} \,_1F_1\left(\begin{smallmatrix} n+1 \\ 2n+2 \end{smallmatrix} \middle| \alpha\right). \tag{8.15}$$

As $\Gamma(m+1) = m!$ for every natural integer m, lemma 8.4 shows that

$$|r_n| \le \mathrm{Max}(1, e^{\mathrm{Re}\,\alpha}) \frac{|\delta|^n |\alpha|^{2n+1}}{n!}. \tag{8.16}$$

Therefore $\lim_{n \to +\infty} r_n = 0$.

Now assume that $e^\alpha \in \mathbb{Q}(i\sqrt{d})$, that is $e^\alpha = \gamma/\delta'$ with $\gamma \in \mathbb{A}_\mathbb{K}$, $\delta' \in \mathbb{N}$. Then (8.15) becomes $q_n \gamma - \delta' p_n = \delta' r_n$, whence $a_n = q_n \gamma - \delta' p_n \in \mathbb{A}_\mathbb{K}$. But $\lim_{n \to +\infty} r_n = 0 \Rightarrow \lim_{n \to +\infty} a_n = 0$, and consequently $a_n = 0$ for every large n by lemma 8.2. Hence, for every n sufficiently large, $r_n = q_n e^\alpha - p_n = 0$ and $r_{n+1} = q_{n+1} e^\alpha - p_{n+1} = 0$.

Thus the system of equations $\begin{cases} q_n x + p_n y = 0 \\ q_{n+1} x + p_{n+1} y = 0 \end{cases}$

has a non trivial solution, namely $x = e^\alpha$, $y = -1$, and its determinant Δ_n is equal to zero. However,

$$\Delta_n = \begin{vmatrix} q_n & p_n \\ q_{n+1} & p_{n+1} \end{vmatrix} = -\begin{vmatrix} D_n Q_n(\alpha) & D_n P_n(\alpha) \\ D_{n+1} Q_{n+1}(\alpha) & D_{n+1} P_{n+1}(\alpha) \end{vmatrix},$$

with the notations of section 8.1 for Padé approximants P_n, Q_n.

Now lemma 6.1 shows that $\Delta_n = -D_n D_{n+1} c_n \alpha^{2n+1} \ne 0$. This contradiction proves theorem 8.5.

Exercises

8.1   Hypergeometric differential equation

1) Prove that the hypergeometric function $_2F_1\left(\begin{smallmatrix} a,b \\ c \end{smallmatrix} \middle| x\right)$ satisfies (8.4).

2) Prove that $f(x) = x^{1-c} \, _2F_1\left(\begin{smallmatrix} a-c+1,b-c+1 \\ 2-c \end{smallmatrix} \middle| x\right)$ also satisfies (8.4).

3) Prove that the general solution of (8.4) on $]0,1[$ is given by (8.5).

8.2   Prove relation (8.6).

8.3   Prove theorem 8.3.

8.4   Euler Beta function and integral representations

Let α, β be positive real numbers. We define *Euler Beta function* by

$$B(\alpha,\beta) = \int_0^1 t^{\alpha-1}(1-t)^{\beta-1}\, dt.$$

1) Prove that $B(\alpha,\beta)$ exists for $\alpha > 0$ and $\beta > 0$, and that
$$B(\alpha,\beta) = B(\beta,\alpha).$$

2) Euler Γ function has been defined in section 3.5, page 30. Prove that

$$\Gamma(\alpha)\Gamma(\beta) = 4\int_{u=0}^{+\infty}\int_{v=0}^{+\infty} e^{-(u^2+v^2)} u^{2\alpha-1} v^{2\beta-1}\, du\, dv.$$

Deduce that $\Gamma(\alpha)\Gamma(\beta) = \Gamma(\alpha+\beta)\cdot 2\int_0^{\frac{\pi}{2}} \cos^{2\alpha-1}\theta \sin^{2\beta-1}\theta\, d\theta$.

Then prove that $B(\alpha,\beta) = \dfrac{\Gamma(\alpha)\Gamma(\beta)}{\Gamma(\alpha+\beta)}$.

3) By using the functional relation $\Gamma(x+1) = x\Gamma(x)$ (formula (3.18)), express $(a)_n$ by means of the Γ function.

Deduce the integral representation of $_1F_1\left(\begin{smallmatrix} a \\ c \end{smallmatrix} \middle| x\right)$ given in lemma 8.3.

4) Prove that Gauss hypergeometric function can be written as

$$_2F_1\left(\begin{smallmatrix} a,b \\ c \end{smallmatrix} \middle| x\right) = \frac{\Gamma(c)}{\Gamma(a)\Gamma(c-a)} \int_0^1 t^{a-1}(1-t)^{c-a-1}(1-tx)^{-b}\, dt,$$

for $c > a > 0$ and $|x| < 1$.

8.5   Prove lemma 8.4.

8.6   Prove lemma 8.5.

8.7   **Diophantine approximation of $\sqrt[3]{17}$, Baker 1964**

1) By using theorem 8.2, prove that

$$
{}_2F_1\left(\begin{matrix} -n,-n+\frac{1}{3} \\ -2n \end{matrix}\middle|\frac{1}{18^3}\right)\times 7\sqrt[3]{17} - {}_2F_1\left(\begin{matrix} -n,-n-\frac{1}{3} \\ -2n \end{matrix}\middle|\frac{1}{18^3}\right)\times 18
$$

$$
= \frac{(-1)^n\left(-n-\frac{1}{3}\right)_{2n+1}}{18^{6n+2}\dbinom{2n}{n}(2n+1)!}\; {}_2F_1\left(\begin{matrix} n+1,n+\frac{2}{3} \\ 2n+2 \end{matrix}\middle|\frac{1}{18^3}\right).
$$

2) Let $u_n = 3n+a$ $(a\in\mathbb{Z})$, and let $\alpha(k)$ be the exponent of 3 in the expansion of $k!$ as a product of primes (see lemma 7.3). Prove that $u_1 u_2 ... u_k$ is divisible by $k!/3^{\alpha(k)}$ and that $\alpha(k)\le k/2$.

3) Let $D_n = 3^{\alpha(n)}(3\cdot 18^3)^n\dbinom{2n}{n}$. Prove that

$$
q_n = 7D_n\cdot {}_2F_1\left(\begin{matrix} -n,-n+1/3 \\ -2n \end{matrix}\middle|\frac{1}{18^3}\right) \quad\text{and}\quad p_n = 18D_n\cdot {}_2F_1\left(\begin{matrix} -n,-n-1/3 \\ -2n \end{matrix}\middle|\frac{1}{18^3}\right)
$$

are integers.

4) By using the result of exercise 8.4, question 4), prove that

$$
\left|{}_2F_1\left(\begin{matrix} n+1,n+\frac{2}{3} \\ 2n+2 \end{matrix}\middle|\frac{1}{18^3}\right)\right| \le \frac{18^2}{7^2\cdot 17^{2/3}}\frac{(2n+1)!}{(n!)^2}\left(\frac{18^3}{4(18^3-1)}\right)^n .
$$

5) We define $r_n = q_n\sqrt[3]{17} - p_n$. Prove that, for every $n\in\mathbb{N}$,

$$
|r_n| \le \frac{1}{3\cdot 7^2\cdot 17^{2/3}}\left(\frac{3\sqrt{3}}{4\cdot 7^3\cdot 17}\right)^n .
$$

6) Deduce that $\sqrt[3]{17}$ is irrational.

8.8   **Padé approximants of $\mathrm{Log}(1-x)$ and irrationality of $\mathrm{Log}\,2$**

1) Find a link between $\mathrm{Log}(1-x)$ and $(1-x)^\alpha$.

2) Deduce the diagonal Padé approximants (Q_n, P_n, R_n) of $\mathrm{Log}(1-x)$.

3) Define $D_n = 2^n \binom{2n}{n} \times \mathrm{LCM}(1,2,...,n)$, $p_n = D_n P_n(\frac{1}{2})$, $q_n = D_n Q_n(\frac{1}{2})$.

Prove that p_n and q_n are integers.

4) Let $r_n = q_n \mathrm{Log}\, 2 + p_n$. By using theorem 7.8, prove that $|r_n| \leq (\frac{3}{4})^n$.

5) Prove that $\mathrm{Log}\, 2$ is irrational.

8.9 Irrationality of $\displaystyle\sum_{n=0}^{+\infty} \frac{1}{2^{2^n} + a_n}$ (Erdös 1964)

Let $a \in \mathbb{Z}$, $a \neq -2^{2^k}$ if $k \in \mathbb{N}$. Define $f(x) = \sum_{n=0}^{+\infty} x^{2^n} / \left(1 + a x^{2^n}\right)$.

1) Prove that f is analytic in $D = \{x \in \mathbb{C} \,/\, |x| < 1\}$ and satisfies the functional equation $f(x^2) = f(x) - x/(1+ax)$ for every $x \in D$.

2) Assume that $f(\frac{1}{2}) \in \mathbb{Q}$. Prove that $f(1/2^{2^n}) \in \mathbb{Q}$, and that there exists a sequence $D_n \in \mathbb{Z}$ satisfying $|D_n| \leq C \cdot 2^{2^n}$ and $D_n f(2^{-2^n}) \in \mathbb{Z}$, where C does not depend on n.

3) Compute a $[1/1]$-Padé approximant of $f(x)$.

4) Prove that $f\left(\frac{1}{2}\right) = \sum_{n=0}^{+\infty} 1 / \left(2^{2^n} + a\right)$ is irrational.

5) More generally, now let $a_n \in \mathbb{Z}$, with $a_n \neq -2^{2^k}$ for $(n,k) \in \mathbb{N}^2$. Assume that, for all $\eta > 1$, $|a_n| \leq \eta^{2^n}$ for large n. For $h \in \mathbb{N}$, define

$$f_h(x) = \sum_{n=0}^{+\infty} x^{2^n} / \left(1 + a_{n+h} x^{2^n}\right).$$

a) Prove that f_h is analytic in D and, for fixed $\eta > 1$, find an upper bound for the coefficients of its Taylor series expansion.

b) Prove that $\sum_{n=0}^{+\infty} 1 / \left(2^{2^n} + a_n\right)$ is irrational by using the $[1/1]$-Padé approximants of f_h.

8.10 q-exponential function

Let $q \in \mathbb{C}$, $|q| > 1$. Define, for all $n \in \mathbb{N}$,

$$n_q! = (q-1)(q^2-1)\cdots(q^n-1) \text{ if } n \geq 1, \quad 0_q! = 1.$$

The q-exponential function E_q is defined by $E_q(x) = \sum_{n=0}^{+\infty} x^n / n_q!$.

It satisfies the functional equation $E_q(qx) = (1+x)E_q(x)$, which yields

$$E_q(x) = \prod_{n=1}^{+\infty}\left(1+xq^{-n}\right).$$

For details, see exercise 5.22, page 75.

It is clear that the q-exponential function is an entire function, and it can be considered a q-analogue of the ordinary exponential function, since

$$\lim_{q\to 1} E_q\left[(q-1)x\right] = e^x.$$

The aim of this exercise is to compute the Padé-approximants of E_q and to prove the irrationality of $E_q(x)$ for every $x \in \mathbb{Q}^*$ and $q \in \mathbb{Z}$, $|q| \geq 2$ (Bundschuh 1969, Borwein 1988).

1) For all integer $n \geq 1$, define $g_n(z) = \dfrac{1}{(z-1)(z-q)\cdots\left(z-q^n\right)}$.

Find a relation between $g_n(qz)$ and $g_n(z)$. Deduce that

$$g_n(z) = \sum_{k=0}^{+\infty} a_n(k)z^k, \quad \text{where } a_n(k) = \frac{(-1)^{n+1}}{q^{\frac{n(n+1)}{2}+nk}} \frac{(n+k)_q!}{k_q!n_q!}.$$

2) For all $n \in \mathbb{N}$, let C_n be the positively oriented circle with center O and radius $|q|^{n+1}$. Consider the complex integral

$$I_n(x) = \frac{1}{2i\pi} \int_{C_n} \frac{E_q(xz)}{z^{n+1}(z-1)(z-q)\cdots\left(z-q^n\right)}\,dz.$$

a) Evaluate $I_n(x)$ by using the residue theorem.

b) Expand $E_q(xz)$ in a power series of x, and prove that the $[n/n]$ Padé approximants of E_q are $Q_n(x)E_q(x) + P_n(x) = I_n(x)$, where

$$P_n(x) = \sum_{k=0}^{n} \frac{(-1)^{n+1}}{q^{\frac{n(n+1)}{2}+nk}} \frac{(n+k)_q!}{k_q!n_q!(n-k)_q!}x^{n-k}$$

$$\tag{8.17}$$

$$Q_n(x) = \sum_{k=0}^{n} \frac{(-1)^{n-k}}{q^{2nk-\frac{k(k-1)}{2}}k_q!(n-k)_q!}(1+x)(1+qx)\cdots\left(1+q^{k-1}x\right)$$

c) Prove that, for $|q| \geq 2$, $|I_n(x)| \leq \dfrac{|q|^{n+1}}{(2n+1)_{|q|}!} E_{|q|}\left(\left|xq^{n+1}\right|\right)|x|^{2n+1}$.

3) Now, we assume that $q \in \mathbb{Z}$, $|q| \geq 2$, and that $x = r/s \in \mathbb{Q}^*$.

a) For every $(n,k) \in \mathbb{N}^2$ such that $0 \leq k \leq n$, we define the q-binomial coefficients by

$$\binom{n}{k}_q = \frac{n_q!}{k_q!(n-k)_q!}.$$

Prove that $\dbinom{n}{k}_q = q^k \dbinom{n-1}{k}_q + \dbinom{n-1}{k-1}_q$.

Deduce that $\dbinom{n}{k}_q \in \mathbb{Z}$, which means that, for every $(n,k) \in \mathbb{N}^2$,

$$(q-1)(q^2-1)\cdots(q^k-1) \mid (q^{n+1}-1)(q^{n+2}-1)\cdots(q^{n+k}-1).$$

b) Prove that the $[n/n]$ Padé-approximants of $E_q(x)$ defined by

$$B_n(x) = n_q! q^{\frac{n(3n+1)}{2}} Q_n(x) \text{ and } A_n(x) = n_q! q^{\frac{n(3n+1)}{2}} P_n(x)$$

have integer coefficients and that, for every fixed $x \neq 0$ satisfying $|x| \leq 1$ and every n sufficiently large,

$$\left|B_n(x)E_q(x) + A_n(x)\right| \leq CE_{|q|}\left(\left|xq^{n+1}\right|\right)|x|^{2n+1},$$

where $C > 0$ does not depend on n nor x.

c) Apply the above results to $x = \dfrac{r}{sq^n}$ and prove that $E_q\left(\dfrac{r}{s}\right) \notin \mathbb{Q}$.

Chapter 9

Algebraic numbers and irrationality measures

In section 9.1, we generalize the notions of rational and quadratic numbers by defining algebraic numbers. In section 9.2, we present the basic properties of algebraic integers. This is an introduction to the algebraic theory of numbers, which will be studied more thoroughly in chapters 10 and 11. Section 9.3 is devoted to the proof of the celebrated Liouville's theorem, which asserts the existence of transcendental (that is non algebraic) numbers. As a matter of fact, Liouville's theorem provides a so-called irrationality measure for every algebraic number. We study this notion in more detail in section 9.4 and show that good diophantine approximations allow to obtain irrationality measures for specific numbers, for example e. In section 9.5, we explain how knowing an irrationality measure of $\sqrt[n]{d}$ allows to solve the diophantine equation $x^n - dy^n = k$ $(n \geq 3)$. Finally, we state without proof the theorems of Thue and Roth, which improve Liouville's theorem.

9.1 Algebraic numbers

Definition 9.1 *Let $\alpha \in \mathbb{C}$. We say that α is algebraic if there exists a non zero polynomial $P \in \mathbb{Z}[x]$ such that $P(\alpha) = 0$.*

For example, $\alpha = 1/\sqrt{2}$ is algebraic since $2\alpha^2 - 1 = 0$, and $\beta = 1 + \sqrt[3]{5}$ is algebraic because $(\beta - 1)^3 - 5 = \beta^3 - 3\beta^2 + 3\beta - 6 = 0$.

Let α be any algebraic number, and let $P \in \mathbb{Z}[x]$, $P \neq 0$, such that $P(\alpha) = 0$ and P has *minimal* degree. By dividing P by its leading coefficient we obtain a *monic* polynomial $P_\alpha \in \mathbb{Q}[x]$, which satisfies $P_\alpha(\alpha) = 0$, P_α of minimal degree. Such a polynomial is unique (exercise 9.1) and is called the *minimal polynomial of α*.

Theorem 9.1 *Let $P \in \mathbb{Q}[x]$ be a monic polynomial satisfying $P(\alpha) = 0$. Then P is the minimal polynomial of α if, and only if, P is irreducible in $\mathbb{Q}[x]$, that is if $P(x) = P_1(x)P_2(x)$, with P_1 and $P_2 \in \mathbb{Q}[x]$, implies $P_1(x) \in \mathbb{Q}$ or $P_2(x) \in \mathbb{Q}$.*

Note that the notion of an irreducible polynomial is a special case of the notion of an irreducible element in a domain $\mathbb{A}$ (definition 5.1). Indeed, the units of $\mathbb{Q}[x]$ are the constant polynomials $P(x) = a \in \mathbb{Q}^*$. For the proof of theorem 9.1, see exercise 9.2.

Example 9.1 Let $\alpha = (a + b\sqrt{d})/c$ be a *quadratic irrational* number, which means that a, b, c, d are rational integers and d is not a square. Then α is a root of $P(x) = (x - \alpha)(x - \alpha^*)$, where $\alpha^* = (a - b\sqrt{d})/c$ is the conjugate of α. Therefore α is a zero of

$$P(x) = x^2 - \frac{2a}{c}x + \frac{a^2 - db^2}{c^2}.$$

This polynomial is irreducible in $\mathbb{Q}[x]$. Indeed, if it was reducible, one could write it as $P(x) = (x + r)(x + r')$, with r, $r' \in \mathbb{Q}$. Hence its roots would be rational, which is not. Thus P is the minimal polynomial of α.

Definition 9.2 *The degree of the minimal polynomial P_α of the algebraic number α is called the degree of α and denoted by $\deg(\alpha)$.*

Thus, if α is rational, α is algebraic of degree 1. If α is quadratic irrational, α is algebraic of degree 2 (see example 9.1).

The p-th root of unity $\omega = e^{2i\pi/p}$, p prime, connected with the Fermat equation $x^p + y^p = z^p$, provides another interesting example. In order to study it, we will need the following result, which is known as *Eisenstein's criterion*.

Lemma 9.1 *Let $P(x) = x^n + a_{n-1}x^{n-1} + \cdots + a_1 x + a_0$, with a_0, a_1, ..., ..., $a_{n-1} \in \mathbb{Z}$. Let p be prime. Assume that p divides a_0, a_1, ..., a_{n-1}, while p^2 does not divide a_0. Then P is irreducible in $\mathbb{Q}[x]$.*

The proof is given in exercise 9.3. It uses the notion of algebraic integer and theorem 9.4. As an application, consider $\beta = 1 + \sqrt[3]{5}$. We know that

$P(\beta) = 0$, with $P(x) = x^3 - 3x^2 + 3x - 6$ (page 134). Therefore, by Eisenstein's criterion, P is irreducible in $\mathbb{Q}[x]$ (take $p = 3$). Hence P is the minimal polynomial of β (theorem 9.1) and $\deg(\beta) = 3$.

Now return to $\omega = e^{2i\pi/p}$, p prime.

Theorem 9.2 *Let $\omega = e^{2i\pi/p}$, p prime. Then ω is algebraic of degree $p-1$. Its minimal polynomial is $P_\omega(x) = x^{p-1} + x^{p-2} + \cdots + x + 1$.*

Definition 9.3 *For p prime, $P_\omega(x) = x^{p-1} + x^{p-2} + \cdots + x + 1$ is called the cyclotomic polynomial of order p.*

Proof of theorem 9.2 It is clear that
$$\omega^{p-1} + \omega^{p-2} + \cdots + \omega + 1 = (\omega^p - 1)/(\omega - 1) = 0.$$
It remains to check that P_ω is irreducible in $\mathbb{Q}[x]$ (theorem 9.1). But P_ω is irreducible if, and only if, $Q(x) = P_\omega(x+1)$ is irreducible. We have

$$Q(x) = \left((x+1)^p - 1\right)/\left((x+1) - 1\right) = \sum_{k=1}^{p} \binom{p}{k} x^{k-1}.$$

For $k = 1, 2, \ldots, p-1$, $p \mid \binom{p}{k}$ since $\binom{p}{k} = p(p-1)\ldots(p-k+1)/k!$ and p does not divide $k!$. As $a_0 = p$, Q is irreducible in $\mathbb{Q}[x]$ by Eisenstein's criterion, and so is P_ω.

9.2   Algebraic integers

We say that α is an *algebraic integer* if it is algebraic and if its minimal polynomial P_α has integer coefficients. In other words, if there exist a_0, $a_1, \ldots, a_{d-1} \in \mathbb{Z}$ such that $\alpha^d + a_{d-1}\alpha^{d-1} + \cdots + a_1\alpha + a_0 = 0$. For example, $\sqrt{2}$, $\sqrt[3]{5}$, $e^{\frac{2i\pi}{p}}$ are algebraic integers. The rational integers are the algebraic integers of degree 1. The quadratic irrational integers (see section 5.2) are the algebraic integers of degree 2.

Theorem 9.3 *Let α be an algebraic number. Then there exists $k \in \mathbb{N}$ such that $\beta = k\alpha$ is an algebraic integer.*

Proof Let $P_\alpha(x) = x^d + a_{d-1}x^{d-1} + \cdots + a_0$ (with $a_0, \ldots, a_{d-1} \in \mathbb{Q}$) be the minimal polynomial of α.

Let $k \in \mathbb{N}$ such that $ka_0 = b_0$, $ka_2 = b_1$, ..., $ka_{d-1} = b_{d-1} \in \mathbb{Z}$. Then

$$k^d \alpha^d + k^{d-1} b_{d-1} \alpha^{d-1} + \cdots + k^{d-1} b_0 = 0.$$

Hence $(k\alpha)^d + b_{d-1}(k\alpha)^{d-1} + kb_{d-2}(k\alpha)^{d-2} + \cdots + k^{d-1} b_0 = 0$, and $\beta = k\alpha$ is an algebraic integer, Q.E.D.

The least positive integer k such that $k\alpha$ is an algebraic integer is called the *denominator* of α and is denoted by $\mathrm{den}(\alpha)$. For instance, if $\alpha = p/q$ is an irreducible rational number with $q > 0$, the minimal polynomial of α is $P_\alpha(x) = x - p/q$, and $\mathrm{den}(\alpha) = q$, as expected.

It is natural to ask if the sum and the product of two algebraic integers is also an algebraic integer. The answer is positive (theorem 9.4 below). In order to show it, we will need a lemma, initially stated by Newton, on symmetric polynomials. We say that the polynomial in n indeterminates $P(x_1, x_2, ..., x_n)$, with coefficients in a domain $\mathbb{A}$, is *symmetric*, if $P(x_{\tau(1)}, x_{\tau(2)}, ..., x_{\tau(n)}) = P(x_1, x_2, ..., x_n)$ for any permutation τ of the indices $1, 2, ..., n$ (that is for any permutation of the indeterminates x_1, x_2, ..., x_n). The following polynomial, for example, is symmetric:

$$P(x_1, x_2, x_3) = x_1^3 + x_2^3 + x_3^3 + 2x_1^2 x_2^2 + 2x_1^2 x_3^2 + 2x_2^2 x_3^2 + 5x_1 x_2 x_3$$

Lemma 9.2   *Let $P \in \mathbb{A}[x_1, x_2, ..., x_n]$ be symmetric. Let*

$$\begin{cases} \sigma_1 = x_1 + x_2 + \cdots + x_n = \sum_i x_i \\ \sigma_2 = \sum_{i<j} x_i x_j \\ \sigma_3 = \sum_{i<j<k} x_i x_j x_k \\ \vdots \\ \sigma_n = x_1 x_2 ... x_n \end{cases} \qquad (9.1)$$

Then $P(x_1, x_2, ..., x_n)$ can be written as a polynomial in the n indeterminates σ_1, $\sigma_2, ..., \sigma_n$, with coefficients in $\mathbb{A}$.

Proof Let α_1 be the highest occurring power of x_1 in P. Among the terms which contain $x_1^{\alpha_1}$, let α_2 be the highest occurring power of x_2. Among the terms which contain $x_1^{\alpha_1} x_2^{\alpha_2}$, let α_3 be the highest occurring power of x_3, and so on. This process defines a *leading term*, of the form $ax_1^{\alpha_1} x_2^{\alpha_2} ... x_n^{\alpha_n}$. Since P is symmetric $\alpha_1 \geq \alpha_2 \geq \cdots \geq \alpha_n$ (indeed, P also contains the term $ax_2^{\alpha_1} x_1^{\alpha_2} ... x_n^{\alpha_n}$, whence $\alpha_2 \leq \alpha_1$, and so on). Now we observe that the leading term of $\sigma_1^{\alpha_1 - \alpha_2} \sigma_2^{\alpha_2 - \alpha_3} ... \sigma_{n-1}^{\alpha_{n-1} - \alpha_n} \sigma_n^{\alpha_n}$ is $x_1^{\alpha_1} x_2^{\alpha_2} ... x_n^{\alpha_n}$. Therefore, by computing $P(x_1, x_2, ..., x_n) - a\sigma_1^{\alpha_1 - \alpha_2} \sigma_2^{\alpha_2 - \alpha_3} ... \sigma_{n-1}^{\alpha_{n-1} - \alpha_n} \sigma_n^{\alpha_n}$, we eliminate all terms of the form $ax_{\tau(1)}^{\alpha_1} x_{\tau(2)}^{\alpha_2} ... x_{\tau(n)}^{\alpha_n}$, and lemma 9.2 follows by induction.

Remark 9.1   This proof is *constructive*. It enables to transform, in practice, $P \in \mathbb{A}[x_1, x_2, ..., x_n]$ into $P \in \mathbb{A}[\sigma_1, \sigma_2, ..., \sigma_n]$. For example, let $P(x_1, x_2, x_3) = x_1^3 + x_2^3 + x_3^3 + 2x_1^2 x_2^2 + 2x_2^2 x_3^2 + 2x_1^2 x_3^2 + 5x_1 x_2 x_3$. First we see that the leading term corresponds to $\alpha_1 = 3$, $\alpha_2 = 0$, $\alpha_3 = 0$, whence

$$P(x_1, x_2, x_3) - \sigma_1^3 = P(x_1, x_2, x_3) - (x_1 + x_2 + x_3)^3$$
$$= 2x_1^2 x_2^2 + 2x_2^2 x_3^2 + 2x_1^2 x_3^2 - 3x_1^2 x_2 - 3x_1 x_2^2$$
$$- 3x_2^2 x_3 - 3x_2 x_3^2 - 3x_1^2 x_3 - 3x_1 x_3^2 - x_1 x_2 x_3.$$

The new leading term corresponds to $\alpha_1 = 2$, $\alpha_2 = 2$, $\alpha_3 = 0$, and

$$P(x_1, x_2, x_3) - \sigma_1^3 - 2\sigma_2^2 = P(x_1, x_2, x_3) - \sigma_1^3 - 2(x_1 x_2 + x_2 x_3 + x_1 x_3)^2$$
$$= -4x_1^2 x_2 x_3 - 4x_1 x_2^2 x_3 - 4x_1 x_2 x_3^2$$
$$- 3x_1^2 x_2 - 3x_1 x_2^2 - 3x_2^2 x_3 - 3x_2 x_3^2 - 3x_1^2 x_3 - 3x_1 x_3^2 - x_1 x_2 x_3.$$

Now $\alpha_1 = 2$, $\alpha_2 = 1$, $\alpha_3 = 1$, whence

$$P(x_1, x_2, x_3) - \sigma_1^3 - 2\sigma_2^2 + 4\sigma_1 \sigma_3$$
$$= -3x_1^2 x_2 - 3x_1 x_2^2 - 3x_2^2 x_3 - 3x_2 x_3^2 - 3x_1^2 x_3 - 3x_1 x_3^2 - x_1 x_2 x_3.$$

Finally, we have $\alpha_1 = 2$, $\alpha_2 = 1$, $\alpha_3 = 0$, which yields

$$P(x_1, x_2, x_3) - \sigma_1^3 - 2\sigma_2^2 + 4\sigma_1 \sigma_3 + 3\sigma_1 \sigma_2 = 8x_1 x_2 x_3.$$

Hence $P(x_1, x_2, x_3) = \sigma_1^3 + 2\sigma_2^2 - 4\sigma_1 \sigma_3 - 3\sigma_1 \sigma_2 + 8\sigma_3$.

Theorem 9.4   Let α and β be two algebraic integers. Then $\alpha + \beta$ and $\alpha\beta$ are algebraic integers.

Proof   Let P_α and P_β be the minimal polynomials of α and β, with degrees d and n respectively. Let $x_1 = \beta$, x_2, ..., x_n be the zeros of P_β in $\mathbb{C}$.

We know that $P_\beta(x) = x^n - \sigma_1 x^{n-1} + \sigma_2 x^{n-2} + \cdots + (-1)^n \sigma_n$, where σ_1, $\sigma_2, ..., \sigma_n$ are defined in (9.1). Therefore σ_1, σ_2, ..., $\sigma_n \in \mathbb{Z}$. Since $x_1 = \beta$, the polynomial $Q(x) \doteq P_\alpha(x - x_1) P_\alpha(x - x_2) ... P_\alpha(x - x_n)$ satisfies $Q(\alpha + \beta) = 0$. Its coefficients are symmetric polynomials in x_1, x_2, ..., x_n, with coefficients in $\mathbb{Z}$. Hence, by lemma 9.2, $Q \in \mathbb{Z}[x]$, which proves that $\alpha + \beta$ is algebraic.

The proof is similar for $\alpha\beta$. Just replace $Q(x)$ by

$$R(x) = (x_1 x_2 ... x_n)^d P_\alpha\left(\frac{x}{x_1}\right) P_\alpha\left(\frac{x}{x_2}\right) ... P_\alpha\left(\frac{x}{x_n}\right).$$

Corollary 9.1 *If α and β are algebraic numbers, so are $\alpha + \beta$ and $\alpha\beta$. More precisely, the set $\overline{\mathbb{Q}}$ of algebraic numbers is a subfield of $\mathbb{C}$.*

Proof See exercise 9.4.

9.3 Transcendental numbers and Liouville's theorem

Definition 9.4 *A complex number α is said to be transcendental if it is not algebraic, that is if $P(\alpha) \neq 0$ for every $P \in \mathbb{Z}[x]$, $P \neq 0$.*

Even though it is easy to give examples of algebraic numbers, it is far from being evident to exhibit transcendental numbers. An elementary example is given in exercise 1.12. But the first transcendental number in history has been constructed by Joseph Liouville (1844), by using the following result, which is known as *Liouville's theorem*.

Theorem 9.5 *Let α be an algebraic number of degree $d \geq 2$. Let $k \in \mathbb{N}$ such that $P = k P_\alpha \in \mathbb{Z}[x]$. Define $C' = \max_{[\alpha-1,\alpha+1]} |P'(x)|$ and $C = \min(1, 1/C')$. Then, for every rational number p/q with $q > 0$,*

$$\left| \alpha - \frac{p}{q} \right| \geq \frac{C}{q^d}. \tag{9.2}$$

Proof For every $p/q \in \mathbb{Q}$, the number $q^d P(p/q)$ is an integer since $P \in \mathbb{Z}[x]$, and it is different from zero since $\deg(\alpha) \geq 2$. Indeed, if it was zero, P_α would have a rational root, which is impossible because it is irreducible in $\mathbb{Q}[x]$. Therefore $\left|q^d\left(P(\alpha)-P(p/q)\right)\right|=\left|q^d P(p/q)\right|\geq 1$.

Now the mean value theorem yields $\max_{[\alpha,p/q]}|P'(x)|.\left|\alpha-\dfrac{p}{q}\right|\geq\dfrac{1}{q^d}$.

If $|\alpha-p/q|\geq 1$, then evidently $|\alpha-p/q|\geq q^{-d}$. Otherwise, we can write $\max_{[\alpha,p/q]}|P'(x)|\leq C'$, and therefore $|\alpha-p/q|\geq q^{-d}/C'$, which proves Liouville's theorem.

Remark 9.2   Liouville's theorem is *explicit*. It enables us to compute C when α is given. For example, let $\alpha=\sqrt[3]{17}$. Then $\alpha^3-17=0$ and $P(x)=x^3-17$ is the minimal polynomial of α, since P is irreducible in $\mathbb{Q}[x]$ by Eisenstein's criterion with $p=17$. Moreover,

$$\max_{[\alpha-1,\alpha+1]}|P'(x)|=3(\sqrt[3]{17}+1)^2=C',$$

whence $C=1/C'\geq 0.026$. Therefore, for any $p/q\in\mathbb{Q}$ with $q>0$,

$$\left|\sqrt[3]{17}-\frac{p}{q}\right|\geq\frac{0.026}{q^3}. \tag{9.3}$$

Definition 9.5 *Let $\alpha\in\mathbb{R}$ be an irrational number. We say that α is a Liouville number if, for every large $n\in\mathbb{N}$, there exists a rational number p/q, with $q\geq 2$, satisfying*

$$\left|\alpha-\frac{p}{q}\right|<\frac{1}{q^n}. \tag{9.4}$$

Example 9.2   Following Liouville, consider the Engel series

$$\alpha=\sum_{k=0}^{+\infty}\frac{1}{10^{k!}}. \tag{9.5}$$

The decimal expansion of α is $\alpha=1,110001000000000000000000100...$
For $n\geq 1$, we have

$$\left|\alpha-\sum_{k=0}^{n}\frac{1}{10^{k!}}\right|=\sum_{k=n+1}^{+\infty}\frac{1}{10^{k!}}\leq\frac{1}{10^{(n+1)!}}\left(1+\frac{1}{10}+\frac{1}{10^2}+\cdots\right)$$

$$\leq\frac{1}{(10^{n!})^n}\cdot\frac{1}{10^{n!}}\cdot\frac{10}{9}<\frac{1}{(10^{n!})^n}.$$

Define $\dfrac{p}{q} = \sum_{k=0}^{n} \dfrac{1}{10^{k!}}, \, q = 10^{n!}$.

It is now clear that (9.4) is true for $n \geq 1$ and that α is a Liouville number (the irrationality of α results from theorem 2.2 or theorem 2.4).

Theorem 9.6 *Any Liouville number is transcendental.*

Proof Let α be a Liouville number. Assume that α is algebraic. Then $d = \deg \alpha \geq 2$, since α is irrational. For every large n, there exists $p/q \in \mathbb{Q}$, with $q \geq 2$, satisfying (9.4). By Liouville's theorem 9.5,

$$\dfrac{C}{q^d} \leq \left| \alpha - \dfrac{p}{q} \right| < \dfrac{1}{q^n}. \text{ Contradiction when } n \to +\infty.$$

9.4 Irrationality measures

Definition 9.6 *Let $\alpha \in \mathbb{R}$ be an irrational number. We say that $\mu > 0$ is an irrationality measure of α if there exists a constant $C > 0$ such that, for any rational p/q with $q > 0$,*

$$\left| \alpha - \dfrac{p}{q} \right| \geq \dfrac{C}{q^\mu}. \tag{9.6}$$

Liouville's theorem 9.5 implies that d is an irrationality measure for any algebraic number of degree d. On the other hand, a Liouville number has no irrationality measure (exercise 9.5). If α has an irrationality measure, the lower bound of all its irrationality measures is called the *best measure of irrationality* of α and is denoted by $\mu(\alpha)$. It can be proved that $\mu(\alpha) \geq 2$ (exercise 9.6). When α is a Liouville number, we put $\mu(\alpha) = +\infty$.

The following theorem will enable us to find irrationality measures by using sequences of good diophantine approximations, such as the ones we obtain by means of Padé approximants (section 8.4, page 126).

Theorem 9.7 *Let $\alpha \in \mathbb{R}$. Assume that there exist constants $a > 0$, $b > 0$, $k > 0$, $\ell \geq \frac{1}{2}$, $h \geq 1$, an increasing function $g : \mathbb{N} \to \mathbb{R}_+^*$, and a sequence p_n/q_n of diophantine approximation of α satisfying*

$$\forall n \in \mathbb{N}, \quad q_n p_{n+1} - q_{n+1} p_n \neq 0, \tag{9.7}$$

$$\cdot \; \forall n \in \mathbb{N}, \quad |q_n| \le k(g(n))^a, \tag{9.8}$$

$$\forall n \in \mathbb{N}, \quad |q_n \alpha - p_n| \le \ell/g(n), \tag{9.9}$$

$$g(0) = 1 \text{ and } \lim_{n \to +\infty} g(n) = +\infty, \tag{9.10}$$

$$\forall n \in \mathbb{N}, \quad g(n+1) \le b(g(n))^h. \tag{9.11}$$

Then α is irrational and, for every rational p/q with $q > 0$,

$$\left| \alpha - \frac{p}{q} \right| \ge \frac{C}{q^\mu}, \tag{9.12}$$

where $C = \dfrac{1}{2kb^{a(h+1)}(2\ell)^{ah^2}}$ and $\mu = ah^2 + 1$.

Proof Let $(p, q) \in \mathbb{Z} \times \mathbb{N}^*$ be given. Let ν be the smallest integer satisfying $q\ell/g(\nu) < 1/2$. Clearly ν exists because $\lim_{n \to +\infty} g(n) = +\infty$. Since $q \ge 1$, $g(0) = 1$ and $\ell \ge 1/2$, we have $\nu \ge 1$. Then $q\ell/g(\nu - 1) \ge 1/2$, whence $g(\nu - 1) \le 2q\ell$. From (9.11) we deduce that $g(\nu) \le b(2q\ell)^h$. And by using (9.11) again we obtain

$$g(\nu) < g(\nu + 1) \le b^{h+1}(2q\ell)^{h^2}. \tag{9.13}$$

Now consider the determinant

$$\Delta_\nu = \begin{vmatrix} q_\nu & p_\nu \\ q_{\nu+1} & p_{\nu+1} \end{vmatrix}.$$

From (9.7) we know that $\Delta_\nu \neq 0$. This means that the vectors (q_ν, p_ν) and $(q_{\nu+1}, p_{\nu+1})$ form a basis of $\mathbb{R}^2$. Consequently, at least one of the two determinants

$$\begin{vmatrix} q_\nu & p_\nu \\ q & p \end{vmatrix} \quad \text{or} \quad \begin{vmatrix} q_{\nu+1} & p_{\nu+1} \\ q & p \end{vmatrix}$$

is different from zero. Let $m = \nu$ or $\nu + 1$, such that $\delta_m = \begin{vmatrix} q_m & p_m \\ q & p \end{vmatrix} \neq 0$.

As $\delta_m \in \mathbb{Z}$, this implies $|\delta_m| \ge 1$, whence $|pq_m - qp_m| \ge 1$. Therefore $|q(q_m \alpha - p_m) - q_m(q\alpha - p)| \ge 1$ and, by the triangle inequality,

$$q|q_m \alpha - p_m| + |q_m||q\alpha - p| \ge 1.$$

Now, by using (9.9) and the definitions of ν and m, we have

$$q\left|q_m\alpha - p_m\right| \le \frac{q\ell}{g(m)} \le \frac{q\ell}{g(v)} < \frac{1}{2}, \quad \text{whence } \left|q_m\right|\left|q\alpha - p\right| > \frac{1}{2}.$$

Moreover, by (9.8) and (9.13), $\left|q_m\right| \le k(g(m))^a \le kb^{a(h+1)}(2q\ell)^{ah^2}$, and we finally obtain

$$\left|q\alpha - p\right| > \frac{1}{2kb^{a(h+1)}(2\ell)^{ah^2}} \frac{1}{q^{ah^2}}, \quad \text{Q.E.D.}$$

Example 9.3 Let us compute an irrationality measure of e. By (8.14), (8.15), (8.16), (8.17), (8.18), we know that

$$\forall n \in \mathbb{N}, \quad \left|q_n e - p_n\right| \le \frac{e}{n!}, \tag{9.14}$$

$$p_n = \frac{(2n)!}{n!} \, {}_1F_1\left(\begin{smallmatrix} -n \\ -2n \end{smallmatrix} \middle| 1\right), \quad q_n = \frac{(2n)!}{n!} \, {}_1F_1\left(\begin{smallmatrix} -n \\ -2n \end{smallmatrix} \middle| -1\right) \tag{9.15}$$

$$q_n p_{n+1} - p_n q_{n+1} \ne 0. \tag{9.16}$$

We apply theorem 9.7 with $g(n) = n!$, which satisfies (see exercise 9.7)

$$\forall n \in \mathbb{N}, \quad (n+1)! \le 3(n!)^{\frac{3}{2}}. \tag{9.17}$$

Now we look for an upper bound for $\left|q_n\right|$. We have

$$\left|{}_1F_1\left(\begin{smallmatrix} -n \\ -2n \end{smallmatrix} \middle| -1\right)\right| \le \sum_{k=0}^{n} \frac{n(n-1)...(n-k+1)}{2n(2n-1)...(2n-k+1)k!} \le \sum_{k=0}^{n} \frac{1}{k!} \le e,$$

and therefore $\left|q_n\right| \le e\dfrac{(2n)!}{n!} = e\dbinom{2n}{n}n!$. By using the upper bound

$$\forall n \in \mathbb{N}, \quad \binom{2n}{n} \le 4^n, \tag{9.18}$$

we deduce (see exercise 9.8 for details) that

$$\forall n \in \mathbb{N}, \quad \left|q_n\right| \le 30(n!)^2. \tag{9.19}$$

We now apply theorem 9.7 with $a = 2$, $b = 3$, $k = 30$, $\ell = e$, $h = 3/2$, and see that, for any rational p/q with $q > 0$,

$$\left|e - \frac{p}{q}\right| \ge \frac{10^{-9}}{q^{5.5}}. \tag{9.20}$$

Consequently, e is not a Liouville number.

Remark 9.3 The irrationality measure $\mu = 5.5$ we have obtained here is far from being the best one. By improving the estimates (9.17) and

(9.19), it can be proved (see exercise 9.9) that, for every $\varepsilon > 0$, there exists C such that, for every $p/q \in \mathbb{Q}$ with $q > 0$, $\left| e - p/q \right| \geq C/q^{2+\varepsilon}$. This means that the best measure of irrationality of e satisfies $\mu(e) \leq 2$. However, the best measure of irrationality of any real number α satisfies $\mu(\alpha) \geq 2$ (see exercise 9.6). Therefore

$$\mu(e) = 2. \tag{9.21}$$

9.5 Diophantine equations and irrationality measures

Theorem 9.8 *Let* $k \in \mathbb{Z}$, $d \in \mathbb{N} - \{0,1\}$ *and* $n \in \mathbb{N} - \{0,1,2\}$, *with* $\sqrt[n]{d} \notin \mathbb{Q}$. *Assume that there exist* $C > 0$ *and* $\mu < n$ *such that, for every rational* p/q *with* $q > 0$,

$$\left| \sqrt[n]{d} - \frac{p}{q} \right| \geq \frac{C}{q^{\mu}}. \tag{9.22}$$

Then the diophantine equation

$$x^n - dy^n = k \tag{9.23}$$

has only finitely many solutions $(x,y) \in \mathbb{Z}^2$. *More precisely, let* (x,y) *be a solution of* (9.23), *such that* $x > 0$, $y > 0$, x *and* y *coprime. Then, either* $y \leq 2\left| k \right|$, *or* $x = P_n$, $y = Q_n$, *where* P_n/Q_n *is a convergent of the regular continued fraction expansion of* $\sqrt[n]{d}$ *satisfying* $Q_n \leq \left(\dfrac{\left| k \right|}{C} \right)^{\frac{1}{n-\mu}}$.

Proof If n is even, we can limit ourselves to the case where $x > 0$, $y > 0$. If n is odd, and if x and y have different signs, (9.23) can be written $x'^n + dy'^n = k'$, $x' > 0$, $y' > 0$, which has only finitely many solutions, and again we only need to consider the case $x > 0$, $y > 0$. Finally, we can assume that x and y are coprime (otherwise we divide the equation by the n-th power of GCD(x, y)). Under these assumptions,

$$\left| x^n - dy^n \right| = \left| x - \sqrt[n]{d}\, y \right| \left(x^{n-1} + x^{n-2} \sqrt[n]{d}\, y + \cdots + (\sqrt[n]{d}\, y)^{n-1} \right)$$

$$\geq \left| x - \sqrt[n]{d}\, y \right| y^{n-1} \geq \left| x/y - \sqrt[n]{d} \right| y^n.$$

Therefore, if (x,y) is a solution of (9.23), then

$$\left|\frac{x}{y} - \sqrt[n]{d}\right| \le \frac{|k|}{y^n} = \frac{|k|}{y} \frac{1}{y^{n-1}}.$$

Thus, if $y > 2|k|$, x/y is a convergent of the regular continued fraction expansion of $\sqrt[n]{d}$ by theorem 4.3. In this case, by using (9.22), we see that $|k|/y^n \ge C/y^\mu$, whence $y^{n-\mu} \le |k|/C$. By assumption, $n - \mu > 0$, and therefore $y \le (|k|/C)^{\frac{1}{n-\mu}}$. The proof of theorem 9.8 is complete.

Remark 9.4 It should be noted that Liouville's theorem 9.5 does not allow to obtain (9.22) with $\mu < n$ (it gives $\mu = n$). We will be able to apply theorem 9.8 whenever we know an improvement of the *Liouville exponent* n. By refining the methods of section 9.4, it is possible to obtain this improvement in special cases. For example, table 9.1 (page 146), which has been given by Bennet in 1997, provides explicit irrationality measures for some numbers of the form $\sqrt[n]{d}$, $n \ge 3$.

Exemple 9.4 Consider the diophantine equation
$$x^3 - 6y^3 = 1. \tag{9.24}$$
If $x \ge 0$ and $y \le 0$, there is only one solution, $x = 1$, $y = 0$. If $x \le 0$ and $y \ge 0$, there is no solution. Therefore we have to solve the equations $x^3 - 6y^3 = \pm 1$, with $x > 0$, $y > 0$, x and y coprime, and theorem 9.8 applies. If $y \le 2|k| = 2$, there is no solution since 7 and 49 are not cubes. Hence $x = P_n$ and $y = Q_n$, where P_n/Q_n is a convergent of the regular continued fraction expansion of $\sqrt[3]{6}$ satisfying $Q_n \le (1/C)^{\frac{1}{3-\mu}}$.. By using the values of C and μ given by table 9.1, we obtain $y = Q_n \le 1193$.
Compute the first terms of the regular continued fraction expansion of $\alpha = \sqrt[3]{6}$. First $\alpha^3 - 6 = 0$ and $\alpha = 1 + 1/\alpha_1$, whence $(1 + 1/\alpha_1)^3 - 6 = 0$, and $5\alpha_1^3 - 3\alpha_1^2 - 3\alpha_1 - 1 = 0$, which implies $\alpha_1 = 1 + 1/\alpha_2$. Replacing yields $2\alpha_2^3 - 6\alpha_2^2 - 12\alpha_2 - 5 = 0$ and $\alpha_2 = 4 + 1/\alpha_3$. Now α_3 fulfills $21\alpha_3^3 - 36\alpha_3^2 - 18\alpha_3 - 2 = 0$, and $\alpha_3 = 2 + 1/\alpha_4$, and so on. Finally we have $\sqrt[3]{6} = [1,1,4,2,7,3,511,1,2,...]$. The first convergents can be computed by using (4.5), and we have $P_0 = 1$, $Q_0 = 1$, $P_1 = 2$, $Q_1 = 1$, $P_2 = 9$, $Q_2 = 5$, $P_3 = 20$, $Q_3 = 11$, $P_4 = 149$, $Q_4 = 82$, $P_5 = 467$, $Q_5 = 257$,

Table 9.1: Explicit irrationality measures $\left|\sqrt[n]{d} - p/q\right| \geq C/q^{-\mu}$

$\sqrt[n]{d}$	C	μ	$\sqrt[n]{d}$	C	μ	$\sqrt[n]{d}$	C	μ
$\sqrt[3]{2}$	0.25	2.47	$\sqrt[3]{62}$	0.04	2.50	$\sqrt[5]{18}$	0.21	4.29
$\sqrt[3]{3}$	0.39	2.76	$\sqrt[3]{63}$	0.02	2.43	$\sqrt[5]{22}$	0.14	4.91
$\sqrt[3]{5}$	0.29	2.80	$\sqrt[3]{65}$	0.02	2.43	$\sqrt[5]{28}$	0.05	3.41
$\sqrt[3]{6}$	0.01	2.35	$\sqrt[3]{66}$	0.04	2.50	$\sqrt[5]{30}$	0.02	3.04
$\sqrt[3]{7}$	0.08	2.70	$\sqrt[3]{67}$	0.06	2.56	$\sqrt[5]{31}$	0.01	2.83
$\sqrt[3]{10}$	0.15	2.45	$\sqrt[3]{68}$	0.08	2.60	$\sqrt[5]{33}$	0.01	2.82
$\sqrt[3]{11}$	0.22	2.91	$\sqrt[3]{70}$	0.12	2.68	$\sqrt[5]{34}$	0.02	3.02
$\sqrt[3]{12}$	0.28	2.95	$\sqrt[3]{76}$	0.10	2.54	$\sqrt[5]{37}$	0.05	3.48
$\sqrt[3]{13}$	0.35	2.86	$\sqrt[3]{78}$	0.03	2.60	$\sqrt[5]{39}$	0.08	2.91
$\sqrt[3]{15}$	0.19	2.54	$\sqrt[3]{83}$	0.10	2.72	$\sqrt[5]{40}$	0.09	3.90
$\sqrt[3]{17}$	0.01	2.22	$\sqrt[3]{84}$	0.37	2.92	$\sqrt[5]{42}$	0.11	4.19
$\sqrt[3]{19}$	0.02	2.30	$\sqrt[3]{90}$	0.09	2.41			
$\sqrt[3]{20}$	0.01	2.23	$\sqrt[3]{91}$	0.01	2.29	$\sqrt[7]{5}$	0.25	4.43
$\sqrt[3]{22}$	0.08	2.31				$\sqrt[7]{10}$	0.38	5.17
$\sqrt[3]{26}$	0.03	2.53	$\sqrt[4]{5}$	0.03	2.77	$\sqrt[7]{11}$	0.40	3.34
$\sqrt[3]{28}$	0.03	2.52	$\sqrt[4]{14}$	0.06	3.78	$\sqrt[7]{12}$	0.42	3.88
$\sqrt[3]{30}$	0.10	2.72	$\sqrt[4]{15}$	0.03	3.27	$\sqrt[7]{13}$	0.44	4.91
$\sqrt[3]{31}$	0.14	2.97	$\sqrt[4]{17}$	0.03	3.24	$\sqrt[7]{17}$	0.03	5.20
$\sqrt[3]{37}$	0.01	2.27	$\sqrt[4]{18}$	0.05	3.67	$\sqrt[7]{23}$	0.43	6.03
$\sqrt[3]{39}$	0.09	2.21	$\sqrt[4]{37}$	0.33	3.34	$\sqrt[7]{45}$	0.27	5.10
$\sqrt[3]{42}$	0.13	2.46	$\sqrt[4]{39}$	0.005	2.52			
$\sqrt[3]{43}$	0.01	2.32				$\sqrt[11]{48}$	0.42	5.05
$\sqrt[3]{44}$	0.22	2.87	$\sqrt[5]{3}$	0.24	3.61			
$\sqrt[3]{52}$	0.26	2.97	$\sqrt[5]{6}$	0.43	3.33	$\sqrt[13]{6}$	0.14	4.22
$\sqrt[3]{58}$	0.12	2.71	$\sqrt[5]{10}$	0.41	3.92	$\sqrt[13]{20}$	0.25	5.87
$\sqrt[3]{60}$	0.08	2.61	$\sqrt[5]{11}$	0.38	4.23			
$\sqrt[3]{61}$	0.06	2.56	$\sqrt[5]{15}$	0.28	4.27	$\sqrt[17]{50}$	0.25	6.96

$P_6 = 238786$, $Q_6 = 131409$. Only the first six can yield a solution of (9.24) since $y = Q_n \leq 1193$. Now it is easy to check that none of the couples $x = P_i$, $y = Q_i$, with $0 \leq i \leq 5$, is a solution of (9.24). Therefore (9.24) has no other solution than $x = 1$, $y = 0$.

9.6 The theorems of Thue and Roth

A lot of work has been carried out in order to improve Liouville exponent in the general case. The first success in this direction has been Thue's theorem (1908):

Theorem 9.9 *Let α be an algebraic number of degree $d \geq 2$. Then there exist a constant $C > 0$ such that, for every $p/q \in \mathbb{Q}$ with $q > 0$,*

$$\left| \alpha - \frac{p}{q} \right| \geq \frac{C}{q^{1 + \frac{d}{2}}}.$$

Theorem 9.9 gives for $d \geq 3$ a *non explicit* improvement of the Liouville exponent. Indeed, the constant C cannot be computed in terms of α. The proof of Thue's theorem uses methods from transcendental number theory, and we must postpone it until exercise 12.17.

Corollary 9.2 *Let $P(X) = a_0 + a_1 X + \cdots + a_d X^d \in \mathbb{Z}[X]$, irreducible in $\mathbb{Q}[X]$, with degree $d \geq 3$. Then the homogeneous diophantine equation*

$$\sum_{i=0}^{d} a_i x^i y^{d-i} = k \tag{9.25}$$

has only finitely many solutions $(x, y) \in \mathbb{Z} \times \mathbb{Z}$.

Proof See exercise 9.10.

In the general case, it is not known how to obtain an upper bound for the solutions (x, y) of (9.25) in function of the a_i's (contrary to theorem 9.8). Thus corollary 9.2 gives no method of effective resolution of equation (9.25).

The best result in diophantine approximation is Roth's theorem (1955):

Theorem 9.10 *Let α be an algebraic number of degree $d \geq 2$. Then, for every $\varepsilon > 0$, there exists a constant $C > 0$ such that $|\alpha - p/q| \geq C/q^{2+\varepsilon}$ for every rational number p/q with $q > 0$.*

In other words, the best irrationality measure of any algebraic number α of degree $d \geq 2$ is $\mu(\alpha) = 2$. Roth's theorem, as Thue's theorem, is not explicit. For details, see for example [18].

Example 9.5   Let $\alpha = 3 \prod_{n=0}^{+\infty} \left(1 + 10^{-4^n}\right)$.

By exercise 2.10, we know that $\alpha \notin \mathbb{Q}$. Moreover, we have seen in exercise 4.15 that there exists an infinite sequence r_n / s_n of rational numbers such that, for every $n \geq 1$, $\left|\alpha - r_n / s_n\right| < 1/\left(2s_n^3\right)$.

Then Liouville's theorem shows that α cannot be algebraic of degree 2. Indeed, if it was, there would be a constant $C > 0$ such that, for every $n \geq 1$, $C/s_n^2 \leq \left|\alpha - r_n / s_n\right| < 1/2s_n^3$. Contradiction when $n \to +\infty$.

Similarly, Thue's theorem shows that α is not algebraic of degree 3, and Roth's theorem applied with $\varepsilon = 1/2$ shows that α is transcendental.

Exercises

9.1   Let α be any algebraic number. Prove that the minimal polynomial of α is unique.

9.2   Proof of theorem 9.1
1) Prove that the minimal polynomial P_α of α is irreducible in $\mathbb{Q}[x]$.
2) Let $P \in \mathbb{Q}[x]$, monic, irreducible in $\mathbb{Q}[x]$, and satisfying $P(\alpha) = 0$. Prove that $P = P_\alpha$.

9.3   Proof of Eisenstein's criterion
1) Let $P \in \mathbb{Z}[x]$. Assume that P is reducible in $\mathbb{Q}[x]$. By using theorem 9.4 , prove that P is reducible in $\mathbb{Z}[x]$.
2) Prove Eisenstein's criterion.

9.4   Field of algebraic numbers
1) Assume α and β algebraic. Prove that $\alpha + \beta$ and $\alpha\beta$ are algebraic.
2) Prove that the set $\overline{\mathbb{Q}}$ of algebraic numbers is a subfield of $\mathbb{C}$.

9.5   Assume that α is a Liouville number. Prove that α has no irrationality measure.

9.6   By using the properties of regular continued fractions, prove that the best irrationality measure of α satisfies $\mu(\alpha) \geq 2$.

9.7   Prove that, for every $n \in \mathbb{N}$, $(n+1)! \leq 3(n!)^{3/2}$.

9.8   Prove that, for every $n \in \mathbb{N}$, $\binom{2n}{n} \leq 4^n$. Deduce (9.19).

9.9   Best irrationality measure of e
1) Let $\eta > 0$. Prove that there exists $C_1 > 0$ (depending on η) such that, for every $n \in \mathbb{N}$, $(n+1)! \leq C_1(n!)^{1+\eta}$.
2) Prove that there exists $C_2 > 0$ (depending on η) such that, for every $n \in \mathbb{N}$, $|q_n| \leq C_2(n!)^{1+\eta}$, where q_n is defined in (9.15).
3) Let $\varepsilon > 0$. Prove that there exists $C > 0$ (depending on ε) such that $|e - p/q| \geq Cq^{-2-\varepsilon}$ for every rational p/q with $q > 0$.

9.10   Prove corollary 9.2.

9.11   Compute the minimal polynomial of $\alpha + \beta$ when
1) $\alpha = \sqrt{2}$ and $\beta = \sqrt[3]{3}$.
2) $\alpha = i$ and $\beta = j = e^{2i\pi/3}$.
3) α and β are the real roots of $x^3 + x + 1$ and $x^3 + x^2 + 1$ respectively.

9.12   Let α be algebraic. Prove that $A = \{P \in \mathbb{Q}[x] / P(\alpha) = 0\}$ is an ideal of $\mathbb{Q}[x]$. Deduce that A is principal (see exercise 6.20) and that $A = \mathbb{Q}[x]P_\alpha$.

9.13   Irrationality measure of $\sqrt[3]{17}$
1) With the notations of exercise 8.7, prove that, for every $n \in \mathbb{N}$,
$$|q_n| \leq 7\left(12\sqrt{3} \cdot 5833\right)^n.$$

2) Deduce that, for every $p/q \in \mathbb{Q}$ with $q > 0$,

$$\left| \sqrt[3]{17} - \frac{p}{q} \right| \geq \frac{5 \cdot 10^{-12}}{q^{2.392}}.$$

9.14   Irrationality measure of $\mathrm{Log}\, 2$

1) With the notations of exercise 8.8, prove that, for every $n \in \mathbb{N}$,

$$|q_n| \leq 36^n.$$

2) Deduce that, for every $p/q \in \mathbb{Q}$ with $q > 0$,

$$\left| \mathrm{Log}\, 2 - \frac{p}{q} \right| \geq \frac{6 \cdot 10^{-8}}{q^{13.46}}.$$

9.15   Solve the diophantine equation $x^4 - 5y^4 = 11$.

9.16   Denote by $(a_n)_{n \in \mathbb{N}}$ the sequence of all integers of the form $2^a 3^b$, arranged in increasing order. Thus $a_0 = 1$, $a_1 = 2$, $a_2 = 3$, $a_3 = 4$, $a_4 = 6$, $a_5 = 8$, $a_6 = 9$, $a_7 = 12$, $a_8 = 16$, $a_9 = 18$, $a_{10} = 24$, $a_{11} = 27$, $a_{12} = 32$. The aim of this exercise is to prove that $\lim_{n \to +\infty} (a_{n+1} - a_n) = +\infty$.

1) Let a, b, $c \in \mathbb{N}^*$ be given. Prove that the equation $ax^3 - by^3 = c$ has only finitely many solutions $(x, y) \in \mathbb{N}^2$.

2) Let $c \in \mathbb{N}$ be given. Prove that the equation $a_{n+1} - a_n = c$, where the unknown is n, has only finitely many solutions.

3) Prove that $\lim_{n \to +\infty} (a_{n+1} - a_n) = +\infty$.

9.17   A variation on theorem 9.7

In this exercise we aim to prove the following result.

Let $\alpha \in \mathbb{R}$. Let $k_0 > 0$, $\ell_0 \geq \frac{1}{2} \geq \ell_1 > 0$, $1 < E_0 \leq E_1$ and $Q > 1$ be real numbers such that there exists a sequence $(p_n, q_n) \in \mathbb{Z}^2$ satisfying

$$\forall n \in \mathbb{N}, \qquad |q_n| \leq k_0 Q^n, \qquad\qquad (*)$$

$$\forall n \in \mathbb{N}, \qquad \ell_1 E_1^{-n} \leq |q_n \alpha - p_n| \leq \ell_0 E_0^{-n}. \qquad (**)$$

Then, for every $(p, q) \in \mathbb{Z} \times \mathbb{N}^*$, $\left| \alpha - \frac{p}{q} \right| \geq \frac{C}{q^\mu}$, with

$$\mu = \frac{\mathrm{Log}(QE_1)}{\mathrm{Log}\, E_0}, \quad C = \frac{\ell_1}{k_0}(QE_1)^{-1-\frac{\mathrm{Log}(\ell_0/\ell_1)}{\mathrm{Log}\, E_0}}.$$

1) Let n be the least integer satisfying $|q_n\alpha - p_n| < \ell_1/q$. Prove that n exists, that $n \geq 1$, that $q_n \neq 0$, and that $n \leq \mathrm{Log}(\ell_1 q/\ell_0)/\mathrm{Log}\, E_0 + 1$.

2) Prove that, if $q_n p - p_n q = 0$, then $\left|\alpha - \dfrac{p}{q}\right| \geq \dfrac{\ell_1 E_1^{-n}}{|q_n|}$.

Deduce that $\left|\alpha - \dfrac{p}{q}\right| \geq \dfrac{C}{q^\mu}$, with C and μ as above.

3) Prove that this result remains true if $q_n p - p_n q \neq 0$.

9.18 Irrationality measures of π^2 and π

Let $\zeta(2) = \displaystyle\sum_{n=1}^{+\infty} \frac{1}{n^2}$. We know by exercise 7.17 that $\zeta(2) = \dfrac{\pi^2}{6}$.

1) For $(r,s) \in \mathbb{N}^2$, define $I_{r,s} = \displaystyle\int_0^1 \int_0^1 \frac{x^r y^s}{1-xy}\, dx\, dy$.

Prove that $I_{r,r} = \zeta(2) - \displaystyle\sum_{k=1}^r \frac{1}{k^2}$.

Prove that $I_{r,s} = \dfrac{1}{r-s}\left(\dfrac{1}{s+1} + \dfrac{1}{s+2} + \cdots + \dfrac{1}{r}\right)$ when $r > s$.

2) Let $P_n(x) = \dfrac{1}{n!}\dfrac{d^n}{dx^n}\left(x^n(1-x)^n\right)$ be the n-th Legendre polynomial.

Prove that $P_n \in \mathbb{Z}[x]$.

3) Define $K_n = \displaystyle\int_0^1 \int_0^1 \frac{P_n(x)(1-y)^n}{1-xy}\, dx\, dy$.

Prove that $K_n = b_n\zeta(2) + a_n$, with $a_n, b_n \in \mathbb{Q}$.

Denote $\delta(n) = \mathrm{LCM}(1,2,\ldots,n)$.

Check that $q_n = [\delta(n)]^2 b_n$ and $p_n = [\delta(n)]^2 a_n$ are integers.

4) By using theorem 7.8, prove that $|q_n| \leq (144)^n$.

5) Prove that $K_n = (-1)^n \displaystyle\int_0^1 \int_0^1 \frac{x^n y^n(1-x)^n(1-y)^n}{(1-xy)^{n+1}}\, dx\, dy$.

6) Prove that $(1-x)(1-y) \le \left(1-\sqrt{xy}\right)^2$ for all $x \ge 0$, $y \ge 0$. Deduce that

$$|K_n| \le \frac{\pi^2}{6} \left(\frac{\sqrt{5}-1}{2}\right)^{5n} \le 1.65 \times (0.0902)^n.$$

7) By noting that $|K_n| \ge \int_\varepsilon^{1-\varepsilon} \int_\varepsilon^{1-\varepsilon} \frac{x^n y^n (1-x)^n (1-y)^n}{(1-xy)^{n+1}} dxdy$ for $\varepsilon \in \left[0, \frac{1}{2}\right]$,

prove that $|K_n| \ge 5 \cdot 10^{-8} (0.0832)^n$.

8) By using the result of exercise 9.17, find an irrationality measure of π^2. Deduce an irrationality measure of π.

9.19   Let α be an algebraic number, P_α its minimal polynomial. Prove that P_α has only single zeros.

9.20   A striking number (following exercise 4.15)

Let $\alpha = 3 \prod_{n=0}^{+\infty} \left(1 + 10^{-4^n}\right)$.

We know that α is transcendental (example 9.5). By using the sequences r_n and s_n defined in exercise 4.15, find an irrationality measure of α. Deduce that α is not a Liouville number.

9.21   A transcendental number (following exercise 1.12)
Find an irrationality measure of the transcendental number

$$\beta = \sum_{n=0}^{+\infty} q^{-2^n} \quad (q \in \mathbb{Z}, \ |q| \ge 2).$$

Deduce that β is not a Liouville number.

9.22   Let $r \in \mathbb{N}$, with $r \ge 3$. Let $q \in \mathbb{Z}$, with $|q| \ge 2$. Let $\gamma = \sum_{n=0}^{+\infty} q^{-r^n}$.

Prove that γ is transcendental and find an irrationality measure.

Chapter 10

Number fields

This chapter aims to provide an introduction to algebraic number theory. In section 10.1, we define number fields, which generalize quadratic fields (chapter 5). Section 10.2 is devoted to the study of conjugates, norms and traces. We define the ring of integers $\mathbb{A}_{\mathbb{K}}$ of a number field $\mathbb{K}$ in section 10.3, and its units in section 10.4. We will be interested, in particular, in the case of cyclotomic fields. Section 10.5 is an introduction to the theory of ideals, which will be studied in more detail in chapter 11. Here, we merely define the discriminant of an ideal of $\mathbb{A}_{\mathbb{K}}$, and, as a special case, the discriminant of the number field $\mathbb{K}$ itself. Finally, we show in section 10.6 how to solve the Fermat equation $x^5 + y^5 = z^5$ by working in the cyclotomic field $\mathbb{K} = \mathbb{Q}(e^{2i\pi/5})$.

10.1 Algebraic number fields

Let $\mathbb{K}$ be a subfield of $\mathbb{C}$. Then $1 \in \mathbb{K}$. Now an easy induction shows that $n \in \mathbb{K}$ for every $n \in \mathbb{N}$. Hence $n \in \mathbb{K}$ for every rational integer n, and finally $n/m \in \mathbb{K}$ for every rational number n/m. Therefore $\mathbb{Q} \subset \mathbb{K}$. As a result (exercise 10.1), $\mathbb{K}$ is a vector space over $\mathbb{Q}$.

Definition 10.1 *Let $\mathbb{K}$ be a subfield of $\mathbb{C}$. We say that $\mathbb{K}$ is a number field if it has finite dimension over $\mathbb{Q}$. We call this dimension the degree of $\mathbb{K}$ and denote it by $d = [\mathbb{K} : \mathbb{Q}]$.*

Now let $\alpha \in \mathbb{C}$ be an algebraic number of degree d. Define

$$\mathbb{K} = \mathbb{Q}(\alpha) = \left\{ a_0 + a_1 \alpha + \cdots + a_{d-1} \alpha^{d-1} \,/\, a_0, a_1, \ldots, a_{d-1} \in \mathbb{Q} \right\}. \quad (10.1)$$

It is easy to check (exercise 10.2) that $1, \alpha, \ldots, \alpha^{d-1}$ are linearly independent over $\mathbb{Q}$, that $\mathbb{K}$ is a subfield of $\mathbb{C}$, and that $\mathbb{K}$ has finite dimension d over $\mathbb{Q}$. Hence $\mathbb{K}$ is a number field of degree d.

Example 10.1 Let $\alpha = 1$. Then $\mathbb{K} = \mathbb{Q}$, of degree 1.

Example 10.2 Let $\alpha = \sqrt{d}$, where $d \in \mathbb{Z}$ is squarefree. Then

$$\mathbb{K} = \mathbb{Q}(\sqrt{d}) = \left\{ a + b\sqrt{d} \mid a, b \in \mathbb{Q} \right\}$$

is a *quadratic field*, of degree 2 (see chapter 5).

Example 10.3 Let $\omega = e^{2i\pi p}$, p odd prime. Then $\mathbb{K} = \mathbb{Q}(\omega)$ is called a *cyclotomic field*. We have $[\mathbb{Q}(\omega) : \mathbb{Q}] = p - 1$, since the minimal polynomial of ω is, by theorem 9.2 and definition 9.3, the cyclotomic polynomial $P_\omega(x) = x^{p-1} + x^{p-2} + \cdots + x + 1$.

Two number fields $\mathbb{K}$ and $\mathbb{L}$ being given, we say that $\mathbb{L}$ is an *extension* of $\mathbb{K}$ if $\mathbb{K} \subset \mathbb{L}$. It is clear, by definition 10.1, that every number field $\mathbb{K}$ is an extension of $\mathbb{Q}$.

Example 10.4 Consider the cyclotomic field $\mathbb{L} = \mathbb{Q}(\omega)$, with $\omega = e^{2i\pi/5}$. We know by exercise 10.3 that $\omega + \omega^4 = 2\cos(2\pi/5) = (-1 + \sqrt{5})/2$. Therefore $\sqrt{5} = 1 + 2\omega + 2\omega^4 \in \mathbb{L}$, and consequently $a + b\sqrt{5} \in \mathbb{L}$ for every $(a, b) \in \mathbb{Q}^2$. Thus the cyclotomic field $\mathbb{L} = \mathbb{Q}(e^{2i\pi/5})$ is an extension of the quadratic field $\mathbb{K} = \mathbb{Q}(\sqrt{5})$. We can be more precise by defining $\theta = 2i\sin(2\pi/5)$. Then $\omega^2 + \omega^3 = 2\cos(4\pi/5) = (-1 - \sqrt{5})/2$, $\omega - \omega^4 = \theta$, $\omega^2 - \omega^3 = 2i\sin(4\pi/5) = (-1 + \sqrt{5})\theta/2$. Therefore, if

$$x = a + b\omega + c\omega^2 + d\omega^3 \quad (a, b, c, d \in \mathbb{Q})$$

is any element of $\mathbb{L}$, then $\overline{x} = a + b\omega^4 + c\omega^3 + d\omega^2$, whence

$$\begin{cases} x + \overline{x} = 2a + b\dfrac{\sqrt{5}-1}{2} - (c+d)\dfrac{\sqrt{5}+1}{2} \\[2mm] x - \overline{x} = b\theta + (c-d)\dfrac{\sqrt{5}-1}{2}\theta \end{cases}$$

This yields

$$x = \frac{1}{2}\left(2a + b\frac{\sqrt{5}-1}{2} - (c+d)\frac{\sqrt{5}+1}{2} \right) + \left(b + (c-d)\frac{\sqrt{5}-1}{2} \right)\theta. \quad (10.2)$$

In other words, x can be written as $\alpha + \beta\theta$, with $\alpha, \beta \in \mathbb{K} = \mathbb{Q}(\sqrt{5})$.

Therefore $\mathbb{L} \subset \mathbb{K}(\theta)$. Conversely, since $\sqrt{5} \in \mathbb{L}$ and $\theta \in \mathbb{L}$, we have $\mathbb{K}(\theta) \subset \mathbb{L}$. Thus $\mathbb{L} = \mathbb{K}(\theta)$. We say that the extension $\mathbb{L}$ has been obtained by *adjoining* to $\mathbb{K}$ an algebraic number θ (here θ is algebraic because $\theta^2 = -(5 + \sqrt{5})/2$, whence $\theta^4 + 5\theta^2 + 5 = 0$).

This adjunction process generalizes easily. If α, β, γ, ... are algebraic numbers, we define $\mathbb{K}$, $\mathbb{L}$, $\mathbb{M}$, ..., by $\mathbb{K} = \mathbb{Q}(\alpha)$, $\mathbb{L} = \mathbb{Q}(\alpha, \beta) = \mathbb{K}(\beta)$, $\mathbb{M} = \mathbb{Q}(\alpha, \beta, \gamma) = \mathbb{L}(\gamma),....$ It can be checked (see exercise 10.4 for details) that $\mathbb{K}$, $\mathbb{L}$, $\mathbb{M}$, ... are indeed number fields.

Conversely, let $\mathbb{K}$ be a number field of degree d. Then every $x \in \mathbb{K}$ is algebraic of degree not greater than d, since the numbers 1, x, $x^2,..., x^d$ are linearly dependent over $\mathbb{Q}$. First choose $\alpha_1 \in \mathbb{K}^*$. Then $\mathbb{Q}(\alpha_1) \subset \mathbb{K}$. If $\mathbb{K} = \mathbb{Q}(\alpha_1)$, there is nothing more to prove. If not, there is $\alpha_2 \in \mathbb{K}$ with $\alpha_2 \notin \mathbb{Q}(\alpha_1)$. Then $\mathbb{Q}(\alpha_1, \alpha_2) \subset \mathbb{K}$ and $[\mathbb{Q}(\alpha_1, \alpha_2) : \mathbb{Q}] > [\mathbb{Q}(\alpha_1) : \mathbb{Q}]$, since $\alpha_2 \notin \mathbb{Q}(\alpha_1)$. By induction, we obtain an increasing sequence of vector subspaces of $\mathbb{K}$. Since $\mathbb{K}$ has finite dimension over $\mathbb{Q}$, there exist α_1, $\alpha_2 ,..., \alpha_n \in \mathbb{K}$ such that $\mathbb{K} = \mathbb{Q}(\alpha_1, \alpha_2,..., \alpha_n)$.

It is a striking result that, in fact, α_1 can be chosen in such a way that $n = 1$. This is known as the *primitive element theorem*:

Theorem 10.1 *Let $\mathbb{K}$ be a number field. Then there exists $\theta \in \mathbb{K}$ such that $\mathbb{K} = \mathbb{Q}(\theta)$.*

Proof Since $\mathbb{K} = \mathbb{Q}(\alpha_1, \alpha_2,..., \alpha_n)$, we only have to prove that, for every $(\alpha, \beta) \in \mathbb{K}^2$, there exists θ such that $\mathbb{Q}(\alpha, \beta) = \mathbb{Q}(\theta)$. We will give a constructive proof by computing θ explicitly. Let P_α and P_β be the minimal polynomials of α and β, $\alpha_1 = \alpha$, $\alpha_2,..., \alpha_n$ be the zeros of P_α, $\beta_1 = \beta$, $\beta_2,..., \beta_p$ be the zeros of P_β. Then the α_i's are all different, and so are the β_j's (see exercise 9.19). Therefore, for every $i = 1$, $2,..., n$ and every $j = 2,..., p$, the equation

$$\alpha_i + x\beta_j = \alpha_1 + x\beta_1 \tag{10.3}$$

has only one solution $x \in \mathbb{C}$. Choose $\gamma \in \mathbb{Z}^*$ such that γ satisfies none of the equations (10.3), and define $\theta = \alpha + \gamma\beta$. Clearly $\theta \in \mathbb{Q}(\alpha, \beta)$, whence $\mathbb{Q}(\theta) \subset \mathbb{Q}(\alpha, \beta)$.

We now prove that $\mathbb{Q}(\alpha, \beta) \subset \mathbb{Q}(\theta)$. For this, it suffices to prove that $\beta \in \mathbb{Q}(\theta)$, since $\alpha = \theta - \gamma\beta$. Let $Q(x) = P_\alpha(\theta - \gamma x)$. Then Q is a polynomial with coefficients in $\mathbb{Q}(\theta)$, and $Q(\beta) = P_\alpha(\alpha) = 0$. Hence β is a common root of Q and P_β. It is the only one. Indeed, if $\beta_j (j \geq 2)$ was a root of Q, then $\theta - \gamma\beta_j$ would be a root of P_α, and therefore equal to one of the α_i's, contrary to the choice of γ. Since P_α and Q belong to $\mathbb{Q}(\theta)[x]$, the same holds for their greatest common divisor $D(x)$, which has degree 1 since P_α and Q have only one common zero β. Hence $D(x) = ax + b$, with $a, b \in \mathbb{Q}(\theta)$, $a \neq 0$, and consequently $D(\beta) = a\beta + b = 0$, which yields $\beta = -b/a \in \mathbb{Q}(\theta)$, Q.E.D.

Example 10.5 Let $\mathbb{K} = \mathbb{Q}\left(\sqrt{3}, \sqrt[3]{2}\right)$. Then $\alpha_1 = \alpha = \sqrt{3}$, $\alpha_2 = -\sqrt{3}$, and $\beta_1 = \beta = \sqrt[3]{2}$, $\beta_2 = j\sqrt[3]{2}$, $\beta_3 = j^2\sqrt[3]{2}$, with $j = e^{2i\pi/3}$. It is clear that $\gamma = 1$ satisfies none of the equations (10.3). Hence $\mathbb{K} = \mathbb{Q}(\sqrt{3} + \sqrt[3]{2})$.

10.2   Conjugates, norms and traces

Let $\mathbb{K}$ be a number field of degree d. We say that $\sigma : \mathbb{K} \to \mathbb{C}$ is a *monomorphism* of $\mathbb{K}$ if it is a field homomorphism which leaves $\mathbb{Q}$ unchanged. In other words,

$$\begin{cases} \sigma(x + y) = \sigma(x) + \sigma(y) & \text{for every } (x, y) \in \mathbb{K}^2 \\ \sigma(xy) = \sigma(x)\sigma(y) & \text{for every } (x, y) \in \mathbb{K}^2 \\ \sigma(r) = r & \text{for every } r \in \mathbb{Q} \end{cases} \qquad (10.4)$$

In particular, if we consider σ as a mapping from the $\mathbb{Q}$-vector space $\mathbb{K}$ to the $\mathbb{Q}$-vector space $\mathbb{C}$, then σ is a linear mapping.

Theorem 10.2 *In every number field $\mathbb{K}$ of degree d, there exist exactly d monomorphisms $\sigma_1, \sigma_2, ..., \sigma_d$. If θ is any primitive element of $\mathbb{K}$,*

these monomorphisms are defined by $\sigma_i(\theta) = \theta_i$, *where* $\theta_1 = \theta$, $\theta_2,..., \theta_d$
are the roots of the minimal polynomial of θ.

Proof As θ is a primitive element of $\mathbb{K}$, we have $\mathbb{K} = \mathbb{Q}(\theta)$ by
theorem 10.1. Let $P(x) = \sum_{i=0}^{d} a_i x^i$ be the minimal polynomial of θ, and
$\theta_1 = \theta$, θ_2, ..., θ_d be the roots of P_θ. These roots are all different (see
exercise 9.19). For every monomorphism σ of $\mathbb{K}$, we have

$$0 = \sum_{i=0}^{d} a_i \theta^i \Rightarrow 0 = \sigma\left(\sum_{i=0}^{d} a_i \theta^i\right) = \sum_{i=0}^{d} \sigma(a_i)\sigma(\theta)^i = \sum_{i=0}^{d} a_i \sigma(\theta)^i$$

since $a_i \in \mathbb{Q}$ for all i. Therefore $\sigma(\theta)$ is a zero of P_θ, that is $\sigma(\theta) = \theta$,
or $\theta_2,...,$ or θ_d. Hence there exist at most d monomorphisms of $\mathbb{K}$.
Conversely, define $\sigma_i : \mathbb{K} \rightarrow \mathbb{C}$ by

$$\sigma_i(a_0 + a_1\theta + \cdots + a_{d-1}\theta^{d-1}) = a_0 + a_1\theta_i + \cdots + a_{d-1}\theta_i^{d-1}.$$

It can be checked (see exercise 10.5) that σ_i is a monomorphism of
$\mathbb{K}$ and that $\sigma_i \neq \sigma_j$ if $i \neq j$. This proves theorem 10.2.

For every $x \in \mathbb{K}$, the numbers $x = \sigma_1(x)$, $\sigma_2(x)$, ..., $\sigma_d(x)$ are called the
conjugates of x.

Example 10.6 Let $\mathbb{K} = \mathbb{Q}(\sqrt{d})$ be a quadratic field. Here $\theta = \sqrt{d}$,
$P_\theta(x) = x^2 - d$, $\theta_1 = \theta = \sqrt{d}$, $\theta_2 = -\sqrt{d}$. Hence $\sigma_1(a + b\sqrt{d}) = a + b\sqrt{d}$
and $\sigma_2(a + b\sqrt{d}) = a - b\sqrt{d}$, which means that σ_1 is the identity in $\mathbb{K}$,
whereas $\sigma_2(x)$ is the conjugate x^* of x as defined in chapter 5.

Let $\mathbb{K}$ be a number field of degree d, and $\sigma_1 = Id$, σ_2, ..., σ_d be the
d monomorphisms of $\mathbb{K}$. For every x in $\mathbb{K}$, define the *norm* of x,
denoted by $N(x)$, and the *trace* of x, denoted by $T(x)$, by

$$N(x) = \sigma_1(x)\sigma_2(x)...\sigma_d(x). \tag{10.5}$$

$$T(x) = \sigma_1(x) + \sigma_2(x) + \cdots + \sigma_d(x). \tag{10.6}$$

If $x \in \mathbb{Q}$, then $\sigma_i(x) = x$ for all i, and therefore $N(x) = x^d$.

Example 10.7 Let $\mathbb{K} = \mathbb{Q}(\sqrt{d})$ be a quadratic field. We have seen in example 10.6 that $\sigma_1\left(a + b\sqrt{d}\right) = a + b\sqrt{d}$ and $\sigma_2(a + b\sqrt{d}) = a - b\sqrt{d}$. Therefore $N(a + b\sqrt{d}) = a^2 - db^2$, $T(a + b\sqrt{d}) = 2a$, which corresponds to the definitions of chapter 5.

Exemple 10.8 Let $\mathbb{K} = \mathbb{Q}(\sqrt[3]{d})$, where d is cubefree. We say that $\mathbb{K}$ is a *cubic field* (which means of degree 3). We even call it a *pure* cubic field, because it has a primitive element of the form $\sqrt[3]{d}$. Here $\theta = \sqrt[3]{d}$, $P_\theta(x) = x^3 - d$, $\theta_1 = \theta$, $\theta_2 = j\theta$, $\theta_3 = j^2\theta$, $j = e^{2i\pi/3}$. Now let $x \in \mathbb{K}$, $x = a + b\theta + c\theta^2$, with $a, b, c \in \mathbb{Q}$. Then

$$\begin{aligned} N(x) &= (a + b\theta + c\theta^2)(a + bj\theta + cj^2\theta^2)(a + bj^2\theta + cj\theta^2) \\ &= (a^2 - bcd + (c^2d - ab)j^2\theta + (b^2 - ac)j\theta^2)(a + bj^2\theta + cj\theta^2). \end{aligned}$$

Therefore $N\left(a + b\sqrt[3]{d} + c\sqrt[3]{d}^2\right) = a^3 + db^3 + d^2c^3 - 3abcd$. $\qquad$ (10.7)

A similar computation shows that

$$T\left(a + b\sqrt[3]{d} + c\sqrt[3]{d}\right) = 3a. \qquad (10.8)$$

Theorem 10.3 *Let $\mathbb{K}$ be a number field of degree d. Then, for every $x \in \mathbb{K}$, $N(x) \in \mathbb{Q}$ and $T(x) \in \mathbb{Q}$. Moreover, for every $(x, y) \in \mathbb{K}^2$,*

$$T(x + y) = T(x) + T(y), \qquad (10.9)$$
$$N(xy) = N(x)N(y), \qquad (10.10)$$
$$N(x) = 0 \Leftrightarrow x = 0. \qquad (10.11)$$

Proof See exercise 10.6.

10.3 Ring of integers of a number field

Let $\mathbb{K}$ be a number field of degree d. We have seen that every $\alpha \in \mathbb{K}$ is algebraic of degree not greater than d. We say that x is an integer of $\mathbb{K}$ if x is an algebraic integer, that is if P_α has rational integer coefficients (see section 9.2). The set $\mathbb{A}_\mathbb{K}$ of all integers of $\mathbb{K}$ is a ring by theorem 9.4. We call it the ring of integers of $\mathbb{K}$.

We observe that, if $\alpha \in \mathbb{A}_\mathbb{K}$, then $N(\alpha) \in \mathbb{Z}$ and $T(\alpha) \in \mathbb{Z}$.

Indeed, $P_\alpha(\alpha) = 0 \Rightarrow P_\alpha(\sigma_i(\alpha)) = 0$ for any monomorphism σ_i of $\mathbb{K}$. Therefore $\sigma_1(\alpha)$, $\sigma_2(\alpha)$, ..., $\sigma_d(\alpha)$ are algebraic integers, and $N(\alpha)$ and $T(\alpha)$ are algebraic integers by theorem 9.4. Since $N(\alpha) \in \mathbb{Q}$ and $T(\alpha) \in \mathbb{Q}$ by theorem 10.3, this implies $N(\alpha) \in \mathbb{Z}$ and $T(\alpha) \in \mathbb{Z}$.

If $\mathbb{K} = \mathbb{Q}(\sqrt{d})$ is a quadratic field, the ring $\mathbb{A}_\mathbb{K}$ of its integers is given by theorem 5.3. Another example, connected with the Fermat equation $x^p + y^p = z^p$, is provided by the cyclotomic field $\mathbb{K} = \mathbb{Q}(e^{2i\pi/p})$:

Theorem 10.4 *Let* $\mathbb{K} = \mathbb{Q}(\omega)$, $\omega = e^{2i\pi/p}$, *p odd prime. Then*
$$\mathbb{A}_\mathbb{K} = \mathbb{Z}(\omega) = \left\{ a_0 + a_1\omega + \cdots + a_{p-2}\omega^{p-2} \mid a_i \in \mathbb{Z} \right\}.$$

Proof First, we will have to compute some traces and norms in $\mathbb{Q}(\omega)$. The minimal polynomial of ω is the cyclotomic polynomial
$$P(x) = 1 + x + x^2 + \cdots + x^{p-1} = (x - \omega)(x - \omega^2)...(x - \omega^{p-1}).$$
Therefore the $p - 1$ monomorphisms of $\mathbb{K} = \mathbb{Q}(\omega)$ are defined by $\sigma_i(\omega) = \omega^i$ for $i = 1, 2, ..., p - 1$. We have
$$P(1) = p = (1 - \omega)(1 - \omega^2)...(1 - \omega^{p-1}) = \sigma_1(1 - \omega)\sigma_2(1 - \omega)...\sigma_{p-1}(1 - \omega).$$
This immediatly yields
$$N(1 - \omega) = p. \tag{10.12}$$
Then we observe that, for every $x \in \mathbb{A}_\mathbb{K}$,
$$T(x(1 - \omega)) \text{ is a multiple of } p. \tag{10.13}$$
Indeed, $T(x(1 - \omega)) = \sum_{i=1}^{p-1} \sigma_i(x)(1 - \omega^i) = (1 - \omega)\sum_{i=1}^{p-1} \sigma_i(x)\left(\sum_{j=0}^{i-1} \omega^j\right).$
Therefore there exists $y \in \mathbb{A}_\mathbb{K}$ such that $T(x(1 - \omega)) = (1 - \omega)y$.

Taking the norms yields $N\big(T\big(x(1 - \omega)\big)\big) = \big(T\big(x(1 - \omega)\big)\big)^d = pN(y)$ by using (10.12). Therefore $p \mid T(x(1 - \omega))$, which proves (10.13).
Now we show that
$$T(\omega^i) = -1 \quad \text{for} \quad i = 1, 2, ..., p - 1. \tag{10.14}$$
Indeed, we have
$$T(\omega^i) = \sum_{j=1}^{p-1} \sigma_j(\omega^i) = \sum_{j=1}^{p-1} \omega^{ij} = \sum_{j=1}^{p-1} \sigma_i(\omega^j)$$
$$= \sigma_i\left(\sum_{j=1}^{p-1} \omega^j\right) = \sigma_i(-1) = -1.$$

Now, let $x = a_0 + a_1\omega + \cdots + a_{p-2}\omega^{p-2}$ $(a_i \in \mathbb{Q})$ be an element of $\mathbb{A}_\mathbb{K}$.
Then $x(1-\omega) = a_0(1-\omega) + a_1(\omega - \omega^2) + \cdots + a_{p-2}(\omega^{p-2} - \omega^{p-1})$, whence

$$T(x(1-\omega)) = a_0(T(1) - T(\omega)) + \cdots + a_{p-2}(T(\omega^{p-2}) - T(\omega^{p-1})).$$

But $T(1) = \sum_{i=1}^{p-1} \sigma_i(1) = \sum_{i=1}^{p-1} 1 = p - 1$. Therefore, $T(x(1-\omega)) = pa_0$ by
(10.14). Thus pa_0 is a rational integer since $x(1-\omega) \in \mathbb{A}_\mathbb{K}$, and this
integer is a multiple of p by (10.13). Consequently, $a_0 \in \mathbb{Z}$.
Now $(x - a_0)\omega^{p-1} = a_1 + a_2\omega + \cdots + a_{p-2}\omega^{p-3} \in \mathbb{A}_\mathbb{K}$, and by arguing the
same way we see that $a_1 \in \mathbb{Z}$. By induction, we deduce that $x \in \mathbb{Z}(\omega)$.
Hence $\mathbb{A}_\mathbb{K} \subset \mathbb{Z}(\omega)$. Since $\mathbb{Z}(\omega) \subset \mathbb{A}_\mathbb{K}$ by theorem 9.4, $\mathbb{A}_\mathbb{K} = \mathbb{Z}(\omega)$.

10.4   Units

Let $\mathbb{A}_\mathbb{K}$ be the ring of integers of the number field $\mathbb{K}$. We say that
$\varepsilon \in \mathbb{A}_\mathbb{K}$ is a *unit* of $\mathbb{A}_\mathbb{K}$ if ε is invertible in $\mathbb{A}_\mathbb{K}$ (see 5.3.1, page 60). We
will generalize theorem 5.4 by using the following lemma.

Lemma 10.1   *Let $\mathbb{K}$ be a number field of degree d, and let $\sigma_1 = Id$,
σ_2, ..., σ_d be the monomorphisms of $\mathbb{K}$. Then, for every $x \in \mathbb{A}_\mathbb{K}$,
$\sigma_2(x)\sigma_3(x)...\sigma_d(x) \in \mathbb{A}_\mathbb{K}$, and therefore x divides $N(x)$ in $\mathbb{A}_\mathbb{K}$.*

Proof   See exercise 10.7. Now, theorem 10.5 generalizes theorem 5.4:

Theorem 10.5   *Let $\mathbb{A}_\mathbb{K}^\times$ be the group of units of the ring of integers $\mathbb{A}_\mathbb{K}$
of the number field $\mathbb{K}$. Then, for every $\varepsilon \in \mathbb{A}_\mathbb{K}$, $\varepsilon \in \mathbb{A}_\mathbb{K}^\times \Leftrightarrow |N(\varepsilon)| = 1$.*

Proof   See exercise 10.8. The group of units of $\mathbb{A}_\mathbb{K}$ is given by theorem
5.5 when $\mathbb{K}$ is an imaginary quadratic field and by theorem 5.7 when $\mathbb{K}$
is a real quadratic field. In the general case, it is a difficult problem to
determine $\mathbb{A}_\mathbb{K}^\times$ for a given number field $\mathbb{K}$. However, in the case of
cyclotomic fields, we have the following result, which is known as
Kummer's lemma:

Lemma 10.2   *Let $\mathbb{K} = \mathbb{Q}(\omega)$, $\omega = e^{2i\pi/p}$, p odd prime. Then the units
of $\mathbb{A}_\mathbb{K}$ are of the form $\varepsilon = \omega^k \eta$, where $k \in \mathbb{N}$ and η is a real unit.*

Proof *Step one.* Let $\varepsilon \in \mathbb{A}_{\mathbb{K}}^{\times}$. We prove that $\varepsilon / \bar{\varepsilon} = \pm \omega^k$, with $k \in \mathbb{N}$.

Put $\varepsilon / \bar{\varepsilon} = a_0 + a_1 \omega + \cdots + a_{p-2} \omega^{p-2}$, with $a_i \in \mathbb{Z}$, $i = 1, \ldots, p-2$.

Since $P_\omega(\omega) = 1 + \omega + \cdots + \omega^{p-1} = 0$, we have for every $m \in \mathbb{Z}$

$$\varepsilon / \bar{\varepsilon} = (a_0 + m) + (a_1 + m)\omega + \cdots + (a_{p-2} + m)\omega^{p-2} + m\omega^{p-1}.$$

Now we choose m such that $\varepsilon / \bar{\varepsilon} = b_0 + b_1 \omega + \cdots + b_{p-1} \omega^{p-1}$, with

$b_i = a_i + m$ for $i \le p-2$, $b_{p-1} = m$, and $-\frac{p}{2} < b_0 + b_1 + \cdots + b_{p-1} < \frac{p}{2}$, and

we define $P(x) = b_0 + b_1 x + \cdots + b_{p-1} x^{p-1}$. We have $\varepsilon / \bar{\varepsilon} = P(\omega)$ and

$\overline{\varepsilon / \varepsilon} = P(\bar{\omega}) = P(\omega^{p-1})$. Dividing $P(x)P(x^{p-1})$ by $x^p - 1$ yields

$$P(x)P(x^{p-1}) = Q(x)(x^p - 1) + R(x). \tag{10.15}$$

But it can be proved (see exercise 10.9 for details) that

$$R(x) = \sum_{r=0}^{p-1} B_r x^r, \quad B_r = \sum_{(i,j) \in E_r} b_i b_j, \tag{10.16}$$

where $E_r = \left\{ (i, j) \in \mathbb{N}^2 / 0 \le i \le p-1, \ 0 \le j \le p-1, \ i - j \equiv r \,(\mathrm{mod}\ p) \right\}$.

Replacing x by ω in (10.15) yields $1 = B_0 + B_1 \omega + \cdots + B_{p-1} \omega^{p-1}$. Hence

ω is a zero of the polynomial $H(x) = (B_0 - 1) + B_1 x + \cdots + B_{p-1} x^{p-1}$.

Therefore H is a multiple of the minimal polynomial P_ω of ω (exercise

9.12). Since $\deg H = p - 1 = \deg P_\omega$, there exists $h \in \mathbb{Z}$ such that

$B_0 - 1 = B_1 = \cdots = B_{p-1} = h$. Taking $x = 1$ in (10.15) yields

$$B_0 + B_1 + \cdots + B_{p-1} = (b_0 + b_1 + \cdots + b_{p-1})^2.$$

Hence $(b_0 + b_1 + \cdots + b_{p-1})^2 = 1 + ph$ and $b_0 + b_1 + \cdots + b_{p-1} \equiv \pm 1 \,(\mathrm{mod}\ p)$.

Taking into account that $-\frac{p}{2} < b_0 + b_1 + \cdots + b_{p-1} < \frac{p}{2}$, we see that

$b_0 + b_1 + \cdots + b_{p-1} = \pm 1$, whence $h = 0$ and $B_0 = 1$, $B_i = 0$ for $i \ge 1$.

But by (10.16), $B_0 = \sum_{i=0}^{p-1} b_i^2$. Therefore $B_0 = 1$ implies that one of the

b_i's is equal to ± 1, whereas the other ones are zero, and $\varepsilon / \bar{\varepsilon} = \pm \omega^k$.

Step two. We prove that, in fact, $\varepsilon / \bar{\varepsilon} = \omega^k$ for some $k \in \mathbb{N}$. We argue by
reduction ad absurdum and assume that there exists $k \in \mathbb{N}$ such that
$\varepsilon / \bar{\varepsilon} = -\omega^k$. Then there exists $s \in \mathbb{N}$ such that $\varepsilon / \bar{\varepsilon} = -\omega^{2s}$ (if k is odd,

replace it by $k+p$). Then $v = \varepsilon\omega^{-s} = -\overline{\varepsilon}\omega^{s}$ is a unit. Using the fact that $\left(\omega, \omega^2, \ldots, \omega^{p-1}\right)$ is a basis of $\mathbb{K}$ over $\mathbb{Q}$, put

$$v = a_0\omega + a_1\omega^2 + \cdots + a_{p-2}\omega^{p-1} = P(\omega).$$

Then $\overline{v} = -v = P(\overline{\omega}) = P(\omega^{-1})$, whence $2v = P(\omega) - P(\omega^{-1}) = (\omega - \omega^{-1})\alpha$ with $\alpha \in \mathbb{A}_{\mathbb{K}}$. Taking the norms yields $N(\omega - \omega^{-1}) \mid N(2v)$. Now $N(2v) = N(2)N(v) = 2^{p-1}$ and $N(\omega - \omega^{-1}) = N(\omega^{-1})N(\omega-1)N(\omega+1) = -pN(\omega+1)$ by (10.12). Therefore $p \mid 2^{p-1}$, contradiction.

Step three. Since there exists $s \in \mathbb{N}$ such that $\varepsilon / \overline{\varepsilon} = \omega^{2s}$, we have $\varepsilon\omega^{-s} = \overline{\varepsilon}\omega^{s}$. Then $\eta = \varepsilon\omega^{-s}$ is a unit satisfying $\eta = \overline{\eta}$. Hence η is a real unit and $\varepsilon = \eta\omega^{s}$, Q.E.D.

Example 10.9 Let $\mathbb{K} = \mathbb{Q}(\omega)$, $\omega = e^{2i\pi/5}$. To find the units of $\mathbb{A}_{\mathbb{K}}$, we look for the real units. We know already by example 10.4 that $\mathbb{K} = \mathbb{Q}\left(\sqrt{5}\right)(\theta)$, with $\theta = 2i\sin(2\pi/5)$. Therefore the real units of $\mathbb{A}_{\mathbb{K}}$ are the units of the ring of integers of $\mathbb{Q}\left(\sqrt{5}\right)$. Hence, by example 5.1,

$$\mathbb{A}_{\mathbb{K}}^{\times} = \left\{\pm\omega^{n}\Phi^{k} / 0 \leq n \leq 4, \ k \in \mathbb{Z}\right\},$$

where $\Phi = (1+\sqrt{5})/2$ is the golden number.

Example 10.10 Let $\mathbb{K} = \mathbb{Q}(\omega)$, $\omega = e^{2i\pi/7}$. It can be proved (see exercise 10.10 for details) that

$$\mathbb{A}_{\mathbb{K}}^{\times} = \left\{\pm\omega^{n}\eta^{k}\mu^{\ell} / 0 \leq n \leq 6, \ (k, \ell) \in \mathbb{Z}^2\right\},$$

where $\eta = \omega + \omega^6 = 2\cos(2\pi/7)$ and $\mu = \omega^3 + \omega^4 = 2\cos(6\pi/7)$.

In fact, the general structure of the group of units of $\mathbb{A}_{\mathbb{K}}$ is very simple. It is given by *Dirichlet units theorem*, which we state here without proof (see for example [27] or [18]).

Theorem 10.6 *Let $\mathbb{K}$ be a number field. Then there exist units η_1, η_2, ..., η_k (called fundamental units) such that every $\varepsilon \in \mathbb{A}_{\mathbb{K}}^{\times}$ can be* *uniquely written as* $\varepsilon = \zeta \eta_1^{n_1} \eta_2^{n_2} ... \eta_k^{n_k}$ *, where* $n_i \in \mathbb{Z}$ *for every i, and* ζ *is a root of unity belonging to* $\mathbb{A}_\mathbb{K}$ *.*

10.5   Discriminants and integral basis

Let $\mathbb{K} = \mathbb{Q}(\theta)$ be a number field of degree d. Then $(1, \theta, ..., \theta^{d-1})$ is a basis of the $\mathbb{Q}$ - vector space $\mathbb{K}$.

More generally, let $(\alpha_1, \alpha_2, ... \alpha_d)$ be any basis of the $\mathbb{Q}$ - vector space $\mathbb{K}$. We define its *discriminant* $\Delta[\alpha_1, \alpha_2, ... \alpha_d]$ by

$$
\begin{cases}
\Delta[\alpha_1, \alpha_2, ... \alpha_d] = \left(D[\alpha_1, \alpha_2, ... \alpha_d] \right)^2, \text{ with} \\
\\
D[\alpha_1, \alpha_2, ... \alpha_d] = \begin{vmatrix} \sigma_1(\alpha_1) & \sigma_1(\alpha_2) ... & \sigma_1(\alpha_d) \\ \sigma_2(\alpha_1) & \sigma_2(\alpha_2) ... & \sigma_2(\alpha_d) \\ \vdots & & \\ \sigma_d(\alpha_1) & \sigma_d(\alpha_2) ... & \sigma_d(\alpha_d) \end{vmatrix}
\end{cases} \tag{10.17}
$$

where the σ_i's are the d monomorphisms of $\mathbb{K}$.

Lemma 10.3   *Let* $\mathbb{K} = \mathbb{Q}(\theta)$ *be a number field of degree d, and let* P_θ *be the minimal polynomial of* θ *. Then*

$$
\Delta\left[1, \theta, ..., \theta^{d-1}\right] = (-1)^{\frac{d(d-1)}{2}} N\left(P_\theta'(\theta) \right).
$$

Proof   We have, by definition,

$$
\Delta\left[1, \theta, ..., \theta^{d-1}\right] = \det\left(\sigma_i(\theta^j) \right)^2 = \begin{vmatrix} 1 & 1 & \cdots & 1 \\ \theta_1 & \theta_2 & \cdots & \theta_d \\ \vdots & \vdots & & \vdots \\ \theta_1^{d-1} & \theta_2^{d-1} & \cdots & \theta_d^{d-1} \end{vmatrix}^2,
$$

where $\theta_1 = \theta$, θ_2, ..., θ_d are the zeros of P_θ. This is a *Vandermonde determinant* (see exercise 10.11), and therefore

$$
\Delta\left[1, \theta, ..., \theta^{d-1}\right] = \prod_{1 \le j < i \le d} (\theta_i - \theta_j)^2.
$$

But $P_\theta(x) = \prod_{i=1}^d (x - \theta_i)$. Therefore $P_\theta'(x) = \sum_{j=1}^d \prod_{i \ne j} (x - \theta_i)$, which yields $P_\theta'(\theta_j) = \prod_{i \ne j} (\theta_j - \theta_i)$.

Multiplying for $j = 1, 2, ..., d$, we get $N(P'_\theta(\theta)) = \prod_{j=1}^{d} \prod_{i \neq j} (\theta_j - \theta_i)$.

In this product, every couple (i, j) with $i \neq j$ contributes for $(\theta_j - \theta_i)(\theta_i - \theta_j) = -(\theta_i - \theta_j)^2$. Hence, comparing with (10.18) yields $\Delta[1, \theta, ..., \theta^{d-1}] = (-1)^s N(P'_\theta(\theta))$, where s is half the number of couples (i, j) satisfying $1 \leq i \leq d$, $1 \leq j \leq d$, $i \neq j$. Since $s = (d^2 - d)/2$, lemma 10.3 is proved.

Lemma 10.4 *Let* $\mathbb{K} = \mathbb{Q}(\theta)$ *be a number field of degree* d. *Let* $(\alpha_1, \alpha_2, ..., \alpha_d)$ *and* $(\beta_1, \beta_2, ..., \beta_d)$ *be two basis of the* $\mathbb{Q}$ - *vector space* $\mathbb{K}$. *Assume* $\alpha_j = \sum_{i=1}^{d} c_{ij} \beta_i$, *with* $c_{ij} \in \mathbb{Q}$. *Then*
$$\Delta[\alpha_1, \alpha_2, ..., \alpha_d] = (\det(c_{ij}))^2 \Delta[\beta_1, \beta_2, ..., \beta_d].$$

Proof See exercise 10.12.

Theorem 10.7 *Let* $(\alpha_1, \alpha_2, ..., \alpha_d)$ *be a basis of the* $\mathbb{Q}$ - *vector space* $\mathbb{K} = \mathbb{Q}(\theta)$. *Then* $\Delta[\alpha_1, ..., \alpha_d] \in \mathbb{Q}^{*+}$. *Moreover, if* $\alpha_i \in \mathbb{A}_{\mathbb{K}}$ *for every* i, *then* $\Delta[\alpha_1, ..., \alpha_d] \in \mathbb{N} - \{0\}$.

Proof See exercise 10.13.

Now let I be an ideal of $\mathbb{A}_{\mathbb{K}}$. We recall from section 5.2 that I is a subgroup of $(\mathbb{A}_{\mathbb{K}}, +)$, satisfying also: $\forall x \in I$, $\forall y \in \mathbb{A}_{\mathbb{K}}$, $xy \in I$.

Theorem 10.8 *Let* $\mathbb{K} = \mathbb{Q}(\theta)$ *be a number field of degree* d. *Let* $I \neq \{0\}$ *be an ideal of* $\mathbb{A}_{\mathbb{K}}$. *Then there exist* α_1, α_2, ..., $\alpha_d \in I$, *linearly independent over* $\mathbb{Q}$, *such that*
$$I = \{n_1 \alpha_1 + n_2 \alpha_2 + \cdots + n_d \alpha_d \,/\, n_1, n_2, ..., n_d \in \mathbb{Z}\}.$$

In other words, every element of the ideal I can be written as a linear combination, with coefficients in $\mathbb{Z}$, of α_1, α_2, ..., α_d. Such a set of linear combinations with coefficients in a domain $\mathbb{A}$ is called an $\mathbb{A}$ - *module*. It fulfills the same axioms as a vector space over a field $\mathbb{K}$, except that the scalars are the elements of $\mathbb{A}$. Theorem 10.8 can also be expressed by saying that every ideal I of $\mathbb{A}_{\mathbb{K}}$ is a $\mathbb{Z}$ - module of rank d (in the case of modules, *rank* replaces *dimension*).

Proof of theorem 10.8 Let $\alpha \in I$, $\alpha \neq 0$. Then, by lemma 10.1,
$$N(\alpha) = \alpha \sigma_2(\alpha)...\sigma_d(\alpha) \in I.$$
Let δ be a denominator of θ. By theorem 9.3, $\delta\theta \in \mathbb{A}_{\mathbb{K}}$, and therefore $\beta = N(\alpha)\delta\theta \in I$. The numbers 1, β, ..., β^{d-1} are linearly independent over $\mathbb{Q}$ since $\delta N(\alpha) \in \mathbb{Z}$. Thus $\left(1, \beta,..., \beta^{d-1}\right)$ is a basis of $\mathbb{K}$, made from elements of I. Among all the basis made from elements of I, we select one, denoted by $\left(\alpha_1, \alpha_2,..., \alpha_d\right)$, such that $\Delta[\alpha_1, \alpha_2,..., \alpha_d]$ is least. This is possible because $\Delta[\alpha_1, \alpha_2,..., \alpha_d] \in \mathbb{N}^*$ by theorem 10.7. Now we show that $B = \left(\alpha_1, \alpha_2,..., \alpha_d\right)$ satisfies the conditions of theorem 10.8. Assume that this is not the case. Then there exists $x \in I$ with $x = x_1\alpha_1 + \cdots + x_d\alpha_d$ (where $x_i \in \mathbb{Q}$ for all i), such that at least one $x_i \notin \mathbb{Z}$. Without loss of generality, we can assume that $x_1 \notin \mathbb{Z}$. Thus $r = x_1 - [x_1] \in]0,1[$. Define $y = x - [x_1]\alpha_1 = r\alpha_1 + x_2\alpha_2 + \cdots + x_d\alpha_d \in I$. Then $B' = \left(y, \alpha_2,..., \alpha_d\right)$ is a basis of $\mathbb{K}$, made from elements of I. The determinant relevant to the change of basis $B \to B'$ is
$$\begin{vmatrix} r & 0 & ... & 0 \\ x_2 & 1 & ... & 0 \\ \vdots & 0 & \ddots & \vdots \\ x_d & 0 & ... & 1 \end{vmatrix} = r.$$
Hence lemma 10.4 yields
$$\Delta[y, \alpha_2,..., \alpha_d] = r^2\Delta[\alpha_1, \alpha_2,..., \alpha_d] < \Delta[\alpha_1, \alpha_2,..., \alpha_d],$$
which contradicts the choice of $\alpha_1, \alpha_2,..., \alpha_d$. Theorem 10.8 is proved.

Now let $\left(\alpha_1, \alpha_2,..., \alpha_d\right)$ and $\left(\beta_1, \beta_2,..., \beta_d\right)$ be two basis of the ideal I, which is a $\mathbb{Z}$ - module by theorem 10.8. We can find $c_{ij} \in \mathbb{Z}$, $c'_{ij} \in \mathbb{Z}$, such that $\alpha_j = \sum_{i=1}^d c_{ij}\beta_i$ and $\beta_j = \sum_{i=1}^d c'_{ij}\alpha_i$.
Thus the square matrices $C = (c_{ij})$ and $C' = (c'_{ij})$ satisfy $C' = C^{-1}$. Taking the determinants yields $\det C \cdot \det C' = 1$. However, $\det C \in \mathbb{Z}$ and $\det C' \in \mathbb{Z}$, whence $\det C = \pm 1$.

Consequently, lemma 10.4 implies that $\Delta[\alpha_1,\alpha_2,...,\alpha_d]=\Delta[\beta_1,\beta_2,...,\beta_d]$. This motivates the following definition.

Definition 10.2 *Let* I *be an ideal of the ring* $\mathbb{A}_\mathbb{K}$ *of integers of a number field* $\mathbb{K}$. *The natural integer* $\Delta(I)=\Delta[\alpha_1,\alpha_2,...,\alpha_d]$, *which does not depend on the basis* $(\alpha_1,\alpha_2,...,\alpha_d)$ *of the* $\mathbb{Z}$ *- module* I, *is called the discriminant of* I.

Theorem 10.8 and definition 10.2 apply, in particular, when $I=\mathbb{A}_\mathbb{K}$. Thus the ring of integers $\mathbb{A}_\mathbb{K}$ of the number field $\mathbb{K}$ is a $\mathbb{Z}$ - module of rank d. Any basis $(\alpha_1,\alpha_2,...,\alpha_d)$ of the $\mathbb{Z}$ - module $\mathbb{A}_\mathbb{K}$ is called an *integral basis* of $\mathbb{K}$. The *discriminant* of $\mathbb{K}$ is defined by

$$\Delta(\mathbb{K})=\Delta(\mathbb{A}_\mathbb{K})=\Delta[\alpha_1,\alpha_2,...,\alpha_d].$$

It does not depend on the choice of the integral basis $(\alpha_1,\alpha_2,...,\alpha_d)$.

Example 10.11 Let $\mathbb{K}=\mathbb{Q}(\sqrt{d})$ be a quadratic field. By theorem 5.3, $\mathbb{A}_\mathbb{K}=\mathbb{Z}(\sqrt{d})$ if $d\equiv 2$ or 3 (mod 4), and $\mathbb{A}_\mathbb{K}=\mathbb{Z}((1+\sqrt{d})/2)$ if $d\equiv 1$ (mod 4). Hence $(1,\sqrt{d})$ is an integral basis of $\mathbb{K}$ if $d\equiv 2$ or 3 (mod 4) and $(1,(1+\sqrt{d})/2)$ is an integral basis of $\mathbb{K}$ if $d\equiv 1$ (mod 4). Thus

$$\text{If } d\equiv 2 \text{ or } 3 \text{ (mod 4)}, \quad \Delta(\mathbb{K})=\begin{vmatrix} 1 & \sqrt{d} \\ 1 & -\sqrt{d} \end{vmatrix}^2=4d. \tag{10.19}$$

$$\text{If } d\equiv 1 \text{ (mod 4)}, \quad \Delta(\mathbb{K})=\begin{vmatrix} 1 & (1+\sqrt{d})/2 \\ 1 & (1-\sqrt{d})/2 \end{vmatrix}^2=d. \tag{10.20}$$

Example 10.12 Let $\mathbb{K}=\mathbb{Q}(\omega)$, $\omega=e^{2i\pi/p}$, p odd prime, be a cyclotomic field. Theorem 10.4 shows that $\mathbb{A}_\mathbb{K}=\mathbb{Z}(\omega)$, and therefore $(1,\omega,...,\omega^{p-2})$ is an integral basis of $\mathbb{K}$. By lemma 10.3,

$$\Delta(K)=(-1)^{(p-1)(p-2)/2}N(P'_\omega(\omega)).$$

However $P_\omega(x)=(1-x^p)/(1-x)$, whence $P'_\omega(\omega)=-p\omega^{p-1}/(1-\omega)$.

Thus, by (10.12), $N\left(P'_\omega(\omega)\right) = N(-p)N(\omega)^{p-1}/N(1-\omega) = (-p)^{p-1}/p$.

Finally, since $(-1)^{p-2} = -1$, we obtain

$$\Delta(\mathbb{K}) = (-1)^{\frac{p-1}{2}} p^{p-2}. \tag{10.21}$$

10.6 Fermat's equation $x^5 + y^5 = z^5$

Theorem 10.9 *Fermat's equation $x^5 + y^5 = z^5$ has no solution if x, y, z are non-zero integers.*

To prove theorem 10.9, we will work in the ring $\mathbb{A}_\mathbb{K}$ of integers of the cyclotomic field $\mathbb{K} = \mathbb{Q}(\omega)$, with $\omega = e^{2i\pi/5}$. We will need two lemmas.

Lemma 10.5 *Let $\omega = e^{2i\pi/5}$, $\mathbb{K} = \mathbb{Q}(\omega)$. Then $\mathbb{A}_\mathbb{K}$ is an euclidean domain. In other words, for every $(a,b) \in \mathbb{A}_\mathbb{K} \times \mathbb{A}_\mathbb{K}^*$, there exists $(q,r) \in \mathbb{A}_\mathbb{K}^2$ such that $a = bq + r$ and $|N(r)| < |N(b)|$.*

Proof See exercice 10.14.

It is readily seen that definition 5.4 and theorem 5.15 (page 68) remain unchanged with "number field" in place of "quadratic field". Therefore lemma 10.5 implies that $\mathbb{A}_\mathbb{K} = \mathbb{Z}(\omega)$ is a unique factorization domain. Hence any element of $\mathbb{A}_\mathbb{K}$ can be uniquely written as a product of prime elements of $\mathbb{A}_\mathbb{K}$.

Now let a, b, $c \in \mathbb{A}_\mathbb{K}$. We say that $a \equiv b \pmod{c}$ if $a - b = cx$ for some $x \in \mathbb{A}_\mathbb{K}$. It is clear that the congruence in $\mathbb{A}_\mathbb{K}$ is a generalization of the congruence in $\mathbb{Z}$ and has the same properties.

Lemma 10.6 *Let $\omega = e^{2i\pi/5}$, $\mathbb{K} = \mathbb{Q}(\omega)$. Let ε be a unit of $\mathbb{A}_\mathbb{K}$, such that $\varepsilon \equiv k \pmod 5$, $k \in \mathbb{Z}$. Then there is a unit η in $\mathbb{A}_\mathbb{K}$, with $\varepsilon = \eta^5$.*

Proof See exercice 10.15.

Now we prove theorem 10.9. We assume that (x,y,z), $x \neq 0$, $y \neq 0$, $z \neq 0$ satisfies $x^5 + y^5 = z^5$, with x, y, z pairwise coprime. By Sophie Germain theorem (exercise 6.13), 5 divides one of the three numbers x, y, z. Without loss of generality, we may assume that $z = 5z'$, $z' \in \mathbb{Z}$.

Step one. Let $\lambda = 1 - \omega \in \mathbb{A}_{\mathbb{K}}$. Then $N(\lambda) = 5$ by (10.12). Since $\lambda \mid N(\lambda)$ by lemma 10.1, we have $x^5 + y^5 = 5 z'^5 = N(\lambda) z'^5 = \lambda^5 (\lambda' z')^5$, with $\lambda' \in \mathbb{A}_{\mathbb{K}}$. Hence we only have to prove that the equation

$$x^5 + y^5 = \varepsilon \lambda^{5k} z^5 \tag{10.22}$$

has no solution $(x, y, z) \in \mathbb{A}_{\mathbb{K}}^*$, x, y, z pairwise coprime, $k \in \mathbb{N}^*$, $\varepsilon \in \mathbb{A}_{\mathbb{K}}^\times$. We will use the method of descent on the integer k. We write (10.22) in the form

$$(x + y)(x + \omega y)(x + \omega^2 y)(x + \omega^3 y)(x + \omega^4 y) = \varepsilon \lambda^{5k} z^5. \tag{10.23}$$

As theorem 5.12 remains evidently true in every ring $\mathbb{A}_{\mathbb{K}}$, $N(\lambda) = 5$ implies that λ is irreducible in $\mathbb{A}_{\mathbb{K}}$. As $\mathbb{A}_{\mathbb{K}}$ is euclidean by lemma 10.5, it is a unique factorization domain by theorem 5.15 and λ is prime in $\mathbb{A}_{\mathbb{K}}$ by theorem 5.14.

Therefore λ divides at least one of the numbers $x + \omega^i y$, $i = 0, 1, 2, 3, 4$. But $(x + \omega^a y) - (x + \omega^b y) = y \omega^a (1 - \omega^{b-a})$ for all $a < b$. Therefore $\lambda = 1 - \omega$ divides $(x + \omega^a y) - (x + \omega^b y)$, which implies that λ divides all the numbers $x + \omega^i y$, $i = 0, 1, 2, 3, 4$. Hence the numbers $(x + \omega^i y)/\lambda$ belong to $\mathbb{A}_{\mathbb{K}}$. Moreover, they are pairwise coprime (see exercise 10.16).

Step two. Now we show that λ^2 divides one of the $x + \omega^i y$ (and only one, since the $(x + \omega^i y)/\lambda$ are pairwise coprime). For this, we observe that $\left(1, \lambda, \lambda^2, \lambda^3\right)$ is an integral basis of $\mathbb{K}$ since 1, $\omega = 1 - \lambda$, ω^2, ω^3 can be written as linear combinations of 1, λ, λ^2, λ^3 with coefficients in $\mathbb{Z}$. Therefore there exist a_i, $b_i \in \mathbb{Z}$ such that $x = a_0 + a_1 \lambda + a_2 \lambda^2 + a_3 \lambda^3$ and $y = b_0 + b_1 \lambda + b_2 \lambda^2 + b_3 \lambda^3$. Thus $x \equiv a_0 + a_1 \lambda \pmod{\lambda^2}$ and $y \equiv b_0 + b_1 \lambda \pmod{\lambda^2}$, $\omega^i \equiv 1 - i\lambda \pmod{\lambda^2}$, $i = 1, 2, 3, 4$, whence

$$x + \omega^i y \equiv a_0 + b_0 + (a_1 + b_1 - i b_0) \lambda \pmod{\lambda^2}. \tag{10.24}$$

We know by step one that $\lambda \mid (x + \omega^i y)$. Therefore $\lambda \mid (a_0 + b_0)$. Since $a_0 + b_0 \in \mathbb{Z}$, taking the norms yields $N(\lambda) = 5 \mid (a_0 + b_0)^5$, which implies that $N(\lambda) \mid (a_0 + b_0)$.

In particular, $(1-\omega)(1-\omega^4)\,|\,(a_0+b_0)$, whence $-\omega^4(1-\omega)^2\,|\,(a_0+b_0)$. Since $-\omega^4 \in \mathbb{A}_\mathbb{K}^\times$, this means that $\lambda^2\,|\,(a_0+b_0)$, and (10.24) becomes

$$(x+\omega^j y)/\lambda \equiv a_1+b_1-ib_0 \quad (\mathrm{mod}\ \lambda). \tag{10.25}$$

But in $\mathbb{Z}$, $b_0 \not\equiv 0$ (mod 5). If not, λ would divide b_0 since $\lambda|5$, λ would divide a_0, and λ would divide x and y. Thus there exists $i \in \{0,1,2,3,4\}$ satisfying $a_1+b_1-ib_0 \equiv 0$ (mod 5). Therefore by (10.25) there exists $i \in \{0,1,2,3,4\}$ such that $\lambda^2\,|\,(x+\omega^j y)$. Even if it means replacing y by $\omega^j y$ in (10.22), we may assume that $\lambda^2\,|\,(x+y)$.

Step three. Since $\lambda^2\,|\,(x+y)$ and $\lambda|(x+\omega^j y)$ for every $i=1$, 2, 3, 4, we must have $k \geq 2$ in (10.23), which can be written as

$$(x+y) \times \frac{x+\omega y}{\lambda} \times \frac{x+\omega^2 y}{\lambda} \times \frac{x+\omega^3 y}{\lambda} \times \frac{x+\omega^4 y}{\lambda} = \varepsilon\lambda^{5k-4}z^5.$$

As λ^2 does not divide any of the $x+\omega^j y$ for $i=1$, 2, 3, 4, λ^{5k-4} must divide $x+y$, and we have

$$\frac{x+y}{\varepsilon\lambda^{5k-4}} \times \frac{x+\omega y}{\lambda} \times \frac{x+\omega^2 y}{\lambda} \times \frac{x+\omega^3 y}{\lambda} \times \frac{x+\omega^4 y}{\lambda} = z^5. \tag{10.26}$$

Now, $\mathbb{A}_\mathbb{K}$ is factorial, and each of the pairwise coprime factors in the left-hand side of (10.26) must be, up to a unit, a fifth power (see section 5.7). In particular,

$$\begin{cases} x+y = e_1\lambda^{5k-4}\alpha^5 \\ x+\omega y = e_2\lambda\beta^5 \\ x+\omega^2 y = e_3\lambda\gamma^5 \end{cases} \tag{10.27}$$

with $(e_1,e_2,e_3) \in \mathbb{A}_\mathbb{K}^{\times 3}$, $(\alpha,\beta,\gamma) \in \mathbb{A}_\mathbb{K}^3$.

None of the numbers α, β, γ is zero (otherwise z would be zero). The last two equations in (10.27) imply that $x = e_3\gamma^5 - \omega e_2\beta^5$ and $y = -\omega^4 e_3\gamma^5 + \omega^4 e_2\beta^5$ (whence β and γ coprime). Replacing in the first equation yields $(1+\omega)e_2\beta^5 - e_3\gamma^5 = \omega e_1\lambda^{5(k-1)}\alpha^5$. After division by $(1+\omega)e_2$, we get

$$\beta^5 + \delta\gamma^5 = \nu\lambda^{5(k-1)}\alpha^5, \tag{10.28}$$

where δ and ν are units of $\mathbb{A}_\mathbb{K}$. Indeed, $(1+\omega) \in \mathbb{A}_\mathbb{K}^\times$ since clearly $(1+\omega)(1+\omega^2 + \omega^4) = 1$.

Step four. Now, if $\theta = a_0 + a_1\omega + a_2\omega^2 + a_3\omega^3 \in \mathbb{A}_\mathbb{K}$ ($a_i \in \mathbb{Z}$), then

$$\theta^5 = \sum_{i_1+i_2+i_3+i_4=5} \frac{5!}{i_1! i_2! i_3! i_4!} a_0^{i_1} (a_1\omega)^{i_2} (a_2\omega^2)^{i_3} (a_3\omega^3)^{i_4},$$

which implies that $\theta^5 \equiv a_0^5 + a_1^5 + a_2^5 + a_3^5$ (mod 5). Hence θ^5 is congruent to some rational integer modulo 5. In particular, $\lambda = 1 - \omega \Rightarrow \lambda^5 \equiv 0$ (mod 5). Now reduce (10.28) modulo 5. We obtain $A + \delta B \equiv 0$ (mod 5), with $(A,B) \in \mathbb{Z}^2$. But $B \not\equiv 0$ (mod 5). Otherwise, $A \equiv 0$ (mod 5), and β and γ would not be coprime. Therefore there exists $C \in \mathbb{Z}$ such that $BC \equiv 1$ (mod 5), whence $\delta \equiv D$ (mod 5), with $D \in \mathbb{Z}$. By lemma 10.6, there exists $\eta \in \mathbb{A}_\mathbb{K}^\times$ such that $\delta = \eta^5$. Replacing in (10.28) yields

$$\beta^5 + (\eta\gamma)^5 = \nu\lambda^{5(k-1)}\alpha^5. \tag{10.29}$$

Comparing to (10.22) shows that theorem 10.9 is proved by descent on k.

Exercises

10.1   Check that any subfield $\mathbb{K}$ of $\mathbb{C}$ is a vector space over $\mathbb{Q}$.

10.2   Let $\alpha \in \mathbb{C}$ be algebraic of degree d. Prove that $\mathbb{K} = \mathbb{Q}(\alpha)$ is a $\mathbb{Q}$ - vector space of dimension d. Deduce that $\mathbb{K}$ is a number field of degree d.

10.3   Compute $\cos(2\pi/5)$.

10.4   Let $\mathbb{K}$ be a number field, and let $\alpha \in \mathbb{C}$ be algebraic. Prove that $\mathbb{L} = \mathbb{K}(\alpha)$ is a number field.

10.5   Let $\mathbb{K} = \mathbb{Q}(\theta)$ be a number field of degree d, and let $\theta_1 = \theta$, $\theta_2, ..., \theta_d$ be the zeros of P_θ, minimal polynomial of θ. Let $\sigma_i : \mathbb{K} \to \mathbb{C}$, defined by $\sigma_i(a_0 + a_1\theta + \cdots + a_{d-1}\theta^{d-1}) = a_0 + a_1\theta_i + \cdots + a_{d-1}\theta_i^{d-1}$.
Prove that σ_i is a monomorphism of $\mathbb{K}$ and that $\sigma_i \neq \sigma_j$ if $i \neq j$.

10.6   Prove formulas (10.9), (10.10), (10.11).

10.7   Prove lemma 10.1.

10.8   Prove theorem 10.5.

10.9   Prove formula (10.16).

10.10   **Units of $\mathbb{K} = \mathbb{Q}(\omega)$, $\omega = e^{2i\pi/7}$**

1) Let $\eta = \omega + \omega^{-1}$, $\mu = \omega^3 + \omega^{-3}$, $\nu = \omega^2 + \omega^{-2}$. Prove that η, μ and ν have the same minimal polynomial. Deduce that η, μ and ν are real units of $\mathbb{A}_\mathbb{K} = \mathbb{Z}(\omega)$.

2) Prove that $\eta^r \mu^s = \pm 1$, with $r, s \in \mathbb{Z}$, implies $r = s = 0$. Deduce that all units of the form $\varepsilon = \pm \eta^r \mu^s$, with $r, s \in \mathbb{Z}$, are different.

3) Let $\varepsilon = E(\omega)$ be a real unit of $\mathbb{A}_\mathbb{K}$. Prove that
$$E(\omega) = a\eta + b\mu + c\nu, \text{ with } a, b, c \in \mathbb{Z},$$
and that there exist real numbers r and s such that $E(\omega) = \pm |\eta|^r |\mu|^s$.

4) Let $E(\omega) = a\eta + b\mu + c\nu$ be a real positive unit such that the real numbers r and s defined in question 3) satisfy $|r| \leq 1/2$, $|s| \leq 1/2$.

   a) Prove that $0.667 \leq |E(\omega)| \leq 1.499$, $0.597 \leq |E(\omega^2)| \leq 1.674$, $0.398 \leq |E(\omega^3)| \leq 2.512$.

   b) Prove that
   $$\begin{cases} 7a = (\eta - 2)E(\omega) + (\mu - 2)E(\omega^3) + (\nu - 2)E(\omega^2), \\ 7b = (\mu - 2)E(\omega) + (\nu - 2)E(\omega^3) + (\eta - 2)E(\omega^2), \\ 7c = (\nu - 2)E(\omega) + (\eta - 2)E(\omega^3) + (\mu - 2)E(\omega^2). \end{cases}$$
Deduce that $|a| \leq 2$, $|b| \leq 1$, $|c| \leq 1$.

c) Prove that $[(a+b+c)^3 - 7(a^2b+b^2c+c^2a+abc)]^2 = 1$.

d) Deduce the only possible values for a, b, c. Then show that $r = s = 0$.

5) Find all units of $\mathbb{K} = \mathbb{Q}(e^{2i\pi/7})$.

10.11 Vandermonde determinant

Prove that $\Delta = \begin{vmatrix} 1 & 1 & \cdots & 1 \\ \theta_1 & \theta_2 & \cdots & \theta_d \\ \vdots & & & \\ \theta_1^{d-1} & \theta_2^{d-1} & \cdots & \theta_d^{d-1} \end{vmatrix} = \prod_{1 \le j < i \le d} (\theta_i - \theta_j)$.

10.12 Prove lemma 10.4.

10.13 Prove theorem 10.7.

10.14 $\mathbb{Z}(e^{2i\pi/5})$ is an euclidean domain

Let $\omega = e^{2i\pi/5}$, and let $\alpha = a_0 + a_1\omega + a_2\omega^2 + a_3\omega^3 \in \mathbb{K} = \mathbb{Q}(\omega)$.

1) Prove that $N(\alpha) = A^2 - 5B^2$, with $A = \frac{5}{4}\sum_{i=0}^{3} a_i^2 - \frac{1}{4}\left(\sum_{i=0}^{3} a_i\right)^2$.

Deduce that $0 \le N(\alpha) \le A^2$.

2) Let $b_i = a_i - [a_i]$ for $i = 0$, 1, 2, 3, $b_4 = 0$. By using the pigeon-hole principle, prove that at least one of the numbers $b_i - b_j$ $(i \ne j)$ satisfies $|b_i - b_j| \le 1/5$. Deduce that there exist $q \in \{1,2,3,4\}$ and $\beta \in A_{\mathbb{K}}$ such that $|N(\alpha\omega^q - \beta)| < 1$.

3) Prove that $\mathbb{Z}(e^{2i\pi/5})$ is an euclidean domain.

10.15 Prove lemma 10.6.

10.16 Prove that the numbers $(x + \omega^i y)/\lambda$ in the proof of theorem 10.9 are pairwise coprime.

10.17 1) Let $\mathbb{K} \subset \mathbb{L} \subset \mathbb{M}$ be three number fields. Prove that
$$[\mathbb{M}:\mathbb{K}] = [\mathbb{M}:\mathbb{L}] \times [\mathbb{L}:\mathbb{K}].$$

2) Let $\mathbb{M} = \mathbb{Q}(\sqrt[3]{2})$. Prove that any element of $\mathbb{M}$ is either rational or algebraic of exact degree 3.

10.18 Let $\mathbb{K} = \mathbb{Q}(\theta)$ be a number field, and let P_θ be the minimal polynomial of θ. Denote by $\langle P_\theta(x) \rangle$ the principal ideal of $\mathbb{Q}[x]$ generated by P_θ, that is

$$\langle P_\theta(x) \rangle = \{ Q \in \mathbb{Q}[x] \ / \ Q(x) = P_\theta(x)R(x), \ R \in \mathbb{Q}[x] \}.$$

Prove that $\mathbb{K}$ and $\mathbb{Q}[x] \big/ \langle P_\theta(x) \rangle$ are isomorphic.

10.19 Ring of integers of $\mathbb{Q}(\sqrt[3]{2})$

1) Let $\mathbb{K} = \mathbb{Q}(\theta)$, $\theta = \sqrt[3]{2}$, and let $x = a + b\theta + c\theta^2$, a, b, $c \in \mathbb{Q}$, be an integer of $\mathbb{K}$. Use the trace and the norm to prove that $3a$, $3b$, $3c \in \mathbb{Z}$.

2) Let σ_1, σ_2, σ_3 be the three monomorphisms of $\mathbb{K}$. Prove that $\sigma_1(x)\sigma_2(x) + \sigma_2(x)\sigma_3(x) + \sigma_1(x)\sigma_3(x) \in \mathbb{Z}$ for every $x \in A_{\mathbb{K}}$. Deduce that $A_{\mathbb{K}} = \mathbb{Z}(\theta)$.

3) Compute the discriminant of $\mathbb{K}$.

10.20 In the cyclotomic field $\mathbb{Q}(\omega)$, $\omega = e^{2i\pi/7}$

Let $\eta = 2\cos(2\pi/7)$. We know by exercise 10.10, question 1, that η is algebraic of degree 3. We put $\mathbb{K} = \mathbb{Q}(\eta)$.

1) Prove that $\mathbb{L} = \mathbb{Q}(\omega)$, $\omega = e^{2i\pi/7}$, is an extension of degree 2 of $\mathbb{K}$.

2) Deduce that $\left(1, \eta, \eta^2\right)$ is an integral basis of $\mathbb{K}$ and compute $\Delta(\mathbb{K})$.

Chapter 11

Ideals

The theory of ideals has been introduced by Kummer (1810-1893) within the scope of his research on Fermat's equation, and later developed by Dedekind (1830-1916). It aims to overcome the fact that the ring $\mathbb{A}_{\mathbb{K}}$ of integers of a number field is not, in general, a unique factorization domain. In section 11.1, we define the fractional ideals and ideals of a number field and the sum and product of two fractional ideals. Section 11.2 is devoted to the study of the arithmetical properties of ideals. In particular, we prove that any ideal can be written, in a unique way, as a product of prime ideals. In section 11.3, we define and study the norm of ideals. In section 11.4, we introduce complements of algebra, which enable us to factorize, in practice, certain principal ideals as a product of prime ideals. We define the class-number in section 11.5, and show how to compute it for an imaginary quadratic field by connecting it with quadratic forms (section 6.5). Finally, we use the class number in order to solve the Mordell equation $y^2 = x^3 + k$ in a number of cases.

11.1   Fractional ideals and ideals of a number field

Let $\mathbb{K} = \mathbb{Q}(\theta)$ be a number field, and let $\mathbb{A}_{\mathbb{K}}$ be the ring of its integers. Let α_1, α_2, ..., $\alpha_n \in \mathbb{K}$. The *fractional ideal generated by* α_1, α_2,..., α_n is defined by

$$A = \langle \alpha_1, \alpha_2,..., \alpha_n \rangle = \{ x_1\alpha_1 + \cdots + x_n\alpha_n \ / \ x_i \in \mathbb{A}_{\mathbb{K}} \}. \qquad (11.1)$$

In other words, A is the $\mathbb{A}_{\mathbb{K}}$-module generated by $\alpha_1,...,$ α_n. We denote by $\mathcal{F}(\mathbb{K})$ the set of all fractional ideals of $\mathbb{K}$.

If the generators α_1, α_2,..., α_n of A belong to $\mathbb{A}_{\mathbb{K}}$, we say simply that A is an *ideal* (meaning an *integral ideal*) of $\mathbb{A}_{\mathbb{K}}$. It is easy to check (see exercise 11.1) that any ideal A of $\mathbb{A}_\mathbb{K}$ is an ideal (in the general sense of (6.2), pages 79-80) of the domain $\mathbb{A}_\mathbb{K}$, which means that

$$\begin{cases} \forall x, y \in A, \quad x - y \in A \\ \forall x \in A, \forall y \in \mathbb{A}_\mathbb{K}, xy \in A \end{cases} \quad (11.2)$$

and that, conversely, any subset A of $\mathbb{A}_\mathbb{K}$ satisfying (11.2) is an ideal in the sense of (11.1). We denote by $\mathfrak{I}(\mathbb{A}_\mathbb{K})$ the set of all ideals of $\mathbb{A}_\mathbb{K}$.

Now let $A = \langle \alpha_1, \alpha_2, ..., \alpha_n \rangle$ and $B = \langle \beta_1, \beta_2, ..., \beta_m \rangle$ be two fractional ideals of $\mathbb{K}$. We define

$$\begin{cases} A + B = \{a + b \,/\, a \in A, b \in B\} \\ AB = \left\{ \sum_{i=1}^q a_i b_i \,/\, q \in \mathbb{N}^*, a_i \in A, b_i \in B \right\} \end{cases} \quad (11.3)$$

Then $A + B$ and AB are fractional ideals of $\mathbb{K}$ (see exercise 11.2) and

$$\begin{cases} A + B = \langle \alpha_1, ..., \alpha_n, \beta_1, ..., \beta_m \rangle \\ AB = \langle \alpha_1 \beta_1, ..., \alpha_1 \beta_m, ..., \alpha_n \beta_1, ..., \alpha_n \beta_m \rangle \end{cases} \quad (11.4)$$

Example 11.1 Consider the quadratic field $\mathbb{K} = \mathbb{Q}(i\sqrt{5})$.

Let $A = \langle 3, 1 + i\sqrt{5} \rangle$, $B = \langle 3, 1 - i\sqrt{5} \rangle$. Then $A + B = \langle 3, 1 + i\sqrt{5}, 1 - i\sqrt{5} \rangle$.

Therefore $2 = (1 + i\sqrt{5}) + (1 - i\sqrt{5}) \in (A + B)$, whence $1 = 3 - 2 \in (A + B)$. Thus $\mathbb{A}_\mathbb{K} = \langle 1 \rangle \subset (A + B)$. As clearly $(A + B) \subset \mathbb{A}_\mathbb{K}$, $A + B = \langle 1 \rangle = \mathbb{A}_\mathbb{K}$. Similarly $AB = \langle 9, 3(1 - i\sqrt{5}), 3(1 + i\sqrt{5}), 6 \rangle$ by (11.4).

Therefore $3 = 9 - 6 \in (AB)$ and $\langle 3 \rangle \subset (AB)$. Now, for every $x \in (AB)$,

$$x = x_1 \cdot 9 + x_2 \cdot 3(1 - i\sqrt{5}) + x_3 \cdot 3(1 + i\sqrt{5}) + x_4 \cdot 6 \ (x_i \in \mathbb{A}_\mathbb{K}).$$

Thus $x = 3(3x_1 + (1 - i\sqrt{5})x_2 + (1 + i\sqrt{5})x_3 + 2x_4) \in \langle 3 \rangle$, and $(AB) \subset \langle 3 \rangle$. So finally $AB = \langle 3 \rangle$. Note that here A and B are ideals of $\mathbb{A}_\mathbb{K}$.

Definition 11.1 *We say that a fractional ideal is principal if it has only one generator, that is if $A = \langle a \rangle$ for some $a \in \mathbb{K}$.*

It is readily seen that $\langle a \rangle \langle b \rangle = \langle ab \rangle$. Thus multiplying elements of $\mathbb{K}$ amounts to multiply the corresponding principal ideals of $\mathbb{K}$.

It is easy to see that $(\mathcal{F}(\mathbb{K}),+)$ is not a group. Although the addition of ideals is commutative and $\langle 0 \rangle$ is the neutral element, only $\langle 0 \rangle$ has an inverse, as it is easy to check. However:

Theorem 11.1 $\mathcal{F}(\mathbb{K})^*$ *is an abelian group under multiplication. The neutral element is* $\langle 1 \rangle = \mathbb{A}_{\mathbb{K}}$.

Proof The reader will check easily that the multiplication of fractional ideals is commutative and associative. Moreover, $\langle 1 \rangle = \mathbb{A}_{\mathbb{K}}$ is the neutral element, since

$$\langle 1 \rangle \cdot \langle \alpha_1, \alpha_2, ..., \alpha_n \rangle = \langle 1 \cdot \alpha_1, ..., 1 \cdot \alpha_n \rangle = \langle \alpha_1, ..., \alpha_n \rangle.$$

The difficult part of the proof consists in proving that every non-zero fractional ideal A has an inverse. See exercise 11.3.

Example 11.2 As in example 11.1, let $\mathbb{K} = \mathbb{Q}(i\sqrt{5})$, $A = \langle 3, 1 + i\sqrt{5} \rangle$, $B = \langle 3, 1 - i\sqrt{5} \rangle$. We have seen that $AB = \langle 3 \rangle$, whence $AB = \langle 3 \rangle \cdot \langle 1 \rangle$. Therefore $AB \langle \frac{1}{3} \rangle = \langle 1 \rangle$ and $A^{-1} = \langle \frac{1}{3} \rangle B = \langle 1, \frac{1-i\sqrt{5}}{3} \rangle$.

Remark 11.1 Multiplication is distributive over addition in $\mathcal{F}(\mathbb{K})$ (see exercise 11.4).

11.2   Arithmetic in $\mathcal{I}(\mathbb{A}_{\mathbb{K}})$

Let $A, B \in \mathcal{I}(\mathbb{A}_{\mathbb{K}})$ be two (integral) ideals. We say that A *divides* B, which we denote by $A \mid B$, if there exists $C \in \mathcal{I}(\mathbb{A}_{\mathbb{K}})$ such that $AC = B$.

Theorem 11.2 *Let* $A, B \in \mathcal{I}(\mathbb{A}_{\mathbb{K}})$. *Then* $A \mid B \Leftrightarrow B \subset A$.

Proof Assume first that $A \mid B$. Then $B = AC$, with $C \in \mathcal{I}(\mathbb{A}_{\mathbb{K}})$. However $AC \subset A$ by (11.2). Therefore $B \subset A$.
Conversely, assume $B \subset A$. If $A = \langle 0 \rangle$, then $B = \langle 0 \rangle$ and clearly $A \mid B$. If not, A is invertible in $\mathcal{F}(\mathbb{K})^*$ by theorem 11.1, and $B \subset A$ yields $BA^{-1} \subset AA^{-1} = \langle 1 \rangle$. Consequently $BA^{-1} \subset \mathbb{A}_{\mathbb{K}}$ and $C = BA^{-1} \in \mathcal{I}(\mathbb{A}_{\mathbb{K}})$. Finally $BA^{-1} = C$ implies $B = AC$, Q.E.D.

Let A, $B \in \mathcal{I}(\mathbb{A}_{\mathbb{K}})$. Their *greatest common divisor* $D = \mathrm{GCD}(A, B)$ is defined by: $D \mid A$, $D \mid B$, and if $D' \mid A$ and $D' \mid B$, then $D' \mid D$.

Theorem 11.3 *Let* A, $B \in \mathcal{I}(\mathbb{A}_{\mathbb{K}})$. *Then* $\mathrm{GCD}(A, B) = A + B$.

Proof We have $(A + B) \supset A$ and $(A + B) \supset B$ since $0 \in A$ and $0 \in B$. Hence $(A + B) \mid A$ and $(A + B) \mid B$ by theorem 11.2. Moreover, if $D' \mid A$ and $D' \mid B$, then $A + B = D'A' + D'B' = D'(A' + B')$, which means that $D' \mid (A + B)$, Q.E.D.

Let A, $B \in \mathcal{I}(\mathbb{A}_{\mathbb{K}})$. We say A and B to be *coprime* if $\mathrm{GCD}(A, B) = \langle 1 \rangle$, in other words if $A + B = \langle 1 \rangle$ by theorem 11.3. This generalizes *Bézout's identity*. Indeed, if A and B are principal ideals of $\mathbb{A}_{\mathbb{K}}$, $A = \langle a \rangle$, $B = \langle b \rangle$, with $a, b \in \mathbb{A}_{\mathbb{K}}$, then $A + B = \{ au + bv \mid (u, v) \in \mathbb{A}_{\mathbb{K}}^2 \}$.

A consequence is *Gauss theorem*:

Theorem 11.4 *Let* A, B, $C \in \mathcal{I}(\mathbb{A}_{\mathbb{K}})$. *Assume that* $A \mid (BC)$ *and that* A *and* B *are coprime. Then* $A \mid C$.

Proof See exercice 11.5.

Definition 11.1 *We say that* $P \in \mathcal{I}(\mathbb{A}_{\mathbb{K}})^*$ *is prime if* $P \neq \langle 1 \rangle$ *and if, for every* $A \in \mathcal{I}(\mathbb{A}_{\mathbb{K}})$, $A \mid P$ *implies* $A = \langle 1 \rangle$ *or* $A = P$.

Now for every $I \in \mathcal{I}(\mathbb{A}_{\mathbb{K}})$, let $\Delta(I)$ be the discriminant of I. We recall from section 10.5 that $\Delta(I)$ is a natural integer.

Lemma 11.1 *Let* A, $B \in \mathcal{I}(\mathbb{A}_{\mathbb{K}})$. *Then* $B \mid A \implies \Delta(B) \mid \Delta(A)$. *Moreover, if* $B \mid A$ *and* $\Delta(B) = \Delta(A)$, *then* $A = B$.

Proof See exercice 11.6.

We can now prove the *fundamental theorem of arithmetic* in $\mathcal{I}(\mathbb{A}_{\mathbb{K}})$.

Theorem 11.5 *Every* $A \in \mathcal{I}(\mathbb{A}_{\mathbb{K}})$, $A \neq \langle 0 \rangle$, $A \neq \langle 1 \rangle$ *is a product of prime ideals. This expression is unique, apart from rearrangement of factors.*

Proof We first prove that A can be expressed as a product of primes.

Let $P_1 \neq \langle 1 \rangle$ be a divisor of A with *least* discriminant. Then P_1 is prime since $Q \mid P_1$ and $Q \neq P_1 \Rightarrow \Delta(Q) < \Delta(P_1)$ by lemma 11.1, whence $Q = \langle 1 \rangle$. Put $A = P_1 A_1$, $A_1 \in \mathfrak{I}(\mathbb{A}_\mathbb{K})$. We have $A_1 \neq A$ since $P_1 \neq \langle 1 \rangle$. Therefore $\Delta(A_1) < \Delta(A)$ by lemma 11.1. If $A_1 \neq \langle 1 \rangle$, we can argue the same way and find $A_2 \in \mathfrak{I}(\mathbb{A}_\mathbb{K})$ such that $\Delta(A_2) < \Delta(A_1)$, and so on. As a sequence of natural integers cannot be decreasing, there exists $n \in \mathbb{N}^*$ such that $A_n = \langle 1 \rangle$, and $A = P_1 P_2 ... P_n$.

Now assume that $A = P_1 P_2 ... P_n = P_1' P_2' ... P_m'$ where the P_i and P_j' are prime. If P_1 and P_1' are coprime, then $P_1 \mid P_2' ... P_m'$ by Gauss theorem 11.4, and the same arguments show that there exists i such that P_1 and P_i' are not coprime. Apart from a rearrangement of factors, we can assume that $D \neq \langle 1 \rangle$ divides P_1 and P_1'. And since P_1 and P_1' are prime, $D = P_1 = P_1'$. After simplification, we get $P_2 ... P_n = P_2' ... P_m'$, and by induction we have $m = n$, $P_i' = P_i$ for every i, which shows that the factorization is unique.

11.3   Norm of an ideal

Let A be an ideal of $\mathbb{A}_\mathbb{K}$. We say that $a \equiv b \pmod{A}$ if $a - b \in A$. This is an equivalence relation in $\mathbb{A}_\mathbb{K}$. The set of equivalence classes is the quotient ring $\mathbb{A}_\mathbb{K}/A$ (see section 6.2).

Lemma 11.2   *Let $A \in \mathfrak{I}(\mathbb{A}_\mathbb{K})$. Then*

$$A \text{ prime} \Leftrightarrow \mathbb{A}_\mathbb{K}/A \text{ is an integral domain.}$$

Proof Assume first that A is prime. Let $x, y \in \mathbb{A}_\mathbb{K}$ such that $xy \equiv 0 \pmod{A}$. Then $xy \in A$, whence $\langle x \rangle \langle y \rangle = \langle xy \rangle \subset A$. This means that $A \mid \langle x \rangle \langle y \rangle$ by theorem 11.2. Now A is prime, and Gauss theorem 11.4 implies that $A \mid \langle x \rangle$ or $A \mid \langle y \rangle$. Therefore $x \in A$ or $y \in A$, in other words $x \equiv 0 \pmod{A}$ or $y \equiv 0 \pmod{A}$. Thus $\mathbb{A}_\mathbb{K}/A$ is an integral domain. Conversely, assume that $\mathbb{A}_\mathbb{K}/A$ is an integral domain. Let $D \in \mathfrak{I}(\mathbb{A}_\mathbb{K})$, with $D \mid A$, $D \neq A$. We want to prove that $D = \langle 1 \rangle$. Put $A = BD$,

$B \in \mathfrak{I}(\mathbb{A}_{\mathbb{K}})$. Let $x \in D - A$. Such an x exists since $A \subset D$ et $D \neq A$. For every y in B, we have $xy \in (BD) = A$, whence $xy \equiv 0 \pmod{A}$. Since $\mathbb{A}_{\mathbb{K}}/A$ is an integral domain, this yields $x \equiv 0 \pmod{A}$ or $y \equiv 0 \pmod{A}$. But $x \notin A$, and therefore $y \in A$. Thus $B \subset A$. On the other hand, $B \mid A$, whence $A \subset B$. This yields $B = A$. Consequently $A = AD$, $D = \langle 1 \rangle$ and A is prime.

Theorem 11.6 *Let $A \in \mathfrak{I}(\mathbb{A}_{\mathbb{K}})$, $A \neq \langle 0 \rangle$. Then $\mathbb{A}_{\mathbb{K}}/A$ is finite.*

Proof We first construct a "diagonal" basis of the $\mathbb{Z}$-module A (see section 10.5). Let $d = [\mathbb{K} : \mathbb{Q}]$, and let $(\alpha_1, \alpha_2, ..., \alpha_d)$ be an integral basis of $\mathbb{K}$. Put $\rho_1 = c_{11}\alpha_1$, where c_{11} is the least positive rational integer such that $c_{11}\alpha_1 \in A$ (c_{11} exists since $|N(x)|\alpha_1 = \pm x\sigma_2(x)...\sigma_d(x)\alpha_1 \in A$ for every $x \in A^*$ by lemma 10.1).

Now we put $\rho_2 = c_{21}\alpha_1 + c_{22}\alpha_2$, where c_{22} is the least positive rational integer such that there exists $c_{21} \in \mathbb{Z}$ satisfying $c_{21}\alpha_1 + c_{22}\alpha_2 \in A$ (c_{22} exists because $|N(x)|(\alpha_1 + \alpha_2) \in A$ for every $x \in A^*$).

By arguing the same way, we construct a sequence $\rho_1, \rho_2, ..., \rho_d$ of elements of A of the form $\rho_i = \sum_{j=1}^{i} c_{ij}\alpha_i$, with the property:

$$\left(x = \sum_{j=1}^{i} x_{ij}\alpha_j \in A, \text{ with } x_{ij} \in \mathbb{Z}, x_{ii} > 0 \right) \Rightarrow (x_{ii} \geq c_{ii} > 0). \quad (11.5)$$

It is clear that $\rho_1, \rho_2, ..., \rho_d$ are linearly independent over $\mathbb{Q}$. Now we show that $A = \mathbb{Z}(\rho_1, \rho_2, ..., \rho_d)$. Let $x \in A$. Then $x = x_1\alpha_1 + \cdots + x_d\alpha_d$ ($x_i \in \mathbb{Z}$). Put $x_d = c_{dd}q_d + r_d$ ($0 \leq r_d < c_{dd}$). We have

$$x = \sum_{i=1}^{d-1} (x_i - q_d c_{di})\alpha_i + q_d\rho_d + r_d\alpha_d.$$

Since $\rho_d \in A$, $x - q_d\rho_d = \sum_{i=1}^{d-1} x_i'\alpha_i + r_d\alpha_d \in A$, whence $r_d = 0$ by (11.5). By arguing the same way with $x - q_d\rho_d$, we see that $x - q_d\rho_d - q_{d-1}\rho_{d-1} \in A$, and so on. Thus by induction $x = \sum_{i=1}^{d} q_i\rho_i$, $q_i \in \mathbb{Z}$, which proves that $(\rho_1, \rho_2, ..., \rho_d)$ is a basis of the $\mathbb{Z}$-module A.

We now show that $E = \{ r_1\alpha_1 + \cdots + r_d\alpha_d \; / \; 0 \le r_i < c_{ii} \}$ is a complete set of residues modulo A, that is that every $x \in \mathbb{A}_\mathbb{K}$ is congruent mod A to an element of E, and that two distinct elements of E are not congruent mod A.

First let $x = x_1\alpha_1 + \cdots + x_d\alpha_d$ $(a_i \in \mathbb{Z})$ be an element of $\mathbb{A}_\mathbb{K}$. The above computations show that $x = \sum_{i=1}^{d} q_i\rho_i + \sum_{i=1}^{d} r_i\alpha_i$, $0 \le r_i < c_{ii}$, $q_i \in \mathbb{Z}$.

Thus $x \equiv \sum_{i=1}^{d} r_i\alpha_i$ (mod A).

Assume now that $\sum_{i=1}^{d} r_i\alpha_i \equiv \sum_{i=1}^{d} r_i'\alpha_i$ (mod A), $0 \le r_i < c_{ii}$, $0 \le r_i' < c_{ii}$ for every i. We may assume that $r_d \le r_d'$. Then $\sum_{i=1}^{d} (r_d' - r_i)\alpha_i \in A$, with $0 \le r_d' - r_d < c_{dd}$. Therefore $r_d' - r_d = 0$ by (11.5). Similarly, by induction we show that $r_i = r_i'$ for every i. Thus E is a complete set of residues mod A and $\mathbb{A}_\mathbb{K}/A$ is finite. Actually, $|\mathbb{A}_\mathbb{K}/A| = |E| = c_{11}c_{22}\ldots c_{dd}$.

Definition 11.2 *Let A be any non zero ideal of $\mathbb{A}_\mathbb{K}$. The norm of A is defined as the number of elements of $\mathbb{A}_\mathbb{K}/A$, that is*

$$N(A) = |\mathbb{A}_\mathbb{K}/A| \in \mathbb{N}^*.$$

Theorem 11.6 immediatly yields the three following corollaries.

Corollary 11.1 *Let $A \in \mathfrak{I}(\mathbb{A}_\mathbb{K})$, $A \ne \langle 0 \rangle$. Then $\Delta(A) = (N(A))^2\Delta(\mathbb{K})$.*

Proof With the notations of theorem 11.6, we have by lemma 10.4
$$\Delta[\rho_1, \rho_2, \ldots, \rho_d] = (\det(c_{ij}))^2 \Delta[\alpha_1, \alpha_2, \ldots, \alpha_d],$$
that is $\Delta(A) = (N(A))^2\Delta(\mathbb{K})$ since $N(A) = |\mathbb{A}_\mathbb{K}/A| = c_{11}c_{22}\ldots c_{dd}$.

Corollary 11.2 *For every $a \in \mathbb{A}_\mathbb{K}^*$, $N(\langle a \rangle) = |N(a)|$.*

In other words, in the case of principal ideals, the norm of ideals has the same value, up to the sign, as the ordinary norm. To prove corollary 11.2, it suffices to observe that, for every $x \in \langle a \rangle$,

$$x = ay = ay_1\alpha_1 + \cdots + ay_d\alpha_d, \quad y_i \in \mathbb{Z}.$$

Therefore $(a\alpha_1, \ldots, a\alpha_d)$ is a basis of the $\mathbb{Z}$ - module $\langle a \rangle$. Hence

$$\Delta(\langle a\rangle) = \begin{vmatrix} \sigma_1(a\alpha_1) & \dots & \sigma_1(a\alpha_d) \\ \sigma_2(a\alpha_1) & \dots & \sigma_2(a\alpha_d) \\ & \vdots & \\ \sigma_d(a\alpha_1) & \dots & \sigma_d(a\alpha_d) \end{vmatrix}^2 = \left(\sigma_1(a)\dots\sigma_d(a)\right)^2 \Delta(\mathbb{K}).$$

Comparing to corollary 11.1 shows that $N(\langle a\rangle) = |N(a)|$, as claimed.

Corollary 11.3 *Let* $A \in \mathfrak{I}(\mathbb{A}_\mathbb{K})$, $A \neq \langle 0\rangle$. *Then* A *is prime if, and only if,* $\mathbb{A}_\mathbb{K}/A$ *is a field* .

This is a direct consequence of lemma 11.2 and theorems 6.5 and 11.6

Remark 11.2 Corollary 11.2 shows that $N(\langle 1\rangle) = 1$. Conversely, if $N(A) = 1$, $\mathbb{A}_\mathbb{K}/A$ has only one element, which is 0. Therefore every $x \in \mathbb{A}_\mathbb{K}$ is congruent to 0 (mod A), whence $x \in A$. Thus $\mathbb{A}_\mathbb{K} \subset A$, and as $A \subset \mathbb{A}_\mathbb{K}$, $A = \mathbb{A}_\mathbb{K} = \langle 1\rangle$. Consequently

$$\forall A \in \mathfrak{I}(\mathbb{A}_\mathbb{K})^*, \quad N(A) = 1 \Leftrightarrow A = \langle 1\rangle = \mathbb{A}_\mathbb{K}. \tag{11.6}$$

Finally, we leave it to the reader to check (exercise 11.7) that *the norm in* $\mathfrak{I}(\mathbb{A}_\mathbb{K})^*$, *as the norm in* $\mathbb{K}$, *is multiplicative*. In other words,

$$\forall A,\ B \in \mathfrak{I}(\mathbb{A}_\mathbb{K})^*, \quad N(AB) = N(A)N(B).$$

Example 11.3 Let $\mathbb{K} = \mathbb{Q}(i\sqrt{5})$, $A = \langle 3, 1+i\sqrt{5}\rangle$, $B = \langle 3, 1-i\sqrt{5}\rangle$. From example 11.1 we know that $AB = \langle 3\rangle$. Therefore $N(A)N(B) = N(\langle 3\rangle)$ $= N(3) = 9$. Hence $N(A) = 1$, 3 or 9. But $N(A) = 1$ is impossible since $A \neq \langle 1\rangle$. Indeed, $A = \langle 1\rangle \Rightarrow B = \langle 3\rangle$, which is impossible because $N(3) = 9$ does not divide $N(1-i\sqrt{5}) = 6$. Similarly $N(A) = 9$ is impossible since $B \neq \langle 1\rangle$. Therefore $N(A) = 3$. And corollary 11.1 and (10.19) now yield $\Delta(A) = 9 \cdot (-20) = -180$

Remark 11.3 Relations (11.7) and (11.6) imply (exercise 11.8) that
$$N(A)\ prime \Rightarrow A\ prime.$$
This is the case, for example, for the ideals A and B in example 11.3. Thus, the factorization of the principal ideal $\langle 3\rangle$ as a product of primes is

$\langle 3 \rangle = \langle 3, 1 + i\sqrt{5} \rangle \cdot \langle 3, 1 - i\sqrt{5} \rangle$. Here it should be remembered (example 5.3) that the ring $\mathbb{A}_\mathbb{K} = \mathbb{Z}(i\sqrt{5})$ is not a unique factorization domain.

11.4   Factorization of $\langle p \rangle$, p prime, as a product of prime ideals

In this section, we prove a theorem which will enable us to find the factorization of $\langle p \rangle$, p prime, $p \in \mathbb{Z}$, when $\mathbb{A}_\mathbb{K} = \mathbb{Z}(\theta)$. This covers, among others, the cases of quadratic fields (theorem 5.3) and cyclotomic fields (theorem 10.4). Some algebraic preliminaries will be necessary.

Let $\mathbb{A}$ and $\mathbb{B}$ be two domains, and let $\varphi : \mathbb{A} \to \mathbb{B}$. Recall that φ is a *ring homomorphism* if, for every $(a,b) \in \mathbb{A}^2$, $\varphi(a+b) = \varphi(a) + \varphi(b)$ and $\varphi(ab) = \varphi(a)\varphi(b)$. The *kernel* of φ is $\ker \varphi = \{a \in \mathbb{A} \ / \ \varphi(a) = 0\}$, and $\ker \varphi = \{0\} \Leftrightarrow \varphi$ injective. Finally, if φ is bijective from $\mathbb{A}$ to $\mathbb{B}$, we say that φ is an *isomorphism* from $\mathbb{A}$ to $\mathbb{B}$. In this case, $\mathbb{A}$ and $\mathbb{B}$ are said to be *isomorphic*, and we denote it by $A \cong B$.

Lemma 11.3   *Let $\mathbb{A}$ and $\mathbb{B}$ be two domains, and let φ be a ring homomorphism from $\mathbb{A}$ to $\mathbb{B}$. Then $\ker \varphi$ is an ideal of $\mathbb{A}$. Moreover, if φ is surjective, then $\mathbb{B} \cong \mathbb{A}/\ker \varphi$.*

Proof   The reader will check (exercise 11.9) that $\ker \varphi$ is an ideal of $\mathbb{A}$. Now define $\Psi : \mathbb{A}/\ker \varphi \to \mathbb{B}$, for every equivalence class $\overline{a}$, by $\Psi(\overline{a}) = \varphi(a)$. We must check that $\Psi(\overline{a})$ does not depend on the choice of a. Now let a and a' satisfying $a \equiv a'$ (mod $\ker \varphi$). We have $a - a' \in \ker \varphi$, whence $\varphi(a - a') = \varphi(a) - \varphi(a') = 0$ and $\varphi(a) = \varphi(a')$. Thus Ψ is well defined as a mapping from $\mathbb{A}/\ker \varphi$ to $\mathbb{B}$. Moreover we have

$$\Psi(\overline{a} + \overline{b}) = \Psi(\overline{a + b}) = \varphi(a + b) = \varphi(a) + \varphi(b) = \Psi(\overline{a}) + \Psi(\overline{b}),$$

and $\Psi(\overline{ab}) = \Psi(\overline{ab}) = \varphi(ab) = \varphi(a)\varphi(b) = \Psi(\overline{a})\Psi(\overline{b})$. Hence Ψ is a ring homomorphism from $\mathbb{A}/\ker \varphi$ to $\mathbb{B}$.

Since φ is surjective, for every $b \in \mathbb{B}$ there is $a \in \mathbb{A}$ such that $b = \varphi(a) = \Psi(\overline{a})$, and Ψ is surjective.

Finally, let $\bar{a} \in \ker \Psi$. Then $\Psi(\bar{a}) = 0 \Rightarrow \varphi(a) = 0 \Rightarrow a \in \ker \varphi \Rightarrow a \equiv 0$ (mod $\ker \varphi$) $\Rightarrow \bar{a} = 0$ and Ψ is injective. Therefore $\mathbb{A}/\ker \varphi \cong \mathbb{B}$.

Now let $\mathbb{K}$ be any commutative field, and let $\mathbb{K}[x]$ be the ring of polynomials with coefficients in $\mathbb{K}$. We know by exercise 6.20 that $\mathbb{K}[x]$ is an euclidean ring. Hence it is a principal ideal domain and a unique factorization domain. The invertible elements (or units) are the constant polynomials, in other words $\mathbb{K}[x]^{\times} = \mathbb{K}$

Let $P \in \mathbb{K}[x]$ be irreducible (see definition 5.1), and let $\langle P(x) \rangle$ be the principal ideal of $\mathbb{K}[x]$ generated by $P(x)$. We will prove the following *adjunction* result:

Lemma 11.4 *Let $\mathbb{K}$ be a commutative field, and let $P \in \mathbb{K}[x]$ be irreducible. Then $\mathbb{K}[x]/\langle P(x) \rangle$ is a commutative field, and there exists α such that $P(\alpha) = 0$ and $\mathbb{K}(\alpha) \cong \mathbb{K}[x]/\langle P(x) \rangle$.*

This result generalizes exercise 10.18. However, in the case where $\mathbb{K} = \mathbb{Q}$, the adjunction of α is not a problem thanks to d'Alembert's theorem.

Now we prove lemma 11.4. Since $\langle P(x) \rangle$ is the principal ideal generated by $P(x)$ in $\mathbb{K}(x)$, we know that $\mathbb{K}[x]/\langle P(x) \rangle$ is a domain. It remains to prove that any non zero element is invertible. Let $Q \in \mathbb{K}[x]$, $Q \not\equiv 0$ (mod P). Then Q is not a multiple of P. Since P is irreducible, P and Q are coprime. Hence there exist $R, S \in \mathbb{K}[x]$ satisfying $P(x)R(x) + Q(x)S(x) = 1$ (Bézout's identity, see exercise 6.20). Thus $Q(x)S(x) \equiv 1$ (mod P), and Q has an inverse in $\mathbb{K}[x]/\langle P(x) \rangle$.

Now denote by α the equivalence class of x modulo $P(x)$. Then we have $P(\alpha) \equiv P(x) \equiv 0$ (mod P), whence $P(\alpha) = 0$.

The mapping $\varphi : \mathbb{K}[x] \to \mathbb{K}(\alpha)$ defined by $\varphi(Q(x)) = Q(\alpha)$ is clearly a surjective homomorphism. Therefore $\mathbb{K}(\alpha) \cong \mathbb{K}[x]/\langle P(x) \rangle$ by lemma 11.3. Indeed, $Q(\alpha) = 0 \Leftrightarrow Q(x) \equiv 0$ (mod $P(x)$) $\Leftrightarrow Q \in \langle P(x) \rangle$. Thus $\mathbb{K}(\alpha)$ is a *commutative field*.

Now let $p \in \mathbb{Z}$ be prime. Denote $\mathbb{F}_p = \mathbb{Z}/p\mathbb{Z}$. Then $\mathbb{F}_p$ is a commutative field with p elements (see example 6.4). For every $P \in \mathbb{Z}[x]$, we denote by $\overline{P} \in \mathbb{F}_p[x]$ the residue of P modulo p.

Theorem 11.7   *Let* $\mathbb{K} = \mathbb{Q}(\theta)$ *be a number field of degree d, such that* $\mathbb{A}_{\mathbb{K}} = \mathbb{Z}(\theta)$. *Let* $p \in \mathbb{Z}$ *be prime. Assume that the minimal polynomial* P_θ *of* θ *factorizes in* $\mathbb{F}_p[x]$ *as the following product of irreducibles:*

$$\overline{P_\theta}(x) = \left(\overline{f_1}(x)\right)^{r_1} \left(\overline{f_2}(x)\right)^{r_2} \ldots \left(\overline{f_k}(x)\right)^{r_k}.$$

Then, for every $i = 1, 2, \ldots, k$, *the ideal* $P_i = \langle p, f_i(\theta) \rangle$ *is prime and*

$$\langle p \rangle = P_1^{r_1} P_2^{r_2} \ldots P_k^{r_k}.$$

Proof   For every $i = 1, 2, \ldots, k$, let α_i such that $\overline{f_i}(\alpha_i) = 0$ (lemma 11.4). Let $v_i : \mathbb{Z}(\theta) \to \mathbb{F}_p(\alpha_i)$, defined by $v_i(P(\theta)) = \overline{P}(\alpha_i)$. Then v_i is a surjective ring homomorphism (see exercise 11.10). Hence $\ker v_i$ is an ideal P_i of $\mathbb{Z}(\theta) = \mathbb{A}_{\mathbb{K}}$ and $\mathbb{A}_{\mathbb{K}}/P_i \cong \mathbb{F}_p(\alpha_i)$ (lemma 11.3). But $\mathbb{F}_p(\alpha_i)$ is a field by lemma 11.4. Thus P_i is a prime ideal (corollary 11.3).
Now we show that $P_i = \langle p, f_i(\theta) \rangle$. First we have $\langle p \rangle \subset P_i$, since

$$P(\theta) \in \langle p \rangle \;\Rightarrow\; P(\theta) = pQ(\theta) \;\Rightarrow\; v_i(P(\theta)) = 0 \;\Rightarrow\; P(\theta) \in \ker v_i = P_i.$$

Moreover, $\langle f_i(\theta) \rangle \subset P_i$, since

$$P(\theta) \in \langle f_i(\theta) \rangle \;\Rightarrow\; P(\theta) = f_i(\theta)Q(\theta) \;\Rightarrow\; v_i(P(\theta)) = \overline{f}_i(\alpha_i)\overline{Q}(\alpha_i) = 0.$$

Therefore $\langle p, f_i(\theta) \rangle \subset P_i$.
Conversely, let $P(\theta) \in P_i$. Then $v_i(P(\theta)) = \overline{P}(\alpha_i) = 0$. Since $\overline{f_i}$ is irreducible in $\mathbb{F}_p[x]$, this implies that $\overline{P}(x) = \overline{f_i}(x)\overline{Q}(x)$. Therefore all coefficients of $P(x) - f_i(x)Q(x) \in \mathbb{Z}[x]$ are multiples of p. Hence

$$P(\theta) = \left(P(\theta) - f_i(\theta)Q(\theta)\right) + f_i(\theta)Q(\theta) \in \langle p, f_i(\theta) \rangle.$$

Therefore $P_i = \langle p, f_i(\theta) \rangle$ and P_i is prime.
Now we show that $\langle p \rangle = P_1^{r_1} P_2^{r_2} \ldots P_k^{r_k}$. Observe that, for $a, b, c \in \mathbb{A}_{\mathbb{K}}$, $\langle a, b \rangle \langle a, c \rangle = \langle a^2, ab, ac, bc \rangle \subset \langle a, bc \rangle$ since a^2, ab, ac are multiples of a.

Therefore $P_1^{r_1} P_2^{r_2} ... P_k^{r_k} \subset \langle p, f_1(\theta)^{r_1} f_2(\theta)^{r_2} ... f_k(\theta)^{r_k} \rangle$ by induction. But

$f_1(\theta)^{r_1} f_2(\theta)^{r_2} ... f_k(\theta)^{r_k} \equiv P_\theta(\theta) \equiv 0$ (mod p), whence $P_1^{r_1} ... P_k^{r_k} \subset \langle p \rangle$.

Thus $\langle p \rangle | P_1^{r_1} ... P_k^{r_k}$ by theorem 11.2, and since the factorization as a product of prime ideals is unique, we obtain

$$\langle p \rangle = P_1^{s_1} P_2^{s_2} ... P_k^{s_k}, \text{ with } s_i \le r_i, \ i = 1, 2, ..., k. \quad (11.8)$$

Now we use the norms. By corollary 11.2, $N(\langle p \rangle) = |N(p)| = p^d$. And

$N(P_i) = |\mathbb{A}_\mathbb{K} / P_i| = |\mathbb{F}_p(\alpha_i)|$ since $\mathbb{A}_\mathbb{K} / P_i \cong \mathbb{F}_p(\alpha_i)$. Moreover, if we put

$d_i = \deg \bar{f}_i$, we see that any element of $\mathbb{F}_p(\alpha_i)$ can be written as

$a_0 + a_1 \alpha_i + \cdots + a_{d_i - 1} \alpha_i^{d_i - 1}$, $a_i \in \mathbb{F}_p$, whence $|\mathbb{F}_p(\alpha_i)| = p^{d_i} = N(P_i)$.

Now taking the norms in (11.8) yields

$$p^d = N(P_1)^{s_1} N(P_2)^{s_2} ... N(P_k)^{s_k} = p^{s_1 d_1 + \cdots + s_k d_k}.$$

Thus $d = s_1 d_1 + \cdots + s_k d_k$. But $d = \deg P_\theta$ implies $d \ge r_1 d_1 + \cdots + r_k d_k$.

Since $r_i \ge s_i$, we deduce that $r_i = s_i$ for all i. Theorem 11.7 is proved.

Example 11.4 Let $\mathbb{K} = \mathbb{Q}(i\sqrt{5})$. Then, by theorem 5.3, $\mathbb{A}_\mathbb{K} = \mathbb{Z}(i\sqrt{5})$.

Theorem 11.7 enables us to find quickly the factorization of $\langle 3 \rangle$ as a product of primes, already laboriously obtained in remark 11.3.

Here $\theta = i\sqrt{5}$, $P_\theta(x) = x^2 + 5$. We reduce P_θ mod 3:

$$\overline{P_\theta}(x) = x^2 - 1 = (x+1)(x-1).$$

Theorem 11.7 applies, the ideals $\langle 3, 1+i\sqrt{5} \rangle$ and $\langle 3, -1+i\sqrt{5} \rangle$ are prime

and $\langle 3 \rangle = \langle 3, 1+i\sqrt{5} \rangle \cdot \langle 3, -1+i\sqrt{5} \rangle$.

11.5 Class group and class number

Let $\mathbb{K}$ be a number field. We know by theorem 11.3 that the non-zero fractional ideals of $\mathbb{K}$ form a group under multiplication, denoted by $\mathcal{F}(\mathbb{K})^*$. The set of non-zero *principal ideals* of $\mathbb{K}$ is a subgroup $\mathcal{P}(\mathbb{K})$ of $\mathcal{F}(\mathbb{K})^*$, since $\langle a \rangle \langle b \rangle = \langle ab \rangle$ and $\langle a \rangle^{-1} = \langle a^{-1} \rangle$. By definition, the (ideal) *class group* of $\mathbb{K}$ is the quotient group

$$\mathcal{C}(\mathbb{K}) = \mathcal{F}(\mathbb{K})^* / \mathcal{P}(\mathbb{K}). \quad (11.9)$$

We say that two fractional ideals A and B are *equivalent* if they are equivalent modulo $\mathcal{P}(K)$, in other words

$$\forall A,\ B \in \mathcal{F}(\mathbb{K})^*, \quad A \sim B \iff \exists \alpha \in \mathbb{K}^* \text{ such that } A = \langle \alpha \rangle B. \quad (11.10)$$

For the (integral) ideals of $\mathbb{A}_\mathbb{K}$, (11.10) becomes

$$\forall A,\ B \in \mathcal{I}(\mathbb{K})^*, \quad A \sim B \iff \exists a,b \in \mathbb{A}_\mathbb{K} \text{ such that } \langle a \rangle A = \langle b \rangle B. \quad (11.11)$$

Indeed, any $\alpha \in \mathbb{K}$ can be written $\alpha = b/a$, $b \in \mathbb{A}_\mathbb{K}$, $a \in \mathbb{N} \subset \mathbb{A}_\mathbb{K}$.

Theorem 11.8 *The class group* $\mathcal{C}(\mathbb{K})$ *is finite.*

To prove theorem 11.8, we will need three lemmas.

Lemma 11.5 *Let* $m \in \mathbb{N}^*$ *be given. Then there exist only finitely many ideals* $A \in \mathcal{I}(\mathbb{A}_\mathbb{K})$ *such that* $N(A) = m$.

Proof Exercise 11.11.

Lemma 11.6 *Let* $(\omega_1, \omega_2, ..., \omega_d)$ *be an integral basis of* $\mathbb{K}$, *and let* $\sigma_1,\ \sigma_2, ...,\ \sigma_d$ *be the monomorphisms of* $\mathbb{K}$. *Define*

$$M = \prod_{j=1}^d \left(|\sigma_j(\omega_1)| + |\sigma_j(\omega_2)| + \cdots + |\sigma_j(\omega_d)| \right). \quad (11.12)$$

Then, for every $A \in \mathcal{I}(\mathbb{A}_\mathbb{K})^*$, *there exists* $\alpha \in A^*$ *such that*

$$|N(\alpha)| \leq M \cdot N(A).$$

Proof Let $k = [N(A)^{1/d}]$ be the integer part of $N(A)^{1/d}$. Then we have $k^d \leq N(A) < (k+1)^d$. Let $a_1, a_2, ..., a_d \in \{0,1,...,k\}$. The number of all numbers $x = a_1\omega_1 + \cdots + a_d\omega_d$ is $(k+1)^d > N(A)$. Since $N(A)$ is the number of equivalence classes mod A, two of these numbers x and x', $x \neq x'$, are congruent mod A by the pigeon-hole principle. Thus $\alpha = x - x' = \sum_{i=1}^d (a_i - a_i')\omega_i \in A^*$, and

$$|N(\alpha)| = \left| \prod_{j=1}^d \sum_{i=1}^d (a_i - a_i')\sigma_j(\omega_i) \right|$$

$$\leq \prod_{j=1}^d \left(\sum_{j=1}^d |a_i - a_i'| |\sigma_j(\omega_i)| \right) \leq k^d M \leq M \cdot N(A).$$

Lemma 11.7 *Let* M *be defined in* (11.12). *For every* $A \in \mathcal{I}(\mathbb{A}_\mathbb{K})^*$, *there exists* $B \in \mathcal{I}(\mathbb{A}_\mathbb{K})^*$ *such that* $A \sim B$ *and* $N(B) \leq M$.

Let δ be a common denominator to all generators of the fractional ideal A^{-1}. Then $C = \langle \delta \rangle A^{-1} \in \mathcal{J}(A_{\mathbb{K}})$ and $AC = \langle \delta \rangle$. By lemma 11.6, there exists $\gamma \in C^*$ such that $|N(\gamma)| \leq M \cdot N(C)$. But

$$\langle \gamma \rangle \subset C \;\Rightarrow\; C \,|\, \langle \gamma \rangle \;\Rightarrow\; \langle \gamma \rangle = BC, \text{ with } B \in \mathcal{J}(A_{\mathbb{K}})^*.$$

Since AC and BC are principal ideals, we have $AC \sim BC$, which yields $A \sim B$ since $\mathcal{C}(\mathbb{K})$ is a multiplicative group. And we see that

$$N(B) = N(\langle \gamma \rangle)/N(C) = |N(\gamma)|/N(C) \leq M.$$

Now theorem 11.8 results easily from lemmas 11.7 and 11.5. Indeed, the number of equivalence classes of fractional ideals is the same as the number of equivalence classes of (integral) ideals (see exercise 11.12).

We denote by $h(\mathbb{K})$ the number of ideal classes (in $\mathbb{K}$ or in $A_{\mathbb{K}}$, which is the same), and call it the *class number* of $\mathbb{K}$. Thus, by definition,

$$h(\mathbb{K}) = |\mathcal{C}(\mathbb{K})|. \tag{11.13}$$

It is generally difficult to compute the class number. However, in the case where $\mathbb{K} = \mathbb{Q}(i\sqrt{d})$ is an imaginary quadratic field, we have the following striking result.

Theorem 11.9 *Let $\mathbb{K} = \mathbb{Q}(i\sqrt{d})$ and let Δ be the discriminant of $\mathbb{K}$. Then $h(\mathbb{K})$ is equal to the number of classes of positive definite quadratic forms with discriminant Δ.*

Proof See exercise 11.13.

Example 11.5 If $d \equiv 1$ or $2 \pmod 4$, then $\Delta(\mathbb{K}) = -4d$, and if $d \equiv 3 \pmod 4$, then $\Delta(\mathbb{K}) = -d$ by (10.19) and (10.20). By using table 6.2, we see that $h(\mathbb{Q}(i)) = 1$, $h(\mathbb{Q}(i\sqrt{2})) = 1$, $h(\mathbb{Q}(i\sqrt{3})) = 1$, $h(\mathbb{Q}(i\sqrt{5})) = 2$, $h(\mathbb{Q}(i\sqrt{6})) = 2$, $h(\mathbb{Q}(i\sqrt{7})) = 1$, and so on.

The following result (Kummer) is fundamental for solving diophantine equations.

Theorem 11.10 *Let $A \in \mathcal{J}(A_{\mathbb{K}})^*$, and let $p \in \mathbb{N}$ be prime. Assume that A^p is principal and that p does not divide $h(\mathbb{K})$. Then A is principal.*

Proof Since A^p is principal, $A^p \sim \langle 1 \rangle$. Therefore the order of the equivalence class of A in the group $\mathcal{C}(\mathbb{K})$ is 1 or p, because p is prime. But we know from example 6.3 that the order of any element of a finite group divides the cardinal of this group. As p does not divide $h(\mathbb{K})$, the order of the equivalence class of A is 1, whence $A \sim \langle 1 \rangle$, and this means that A is principal.

Theoreme 11.11 $\mathbb{A}_{\mathbb{K}}$ *is a principal ideal domain if and only if* $h(\mathbb{K}) = 1$.

Proof If $\mathbb{A}_{\mathbb{K}}$ is a principal ideal domain, every ideal of $\mathbb{A}_{\mathbb{K}}$ is principal, therefore equivalent to $\langle 1 \rangle$ and $h(\mathbb{K}) = 1$.

Conversely, if $h(\mathbb{K}) = 1$, every $A \in \mathcal{I}(\mathbb{A}_{\mathbb{K}})^*$ is equivalent to $\langle 1 \rangle$ since $\mathcal{C}(\mathbb{K})$ has only one element, and therefore A is principal.

Corollary 11.4 *The ring of integers of* $\mathbb{K} = \mathbb{Q}\left(i\sqrt{d}\right)$ *is principal, and therefore a unique factorization domain, for*
$$d = 1,\ 2,\ 3,\ 7,\ 11,\ 19,\ 43,\ 67,\ 163.$$

Proof This results directly from theorem 11.9 and table 6.2 for $d = 1$, 2, 3, 7, 11, 19, 43 and 67. For $d = 163$, see exercise 11.14.

It should be noticed that $\mathbb{A}_{\mathbb{K}}$ is euclidean, and therefore principal, if $d = 1$, 2, 3, 7, 11 by theorem 5.16, but it is not if $d = 19$, 43, 67, 163.

It can be proved that corollary 11.4 gives *all* the imaginary quadratic fields $\mathbb{K}$ such that $\mathbb{A}_{\mathbb{K}}$ is a unique factorization domain. This result has been obtained independently by Baker and Stark in 1966.

11.6   Application to Mordell's equation $y^2 = x^3 + k$

Theorem 11.12 *Let k be a squarefree rational integer satisfying $k < -1$ and $k \equiv 2$ or 3 (mod 4). Assume that the class number $h\left(\mathbb{Q}(\sqrt{k})\right)$ is not divisible by 3. Then Mordell's equation $y^2 = x^3 + k$ is soluble if, and only if, $k = \pm 1 - 3a^2$, $a \in \mathbb{N}^*$. In this case, the solutions are $x = a^2 - k$, $y = \pm a(a^2 + 3k)$.*

Proof Let $(x, y) \in \mathbb{Z}^2$ be a solution of $y^2 = x^3 + k$. First notice that x and y must be coprime. Indeed, if any prime p divides x and y, then $p^2 | k$, which is impossible since k is squarefree. Moreover, x must be odd. If not, we would have $y^2 \equiv k \pmod 4$, which is impossible because the only squares mod 4 are 0 and 1. Now we write Mordell's equation as $(y + \sqrt{k})(y - \sqrt{k}) = x^3$ and we transform it by introducing principal ideals in $\mathbb{K} = \mathbb{Q}(\sqrt{k})$. We obtain

$$\langle y + \sqrt{k} \rangle \langle y - \sqrt{k} \rangle = \langle x \rangle^3. \tag{11.14}$$

Now assume that $\langle y + \sqrt{k} \rangle$ and $\langle y - \sqrt{k} \rangle$ are not coprime. Let P be a prime ideal which divides $\langle y + \sqrt{k} \rangle$ and $\langle y - \sqrt{k} \rangle$. Then $y + \sqrt{k} \in P$, $y - \sqrt{k} \in P$, $2y \in P$, whence $P | \langle 2y \rangle$. Moreover, by (11.14), $P | \langle x \rangle$, since it is one of the prime factors of $\langle x \rangle^3$. Taking the norms yields $N(P) | N(2y) = 4y^2$, $N(P) | N(x) = x^2$. Since x is odd, $N(P)$ is also odd, and therefore $N(P) | y^2$, $N(P) | x^2$. Now $P \neq \langle 1 \rangle$, whence $N(P) \neq 1$ by (11.6). Thus x and y are not coprime, contradiction.

So the principal ideals $\langle y + \sqrt{k} \rangle$ and $\langle y - \sqrt{k} \rangle$ are coprime. As the factorization as a product of prime ideals is unique, (11.14) yields

$$\langle y + \sqrt{k} \rangle = A^3, \ A \in \mathfrak{I}(\mathbb{A}_{\mathbb{K}}). \tag{11.15}$$

Therefore A^3 is principal. Since 3 does not divide $h(\mathbb{K})$, A is principal (theorem 11.10). Hence there exists $(a, b) \in \mathbb{Z}^2$ such that $A = \langle a + b\sqrt{k} \rangle$. Indeed, here $\mathbb{A}_{\mathbb{K}} = \mathbb{Z}(\sqrt{k})$ by theorem 5.3 since $d \equiv 2$ or $3 \pmod 4$. Now (11.15) yields $\langle y + \sqrt{k} \rangle = \langle \left(a + b\sqrt{k} \right)^3 \rangle$. However:

For any number field $\mathbb{K}$, *the equality* $\langle \alpha \rangle = \langle \beta \rangle$, *with* $(\alpha, \beta) \in \mathbb{A}_{\mathbb{K}}^2$, *implies that* $\alpha = \varepsilon \beta$, *where* ε *is a unit of* $\mathbb{A}_{\mathbb{K}}$ (see exercise 11.15).

In our case the only units of $\mathbb{A}_{\mathbb{K}}$ are 1 and -1 by theorem 5.5. Therefore $y + \sqrt{k} = \pm \left(a + b\sqrt{k} \right)^3$, and we obtain

$$y = \pm a(a^2 + 3kb^2), \quad 1 = \pm b(3a^2 + kb^2) \tag{11.16}$$

The second equation yields $b = \pm 1$ and $k = \pm 1 - 3a^2$.

Therefore $y = \pm a(a^2 + 3k)$. Replacing in Mordell's equation yields

$$x^3 = a^6 + 6a^4 k + 9a^2 k^2 - k. \tag{11.17}$$

But $(k + 3a^2)^2 = 1$, and therefore $k = k^3 + 6a^2 k^2 + 9a^4 k$. Now (11.17) becomes $x^3 = a^6 - 3a^4 k + 3a^2 k^2 - k^3$, that is $x = a^2 - k$.

Remark 11.4   This proof highlights the two main problems which arise when using ideals for solving diophantine equations:

a)   It is necessary to go from the *ideals* of $\mathbb{A}_{\mathbb{K}}$ to the *elements* of $\mathbb{A}_{\mathbb{K}}$. The main tool for this is theorem 11.10.

b)   It is necessary to have some knowledge about the *units* of $\mathbb{A}_{\mathbb{K}}$.

These two problems have been raised by Kummer as early as 1847. The interested reader will find an introduction to Kummer's work in the first chapter of [14]. For a detailled study, see [7] or [23].

The proof of theorem 10.9 we have given is a special case of Kummer's method. It uses the fact that $\mathbb{Q}(e^{2i\pi/5})$ is an euclidean domain (lemma 10.5), which implies that $h\left(\mathbb{Q}(e^{2i\pi/5})\right) = 1$. In this case, it is not necessary to invoke the ideal theory. However, we must still know something about the units of $\mathbb{Q}(e^{2i\pi/5})$ (see lemma 10.6).

Exercises

11.1   Prove that $A \in \mathfrak{J}(\mathbb{A}_{\mathbb{K}})$ if and only if A satisfies (11.2).

11.2   Prove (11.4).

11.3   **Proof of theorem 11.1**

1) Let $P(x) = a_0 + a_1 x + \cdots + a_n x^n$, $a_n \neq 0$, where the a_i's are algebraic integers. Let $\rho \in \mathbb{C}$ be a root of P. Prove by induction on n that the coefficients of $P(x)/(x - \rho)$ are algebraic integers.

2) Let $A(x) = a_0 + a_1 x + \cdots + a_n x^n$, $B(x) = b_0 + \cdots + b_m x^m$, where the a_i's and the b_i's are algebraic integers, such that $a_n b_m \neq 0$. Define $C(x) = A(x)B(x) = c_0 + \cdots + c_\ell x^\ell$. By theorem 9.4, we know that the c_i's are algebraic integers. Assume that there exists an algebraic integer δ such that c_i/δ is an algebraic integer for all $i = 0, 1, \ldots, \ell$. Prove that $a_i b_j/\delta$ is an algebraic integer for every (i, j), $0 \leq i \leq n$, $0 \leq j \leq m$.

3) Let $A = \langle \alpha_0, \ldots, \alpha_n \rangle \in \mathcal{I}(\mathbb{A}_{\mathbb{K}})$, $\alpha_n \neq 0$. Let $\sigma_1 = Id$, $\sigma_2, \ldots, \sigma_d$ be the monomorphisms of $\mathbb{K}$. Define $f(x) = \alpha_0 + \alpha_1 x + \cdots + \alpha_n x^n$, and
$$g(x) = \prod_{k=2}^{d} (\sigma_k(\alpha_0) + \sigma_k(\alpha_1)x + \cdots + \sigma_k(\alpha_n)x^n) = \beta_0 + \cdots + \beta_{(d-1)n} x^{(d-1)n}.$$
a) Prove that $f(x)g(x) = \gamma_0 + \cdots + \gamma_{nd} x^{nd} \in \mathbb{Z}[x]$ and that $g(x) \in \mathbb{A}_{\mathbb{K}}[x]$.
b) Let $B = \langle \beta_0, \ldots, \beta_{(d-1)n} \rangle$, $a = \mathrm{GCD}(\gamma_0, \ldots, \gamma_{nd})$. Prove that $\langle a \rangle \subset AB$.
c) Prove that $a \mid \alpha_i \beta_j$ for every couple (i, j). Deduce that $AB = \langle a \rangle$.
4) Prove that every fractional ideal has an inverse.

11.4 Prove that multiplication is distributive over addition in $\mathcal{F}(\mathbb{K})$.

11.5 Prove theorem 11.4 (Gauss theorem).

11.6 Prove lemma 11.1 by using the results of section 10.5.

11.7 Multiplicativity of the norm in $\mathcal{I}(\mathbb{A}_{\mathbb{K}})^*$

1) Let $A, P \in \mathcal{I}(\mathbb{A}_{\mathbb{K}})^*$ be two non zero ideals, with P prime. Prove that there exists $\alpha \in A$ such that $\alpha \notin (AP)$, and that there exists an ideal B such that $AB = \langle \alpha \rangle$, $P \nmid B$, $P + B = \langle 1 \rangle$. Deduce that $A = \langle \alpha \rangle + AP$.

2) Let $(a_1, a_2, \ldots, a_{N(A)})$ be a complete system of residues modulo A, and $(p_1, p_2, \ldots, p_{N(P)})$ be a complete system of residues modulo P. Prove that $(a_1 + \alpha p_1, \ldots, a_1 + \alpha p_{N(P)}, a_2 + \alpha p_1, \ldots, a_{N(A)} + \alpha p_1, \ldots, a_{N(A)} + \alpha p_{N(P)})$ is a complete system of residues modulo (AP).

3) Prove that the norm is multiplicative in $\mathcal{I}(\mathbb{A}_{\mathbb{K}})^*$

11.8 Prove that $N(A)$ prime $\Rightarrow A$ prime.

11.9 Assume that $\mathbb{A}$ and $\mathbb{B}$ are two domains, and let $\varphi : \mathbb{A} \to \mathbb{B}$ be a ring homomorphism. Prove that $\ker \varphi$ is an ideal of $\mathbb{A}$.

11.10 Prove that v_i, in the proof of theorem 11.7, is a surjective ring homomorphism.

11.11 Let $A \in \mathcal{J}(\mathbb{A}_{\mathbb{K}})^*$. By starting from the basis $(\rho_1, \rho_2, ..., \rho_d)$ of the $\mathbb{Z}$ - module A considered in the proof of theorem 11.6, prove that there exists a basis $(\omega_1, \omega_2, ..., \omega_d)$ of the $\mathbb{Z}$ - module A such that $\omega_i = c'_{i1}\alpha_1 + \cdots + c'_{i,i-1}\alpha_{i-1} + c_{ii}\alpha_i$, where $0 \le c'_{ij} < c_{jj}$ for all $j < i$. Deduce that there are only finitely many ideals A with a given norm.

11.12 Prove that the number of equivalence classes in $\mathcal{J}(\mathbb{A}_{\mathbb{K}})$ is the same that the number of equivalence classes in $\mathcal{F}(\mathbb{K})$.

11.13 Proof of theorem 11.9
Let $\mathbb{K} = \mathbb{Q}(\sqrt{d})$, $d < 0$, be an imaginary quadratic field with discriminant $\Delta = \Delta(\mathbb{K})$.
1) Let $A \in \mathcal{J}(\mathbb{A}_{\mathbb{K}})$. Prove that there is a basis of the $\mathbb{Z}$ - module A of the form $(\alpha, \theta\alpha)$, with $\alpha, \theta \in \mathbb{K}$, $\mathrm{Im}(\theta) > 0$.
2) Let $A \in \mathcal{J}(\mathbb{A}_{\mathbb{K}})$, with base $(\alpha, \theta\alpha)$, $\alpha, \theta \in \mathbb{K}$, $\mathrm{Im}(\theta) > 0$. Consider the quadratic form $Q(x, y) = (\alpha x + \theta\alpha y)(\overline{\alpha}x + \overline{\theta\alpha}y) = px^2 + qxy + ry^2$, where $\overline{\alpha}$ denotes the complex conjugate of α, which is equal here to the algebraic conjugate α^* of α in $\mathbb{K}$.
    a) Prove that $p, q, r \in \mathbb{Z}$.
    b) Prove that $q^2 - 4pr = N(A)^2 \Delta$.
    c) Prove that $N(A)$ divides p, q and r.
3) We put $p = N(A)a$, $q = N(A)b$, $r = N(A)c$, and we associate to A the quadratic form $R(x, y) = ax^2 + bxy + cy^2$.
    a) Check that the discriminant of R is Δ and that
$$R(x, y) = a(x + \theta y)(x + \overline{\theta}y).$$

b) Let $A_1 = (\alpha_1, \theta_1\alpha_1)$ and $A_2 = (\alpha_2, \theta_2\alpha_2)$ be two equivalent ideals in $\mathbb{A}_\mathbb{K}$, and let $R_1 = (a_1, b_1, c_1)$ and $R_2 = (a_2, b_2, c_2)$ be the associated quadratic forms. Prove that there exist rational integers s, t, u, v, and $\lambda_1, \lambda_2 \in \mathbb{A}_\mathbb{K}$, satisfying $sv - ut = 1$ and

$$\begin{cases} \lambda_1\alpha_1 = \lambda_2(s\alpha_2 + t\theta_2\alpha_2) \\ \lambda_1\theta_1\alpha_1 = \lambda_2(u\alpha_2 + v\theta_2\alpha_2) \end{cases}$$

c) Deduce that $A \to R(x, y)$ defines a mapping Ψ from the set of equivalence classes of ideals in $\mathbb{A}_\mathbb{K}$ to the set of equivalence classes of quadratic forms with discriminant Δ.

4) Prove that Ψ is surjective.

5) Prove that Ψ is injective and conclude.

11.14 Prove that $h\left(\mathbb{Q}(i\sqrt{163})\right) = 1$.

11.15 Let $\mathbb{K}$ be a number field. Prove that, for every $(\alpha, \beta) \in \mathbb{A}_\mathbb{K}^{*2}$,

$$\langle \alpha \rangle = \langle \beta \rangle \iff \exists \varepsilon \in \mathbb{A}_\mathbb{K}^{\times} \text{ such that } \alpha = \varepsilon\beta.$$

11.16 Let $\mathbb{K} = \mathbb{Q}(i\sqrt{13})$. Factorize the principal ideal $\langle 10 \rangle$ as a product of prime ideals in $\mathbb{A}_\mathbb{K}$. Check the result by multiplying the prime ideals.

11.17 Let $\mathbb{K} = \mathbb{Q}(\sqrt{51})$. Let $A = \langle 2, 1 + \sqrt{51} \rangle$.

1) Determine the fractional ideal A^{-1}.

2) Prove that A is prime in $\mathbb{A}_\mathbb{K}$.

3) Compute $N(A)$.

11.18 Let $\mathbb{K}$ be a number field, and $A \in \mathcal{I}(\mathbb{A}_\mathbb{K})$. Prove that $N(A) \in A$.

11.19 Let $\mathbb{K} = \mathbb{Q}(\omega)$, $\omega = e^{2i\pi/5}$. Express the ideals $\langle 3 \rangle$ and $\langle 5 \rangle$ as a product of prime ideals.

11.20 Let $\mathbb{K}$ be a number field, and let M be defined in lemma 11.6, formula (11.12). Assume that, for every prime $p \in \mathbb{Z}$ satisfying $p \le M$, every prime ideal which divides $\langle p \rangle$ is principal. Prove that $h(\mathbb{K})=1$, which means that $\mathbb{A}_{\mathbb{K}}$ is principal.

11.21 Use the result of exercise 11.20 to prove that the ring of integers $\mathbb{A}_{\mathbb{K}} = \mathbb{Z}(\sqrt{7})$ of the quadratic field $\mathbb{K} = \mathbb{Q}(\sqrt{7})$ is principal.

11.22 Solve the diophantine equation $y^2 = x^3 - 13$.

Chapter 12

Introduction to transcendence methods

Hermite has been the first to prove the transcendence of a classical number, namely e, in 1873. By improving Hermite's method, Lindemann succeeded in proving the transcendence of π in 1882, thus showing the impossibility of the quadrature of the circle. This last chapter is an introduction to the transcendence methods developed in the 1930's by Gelfond, Mahler, Schneider and Siegel. Sections 12.1 and 12.2 are devoted to preliminaries: algebraic functions, house of an algebraic number, size inequality. Mahler's method is the simplest transcendence method and is presented in section 12.3. It allows us to describe, in section 12.4, the main steps of a transcendence proof. Next, we prove Hermite-Lindemann theorem (which implies the transcendence of e and π) in section 12.5 as an illustration of Gelfond's method. In section 12.6, we prove the transcendence of a^b by Schneider's method, and finally give an idea of Siegel-Shidlovsky method in section 12.6.

12.1   Algebraic and transcendental functions

Definition 12.1 *Let* $\mathbb{K}$ *be any commutative field, and let* $\mathbb{A}$ *be any commutative ring containing* $\mathbb{K}$. *We say that* $\alpha \in \mathbb{A}$ *is algebraic over* $\mathbb{K}$ *if there exists* $P \in \mathbb{K}[x]$ *such that* $P(\alpha) = 0$ *and* $P \neq 0$.

For instance, algebraic numbers are the elements of $\mathbb{C}$ which are algebraic over $\mathbb{Q}$.

Now let Ω be any open subset of the complex plane, and let f be an analytic function in Ω. We say that f is *algebraic* if it is algebraic over the field $\mathbb{K} = \mathbb{C}(x)$ of the rational functions with complex coefficients, in other words if there exist polynomials $P_0, P_1, \ldots, P_d \in \mathbb{C}[x]$, not all zero, such that for every $x \in \Omega$,

$$P_d(x)\big(f(x)\big)^d + P_{d-1}(x)\big(f(x)\big)^{d-1} + \cdots + P_0(x) = 0. \qquad (12.1)$$

Example 12.1 Every rational function is algebraic of degree 1.

Example 12.2 $f(x) = (1+x)^{1/2}$ is algebraic of degree 2, since

$$\big(f(x)\big)^2 - (1+x) = 0.$$

When an element of $\mathbb{A}$ is not algebraic over $\mathbb{K}$, we say that it is *transcendental* over $\mathbb{K}$. We have already given, in exercise 1.12 and in chapter 9, transcendence proofs for certain complex numbers. However, the easiest proofs of transcendence deal with the transcendence of functions. For instance, we have the two following lemmas.

Lemma 12.1 *The exponential function $f(x) = e^x$ is transcendental.*

Proof Suppose it to be algebraic, and write

$$P_d(x)e^{dx} + P_{d-1}(x)e^{(d-1)x} + \cdots + P_0(x) = 0, \qquad (12.2)$$

where we choose d to be least, whence $P_d \neq 0$. Then, for any $x \in \mathbb{C}$,

$$P_d(x) = -P_{d-1}(x)e^{-x} - \cdots - P_0(x)e^{-dx}. \qquad (12.3)$$

Let $x \in \mathbb{R}$, $x \to +\infty$. Then the right-hand side of (12.3) vanishes, which implies $\lim_{x \to +\infty} P_d = 0$, a contradiction.

Lemma 12.2 *Let $\Omega = \{x \in \mathbb{C} \ / \ |x| < 1\}$. Then the function*

$$f(x) = \sum_{n=0}^{+\infty} x^{2^n}, \qquad (12.4)$$

which is analytic in Ω, is transcendental.

Proof First we observe that f satisfies a very simple functional equation, namely

$$f(x^2) = \sum_{n=0}^{+\infty} x^{2^{n+1}} = f(x) - x. \qquad (12.5)$$

Now assume that f is algebraic. Then, for every $x \in \Omega$,

$$\big(f(x)\big)^d + Q_{d-1}(x)\big(f(x)\big)^{d-1} + \cdots + Q_0(x) = 0, \qquad (12.6)$$

where the Q_i's are rational functions with complex coefficients, and d is chosen to be least. In (12.6), replace x by x^2 and use (12.5). We get

$$\big(f(x) - x\big)^d + Q_{d-1}(x^2)\big(f(x) - x\big)^{d-1} + \cdots + Q_0(x^2) = 0.$$

Expanding the left-hand member yields

$$\left(f(x)\right)^{d} + \left(Q_{d-1}(x^{2}) - dx\right)\left(f(x)\right)^{d-1} + \cdots = 0.$$

Substract this to (12.6). Since d is least, we get $Q_{d-1}(x) = Q_{d-1}(x^{2}) - dx$. Put $Q_{d-1}(x) = A(x)/B(x)$, with A and B coprime. We have

$$A(x)B(x^{2}) = A(x^{2})B(x) - dxB(x)B(x^{2}). \qquad (12.7)$$

Thus $B(x^{2})|A(x^{2})B(x)$. As $B(x^{2})$ and $A(x^{2})$ are coprime, this implies $B(x^{2})|B(x)$. Looking at the degree of B yields $B(x) = b \in \mathbb{C}^{*}$ and (12.7) becomes $A(x) = A(x^{2}) - bdx$. If $\deg A \geq 1$, this is impossible. Therefore $\deg A = 0$ and $A(x) = a \in \mathbb{C}$, which implies $b = 0$, a contradiction.

12.2   House of an algebraic number

Most of irrationality proofs rest on the fact that a sequence of positive integers cannot vanish (see section 1.6). Or, which is all the same,

$$\alpha \in \mathbb{Z} - \{0\} \;\Rightarrow\; |\alpha| \geq 1. \qquad (12.8)$$

We have seen in remark 8.3 that this property does not remain true if we replace $\mathbb{Z}$ by the ring $\mathbb{A}_{\mathbb{K}}$ of integers of a number field $\mathbb{K}$, except if $\mathbb{K}$ is an imaginary quadratic field (see lemma 8.2). However, in transcendence proofs, we will have to use the most general number fields. This explains the following definition:

Definition 12.2 *Let* $\alpha \in \mathbb{C}$ *be algebraic of degree* d, *and let* $\alpha_{1} = \alpha$, $\alpha_{2}, \ldots,$ α_{d} *be the conjugates of* α, *that is the zeros of its minimal polynomial* P_{α}. *The house of* α *is the positive real number defined by*

$$\overline{|\alpha|} = \max_{i} |\alpha_{i}|. \qquad (12.9)$$

Theorem 12.1   *Let* α *be an algebraic number, and* $\mathbb{K}$ *be any number field containing* α. *Let* $\sigma_{1}, \ldots,$ σ_{m} *be the monomorphisms of* $\mathbb{K}$. *Then*

$$\overline{|\alpha|} = \max_{i} |\sigma_{i}(\alpha)|. \qquad (12.10)$$

Proof   See exercice 12.1.

Theorem 12.2   *Let* α *and* β *be two algebraic numbers. Then*

$$\overline{|\alpha + \beta|} \leq \overline{|\alpha|} + \overline{|\beta|} \qquad (12.11)$$

$$\overline{|\alpha\beta|} \leq \overline{|\alpha|}.\overline{|\beta|} \qquad (12.12)$$

Proof   See exercise 12.2.

Theorem 12.3   *Let α be a non-zero algebraic integer. Then $\overline{|\alpha|} \geq 1$.*

Proof   See exercice 12.3.

Theorem 12.3 plainly generalizes (12.8). However, for transcendence proofs, we will rather use the *size inequality*, which applies to any algebraic number α (not only to algebraic integers), and involves, in addition to the house of α, the denominator den (α) (see theorem 9.3):

Theoreme 12.4   (Size inequality)
Let α be a non-zero algebraic number of degree d. Then

$$|\alpha| \geq \left(\overline{|\alpha|}\right)^{-d+1} \left(\text{den}\,(\alpha)\right)^{-d}. \tag{12.13}$$

Proof   See exercice 12.4.

12.3   Mahler's transcendence method (1929)

It enables to prove the transcendence of values, at algebraic points, of functions satisfying functional equations similar to (12.5). We will show how it works by using the example of the function f defined in (12.4).

Theorem 12.5   *Let α be a non-zero algebraic number, with $|\alpha| < 1$. Then $f(\alpha) = \sum_{n=0}^{+\infty} \alpha^{2^n}$ is transcendental.*

Proof   *Step one.* It is easy to see that $f(x)$, $\left(f(x)\right)^2$, $\left(f(x)\right)^3$,..., are power series with rational integer coefficients. Put

$$\left(f(x)\right)^k = \sum_{n=0}^{+\infty} b_{kn} x^n, \quad b_{kn} \in \mathbb{Z}.$$

Let $m \in \mathbb{N}^*$ be fixed (we will choose its value later). Then there exist $P_0, P_1,..., P_m \in \mathbb{Z}[x]$, not all zero, denoted by

$$P_i(x) = \sum_{n=0}^{m} a_{in} x^n, \quad i = 0, 1,..., m,$$

and a power series $g_m(x)$, such that

$$P_m(x)\left(f(x)\right)^m + P_{m-1}(x)\left(f(x)\right)^{m-1} + \cdots + P_0(x) = x^{m^2} g_m(x). \tag{12.14}$$

Indeed, in order to fulfill (12.14), we have to choose the $P_i(x)$ in such a way that, in the sums and products of the left-hand side, the coefficient of x^n vanishes for $n = 0, 1, 2,..., m^2 - 1$, in other words

$$\sum_{i=0}^{m} \sum_{k=0}^{\min(m,n)} a_{ik} b_{i,n-k} = 0, \quad n = 0, 1,..., m^2 - 1. \qquad (12.15)$$

This is a system of m^2 linear equations with $(m+1)^2$ unknowns (the a_{ik}'s), with coefficients in $\mathbb{Z}$ (the $b_{i,n-k}$'s). As the number of unknowns is greater than the number of equations, this system has a non-zero solution in the field $\mathbb{Q}$, and therefore in $\mathbb{Z}$ after multiplication by a common denominator (compare to lemma 1.1). This yields (12.14).

Moreover, g_m is not identically zero. If it was, f would be algebraic since at least one of the P_i's is not zero, contradiction with lemma 12.2. Therefore we can write

$$g_m(x) = x^\sigma h_m(x), \quad \sigma \geq 0, \quad h_m(0) \neq 0. \qquad (12.16)$$

Step two. Suppose that $f(\alpha)$ is algebraic. Let $\mathbb{K} = \mathbb{Q}(\alpha, f(\alpha))$, $[\mathbb{K} : \mathbb{Q}] = d$. Let $a = \mathrm{den}\,(\alpha)$ be the denominator of α, in such a way that $\alpha = \beta/a$, with $\beta \in \mathbb{A}_{\mathbb{K}}$. Similarly, if $b = \mathrm{den}\,(f(\alpha))$, then we can write $f(\alpha) = \gamma/b$, with $\gamma \in \mathbb{A}_{\mathbb{K}}$.

Taking into account the fact that, for every integer $n \geq 1$,

$$f(x^{2^n}) = f(x) - \sum_{k=0}^{n-1} x^{2^k}, \qquad (12.17)$$

which results from (12.5) by an easy induction, we see that

$$\forall n \in \mathbb{N}^*, \quad f(\alpha^{2^n}) = A_n \Big/ \left(ba^{2^{n-1}} \right), \quad \text{with } A_n \in \mathbb{A}_{\mathbb{K}}.$$

Replacing x by α^{2^n} in (12.14) and (12.16) yields

$$\sum_{i=0}^{m} P_i \left(\frac{\beta^{2^n}}{a^{2^n}} \right) \left(\frac{A_n}{ba^{2^{n-1}}} \right)^i = \alpha^{(m^2+\sigma)2^n} h_m(\alpha^{2^n}). \qquad (12.18)$$

Denote by B_n the left-hand member of (12.18). Then $B_n \in \mathbb{K}$. Since the P_i's have integer coefficients and $\deg P_i \leq m$, we see that

$$\mathrm{den}\,(B_n) \leq a^{m2^n} b^m a^{m2^{n-1}} \leq b^m a^{m2^{n+1}}. \qquad (12.19)$$

Step three. Let $n \to +\infty$. Since $h_m(0) \neq 0$, (12.18) yields

$$B_n \underset{+\infty}{\sim} h_m(0)\alpha^{(m^2+\sigma)2^n}. \qquad (12.20)$$

This implies that, for every large n,
$$B_n \neq 0, \tag{12.21}$$
and there exists $c_1 > 0$, independant on n, such that, for any $n \in \mathbb{N}$,
$$|B_n| \leq c_1 |\alpha|^{m^2 2^n}. \tag{12.22}$$

Step four. We now look for an upper bound for $\overline{|B_n|}$. By theorem 12.2, we can write, since $a_{ij} \in \mathbb{Z}$ for every i, j,

$$\overline{|B_n|} \leq \sum_{i=0}^{m} \left(\sum_{j=0}^{m} \overline{|a_{ij}|} \, \overline{|\alpha|}^{j 2^n} \right) \left(\overline{|f(\alpha^{2^n})|} \right)^i.$$

Put $c_2 = \max_{i,j} |a_{ij}|$, $c_3 = \max \left(1, \overline{|\alpha|}, \overline{|f(\alpha)|} \right)$. Then, by exercice 12.5,

$$\overline{|B_n|} \leq c_2 (m+1)^2 c_3^{m 2^{n+1}}. \tag{12.23}$$

Conclusion. Since $B_n \neq 0$ by (12.21), we can apply the size inequality (12.13). We have $\deg(B_n) \leq d$ since $B_n \in \mathbb{K}$. Taking (12.19), (12.22) and (12.23) into account, we obtain

$$c_1 |\alpha|^{m^2 2^n} \geq \left(c_2 (m+1)^2 c_3^{m 2^{n+1}} \right)^{-d+1} \left(b^m a^{m 2^{n+1}} \right)^{-d}.$$

Taking the logarithms yields
$$-md \operatorname{Log} b + (-d+1) \operatorname{Log} \left(c_2 (m+1)^2 \right)$$
$$+ 2^n \left[-m^2 \operatorname{Log} |\alpha| + 2m \left((-d+1) \operatorname{Log} c_3 - d \operatorname{Log} a \right) \right] \leq \operatorname{Log} c_1. \tag{12.24}$$

Since $|\alpha| < 1$, we can choose m such that
$$-m^2 \operatorname{Log} |\alpha| + 2m \left((-d+1) \operatorname{Log} c_3 - d \operatorname{Log} a \right) > 0.$$

Then the left-hand member of (12.24) tends to $+\infty$ when $n \to +\infty$, which is impossible since it is bounded. Theorem 12.5 is proved.

For details about Mahler's method, see Nishioka's book [19].

Remark 12.1 Exercise 1.12 consists in a very elementary proof of the transcendence of $f(1/2)$.

12.4 Methodological remarks and Siegel's lemma

The proof of theorem 12.5 allows us to highlight the main five steps in a transcendence proof.

a) Construction of an auxiliary function, generally a non explicit one (contrary to what happens in most irrationality proofs). Here, this auxiliary function is $x^{m^2} g_m(x)$ (see (12.14)).

b) Display of an algebraic number β connected with this auxiliary function. Here, the number B_n.

c) Computation of upper bounds for $|\beta|$, den (β), $\overline{|\beta|}$.

d) Proof that $\beta \neq 0$. When using Mahler's method, this part of the proof is simple thanks to (12.20). But generally, it is the most difficult part. One has to prove preliminary results known as *zero lemmas*.

e) Use of the size inequality (12.13), which leads to a contradiction.

Now return to the first step. Generally, it consists in solving a system of linear equations with coefficients in the ring of integers $\mathbb{A}_{\mathbb{K}}$ of a number field $\mathbb{K}$. For example, in the proof of theorem 12.5, $\mathbb{A}_{\mathbb{K}} = \mathbb{Z}$. When applying Mahler's method, this turns out to be a simple task, because one only has to assert the *existence* of the solutions $a_{ik} \in \mathbb{Z}$. We do not need to *estimate* them. In all other methods, it will be necessary to use the so-called *Siegel's lemma* (1930):

Lemma 12.3 *Let* (A_{ij}), $1 \leq i \leq m$, $1 \leq j \leq n$, *be rational integers with* $n > m$. *Let* $A \in \mathbb{N}$ *such that* $\max_{i,j} |A_{ij}| \leq A$.

Then there exists $(x_1, x_2, ..., x_n) \in \mathbb{Z}^n$ *satisfying the two conditions*

$$0 < \max_i |x_i| \leq (nA)^{\frac{m}{n-m}}, \text{ and} \qquad (12.25)$$

$$\begin{cases} A_{11}x_1 + A_{12}x_2 + \cdots + A_{1n}x_n = 0 \\ A_{21}x_1 + A_{22}x_2 + \cdots + A_{2n}x_n = 0 \\ \vdots \\ A_{m1}x_1 + A_{m2}x_2 + \cdots + A_{mn}x_n = 0 \end{cases} \qquad (12.26)$$

Proof We will use the pigeon-hole principle. For $1 \leq j \leq m$, let $-V_j$ (respectively W_j) be the sum of all $A_{ij} \leq 0$ (respectively $A_{ij} \geq 0$). Note that the A_{ij}'s are the coefficients of the j-th equation in (12.26). Then clearly $V_j + W_j \leq nA$. Now let $X \in \mathbb{N}$. To each lattice point $(x_1,...,x_n)$ $\in \mathbb{Z}^n$ satisfying $0 \leq x_i \leq X$ ($1 \leq i \leq n$), we associate the lattice point $(y_1,...,y_m) \in \mathbb{Z}^n$ defined by $y_j = \sum_{i=1}^{n} A_{ij} x_i$ ($1 \leq j \leq m$).

Then $-V_j X \leq y_j \leq W_j X$ for every $j = 1, ..., m$.

Thus the set E of all $(x_1,...,x_n)$ has $(X+1)^n$ elements, whereas the set F of all $(y_1,...,y_m)$ has at most $(nAX+1)^m$ elements.

Choose $X = \left[(nA)^{\frac{m}{n-m}} \right]$. Then $(X+1)^{n-m} > (nA)^m$. Consequently

$$(X+1)^n > (X+1)^m (nA)^m \geq (nAX+1)^m.$$

Therefore $|E| > |F|$, and the mapping $(x_1,...,x_n) \to (y_1,...,y_n)$ is not injective from E to F. Hence there must exist two points $(x_1',...,x_n')$ and $(x_1'',...,x_n'')$ in E, at least, which have the same image $(y_1,...,y_m)$. Then, for all $j = 1, ..., m$, $\sum_{i=1}^{n} A_{ij}(x_i' - x_i'') = 0$, and $x_i = x_i' - x_i''$ fulfills (12.25) and (12.26), Q.E.D.

Refining these arguments leads to *Siegel's lemma in number fields*:

Lemma 12.4   *Let* $\mathbb{K}$ *be a number field of degree* d. *Let* (A_{ij}), $1 \leq i \leq m$, $1 \leq j \leq n$, *be integers of* $\mathbb{K}$, *with* $n > dm$.

Let $A \in \mathbb{N}$ *satisfying* $\max_{i,j} \overline{|A_{ij}|} \leq A$.

Then there exist a real number $M > 0$, *independent on* n *and* m, *and a solution* $(x_1, x_2,..., x_n) \in \mathbb{Z}^n$ *of* (12.26) *satisfying*

$$0 < \max_i |x_i| \leq (nMA)^{\frac{dm}{n-dm}}.$$

Proof   See exercice 12.6.

12.5   Hermite-Lindemann theorem

Theorem 12.6   *Let* $\alpha \in \mathbb{C}^*$ *be algebraic. Then* e^α *is transcendental.*

Corollary 12.1   *e and* π *are transcendental.*

The corollary is evident for e. If π was algebraic, so would be $i\pi$ since i is algebraic. Hence $e^{i\pi} = -1$ would be transcendental, contradiction.

Now we prove Hermite-Lindemann theorem by using Gelfond's method (1934).

Let n be an integer. First we construct an auxiliary function of the form

$$\begin{cases} F_n(x) = \sum_{i=0}^{p} \sum_{j=0}^{q} a_{ij} x^i e^{jx}, \\ a_{ij} \in \mathbb{Z}, \quad p = [n/\mathrm{Log}\, n], \quad q = \left[(\mathrm{Log}\, n)^2 \right]. \end{cases} \tag{12.27}$$

Suppose that α and e^α are both algebraic. Put $\mathbb{K} = \mathbb{Q}(\alpha, e^\alpha)$ and $d = \deg \mathbb{K}$. Let a be a common denominator of α and e^α. Then, by exercise 12.7, there exist a_{ij} in (12.27) such that

$$F_n(0) = F_n'(0) = \cdots = F_n^{(n-1)}(0) = 0, \tag{12.28}$$

$$F_n(\alpha) = F_n'(\alpha) = \cdots = F_n^{(n-1)}(\alpha) = 0, \tag{12.29}$$

$$0 < \max_{i,j} |a_{ij}| \le e^n \quad \text{for } n \text{ sufficiently large.} \tag{12.30}$$

Since the exponential function is transcendental by lemma 12.1 and the a_{ij}'s are not all zero, the function F_n is not identically zero. Therefore there exists an integer k such that $F_n^{(k)}(0) \ne 0$. Define m as the largest integer satisfying

$$F_n(0) = \cdots = F_n^{(m-1)}(0) = F_n(\alpha) = \cdots = F_n^{(m-1)}(\alpha) = 0. \tag{12.31}$$

Clearly $m \ge n$, and there exists $r \in \{0,1\}$ such that

$$\beta = F_n^{(m)}(r\alpha) \ne 0. \tag{12.32}$$

Now we look for an upper bound for $|\beta|$, in view of applying the size inequality. Let us consider

$$G_n(x) = \frac{F_n(x)}{x^m (x - \alpha)^m}. \tag{12.33}$$

Then G_n is an entire function by (12.31). By using the *maximum modulus principle*, we get (see exercise 12.8 for details)

$$|\beta| \le m^{-\frac{m}{6}}. \tag{12.34}$$

Moreover, it can be checked (see exercise 12.9) that

$$\mathrm{den}(\beta) \le a^{\frac{m}{\mathrm{Log}\, m} + (\mathrm{Log}\, m)^2}, \tag{12.35}$$

$$|\overline{\beta}| \le e^{3m \mathrm{Log}(\mathrm{Log}\, m)}. \tag{12.36}$$

By taking the logarithms, the size inequality (12.3) yields

$$\frac{m}{6} \operatorname{Log} m \le d \left(\frac{m}{\operatorname{Log} m} + (\operatorname{Log} m)^2 \right) \operatorname{Log} a + 3m(d-1) \operatorname{Log}(\operatorname{Log} m).$$

This is inconsistant when n (and therefore m) is sufficiently large. This contradiction proves Hermite-Lindemann theorem.

12.6   Gelfond-Schneider theorem

Let α be any non-zero complex number. Then there exists one and only one value of the argument of α, denoted by $\operatorname{Arg}\alpha$, satisfying $-\pi < \operatorname{Arg}\alpha \le \pi$. Recall that the *principal value of the complex logarithm* is defined on $\mathbb{C}^*$ by

$$\operatorname{Log}\alpha = \operatorname{Log}|\alpha| + i\operatorname{Arg}\alpha. \tag{12.37}$$

For every $\alpha \in \mathbb{C}^*$ and every $z \in \mathbb{C}$, α^z is defined by

$$\alpha^z = e^{z \operatorname{Log}\alpha}. \tag{12.38}$$

It is clear that $f_\alpha(z) = \alpha^z$ is an entire function. Moreover, if α is different from 1, it is a transcendental function. To show it, it suffices to suppose it is algebraic, replace z by $x/\operatorname{Log}\alpha$ and use lemma 12.1.

Proving the transcendence of α^β when α and β are both algebraic was one of the famous problems set by Hilbert at the international congress of mathematics in Paris in 1900. It has been solved independently by Gelfond and Schneider in 1934:

Theorem 12.7   *Let α be an algebraic number different from 0 or 1. Let β be an algebraic irrational number. Then α^β is transcendental.*

Corollary 12.2 e^π *is transcendental.*

Indeed, $(-1)^{-i} = e^{-i\operatorname{Log}(-1)} = e^\pi$ is transcendental since -1 and $-i$ are algebraic, while i is irrational. Note that assuming β to be irrational is necessary in theorem 12.7 since, for example, $\sqrt{2} = 2^{1/2}$ is algebraic.

Now we prove theorem 12.7 by using Schneider's method. We argue by *reductio ad absurdum* and assume that $\gamma = \alpha^\beta$ is algebraic. Let $\mathbb{K}$ denote the number field $\mathbb{K} = \mathbb{Q}(\alpha, \beta, \gamma)$. Let $d = \deg \mathbb{K}$, and let M be the number defined in lemma 12.4, which depends only on the choice of an integral basis of $\mathbb{K}$ (see exercise 12.6).

Define $\lambda = \mathrm{den}(\alpha) \times \mathrm{den}(\beta) \times \mathrm{den}(\gamma)$, $\mu = \overline{|\alpha|}\,\overline{|\gamma|}\left(1 + \overline{|\beta|}\right)$, $\nu = e^{2|\mathrm{Log}\,\alpha|}$.

Elementary limit computations show that there exists an integer N such that, for every $n \geq N$,

$$
\left.\begin{array}{ll}
\left(1 + |\beta|\right)(n+1)^5 \leq n^6 \leq \dfrac{1}{2}n^7, & \left(\lambda\mu\right)^{n^8} n^{5n^8} \leq n^{6n^8}, \\[2mm]
\left[Mn^{11}n^{6n^8}\right]^{\frac{d}{n-d}} \leq e^{n^8}, & n^{11}e^{n^8}e^{7n^8 \mathrm{Log}\,n}e^{n^{10}|\mathrm{Log}\,\alpha|} \leq \nu^{n^{10}}, \\[2mm]
(4\nu)^{n^{10}} n^{-n^{10}} \leq e^{-n^{10}}, & \lambda^{(n+1)^8} \leq e^{n^9}, \\[2mm]
n^{11}e^{n^8} \mu^{(n+1)^8}(n+1)^{5n^8} \leq e^{n^9}, & n^{10} > (2d-1)n^9.
\end{array}\right\} \quad (12.39)
$$

Now we choose and fix an integer N fulfilling conditions (12.39).

The *first step* of the proof consists in constructing a polynomial

$$
P(x,y) = \sum_{i=0}^{N^8-1} \sum_{j=0}^{N^3-1} a_{ij} x^i y^j \in \mathbb{Z}(x,y) \tag{12.40}
$$

in such a way that the auxiliary function

$$
F(z) = P\left(z, \alpha^z\right) = \sum_{i=0}^{N^8-1} \sum_{j=0}^{N^3-1} a_{ij} z^i \alpha^{jz} \tag{12.41}
$$

satisfies $F(k + \beta m) = 0$ for any (k,m) with $0 \leq k < N^5$ and $0 \leq m < N^5$. Siegel's lemma shows that it is possible to find $P(x,y)$ fulfilling these requirements, with moreover $0 < \max_{i,j} |a_{ij}| \leq e^{N^8}$ (see exercice 12.16).

The *second step* consists in proving that, for every integer $n \geq N$,

$$
\begin{cases}
(I)_n & F(k + m\beta) = 0, & 0 \leq k, m < n^5 \\[2mm]
(II)_n & \max_{|z| \leq n^6} |F(z)| \leq e^{-n^{10}}.
\end{cases} \tag{12.42}
$$

For this, we proceed by induction on n.

a) First, we observe that $(I)_N$ is the conclusion of the first step.

b) Second, we show that, for every $n \geq N$, $(I)_n \Rightarrow (II)_n$. To do this, we observe that

$$\max_{|z|\le n^7} \left| F(z) \right| \le \sum_{i=0}^{N^8-1} \sum_{j=0}^{N^3-1} \left| a_{ij} \right| n^{7i} e^{j\,\mathrm{Re}(z\,\mathrm{Log}\,\alpha)}.$$

Since $\mathrm{Re}\left(z\,\mathrm{Log}\,\alpha\right) \le \left| z\,\mathrm{Log}\,\alpha \right| \le n^7 \left| \mathrm{Log}\,\alpha \right|$ and $\max_{i,j} \left| a_{ij} \right| \le e^{N^8}$, we get

$$\max_{|z|\le n^7} \left| F(z) \right| \le N^8 N^3 e^{N^8} e^{7N^8 \,\mathrm{Log}\,n} e^{N^3 n^7 \left| \mathrm{Log}\,\alpha \right|}$$

$$\le n^{11} e^{n^8} e^{7n^8 \,\mathrm{Log}\,n} e^{n^{10} \left| \mathrm{Log}\,\alpha \right|} \le v^{n^{10}}. \tag{12.43}$$

As in (12.33), define $G(z) = \dfrac{F(z)}{\prod_{k=0}^{n^5-1} \prod_{m=0}^{n^5-1} (z - k - m\beta)}$.

Since β is irrational, all the numbers $k + m\beta$ are different. Since (I_n) holds, they are zeros of F, and consequently G is an entire function. We now apply the maximum modulus principle in the disk $|z| \le n^7$. Taking (12.43) into account, we obtain

$$|z| \le n^6 \Rightarrow \left| G(z) \right| \le \max_{|z|=n^7} \left| G(z) \right| \le \frac{v^{n^{10}}}{\prod_{k=0}^{n^5-1} \prod_{m=0}^{n^5-1} \left| n^7 - |k + m\beta| \right|}.$$

Therefore, for every complex number z satisfying $|z| \le n^6$, we have

$$\left| F(z) \right| \le v^{n^{10}} \frac{\prod_{k=0}^{n^5-1} \prod_{m=0}^{n^5-1} \left(n^6 + |k + m\beta| \right)}{\prod_{k=0}^{n^5-1} \prod_{m=0}^{n^5-1} \left| n^7 - |k + m\beta| \right|}$$

$$\le v^{n^{10}} \frac{\prod_{k=0}^{n^5-1} \prod_{m=0}^{n^5-1} \left(n^6 + (1+|\beta|) n^5 \right)}{\prod_{k=0}^{n^5-1} \prod_{m=0}^{n^5-1} \left| n^7 - (1+|\beta|) n^5 \right|} \le v^{n^{10}} \frac{\prod_{k=0}^{n^5-1} \prod_{m=0}^{n^5-1} \left(2n^6 \right)}{\prod_{k=0}^{n^5-1} \prod_{m=0}^{n^5-1} \left(n^7/2 \right)}.$$

Hence $\max_{|z|\le n^6} \left| F(z) \right| \le (4v)^{n^{10}} n^{-n^{10}} \le e^{-n^{10}}$. Therefore $(I)_n \Rightarrow (II)_n$.

c) Finally, we show that $(II)_n \Rightarrow (I)_{n+1}$. Let k and m be two integers such that $0 \le k, m < (n+1)^5$. Consider the algebraic number

$$F(k + \beta m) = \sum_{i=0}^{N^8-1} \sum_{j=0}^{N^3-1} a_{ij} (k + \beta m)^i \alpha^{jk} \gamma^{jm}.$$

Since $|k + m\beta| \le (1 + |\beta|)(n+1)^5 \le n^6$, $(II)_n$ yields $\left| F(k + \beta m) \right| \le e^{-n^{10}}$. An estimate of the denominator of $F(k + \beta m)$ is

$$\mathrm{den}\big(F\left(k+\beta m\right)\big) \le \left(\mathrm{den}\beta\right)^{N^8}\left(\mathrm{den}\,\alpha\,\mathrm{den}\,\gamma\right)^{N^3(n+1)^5} \le \lambda^{(n+1)^8} \le e^{n^9}.$$

An estimate of the house of $F\left(k+\beta m\right)$ is

$$\overline{\left|F\left(k+\beta m\right)\right|} \le N^8 N^3 e^{N^8}\left(1+\overline{|\beta|}\right)^{N^8}(n+1)^{5N^8}\left(\overline{|\alpha|}.\overline{|\gamma|}\right)^{N^3(n+1)^5}$$

$$\le n^{11} e^{n^8} \mu^{(n+1)^8}(n+1)^{5n^8} \le e^{n^9}.$$

If $F\left(k+\beta m\right)$ was not zero, then by the size inequality we would have $e^{(1-2d)n^9} \le e^{-n^{10}}$, contradiction with the last condition in (12.39). Hence $(II)_n \Rightarrow (I)_{n+1}$, and $(I)_n$ and $(II)_n$ hold for all $n \ge N$, Q.E.D.

The *third step* is the conclusion:

Let $z \in \mathbb{C}$. Then, for every $n \ge N$ satisfying $n^6 \ge |z|$, $(II)_n$ holds, whence $\left|F\left(z\right)\right| \le e^{-n^{10}}$. Letting $n \to +\infty$ yields

$$F\left(z\right) = \sum_{j=0}^{N^3-1}\left(\sum_{i=0}^{N^8-1} a_{ij} z^i\right)\left(\alpha^z\right)^j = 0.$$

Since the a_{ij}'s are not all zero, this means that $f\left(z\right) = \alpha^z$ is algebraic, and this contradiction proves Gelfond-Schneider theorem.

Remark 12.1 Both Gelfond and Schneider's methods make use of an auxiliary function with "many zeros". The difference is that Gelfond's method involves two repeated zeros with high multiplicity (and the property $(e^z)' = e^z$), whereas Schneider's method rests on simple zeros (and the property $e^a e^b = e^{a+b}$). The interested reader should refer to [28] for a proof of theorem 12.7 by Gelfond's method.

12.7 Siegel-Shidlovski method

It has been introduced by Siegel in 1930 to deal with Bessel functions, and subsequently improved by Shidlovski from 1954 onwards. It enables to prove transcendence results for the values, at algebraic points, of the so-called *E*-functions. We say that an analytic function

$$f(x) = \sum_{n=0}^{+\infty} a_n \frac{x^n}{n!}$$

is an *E*-function if it satisfies the three following conditions:

a. $a_n \in \mathbb{K}$ for all $n \in \mathbb{N}$, where $\mathbb{K}$ is a given number field.

b. For every $\varepsilon > 0$, $\overline{|a_n|} = 0(n^{\varepsilon n})$.

c. For every $\varepsilon > 0$, $\text{den}(a_n) = 0(n^{\varepsilon n})$.

By way of example, we state here without proof two typical results.

Theorem 12.8 (Siegel 1930)

Let $J_0(x) = \sum_{n=0}^{+\infty} \frac{(-1)^n}{(n!)^2} \left(\frac{x}{2}\right)^2$ *be the Bessel function of order* 0.

Let α *be any non-zero algebraic number.*
Then $J_0'(\alpha)/J_0(\alpha)$ *is transcendental.*

Corollary 12.3 *The real number* β *defined by the regular continued fraction* $\beta = [1, 2, 3, \ldots, n, \ldots]$ *is transcendental.*

The proof of collorary 12.3 rests on remark 3.2. See exercise 12.10.

Theorem 12.9 (Shidlovski 1954)
Consider the confluent hypergeometric function

$$_1F_1\left({1 \atop c}\Big|x\right) = \sum_{n=0}^{+\infty} \frac{x^n}{(c)_n} \ , \quad c \in \mathbb{Q} - \mathbb{Z}^-.$$

Let $\alpha \in \mathbb{C}^*$ *be algebraic. Then* $_1F_1\left({1 \atop c}\Big|\alpha\right)$ *is transcendental.*

For details on Siegel-Shidlovski method, see [2], [17], [26].

Exercises

12.1   Let α be an algebraic number of degree d, and let $\mathbb{L} = \mathbb{Q}(\alpha)$.
Let $\mathbb{K}$ be a number field containing α. Put $\mathbb{K} = \mathbb{L}(\theta)$, $[\mathbb{K}:\mathbb{Q}] = m$,
$[\mathbb{K}:\mathbb{L}] = n$. We know by exercise 10.17 that

$$B = \left(\alpha^i \theta^j \ / \ 0 \le i \le d-1 \ ; \ 0 \le j \le n-1\right)$$

is a basis of the $\mathbb{Q}$ - vector space $\mathbb{K}$. Prove that φ_k defined by $\varphi_k(\alpha^i \theta^j) = \alpha_k^i \theta^j$ is a monomorphism of $\mathbb{K}$. Then prove theorem 12.1.

12.2 Prove theorem 12.2.

12.3 House of an algebraic number

1) Let $\alpha \neq 0$ be an algebraic integer. Prove that $\overline{|\alpha|} \geq 1$ (theorem 12.3).

2) Let $\alpha \neq 0$ be an algebraic integer of degree n. Assume that $\overline{|\alpha|} = 1$. For any integer $i \geq 1$, let $P_i(x) = a_{0,i} + a_{1,i} x + \cdots + a_{d_i-1,i} x^{d_i-1} + x^{d_i} \in \mathbb{Z}[x]$ be the minimal polynomial of α^i.

Prove that $|a_{j,i}| \leq \binom{n}{j}$ for every $j = 0, 1, \cdots, d_i - 1$.

3) Deduce the following theorem: *If $\alpha \neq 0$ is an algebraic integer which is not a root of unity, then $\overline{|\alpha|} > 1$.*

12.4 Prove the size inequality by using the norm in $\mathbb{K} = \mathbb{Q}(\alpha)$.

12.5 Prove formula (12.23).

12.6 Prove Siegel's lemma 12.4 by using an integral basis of $\mathbb{K}$.

12.7 By using Siegel's lemma 12.4, check (12.28), (12.29), (12.30).

12.8 We recall the **maximum modulus principle**: *If f is analytic in* $D = \{z \in \mathbb{C} \ / \ |z| < R\}$, *and* $0 < r < R$, *then* $|x| \leq r \Rightarrow |f(x)| \leq \max_{|z|=r} |f(z)|$.

By applying the maximum modulus principle to G_n defined in (12.33), in the disk $|x| \leq m^{2/3}$, prove the estimate (12.34).

12.9 Prove (12.35) and (12.36).

12.10 Prove corollary 12.3.

12.11 Let $q \in \mathbb{C}$, $|q| > 1$. Prove that the Tchakaloff function, defined in section 1.4, is transcendental.

12.12   A result of Lucas, 1870

Prove that $\displaystyle\sum_{n=1}^{+\infty}\frac{x^{2^n}}{1-x^{2^{n+1}}}=\frac{x^2}{1-x^2}$ if $|x|<1$.

Deduce that $\displaystyle\sum\frac{1}{F_{2^n}}=\frac{7-\sqrt5}{2}$ (here F_n is the Fibonacci sequence).

12.13   Let $a\in\mathbb{Z}^*$. Prove that $\displaystyle f_a(x)=\sum_{n=0}^{+\infty}\frac{x^{2^n}}{1+a x^{2^n}}$ is transcendental.

Deduce that $\displaystyle\theta=\sum_{n=0}^{+\infty}1\!\big/\!\big(2^{2^n}+a\big)$ is transcendental.

12.14   Transcendence of e, adapted from Hermite's proof, 1873

1) Let $f\in\mathbb{Z}[x]$. Prove that $n!$ divides all the coefficients of $f^{(n)}$.

2) Let f be a polynomial of degree N. Put

$$F(x)=f(x)+f'(x)+\cdots+f^{(N)}(x).$$

Prove *Hermite's formula*: $\displaystyle e^x\int_0^x f(t)e^{-t}dt=F(0)e^x-F(x)$.

3) By using the polynomial $\displaystyle f(x)=\frac{1}{(n-1)!}x^{n-1}\big((x-1)...(x-d)\big)^n$ for prime n, with $n\to+\infty$, prove that e is transcendental.

12.15   Quadrature of the circle

1) We say that a point A of the plane is constructible (with ruler and compass) if it can be obtained from two given points by a finite number of operations involving only intersections of straight lines and circles.

a) Prove that every point with rational coordinates is constructible.

b) Prove that the coordinates of any constructible point are algebraic numbers.

2) The *quadrature of the circle* is a famous problem, which goes back to ancient Greek geometry. Starting from a given circle (C) of radius 1, it consists in constructing with ruler and compass a square which has the same area as (C). Prove that this is impossible.

12.16   Construct the polynomial $P(x,y)$ of the first step of the proof of Gelfond-Schneider theorem.

12.17 Proof of Thue's theorem

1) Prove that the two following statements of Thue's theorem (theorem 9.9) are equivalent:

Statement 1: If α is algebraic of degree $d \geq 2$, then for every $\varepsilon > 0$ the inequation $\left| \alpha - p/q \right| < q^{-\frac{d}{2}-1-\varepsilon}$ (I) has only finitely many rational solutions p/q with $q > 0$.

Statement 2: If α is algebraic of degree $d \geq 2$, then for every $\varepsilon > 0$ the inequation $\left| \alpha - p/q \right| < q^{-\frac{d}{2}-1-\varepsilon}$ (I) has only finitely many *irreducible* rational solutions p/q with $q > 0$.

In what follow, we assume that α is algebraic of degree $d \geq 3$ and that there exists $\varepsilon > 0$ such that inequation (I) has infinitely many *irreducible* rational solutions p/q with $q > 0$.

We put $\rho = \dfrac{d}{2} + 1 + \varepsilon$.

2) Let $\lambda \in \mathbb{R}$, with $0 < \lambda < 1$, and let $m \in \mathbb{N}$, with $m \geq 2$.

We define $n = \left[(1+\lambda)\dfrac{dm}{2} \right]$.

Prove that there exist $P(x) = \sum_{i=0}^{n} a_i x^i$ and $Q(x) = \sum_{i=0}^{n} b_i x^i$, with coefficients in $\mathbb{Z}$, such that $A(x, y) = P(x) - yQ(x)$ satisfies

$$\frac{\partial^h A}{\partial x^h}(\alpha, \alpha) = 0 \quad \text{for every } h = 0,\ 1, \cdots,\ m-1,$$

and also $\max_{i,j}\left(|a_i|, |b_j| \right) \leq \beta^{m/\lambda}$, where β is a constant which depends only on α (but not on m nor λ).

3) Put $W(x) = P(x)Q'(x) - P'(x)Q(x) = \sum_{i=0}^{2n-1} c_i x^i$. Prove that:

a) α is a root of order at least $m-1$ of W.

b) W is not zero.

c) $\max_{i}\left(\text{Log}|c_i| \right) \leq \gamma \dfrac{m}{\lambda}$, where γ is a constant which depends only on α.

4) Let p_0/q_0 be an irreducible solution of (I), such chosen that $\text{Log}\, q_0 \geq \gamma \lambda^{-2}$. Assume that p_0/q_0 is a zero of W.

Prove that its multiplicity is at most $w = \dfrac{\gamma m}{\lambda \operatorname{Log} q_0} \leq \lambda m$.

5) For every solution p_k/q_k of (I) with $k \geq 1$ and $q_k \geq q_0^2$, we choose in the above construction $m = m(k) = \left\lceil \dfrac{\operatorname{Log} q_k}{\operatorname{Log} q_0} \right\rceil$.

Prove that there exists an integer j, $0 \leq j \leq w+1 \leq \lambda m + 1$, such that

$$\mu_{k,j} = \frac{1}{j!} \frac{\partial^j A}{\partial x^j}\left(\frac{p_0}{q_0}, \frac{p_k}{q_k}\right) = \frac{1}{j!}\left[P^{(j)}\left(\frac{p_0}{q_0}\right) - \frac{p_k}{q_k} Q^{(j)}\left(\frac{p_0}{q_0}\right)\right]$$

is not zero. For every $k \geq 1$, we fix such a $j = j(k)$, and put $v_k = \mu_{k,j}$.

6) By using Taylor's formula, prove that $|v_k| \leq \delta^{\frac{m}{\lambda}} q_0^{(\lambda-1)\rho m}$, where δ depends only on α.

7) Prove that $|v_k| \geq q_0^{-\left(\frac{1+\lambda}{2}d+1\right)m-1}$ and conclude.

12.18 Let $a, b \in \mathbb{N}$, $a \geq 2$, $b \geq 2$ be coprime rational integers.
1) Prove that $\theta = \operatorname{Log} a / \operatorname{Log} b$ is irrational.
2) Prove that θ is transcendental.

Solutions to the exercises

Solutions to the exercises of chapter 1

Exercice 1.1   Since 0 is a repeated zero with multiplicity n of P, we have $P(0) = P'(0) = \cdots = P^{(n-1)}(0) = 0$. For $k \geq n$, the constant term in $P^{(k)}(x)$ is $P^{(k)}(0) = \dfrac{k!}{n!}\binom{n}{k-n} a^{2n-k}(-b)^{k-n}$, since it is equal to the k-th derivative of the term of degree k of $P(x)$, which is given by the expansion of $(a-bx)^n$. Since $k \geq n$, $n! \mid k!$, whence $P^{(k)}(0) \in \mathbb{Z}$ for every $k \geq n$. Therefore $F(0) \in \mathbb{Z}$. Similarly $F(\pi) \in \mathbb{Z}$ and $F(0) + F(\pi)$ is an integer.

Exercice 1.2   We look for $Q(x) = \sum_{i=0}^{q} \alpha_i x^i$, $P(x) = \sum_{i=0}^{p} \beta_i x^i$. Expand $Q(x)f(x) + P(x)$ and order it with respect to the powers of x. Since the terms with degrees 0, 1, ..., $r-1$ have to vanish, we see that the α_i's and β_i's must satisfy the system

$$\begin{cases} a_0\alpha_0 + \beta_0 = 0 \\ a_1\alpha_0 + a_0\alpha_1 + \beta_1 = 0 \\ \vdots \\ a_{r-1}\alpha_0 + a_{r-2}\alpha_1 + \cdots = 0. \end{cases}$$

This is a system with $p + q + 2$ unknowns and r equations. Since $p + q + 2 > r$, it has a non-zero solution $\left(\alpha_0, \alpha_1, ..., \alpha_q, \beta_0, \beta_1, ..., \beta_p\right)$ by the rank theorem. If all the α_i's were zero, we would have $P \neq 0$ and $P(x) = x^r g(x)$, which is clearly impossible since $r > p$. Therefore $Q \neq 0$, as claimed.

Exercise 1.3   Suppose that $\alpha = a/b$, with $a \in \mathbb{Z}$, $b \in \mathbb{Z}$. Then

$$0 < |\alpha - p_n/q_n| \leq \varepsilon(n)/q_n \;\Rightarrow\; 0 < |aq_n - bp_n| \leq b\varepsilon(n).$$

Therefore $\lim_{n \to +\infty} |aq_n - bp_n| = 0$. However, $|aq_n - bp_n| \in \mathbb{N}$, and a sequence of positive integers cannot tend to 0, since it is not less than 1. This proves that α is irrational.

Exercise 1.4   Assume that $\sqrt[3]{2} + \sqrt{5} = p/q$, with p, q integers. Then

$$\left(\sqrt{5}-\frac{p}{q}\right)^3 = -2, \quad \text{whence} \quad \left(5+3\frac{p^2}{q^2}\right)\sqrt{5} = 15\frac{p}{q}+\frac{p^3}{q^3}-2.$$

Hence $\sqrt{5} \in \mathbb{Q}$, which contradicts theorem 1.1. Thus $\alpha = \sqrt[3]{2}+\sqrt{5}$ is irrational.

Exercise 1.5   Assume that $\log_{10} 2 = a/b$, $a,b \in \mathbb{N}$, $b \neq 0$. Since $10^{\log_{10} 2} = 2$, we have $10^{a/b} = 2$, whence $10^a = 2^b$. As $a \geq 1$, 10^a is divisible by 5, and therefore so is 2^b. Contradiction.

Exercise 1.6   By hypothesis, $P(x) = \left(\frac{p}{q}-x\right)^d Q(x)$, with $Q \in \mathbb{Q}[x]$.

Now we compute $Q(x)$. For every x satisfying $|x| < p/q$,

$$Q(x) = \frac{q^d}{p^d}\left(1-\frac{qx}{p}\right)^{-d} P(x) = \frac{q^d}{p^d}\left(\sum_{k=0}^{+\infty}\binom{d+k-1}{k}\frac{q^k}{p^k}x^k\right)\left(\sum_{k=0}^{n}a_k x^k\right).$$

As $\deg Q = n-d$, we obtain $Q(x) = \sum_{k=0}^{n-d}\left[\sum_{m=0}^{k}\binom{d+m-1}{m}\frac{q^{d+m}}{p^{d+m}}a_{k-m}\right]x^k.$

Looking at the terms of higher degree in $P(x) = (p/q-x)^d Q(x)$ yields

$$p^n a_n = (-1)^d q^d \sum_{m=0}^{n-d}\binom{d+m-1}{m}p^{n-d-m}q^m a_{n-d-m}.$$

Since the binomial coefficients and the a_i's are integers, we deduce that q^d divides $p^n a_n$. Since p and q coprime, q^d divides a_n.

Application: Suppose that $\cos(\pi/n) = p/q$, $p,q \in \mathbb{N}$, p and q coprime. We know by de Moivre's formula that, for every $\theta \in \mathbb{R}$ and $r \in \mathbb{N}$,

$$\cos r\theta + i \sin r\theta = (\cos\theta + i\sin\theta)^r = \sum_{k=0}^{r}\binom{r}{k}i^k \sin^k\theta\cos^{r-k}\theta.$$

Identifying the real parts yields

$$\cos r\theta = \sum_{m=0}^{r/2}\binom{r}{2m}(-1)^m\left(1-\cos^2\theta\right)^m\cos^{r-2m}\theta. \tag{*}$$

Now we distinguish three cases:

a. n is odd, $n = 2k+1$, $k \geq 2$. By replacing r by n and θ by π/n in $(*)$, we see that $\cos(\pi/n)$ is a zero of $G(x) = \sum_{m=0}^{k}\binom{n}{2m}(-1)^m\left(1-x^2\right)^m x^{n-2m}+1.$

Here G has degree n, and the coefficient of x^n is

$$a_n = \sum_{m=0}^{k} \binom{n}{2m} = \frac{1}{2}\left((1+1)^n + (1-1)^n\right) = 2^{n-1}.$$

But q divides a_n, and therefore $q = 2^h$, with $0 \leq h \leq n-1$. Similarly, the constant term of G is $a_0 = G(0) = 1$, and if we apply the main result of this exercise to the polynomial $x^n G(1/x)$, we see that p divides $a_0 = 1$, whence $p = 1$. But the equation $\cos(\pi/n) = 1/2^h$ has for solution $n = 3$, $h = 1$, and no other one if $n \geq 4$, since the sequence $\cos(\pi/n)$ is increasing. Thus if n is odd and greater than 3, $\cos(\pi/n)$ is irrational.

b. n is a multiple of 4, $n = 4k$, $k \geq 1$. We replace r by $2k$ and θ by π/n in $(*)$ and see that $\cos(\pi/n)$ is a zero of $H(x) = \sum_{m=0}^{k} \binom{2k}{2m}(-1)^m(1-x^2)^m x^{2k-2m}$.

As in case a, H has degree $2k$, and the coefficient of x^{2k} is

$$a_{2k} = \sum_{m=0}^{k} \binom{2k}{2m} = \frac{1}{2}\left((1+1)^{2k} + (1-1)^{2k}\right) = 2^{2k-1}.$$

Consequently $q = 2^h$, with $0 \leq h \leq 2k-1$. Moreover, the constant term of H is $a_0 = H(0) = (-1)^k$, which implies $p = 1$. Therefore $\cos(\pi/n)$ is irrational if n is a multiple of 4.

c. $n = 2(2k+1)$, with $k \in \mathbb{N}$, $k \geq 1$.

If $\cos(\pi/n) \in \mathbb{Q}$, then $\cos(\pi/(2k+1)) = \cos(2\pi/n) = 2\cos^2(\pi/n) - 1 \in \mathbb{Q}$.

From part a), we know that this is only possible if $k = 1$. But $\cos(\pi/6) = \sqrt{3}/2$ is irrational by theorem 1.1. Hence $\cos(\pi/n)$ is still irrational in this case.

Consequently, $\cos(\pi/n)$ is irrational for $n \geq 4$, and rational for $n = 1$, 2, 3.

Exercise 1.7   *First proof.* We construct a diophantine approximation of $\alpha = \sum_{n=0}^{+\infty} 1/m^{n^2}$ by using the partial sums of the series which defines α (thus using the same method as for proving the irrationality of e).

$$\alpha - \sum_{k=0}^{n} \frac{1}{m^{k^2}} = \sum_{k=n+1}^{+\infty} \frac{1}{m^{k^2}} = \frac{1}{m^{(n+1)^2}} + \frac{1}{m^{(n+2)^2}} + \frac{1}{m^{(n+3)^2}} + \cdots, \text{ whence}$$

$$0 < \alpha - \sum_{k=0}^{n} \frac{1}{m^{k^2}} < \frac{1}{m^{(n+1)^2}}\left(1 + \frac{1}{m} + \frac{1}{m^2} + \frac{1}{m^3} + \cdots\right).$$

Therefore $0 < m^{n^2}\alpha - m^{n^2}\sum_{k=0}^{n} \frac{1}{m^{k^2}} < \frac{1}{m^{2n}} \times \frac{1}{m-1}$.

If $\alpha = a/b$, $a, b \in \mathbb{N}$, then $\beta_m = m^{n^2} a - b m^{n^2} \sum_{k=0}^n 1/m^{k^2}$ is a non-zero integer.

Moreover, as $0 < m^{n^2} a - b m^{n^2} \sum_{k=0}^n 1/m^{k^2} < \dfrac{b}{(m-1)m^{2n}}$ we have $\lim\limits_{m \to +\infty} \beta_m = 0$.

This contradiction proves that $\alpha \notin \mathbb{Q}$.

Second proof. In the definition of Tschakaloff function (1.4), replace q by m^2 and x by m. We obtain

$$T_{m^2}(m) = \sum_{k=0}^{+\infty} \frac{1}{m^{k^2}} = \alpha.$$

Therefore α is irrational by theorem 1.4.

Exercise 1.8   1) We have $\sqrt{5} = 2\Phi - 1$, whence $5 = 4\Phi^2 - 4\Phi + 1$ and $\Phi^2 - \Phi - 1 = 0$. Thus the golden number Φ is indeed irrational quadratic. Note that it is irrational because, if not, $\sqrt{5}$ would be rational. Alternatively, one can also use the result of exercise 1.6.

2) Suppose that $ae^2 + be + c = 0$, a, b, $c \in \mathbb{Z}$, $a \neq 0$. Then $ae + b + c/e = 0$, whence $a \sum_{n=0}^{+\infty} 1/n! + b + c \sum_{n=0}^{+\infty} (-1)^n / n! = 0$. Now introduce the partial sums:

$$\sum_{k=0}^n \frac{a + c(-1)^k}{k!} + b = - \sum_{k=n+1}^{+\infty} \frac{a + c(-1)^k}{k!} \implies \left| \sum_{k=0}^n \frac{a + c(-1)^k}{k!} + b \right| \le (|a| + |c|) \sum_{k=n+1}^{+\infty} \frac{1}{k!}$$

$$\implies \left| n! \sum_{k=0}^n \frac{a + c(-1)^k}{k!} + bn! \right| \le (|a| + |c|) \frac{2}{n+1},$$

by using the estimate of R_n we have established in theorem 1.2. Hence the non negative integer $\tau_n = \left| n! \sum_{k=0}^n \left(a + c(-1)^k \right) / k! + bn! \right|$ tends to 0 when $n \to +\infty$. Therefore $\tau_n = 0$ for n sufficiently large, in which case

$$n!(a + c) + \frac{n!}{1!}(a - c) + \cdots + \frac{n!}{(n-1)!}(a + c(-1)^{n-1}) + (a + c(-1)^n) + bn! = 0.$$

Thus $n \mid (a + c(-1)^n)$ for n sufficiently large, which implies that $a + c(-1)^n = 0$ for n sufficiently large. If n is even, this yields $a + c = 0$, and if n is odd, $a - c = 0$. Therefore $a = 0$, contradiction. The number e is not quadratic.

Exercise 1.9   a) We have

$$f(x) - g(x) = \sum_{n=0}^{+\infty} \left[\frac{x^{2^n}}{1 - x^{2^n}} - \frac{x^{2^n}}{1 + x^{2^n}} \right] = \sum_{n=0}^{+\infty} \frac{2x^{2^{n+1}}}{1 - x^{2^{n+1}}} = 2 \sum_{n=1}^{+\infty} \frac{x^{2^n}}{1 - x^{2^n}} = 2 \left(f(x) - \frac{x}{1-x} \right).$$

b) Introduce the partial sums as in exercises 1.7 and 1.8. We have

$$f\left(\frac{1}{2}\right)=\sum_{k=0}^{n}\frac{1}{2^{2^{k}}-1}+\sum_{k=n+1}^{+\infty}\frac{1}{2^{2^{k}}-1}.$$

We observe that $2^{2^{k}}-1$ divides $2^{2^{n}}-1$ for every $k\leq n$, since

$$2^{2^{n}}-1=(2^{2^{k}})^{2^{n-k}}-1=(2^{2^{k}}-1)((2^{2^{k}})^{2^{n-k}-1}+(2^{2^{k}})^{2^{n-k}-2}+\cdots+1).$$

Now we have $(2^{2^{n}}-1)f\left(\dfrac{1}{2}\right)-(2^{2^{n}}-1)\displaystyle\sum_{k=0}^{n}\frac{1}{2^{2^{k}}-1}=(2^{2^{n}}-1)\sum_{k=n+1}^{+\infty}\frac{1}{2^{2^{k}}-1}.$

Moreover $0<\displaystyle\sum_{k=n+1}^{+\infty}\frac{1}{2^{2^{k}}-1}\leq\sum_{k=n+1}^{+\infty}\frac{1}{\frac{1}{2}2^{2^{k}}}\leq\frac{1}{2^{2^{n+1}-1}}\left(1+\frac{1}{2}+\frac{1}{2^{2}}+\cdots\right)\leq\frac{1}{2^{2^{n+1}-2}}.$

Therefore $0<(2^{2^{n}}-1)f\left(\dfrac{1}{2}\right)-(2^{2^{n}}-1)\displaystyle\sum_{k=0}^{n}\frac{1}{2^{2^{k}}-1}\leq\frac{2^{2^{n}}-1}{2^{2^{n+1}-2}}\leq2^{-2^{n}+2}.$

Thus $0<a_{n}f\left(1/2\right)+b_{n}\leq\varepsilon(n)$, with $\lim_{n\to+\infty}\varepsilon(n)=0$, $a_{n},b_{n}\in\mathbb{Z}$. This proves that $f\left(1/2\right)$ is irrational by theorem 1.5.

c) Since $f\left(1/2\right)=-g\left(1/2\right)+2$ by question a), $g\left(1/2\right)=\sum_{n=0}^{+\infty}1/F_{n}\notin\mathbb{Q}.$

Exercise 1.10   Function f_{a} fulfills a functional equation similar to (1.4):

$$f_{a}(qx)=1+\sum_{n=1}^{+\infty}\frac{(1+a)(1+aq)...(1+aq^{n-1})}{q^{n(n-1)/2}}x^{n}=1+\sum_{k=0}^{+\infty}\frac{(1+a)(1+aq)...(1+aq^{k})}{q^{k(k+1)/2}}x^{k+1}$$

$$=1+\sum_{k=0}^{+\infty}\frac{(1+a)...(1+aq^{k-1})}{q^{k(k+1)/2}}x^{k+1}+\sum_{k=0}^{+\infty}\frac{(1+a)...(1+aq^{k-1})}{q^{k(k+1)/2}}aq^{k}x^{k+1}.$$

This yields $f_{a}(qx)=1+xf_{a}(x)+axf_{a}(qx)$, whence $(1-ax)f_{a}(qx)=1+xf_{a}(x)$. As for Tchakaloff function, if $x=\alpha/\beta\in\mathbb{Q}$ and $f_{a}(x)=\mu/\nu\in\mathbb{Q}$, this implies that, for every $n\in\mathbb{N}$, $f_{a}(x/q^{n})=A_{n}/(\nu\alpha^{n})$, where $A_{n}\in\mathbb{Z}$. Now the proof is the same as the one of theorem 1.4.

Exercise 1.11   1) Since $\sqrt{d}\notin\mathbb{Q}$, by theorem 1.6 there exist infinitely many couples $(x,y)\in\mathbb{Z}^{2}$, with $y\neq0$, such that $0<\left|x-y\sqrt{d}\right|<1/\left|y\right|$. Therefore

$$0<\left|x^{2}-dy^{2}\right|<\frac{\left|x+y\sqrt{d}\right|}{\left|y\right|}\leq\frac{\left|x\right|+\left|y\right|\sqrt{d}}{\left|y\right|}.$$

But $\left|x-y\sqrt{d}\right|<1/\left|y\right|\leq1\Rightarrow\left|\left|x\right|-\left|y\right|\sqrt{d}\right|\leq1\Rightarrow-1\leq\left|x\right|-\left|y\right|\sqrt{d}\leq1$
$\Rightarrow\left|x\right|\leq1+\left|y\right|\sqrt{d}\leq\left|y\right|+\left|y\right|\sqrt{d}$. Hence $0<\left|x^{2}-dy^{2}\right|<1+2\sqrt{d}$.

2) By question 1, the numbers $x^{2}-dy^{2}$ are bounded.

Since x, y and d are integers, it can take only finitely many values. Hence, by the pigeon-hole principle, there exists an integer k, $0 < |k| < 1 + 2\sqrt{d}$, such that the equation $x^2 - dy^2 = k$ has infinitely many solutions. Now, we reduce these solutions mod k. Since there are only finitely many residues, by applying again the pigeon-hole principle we see that there exist k, m and n such that the system $x^2 - dy^2 = k$, $x \equiv m$ (mod k), $y \equiv n$ (mod k) has infinitely many solutions, with $0 < |k| < 1 + 2\sqrt{d}$.

3) We have $(x' - y'\sqrt{d})(x'' + y''\sqrt{d}) = x'x'' - dy'y'' + (x'y'' - y'x'')\sqrt{d}$.

Moreover $x'x'' - dy'y'' \equiv m^2 - dn^2$ (mod k) $\equiv x'^2 - dy'^2$ (mod k).

Now $x'^2 - dy'^2 = k \equiv 0$ (mod k). Therefore $x'x'' - dy'y'' \equiv 0$ (mod k), which means that $x'x'' - dy'y''$ is a multiple of k, and we have $x'x'' - dy'y'' = k\xi$, $\xi \in \mathbb{Z}$. Similarly $x'y'' - y'x'' \equiv mn - mn \equiv 0$ (mod k), whence $x'y'' - y'x'' = k\eta$, $\eta \in \mathbb{Z}$. Therefore

$$(x' - y'\sqrt{d})(x'' + y''\sqrt{d}) = x'x'' - dy'y'' + (x'y'' - y'x'')\sqrt{d} = k(\xi + \eta\sqrt{d}),$$
$$(x' + y'\sqrt{d})(x'' - y''\sqrt{d}) = x'x'' - dy'y'' - (x'y'' - y'x'')\sqrt{d} = k(\xi - \eta\sqrt{d}).$$

4) Multiplying these two equalities yields
$$k^2(\xi^2 - d\eta^2) = (x'^2 - dy'^2)(x''^2 - dy''^2) = k^2.$$

Hence $\xi^2 - d\eta^2 = 1$. It remains to prove that one can choose the couples (x', y') and (x'', y'') in such a way that $\eta \neq 0$. But $\eta = 0 \Rightarrow x'/y' = x''/y''$, which is impossible because the rational numbers x/y given by theorem 1.6 are all different. Therefore the Pell equation has a non trivial solution.

Remark: We will solve Pell's equation in chapters 4 and 5.

Exercise 1.12   1) We have $\alpha = \sum_{n=0}^{+\infty} b_1(n) q^{-n}$, where $b_1(n) = 1$ if $n = 2^m$ and $b_1(n) = 0$ if not. Now an easy computation shows that

$$b_h(n) = \sum_{m_1 + m_2 + \cdots + m_h = n} b_1(m_1) b_1(m_2) \cdots b_1(m_h).$$

a. If the representation of n in base 2 has at least $h + 1$ digits 1, it is impossible to find h exact powers of 2 such that their sum is n. This is evidently true if all these powers are distinct, since the expansion of n in base 2 is unique. This remains true if some of them are identical, since $2^i + 2^i = 2^{i+1}$. Therefore at least one of the $b_1(m_i)$ is zero in the above sum. Hence $b_h(n) = 0$.

b. If the representation of n in base 2 has exactly h digits 1, then there exist exactly h distinct powers of 2, $m_1 < m_2 < \cdots < m_h$, such that their sum is n. Hence $b_1(m_1)b_1(m_2)\cdots b_1(m_h) = 1$. Taking into account all possible permutations, we see that $b_h(n) = h!$

2) a) The representation of n_k in base 2 has exactly d digits 1. Consequently $b_d(n_k) = d!$, and $b_h(n_k) = 0$ for every $h = 1, \cdots, d-1$ by question 1.

b) It is clear that the representation of $n_k + 1$, $n_k + 2$, $\cdots$, $n_k + 2^{k-1}$ in base 2 has at least $d+1$ digits 1. Consequently

$$b_h(n_k + 1) = b_h(n_k + 2) = \cdots = b_h(n_k + 2^{k-1}) = 0 \text{ for every } h = 1, 2, \cdots, d.$$

c) We have $n_k - 1 = (2^k - 1) + 2^{k+1} + \cdots + 2^{k+d-1} = 1 + \cdots + 2^{k-1} + 2^{k+1} + \cdots + 2^{k+d-1}$.

Since $k \geq 2$, we see that $b_h(n_k - 1) = 0$ for every $h = 1, 2, \cdots, d$.

Moreover, if $m \leq 2^{k-2} - 1 = 1 + 2 + \cdots + 2^{k-3}$, then substracting m to $n_k - 1$ leaves the powers 2^{k-2} and 2^{k-1} unchanged. Thus

$$b_h(n_k - 1) = b_h(n_k - 2) = \cdots = b_h(n_k - 2^{k-2}) = 0 \text{ for every } h = 1, 2, \cdots, d.$$

3) Assume α is algebraic. Then there exist $a_0, a_1, \cdots, a_d \in \mathbb{Z}$, with $a_d \neq 0$, satisfying $\sum_{n=0}^{+\infty} \left(\sum_{h=1}^{d} a_h b_h(n) \right) q^{-n} + a_0 = 0$.

Using the numbers n_k defined in question 2 and taking 2b and 2c into account yields

$$\sum_{n=0}^{n_k - 2^{k-2} - 1} \left(\sum_{h=1}^{d} a_h b_h(n) \right) q^{-n} + \sum_{h=1}^{d} a_h b_h(n_k) q^{-n_k} + a_0 = - \sum_{n=n_k + 2^{k-1} + 1}^{+\infty} \left(\sum_{h=1}^{d} a_h b_h(n) \right) q^{-n}.$$

Put $M = |a_1| + \cdots + |a_d|$. Then 2a and the estimate $b_h(n) \leq d!$ yield

$$\left| \sum_{n=0}^{n_k - 2^{k-2} - 1} \left(\sum_{h=1}^{d} a_h b_h(n) \right) q^{-n} + a_d d! q^{-n_k} + a_0 \right| \leq M d! \frac{|q|^{-n_k - 2^{k-1}}}{|q| - 1}.$$

Multiply by $|q|^{n_k}$, in such a way that the left-hand side becomes an integer A_k:

$$A_k = \left| \sum_{n=0}^{n_k - 2^{k-2} - 1} \left(\sum_{h=1}^{d} a_h b_h(n) \right) q^{-n + n_k} + a_d d! + a_0 q^{n_k} \right| \leq M d! \frac{|q|^{-2^{k-1}}}{|q| - 1}.$$

This inequality yields $\lim_{k \to \infty} A_k = 0$, whence $A_k = 0$ for k sufficiently large. Therefore, for k sufficiently large,

$$a_d d! = - \sum_{n=0}^{n_k - 2^{k-2} - 1} \left(\sum_{h=1}^{d} a_h b_h(n) \right) q^{-n + n_k} - a_0 q^{n_k}.$$

This implies that $q^{2^{k-2}}$ divides $a_d d!$ for k sufficiently large.

This contradiction proves that α cannot be algebraic. Therefore it is transcendental. This result is proved here by elementary means. It will be generalized in chapter 12 by means of Mahler's method (theorem 12.5).

Solutions to the exercises of chapter 2

Exercise 2.1   Since $u_0 = \left[1/x_0 \right] + 1$, we have $u_0 \le 1/x_0 + 1 < u_0 + 1$, whence $1 < u_0 x_0 \le 1 + x_0$. Consequently $0 < x_1 \le x_0 \le 1$. Similarly $0 < x_2 \le x_1 \le 1$, and an easy induction shows that $0 < x_{n+1} \le x_n \le 1$ for any $n \in \mathbb{N}$.

2) The integer part is a non decreasing function. Moreover, $x_n > 0$ and (x_n) is non increasing. Hence $u_n = \left[1/x_0 \right] + 1 \implies u_{n+1} \ge u_n, \forall n \in \mathbb{N}$.

But $u_0 = \left[1/x_0 \right] + 1 \ge 2$ because $1/x_0 \ge 1$, whence $u_n \ge 2, \forall n \in \mathbb{N}$.

3) We have $x_0 = \dfrac{1}{u_0} + \dfrac{x_1}{u_0} = \dfrac{1}{u_0} + \dfrac{1}{u_0}\left(\dfrac{1}{u_1} + \dfrac{x_2}{u_1} \right) = \dfrac{1}{u_0} + \dfrac{1}{u_0 u_1} + \dfrac{x_2}{u_0 u_1}$.

By induction, we obtain $x_0 = \dfrac{1}{u_0} + \dfrac{1}{u_0 u_1} + \cdots + \dfrac{1}{u_0 u_1 \ldots u_n} + \dfrac{x_{n+1}}{u_0 u_1 \ldots u_n}$.

Now $0 < x_{n+1} \le 1$ for all $n \in \mathbb{N}$ and $u_0 u_1 \ldots u_n \ge 2^{n+1}$ (questions 1 and 2).

Therefore $\lim\limits_{n \to \infty} \dfrac{x_{n+1}}{u_0 u_1 \ldots u_n} = 0$ and $x_0 = \sum\limits_{n=0}^{+\infty} \dfrac{1}{u_0 u_1 \ldots u_n}$.

4) Suppose that $x_0 = \sum_{n=0}^{+\infty} 1/(v_0 v_1 \ldots v_n)$, with v_n non decreasing, $v_n \in \mathbb{N} - \{0,1\}$.

Then $x_0 = \dfrac{1}{v_0} + \dfrac{1}{v_0} \sum\limits_{n=1}^{+\infty} \dfrac{1}{v_1 \ldots v_n} = \dfrac{1}{v_0} + \dfrac{x_1'}{v_0}$, and $0 < x_1' \le \sum\limits_{n=0}^{+\infty} \dfrac{1}{v_0 \ldots v_n} = x_0$ since (v_n) is non decreasing. Consequently $v_0 = 1/x_0 + x_1'/x_0 \implies 1/x_0 < v_0 \le 1/x_0 + 1$. This means that $v_0 = \left[1/x_0 \right] + 1 = u_0$, and now $x_1' = v_0 x_0 - 1$ yields $x_1' = x_1$. By induction, the same reasoning shows that $v_n = u_n$ for every $n \in \mathbb{N}$.

Exercise 2.2   We approximate x_0 by its partial sums:

$$u_0 u_1 \ldots u_n x_0 - u_0 u_1 \ldots u_n \sum_{k=0}^{n} \frac{1}{u_0 u_1 \ldots u_k} = \sum_{k=1}^{+\infty} \frac{1}{u_{n+1} \ldots u_{n+k}}.$$

As $u_n \ge 2$ for every $n \in \mathbb{N}$, we obtain

$$0 < u_0 u_1 \ldots u_n x_0 - u_0 u_1 \ldots u_n \sum_{k=0}^{n} \frac{1}{u_0 u_1 \ldots u_k} \le \frac{1}{u_{n+1}}\left(1 + \frac{1}{2} + \frac{1}{2^2} + \cdots \right) = \frac{2}{u_{n+1}}.$$

Since $u_n \to +\infty$, x_0 is irrational by theorem 1.5.

Exercise 2.3   1) For every $n \in \mathbb{N}$, we have $q_n \le \alpha_n / (\alpha_n - 1) < q_n + 1$, whence $\alpha_n q_n - q_n \le \alpha_n < \alpha_n q_n - q_n + \alpha_n - 1$. Consequently, for every $n \in \mathbb{N}$,

$$\alpha_n > 1 + 1/q_n \quad \text{and} \quad \alpha_n \le 1 + 1/(q_n - 1) \quad \text{if} \quad q_n \ge 2. \tag{*}$$

If $q_n = 1$, the inequality $q_{n+1} \ge q_n^2$ is obvious, therefore we assume $q_n \ge 2$.

From (2.8) we have $\alpha_n = \alpha_{n+1}(1 + 1/q_n)$. Therefore (*) implies

$$(1 + 1/q_{n+1})(1 + 1/q_n) < 1 + 1/(q_n - 1).$$

Expanding yields $q_n^2 - 1 < q_{n+1}$. Since q_n are q_{n+1} integers, we have $q_n^2 \le q_{n+1}$.

2) Assume that $q_n = 1$ for every $n \in \mathbb{N}$. Then the estimate $\alpha_n > 1$ and (2.9) imply that $\alpha_0 > 2^{n+1}$. Contradiction when $n \to +\infty$.

3) Let N such that $q_N \ge 2$. Then $q_{n+1} \ge q_n^2$ yields $q_n \ge q_N^{2^{n-N}} \ge 2^{2^{n-N}}$, $\forall n \ge N$. As $1 + 1/q_n < \alpha_n \le 1 + 1/(q_n - 1)$ by question 1, we obtain $\lim_{n \to +\infty} \alpha_n = 1$. Hence (2.9) implies $\lim_{n \to +\infty} \prod_{k=0}^{n} (1 + 1/q_k) = \alpha_0$, which means that the infinite product converges to α_0.

4) Since $q'_{n+1} \ge q_n'^2$, we have $\alpha_0 \le \prod_{n=0}^{+\infty} (1 + 1/q_0'^{2^n})$. But this infinite product can be computed the following way for $q_0' \ge 2$:

$$\prod_{n=0}^{+\infty} (1 + 1/q_0'^{2^n}) = \prod_{n=0}^{+\infty} (1 - 1/q_0'^{2^{n+1}}) \Big/ \prod_{n=0}^{+\infty} (1 - 1/q_0'^{2^n}) = 1/(1 - 1/q_0') = q_0'/(q_0' - 1).$$

Therefore $(q_0' - 1)\alpha_0 \le q_0'$ if $q_0' \ge 2$. As this remains true if $q_0' = 1$, we have $(q_0' - 1)\alpha_0 \le q_0'$ for every $q_0' \in \mathbb{N} - \{0\}$. Hence $q_0' \le \alpha_0 / (\alpha_0 - 1)$. However $\alpha_0 > 1 + 1/q_0'$ implies that $q_0' + 1 > \alpha_0 / (\alpha_0 - 1)$, whence

$$q_0' = \left[\alpha_0 / (\alpha_0 - 1) \right] = q_0.$$

Now we put $\alpha_0 = (1 + 1/q_0)\alpha_1$, and have $\alpha_1 = \prod_{n=1}^{+\infty} (1 + 1/q_n')$. Arguing the same way shows, by induction, that $q_n' = q_n$ for every $n \in \mathbb{N}$.

Exercise 2.4   1) We can argue as in exercise 3, question 4. If x is not a root of unity, we have

$$(1-x)\prod_{k=0}^{n-1}(1 + x^{2^k}) = (1-x)\frac{\prod_{k=0}^{n-1}(1 - x^{2^{k+1}})}{\prod_{k=0}^{n-1}(1 - x^{2^k})} = (1-x)\frac{1 - x^{2^n}}{1 - x} = 1 - x^{2^n}.$$

By continuity, this remains true for every complex number x. Therefore, for every $x \in \mathbb{C}$ such that $|x| < 1$, $\prod_{k=0}^{+\infty}(1 + x^{2^k}) = 1/(1-x)$. Now if $q_{n+1} = q_n^2$ for every $n \ge N$, this yields

$$\alpha_N = \prod_{n=N}^{+\infty} \left(1 + 1/q_n\right) = \prod_{n=0}^{+\infty} \left(1 + 1/q_N^{2^n}\right) = q_N/(q_N - 1) \in \mathbb{Q}.$$

Hence $\alpha_0 = (1 + 1/q_0)...(1 + 1/q_{N-1})\alpha_N \in \mathbb{Q}$.

2) If α_0 is rational, so is α_n for every $n \in \mathbb{N}$. Put $\alpha_n = a_n/b_n$, with a_n and b_n coprime and $a_n > b_n$ (since $\alpha_n > 1$). The equality $\alpha_n = (1 + 1/q_n)\alpha_{n+1}$ yields

$$\frac{a_{n+1}}{b_{n+1}} = \frac{q_n a_n}{(q_n + 1)b_n}.$$

Since a_{n+1} and b_{n+1} are coprime, there exists $d_n \in \mathbb{N}^*$ such that $q_n a_n = a_{n+1} d_n$ and $(q_n + 1)b_n = b_{n+1}d_n$. Substracting yields

$$q_n(a_n - b_n) - b_n = d_n(a_{n+1} - b_{n+1}) \geq a_{n+1} - b_{n+1}.$$

But $q_n \leq \alpha_n/(\alpha_n - 1) \Rightarrow q_n(a_n - b_n) \leq a_n$, and therefore $a_n - b_n \geq a_{n+1} - b_{n+1}$. As sequence $(a_n - b_n)$ is a sequence of positive integers, this implies the existence of $k \in \mathbb{N}$ such that $a_n - b_n = k$ for $n \geq N$. Replacing in the relation $q_n(a_n - b_n) - b_n = d_n(a_{n+1} - b_{n+1})$ yields $q_n k - b_n = d_n k$. Hence b_n is divisible by k, and so is a_n since $a_n - b_n = k$. Now a_n and b_n are coprime, whence $k = 1$, and $a_n = \alpha_n/(\alpha_n - 1)$. By the definition of q_n, we now have $q_n = a_n$ for $n \geq N$. Since $q_n a_n = a_{n+1}d_n$, this implies that $q_n^2 = q_{n+1}d_n \geq q_{n+1}$ for $n \geq N$. But $q_{n+1} \geq q_n^2$, and therefore $q_{n+1} = q_n^2$ for every $n \geq N$.

Exercise 2.5   We have $\alpha = \dfrac{1}{p} \times \dfrac{1}{1 - 1/p} = \displaystyle\sum_{n=1}^{+\infty} \dfrac{1}{p^n}$ for $p \in \mathbb{N}$, $p \geq 2$. (*)

This gives the p-adic expansion of α if $p \geq 3$. However, if $p = 2$, the "digits" c_n satify $c_n = 1 = p - 1$ for every $n \in \mathbb{N}^*$, and therefore (*) is not the 2-adic expansion of α. As a matter of fact, if $p = 2$, the 2-adic expansion of $\alpha = 1$ is $\alpha = 1 + \sum_{n=1}^{+\infty} 0 \times 2^{-n} = 1,0000...$

Exercise 2.6   Applying the algorithm of the 7-adic expansion yields

$$\beta = \frac{35}{9} = \frac{27 + 8}{9} = 3 + \frac{8}{9} = 3 + \frac{1}{7}\frac{56}{9} = 3 + \frac{1}{7}\frac{54 + 2}{9} = 3 + \frac{6}{7} + \frac{1}{7^2}\frac{14}{9}$$

$$= 3 + \frac{6}{7} + \frac{1}{7^2}\left(1 + \frac{5}{9}\right) = 3 + \frac{6}{7} + \frac{1}{7^2} + \frac{1}{7^3}\frac{35}{9}.$$

As we obtain again $35/9$, the expansion is now periodic, and the 7-adic expansion of β is $\beta = 3,61361361361... = 3,6\overline{1361}$.

Exercise 2.7   If $e^{\sqrt{2}}$ was rational, so would be $\cosh\sqrt{2} = (e^{\sqrt{2}} + e^{-\sqrt{2}})/2$. But

$$\cosh\sqrt{2} = 1 + \frac{(\sqrt{2})^2}{2!} + \frac{(\sqrt{2})^4}{4!} + \frac{(\sqrt{2})^6}{6!} + \cdots + \frac{(\sqrt{2})^{2n}}{(2n)!} + \cdots$$

$$= 1 + \frac{1}{1} + \frac{1}{1(3\cdot 2)} + \frac{1}{1(3\cdot 2)(5\cdot 3)} + \cdots + \frac{1}{1(3\cdot 2)\ldots(2n-1)n} + \cdots.$$

This is the Engel series expansion of $\cosh\sqrt{2}$, with $u_n = (2n-1)n$ for $n \geq 1$. Hence $\cosh\sqrt{2} \notin \mathbb{Q}$ by theorem 2.4, and $e^{\sqrt{2}}$ is irrational.

Exercise 2.8   1) Let $x_n = t_n - \sqrt{t_n^2 - 1}$ be the least positive root of the equation $X^2 - 2t_n X + 1 = 0$. Then evidently x_n is also the least positive root of the equation $t_n X^2 - 2t_n^2 X + t_n = 0$. Therefore

$$x_n = \left(2t_n^2 - \sqrt{4t_n^4 - 4t_n^2}\right)\Big/2t_n = \left(2t_n^2 - 1 + 1 - \sqrt{(2t_n^2 - 1)^2 - 1}\right)\Big/2t_n.$$

Since $t_{n+1} = 2t_n^2 - 1$, we obtain $x_n = t_n - \sqrt{t_n^2 - 1} = \left(1 + t_{n+1} - \sqrt{t_{n+1}^2 - 1}\right)\Big/2t_n$.

2) Hence $x_n = (1 + x_{n+1})/2t_n$, which yields by induction

$$x_0 = \frac{1}{2t_0} + \frac{1}{2t_0 2t_1} + \cdots + \frac{1}{2t_0 \ldots 2t_n} + \frac{x_{n+1}}{2t_0 \ldots 2t_n}.$$

However, $x_n = 1\Big/\left(t_n + \sqrt{t_n^2 - 1}\right) \leq 1/t_n$. Since t_n is increasing and tends to $+\infty$, we see that $\lim_{n\to+\infty} x_{n+1}/(2t_0 \ldots 2t_n) = 0$, whence

$$x_0 = t_0 - \sqrt{t_0^2 - 1} = \frac{1}{2t_0} + \frac{1}{2t_0 2t_1} + \cdots + \frac{1}{2t_0 \ldots 2t_n} + \cdots.$$

The conclusion follows from the uniqueness of Engel's series expansion.

3) Replace t_0 by $t_0 = 2$ $\left(\text{whence } \sqrt{t_0^2 - 1} = 3\right)$, $t_0 = 3$ $\left(\text{whence } \sqrt{t_0^2 - 1} = 2\sqrt{2}\right)$, $t_0 = 8$ $\left(\text{whence } \sqrt{t_0^2 - 1} = 3\sqrt{7}\right)$, $t_0 = 9$ $\left(\text{whence } \sqrt{t_0^2 - 1} = 4\sqrt{5}\right)$.

In fact, it is possible to express $\sqrt{n}$ by means of an Engel series if n is not a perfect square. This is connected with Pell's equation (see exercise 4.13).

Exercise 2.9   1) Since $q_n = [\alpha_n] + 1$, we have $q_n - 1 \leq \alpha_n < q_n$, whence

$$\frac{1}{\alpha_n} = \frac{1}{q_n} + \frac{1}{\alpha_{n+1}} \implies \frac{1}{q_n - 1} > \frac{1}{q_n} + \frac{1}{q_{n+1}}.$$

Therefore $q_{n+1} > q_n^2 - q_n$. Since q_n is an integer, this yields $q_{n+1} \geq q_n^2 - q_n + 1$.

2) By question 1, we have $q_{n+1} - q_n \geq (q_n - 1)^2$. Since $q_0 \geq 2$ ($\alpha_0 \geq 1$), by induction we see easily that (q_n) is increasing. Therefore $q_n \to +\infty$, and so does α_n, since $q_n - 1 \leq \alpha_n$. However

$$\gamma = \frac{1}{\alpha_0} = \frac{1}{q_0} + \frac{1}{\alpha_1} = \frac{1}{q_0} + \frac{1}{q_1} + \frac{1}{\alpha_2} = \cdots = \frac{1}{q_0} + \frac{1}{q_1} + \cdots + \frac{1}{q_n} + \frac{1}{\alpha_{n+1}}.$$

Hence $\gamma = \sum_{n=0}^{+\infty} 1/q_n$. It remains to prove that the expansion is unique.

For this, assume that $\gamma = \sum_{n=0}^{+\infty} 1/q'_n$, with $q'_{n+1} \geq q'^2_n - q'_n + 1$.

Put $\gamma = 1/\alpha_0$. Then $1/\alpha_0 = 1/q'_0 + 1/\alpha'_1$, whence $1/\alpha_0 > 1/q'_0$ and $q'_0 > \alpha_0$.

Moreover $q'_{n+1} - 1 \geq q'_n(q'_n - 1)$, and therefore $\dfrac{1}{q'_{n+1} - 1} \leq \dfrac{1}{q'_n - 1} - \dfrac{1}{q'_n}$.

If we add these equalities for n from 0 to $+\infty$, we obtain $1/\alpha_0 \leq 1/(q'_0 - 1)$, whence $\alpha_0 \geq q'_0 - 1$. Thus $q'_0 \leq \alpha_0 + 1 < q'_0 + 1$, and $q'_0 = q_0$. Consequently $\alpha'_1 = \alpha_1$, and by induction we now see that $q'_n = q_n$ for every $n \in \mathbb{N}$.

3) First we prove that γ is rational if $q_{n+1} = q_n^2 - q_n + 1$ for $n \geq N$. In this case

$$\frac{1}{q_{n+1} - 1} = \frac{1}{q_n - 1} - \frac{1}{q_n} \text{ for } n \geq N.$$

Adding these equalities for $n = N$ to $+\infty$ yields $\dfrac{1}{\alpha_N} = \sum_{n=N}^{+\infty} \dfrac{1}{q_n} = \dfrac{1}{q_N - 1}$.

Therefore $1/\alpha_N$ is rational, and so is γ.

Assume now that $\gamma \in \mathbb{Q}$, $\gamma = p/q$. The equality $\gamma = 1/q_0 + \cdots + 1/q_{n-1} + 1/\alpha_n$ shows that $1/\alpha_n = k_n/q q_0 \ldots q_{n-1}$, with $k_n \in \mathbb{N}$. Since $1/\alpha_n = 1/q_n + 1/\alpha_{n+1}$, this implies that

$$q q_0 \ldots q_{n-1} + k_{n+1} = k_n q_n \quad (*).$$

However, by the definition of q_n we have $\alpha_{n+1} \geq q_{n+1} - 1 \geq q_n^2 - q_n$. Hence

$$\frac{q q_0 \ldots q_n}{k_{n+1}} \geq q_n(q_n - 1) \text{ and } q q_0 \ldots q_{n-1} \geq k_{n+1}(q_n - 1).$$

Replacing in (*) yields $k_n q_n \geq k_{n+1} q_n$. Therefore (k_n) is not increasing, and there exists an integer k such that $k_n = k$ for every $n \geq N$. This yields

$$\frac{1}{\alpha_{n+1}} = \frac{k}{q q_0 \ldots q_n} = \frac{1}{q_n} \frac{1}{\alpha_n} = \frac{1}{q_n}\left(\frac{1}{q_n} + \frac{1}{\alpha_{n+1}}\right).$$

Hence $\alpha_{n+1} = q_n(q_n - 1)$. But $\alpha_{n+1} \geq q_{n+1} - 1$, and therefore $q_{n+1} \leq q_n^2 - q_n + 1$ for $n \geq N$. Since $q_{n+1} \geq q_n^2 - q_n + 1$, we obtain $q_{n+1} = q_n^2 - q_n + 1$ for $n \geq N$, Q.E.D.

4) Put $q_n = 2^{2^n} - a$. Then

$$q_{n+1} - q_n^2 + q_n = 2^{2^{n+1}} - a - (2^{2^{n+1}} - 2a2^{2^n} + a^2) + 2^{2^n} - a = (2a+1)2^{2^n} - 2a - a^2.$$

Since $a \geq 0$, for $n \geq N$ we have $q_{n+1} - q_n^2 + q_n \geq 2$. Hence $\sum_{n=N}^{+\infty} 1/q_n$ is a Sylvester series whose sum is an irrational number.

Remark This result generalizes the proof of the irrationality of $\sum_{n=1}^{+\infty} 1/(2^{2^n} - 1)$ given in exercise 1.9. It remains true for $a \in \mathbb{Z}$ (exercise 8.9). As a matter of fact, all these numbers are transcendental. It can be proved by using Mahler's method (exercise 12.13).

Exercise 2.10   1) $\alpha = 3,300330000000000330033000...$

2) For any $n \geq 1$, the numbers 3.4^n, $3.4^n + 1$, $3.4^n + 2$, $\cdots$, $3.4^n + 4^{n-1}$ have at least a digit different from 0 or 1 in their representation in base 4. Thus the decimal representation of α contains arbitrarily long sequences of consecutive zeros. As it also contains infinitely many digits 3, it cannot be periodic and α is irrational by theorem 2.2.

3) Clearly $f(x) = \lim_{n \to \infty} P_n(x)$, with $P_n(x) = \prod_{k=0}^{n} \left(1 + x^{4^k}\right) = \sum_{k=0}^{N} a_k x^k$, $a_k = 1$ if $k \in E$, $a_k = 0$ if $k \notin E$, $N = \frac{1}{3}(4^{n+1} - 1)$. Thus $f(x) = \sum_{k=0}^{+\infty} a_k x^k = \sum_{n \in E} x^n$.

4) $f(x)g(x) = \prod_{n=0}^{+\infty} \left(1 + x^{4^n}\right)\left(1 + x^{2.4^n}\right) = \prod_{n=0}^{+\infty} \left(1 + x^{4^n} + x^{2.4^n} + x^{3.4^n}\right)$.

Thus $f(x)g(x) = \prod_{n=0}^{+\infty} \frac{1 - x^{4^{n+1}}}{1 - x^{4^n}} = \lim_{n \to +\infty} \frac{1 - x^4}{1 - x} \frac{1 - x^{4^2}}{1 - x^4} \cdots \frac{1 - x^{4^{n+1}}}{1 - x^{4^n}} = \frac{1}{1 - x}$.

5) For $x = \frac{1}{10}$, this yields $\frac{1}{3}\alpha g\left(\frac{1}{10}\right) = \frac{10}{9}$. Therefore $\frac{1}{\alpha} = \frac{3}{10} g\left(\frac{1}{10}\right)$.

By arguing as in question 3, we see now that $g(x) = \sum_{n \in 2E} x^n$.

Therefore $\frac{1}{\alpha} = 3 \sum_{n \in 2E} 10^{-n-1} = 0,3030000030300...$

Solutions to the exercises of chapter 3

Exercise 3.1   By theorem 3.1, we have

$$P_n Q_{n-1} - P_{n-1} Q_n = (b_n P_{n-1} + a_n P_{n-2}) Q_{n-1} - P_{n-1}(b_n Q_{n-1} + a_n Q_{n-2})$$
$$= -a_n (P_{n-1} Q_{n-2} - P_{n-2} Q_{n-1}).$$

Therefore, by induction,

$$P_n Q_{n-1} - P_{n-1} Q_n = (-1)^n a_1 a_2 \ldots a_n (P_0 Q_{-1} - P_{-1} Q_0) = (-1)^{n-1} a_1 a_2 \ldots a_n.$$

Hence (3.4) is proved. And we have

$$R_n - R_{n-1} = \frac{P_n}{Q_n} - \frac{P_{n-1}}{Q_{n-1}} = \frac{P_n Q_{n-1} - P_{n-1} Q_n}{Q_n Q_{n-1}} = \frac{(-1)^{n-1} a_1 a_2 \ldots a_n}{Q_n Q_{n-1}}.$$

Exercise 3.2   Theorem 3.3 is an easy consequence of (3.5), since

$$R_n - R_0 = \sum_{k=1}^{n} (R_k - R_{k-1}) = \sum_{k=1}^{n} \frac{(-1)^{k-1} a_1 a_2 \ldots a_k}{Q_k Q_{k-1}}.$$

Exercise 3.3   Compute the denominators of the convergents of

$$R = \frac{a_0}{b_0} \ \frac{a_1 b_0}{-a_1 + b_1} \ \frac{a_2 b_1}{-a_2 + b_2} - \cdots.$$

By theorem 3.1, we see that $Q_0 = 1$, $Q_1 = b_0$, $Q_2 = (a_1 + b_1)Q_1 - a_1 b_0 Q_0 = b_1 b_0$, $Q_3 = (a_2 + b_2)Q_2 - a_2 b_1 Q_1 = b_2 b_1 b_0$. Now an easy induction shows that $Q_n = b_0 b_1 \ldots b_{n-1}$. Hence, by theorem 3.2, we know that R is convergent, and that its value is $\sum_{n=1}^{+\infty} (-1)^{n-1} A_1 \ldots A_n / Q_n Q_{n-1}$, with $A_1 = a_0$, $A_n = -a_{n-1} b_{n-2}$ for $n \geq 2$. This proves theorem 3.4.

Exercise 3.4   1) Prove that (v_n) is a Cauchy sequence. We have

$$v_{n+p} - v_n = \prod_{k=0}^{n} (1 + u_k) \times [(1 + u_{n+1})(1 + u_{n+2}) \ldots (1 + u_{n+p}) - 1]$$

$$= \prod_{k=0}^{n} (1 + u_k) \times \left(\sum_{i=1}^{p} u_{n+i} + \sum_{i=1}^{p-1} \sum_{j=i+1}^{p} u_{n+i} u_{n+j} + \cdots + u_{n+1} u_{n+2} \ldots u_{n+p} \right).$$

Therefore $\left| v_{n+p} - v_n \right| \leq \prod_{k=0}^{n} (1 + |u_k|) \left(\sum_{i=1}^{p} |u_{n+i}| + \sum_{i=1}^{p-1} \sum_{j=i+1}^{p} |u_{n+i}||u_{n+j}| + \cdots \right).$

In other words, $\left| v_{n+p} - v_n \right| \leq \left| w_{n+p} - w_n \right|$, where $w_n = \prod_{k=0}^{n} (1 + |u_k|)$.

But $\operatorname{Log} w_n = \sum_{k=0}^{n} \operatorname{Log}(1 + |u_k|) \leq \sum_{k=0}^{n} |u_k|$ since $\operatorname{Log}(1+x) \leq x$ for $x > -1$. Hence (w_n), which is increasing and bounded above by $\sum_{k=0}^{+\infty} |u_k|$, is convergent, and is therefore a Cauchy sequence. For any $\varepsilon > 0$, there exists N such that $n \geq N$ and $p \geq 0 \Rightarrow \left| w_{n+p} - w_n \right| \leq \varepsilon$. This yields $\left| v_{n+p} - v_n \right| \leq \varepsilon$. Consequently (v_n) is a Cauchy sequence, hence convergent in $\mathbb{C}$.

2) Since $\sum_{n=0}^{+\infty} |u_n|$ is convergent, we have $\lim_{n \to +\infty} u_n = 0$. Let $N \in \mathbb{N}$ such that $u_n \leq \frac{1}{2}$ for every $n \geq N$. We can write $\left| \prod_{k=N}^{n} (1 + u_k) \right| \geq \prod_{k=N}^{n} (1 - |u_k|)$. But

$\prod_{k=N}^{n}(1-|u_k|)$ has a non-zero finite limit when $n \to +\infty$. Indeed, $\text{Log}\left(\prod_{k=N}^{n}(1-|u_k|)\right) = \sum_{k=N}^{n}\text{Log}(1-|u_k|)$ has a finite limit ℓ, since the series $\sum_{k=N}^{+\infty}\text{Log}(1-|u_k|)$ is convergent by comparison to $-\sum_{k=N}^{+\infty}|u_k|)$. Therefore $\prod_{k=N}^{n}(1-|u_k|) \to e^{\ell} \neq 0$. This yields $\left|\prod_{k=N}^{+\infty}(1+u_k)\right| \geq e^{\ell}$, and consequently $\prod_{k=N}^{+\infty}(1+u_k) \neq 0$. Thus $\prod_{k=0}^{+\infty}(1+u_k)$ is zero if and only if one among the factors $(1+u_0), (1+u_1),..., (1+u_{N-1})$ is zero.

Exercise 3.5 By the change of variable $x = \sqrt{t}$, we see that

$$\Gamma(1/2) = \int_0^{+\infty} t^{-\frac{1}{2}} e^{-t} dt = 2\int_0^{+\infty} e^{-x^2} dx.$$

Now compute *the Gauss integral* $I = \int_0^{+\infty} e^{-x^2} dx$. We have

$$I^2 = \int_0^{+\infty} e^{-x^2} dx \int_0^{+\infty} e^{-y^2} dy = \int_0^{+\infty}\int_0^{+\infty} e^{-x^2} e^{-y^2} dxdy.$$

Introducing polar coordinates yields

$$I^2 = \int_{\theta=0}^{\pi/2}\int_{r=0}^{+\infty} e^{-r^2} rdrd\theta = [\theta]_0^{\pi/2} \cdot \left[-\frac{1}{2}e^{-r^2}\right]_0^{+\infty} = \frac{\pi}{4}.$$

Therefore $I = \sqrt{\pi}/2$ and $\Gamma(1/2) = \sqrt{\pi}$.

Exercise 3.6 By definition $J_{1/2}(x) = \frac{x^{1/2}}{\sqrt{2}}\sum_{k=0}^{+\infty}\frac{(-1)^k}{k!\Gamma(\frac{1}{2}+k+1)}\frac{x^{2k}}{4^k}$. Now

$$\Gamma\left(\frac{1}{2}+k+1\right) = \left(\frac{1}{2}+k\right)\Gamma\left(\frac{1}{2}+k\right) = \left(\frac{1}{2}+k\right)\left(\frac{1}{2}+k-1\right)\Gamma\left(\frac{1}{2}+k-1\right)$$

$$= \cdots = \left(\frac{1}{2}+k\right)\left(\frac{1}{2}+k-1\right)\cdots\frac{1}{2}\Gamma\left(\frac{1}{2}\right) = \frac{1}{2^{k+1}}(2k+1)(2k-1)\cdots3\cdot1\cdot\sqrt{\pi}.$$

Hence $J_{1/2}(x) = \frac{x^{-1/2}}{\sqrt{2\pi}}\sum_{k=0}^{+\infty}\frac{2(-1)^k x^{2k+1}}{k!2^k(2k+1)(2k-1)\cdots3\cdot1}$

$$= \sqrt{\frac{2}{\pi}}x^{-1/2}\sum_{k=0}^{+\infty}\frac{(-1)^k x^{2k+1}}{(2k+1)!} = \sqrt{\frac{2}{\pi}}x^{-1/2}\sin x.$$

The proof is similar for $J_{-1/2}(x)$.

Exercise 3.7 We look for solutions of (3.23) in the form

$$y = x^\nu\sum_{n=0}^{+\infty}a_n x^n = \sum_{n=0}^{+\infty}a_n x^{n+\nu}.$$

Deriving term by term and replacing in (3.23) yields, for every $n \in \mathbb{N}$,

$$(v+n+2)(v+n+1)a_{n+2} + (v+n+2)a_{n+2} + a_n - v^2 a_{n+2} = 0.$$

For even indices, this relation can be written as $a_{2p+2} = \dfrac{-a_{2p}}{4(p+1)(v+p+1)}$.

Choosing $a_0 = 1/\left(2^v \Gamma(v+1)\right)$ yields Bessel function J_v.

Exercise 3.8   By using repeatedly the property $\Gamma(v+1) = v.\Gamma(v)$, we have

$$J_{v-1}(x) + J_{v+1}(x) = \left(\frac{x}{2}\right)^{v-1} \sum_{n=0}^{+\infty} \frac{(-x^2/4)^n}{n!\Gamma(v+n)} + \left(\frac{x}{2}\right)^{v+1} \sum_{n=0}^{+\infty} \frac{(-x^2/4)^n}{n!\Gamma(v+n+2)}$$

$$= \left(\frac{x}{2}\right)^{v-1} \left(\sum_{n=0}^{+\infty} \frac{(-x^2/4)^n}{n!\Gamma(v+n)} - \sum_{n=0}^{+\infty} \frac{(-x^2/4)^{n+1}}{n!\Gamma(v+n+2)} \right)$$

$$= \left(\frac{x}{2}\right)^{v-1} \left(\frac{1}{\Gamma(v)} + \sum_{n=1}^{+\infty} \left(\frac{1}{n!\Gamma(v+n)} - \frac{1}{(n-1)!\Gamma(v+n+1)} \right) \left(\frac{-x^2}{4} \right)^n \right)$$

$$= \frac{2}{x}\left(\frac{x}{2}\right)^{v} \left(\frac{v}{\Gamma(v+1)} + \sum_{n=1}^{+\infty} \frac{v}{n!\Gamma(v+n+1)} \left(\frac{-x^2}{4} \right)^n \right) = \frac{2v}{x} J_v(x).$$

Exercise 3.9 $J_{v+n}(x) = \left(\dfrac{x}{2}\right)^{v+n} \displaystyle\sum_{k=0}^{+\infty} \dfrac{(-1)^k}{k!\Gamma(v+n+k+1)} \left(\dfrac{x}{2}\right)^{2k}$

$$= \left(\frac{x}{2}\right)^{v+n} \frac{1}{\Gamma(v+n+1)} (1 + R_n(x)),$$

with $R_n(x) = \displaystyle\sum_{k=1}^{+\infty} \dfrac{(-x^2/4)^k}{k!(v+n+k)(v+n+k-1)...(v+n+1)}$.

For n sufficiently large, $v+n > \dfrac{x^2}{4}$ and $\left| R_n(x) \right| \le \displaystyle\sum_{k=1}^{+\infty} \left(\dfrac{x^2}{4(v+n)} \right)^k$.

Therefore $\left| R_n(x) \right| \le \dfrac{x^2/4(v+n)}{1 - x^2/4(v+n)}$. Consequently $\lim_{n \to +\infty} R_n(x) = 0$ and

$$J_{v+n}(x) \underset{n \to +\infty}{\sim} \left(\frac{x}{2}\right)^{v+n} \frac{1}{\Gamma(v+n+1)}.$$

In particular, $J_{v+n}(x) \ne 0$ for n sufficiently large. Now suppose that there exists $x \ne 0$ such that $J_v(x) = J_{v-1}(x) = 0$. Then by (3.24), $J_{v+1}(x) = 0 = J_{v+2}(x) = \cdots = J_{v+n}(x)$ for every n. Hence $J_v(x)$ and $J_{v-1}(x)$ are not both zero.

Exercise 3.10   Start from the *harmonic alternating series*

$$\text{Log}\,2 = \sum_{n=1}^{+\infty} \frac{(-1)^{n-1}}{n} = \sum_{n=0}^{+\infty} \frac{a_0 a_1 \dots a_n}{b_0 b_1 \dots b_n},$$

with $a_0 = 1$, $a_n = -n$ for $n \geq 1$, $b_n = n+1$ for all n, and apply theorem 3.4.

Exercise 3.11   1) (F_n) is a second order linear recurring sequence, with characteristic equation $r^2 = r+1$, whose solutions are the golden number $\Phi = (1+\sqrt{5})/2$, and $\Psi = -1/\Phi = (1-\sqrt{5})/2$. Therefore $F_n = A\Phi^n + B\Psi^n$. As $F_0 = 0$ and $F_1 = 1$, we obtain $F_n = (\Phi^n - \Psi^n)/\sqrt{5}$.

2) Let $R_n = P_n/Q_n$ be the n-th convergent of this continued fraction. Then $Q_{-1} = 0$, $Q_0 = 1$ and $Q_{n+1} = Q_n + Q_{n-1}$ by theorem 3.1. Therefore $Q_n = F_{n+1}$ for every $n \geq -1$. Consequently Q_n is increasing and tends to $+\infty$ (we even have $F_n \underset{+\infty}{\sim} \Phi^n/\sqrt{5}$). Hence the series $\sum_{n=1}^{+\infty} (-1)^{n-1}/Q_n Q_{n-1}$ is convergent by the alternating series test. Now theorem 3.3 shows that the continued fraction α is convergent. As plainly $\alpha = 1 + 1/\alpha$, we have $\alpha^2 = \alpha + 1$, whence $\alpha = \Phi$ or $\alpha = -1/\Psi$. But $\alpha > 0$, and therefore $\alpha = \Phi$, which proves that

$$\Phi = \frac{1+\sqrt{5}}{2} = 1 + \frac{1}{1+} \frac{1}{1+} \cdots \frac{1}{1+} \cdots$$

3) Theorem 3.3 immediatly yields $\Phi = 1 + \sum_{n=1}^{+\infty} \frac{(-1)^{n-1}}{Q_n Q_{n-1}} = 1 + \sum_{n=1}^{+\infty} \frac{(-1)^{n-1}}{F_{n+1} F_n}$.

Therefore $\sum_{n=1}^{+\infty} \frac{(-1)^{n-1}}{F_n F_{n+1}} = \Phi - 1 = \frac{\sqrt{5}-1}{2}$.

Exercise 3.12   1) a) By theorem 3.1, we know that $Q_n = c_n Q_{n-1} + Q_{n-2}$. Since $c_n > 0$, this implies $Q_n > Q_{n-2}$. Hence $Q_n Q_{n-1} > Q_{n-1} Q_{n-2}$, and the sequence $Q_n Q_{n+1}$ is increasing. On the other hand, if $Q_{n-2} \leq Q_{n-1}$, then $Q_n = c_n Q_{n-1} + Q_{n-2}$ implies that $Q_n \leq (1+c_n)Q_{n-1}$. If, on the contrary, $Q_{n-1} \leq Q_{n-2}$, it implies $Q_n \leq (1+c_n)Q_{n-2}$ and, *a fortiori*, $Q_n \leq (1+c_n)(1+c_{n-1})Q_{n-2}$.
Hence, by induction, starting from Q_n, we will have two cases,

$$Q_n \leq (1+c_n)(1+c_{n-1})\dots(1+c_1)Q_0 \quad \text{or} \quad Q_n \leq (1+c_n)(1+c_{n-1})\dots(1+c_2)Q_1.$$

Therefore $Q_n \leq C \prod_{k=1}^{n}(1+c_k)$, with $C = Q_0$ or $C = Q_1/(1+c_1)$.

b) The equality $Q_n = c_n Q_{n-1} + Q_{n-2}$ implies $Q_n - Q_{n-2} = c_n Q_{n-1}$. If n is even, with $n = 2p$, this yields $Q_{2p} - Q_{2p-2} = c_{2p} Q_{2p-1}$. If n is odd, with $n = 2p+1$,

$Q_{2p+1} - Q_{2p-1} = c_{2p+1}Q_{2p}$. The sequence Q_{2p+1} is therefore increasing, whence $Q_{2p+1} \geq Q_1 = c_1$. Replacing in $Q_{2p} - Q_{2p-2} = c_{2p}Q_{2p-1}$ yields $Q_{2p} - Q_{2p-2} \geq c_{2p}c_1$, and therefore $Q_{2p} - Q_0 \geq c_1(c_2 + c_4 + \cdots + c_{2p})$ by addition. This proves the first inequality since $Q_0 = 1$. The proof of the second inequality is similar.

c) If the series $\sum_{n=1}^{+\infty} c_n$ is convergent, then

$$Q_n \leq C\prod_{k=1}^{n}(1+c_k) \quad \Rightarrow \quad \operatorname{Log}Q_n \leq \operatorname{Log}C + \sum_{k=1}^{n}\operatorname{Log}(1+c_k) \leq \operatorname{Log}C + \sum_{k=1}^{n}c_k.$$

Thus Q_n is bounded above, $Q_n \leq M$, say, for every $n \in \mathbb{N}$. Therefore the sequence Q_nQ_{n-1} does not tend to $+\infty$, which proves that the series

$$\sum_{n=1}^{+\infty}(-1)^{n-1}a_1 \ldots a_n \big/ Q_nQ_{n-1} = \sum_{n=1}^{+\infty}(-1)^{n-1} \big/ Q_nQ_{n-1}$$

is divergent since its terms do not tend to 0. Hence the continued fraction is divergent by theorem 3.3.

d) We have just shown that $\dfrac{1}{c_1} + \dfrac{1}{c_2} + \cdots + \dfrac{1}{c_n} + \cdots$ convergent $\Rightarrow \sum_{n=1}^{+\infty} c_n$ divergent.

It remains to prove that $\sum_{n=1}^{+\infty} c_n$ divergent $\Rightarrow \dfrac{1}{c_1} + \dfrac{1}{c_2} + \cdots + \dfrac{1}{c_n} + \cdots$ convergent.

Suppose that $\sum_{n=1}^{+\infty} c_n$ is divergent. As $c_n > 0$, at least one of the two series $c_2 + c_4 + \cdots + c_{2p} + \cdots$ or $c_1 + c_3 + \cdots + c_{2p+1} + \cdots$ must be divergent. Hence, by question b), at least one of the two sequences Q_{2p} or Q_{2p+1} tends to $+\infty$. Thus $\lim_{n \to +\infty} Q_nQ_{n-1} = +\infty$. On the other hand, Q_nQ_{n-1} is increasing (question a). Therefore the series $\sum_{n=1}^{+\infty}(-1)^{n-1}a_1 \ldots a_n \big/ Q_nQ_{n-1} = \sum_{n=1}^{+\infty}(-1)^{n-1} \big/ Q_nQ_{n-1}$ is convergent by the alternating series test, and so is the continued fraction.

2) The continued fraction $\dfrac{a_1}{b_1} + \dfrac{a_2}{b_2} + \cdots$ is equivalent to $\dfrac{1}{c_1} + \dfrac{1}{c_2} + \cdots$, with

$$c_{2n} = \frac{a_1a_3 \ldots a_{2n-1}}{a_2a_4 \ldots a_{2n}}b_{2n} \quad, \quad c_{2n+1} = \frac{a_2a_4 \ldots a_{2n}}{a_1a_3 \ldots a_{2n+1}}b_{2n+1}.$$

Therefore $c_{2n}c_{2n+1} = b_{2n}b_{2n+1}\big/a_{2n+1}$ and $c_{2n-1}c_{2n} = b_{2n-1}b_{2n}\big/a_{2n}$.

Now the classical inequality $\sqrt{ab} \leq (a+b)/2$ yields

$$c_{2n} + c_{2n+1} \geq 2\sqrt{b_{2n}b_{2n+1}/a_{2n+1}} \quad \text{and} \quad c_{2n-1} + c_{2n} \geq 2\sqrt{b_{2n-1}b_{2n}/a_{2n}}.$$

Adding these inequalities yields

$$c_2 + c_3 + c_4 + \cdots + c_{2n} + c_{2n+1} \geq 2\left(\sqrt{b_2b_3/a_3} + \sqrt{b_4b_5/a_5} + \cdots + \sqrt{b_{2n}b_{2n+1}/a_{2n+1}}\right),$$

$$c_1 + c_2 + c_3 + \cdots + c_{2n-1} + c_{2n} \geq 2\left(\sqrt{b_1b_2/a_2} + \sqrt{b_3b_4/a_4} + \cdots + \sqrt{b_{2n-1}b_{2n}/a_{2n}}\right).$$

As the series whose general term is $u_n = \sqrt{b_n b_{n+1}/a_{n+1}}$ is divergent, at least one of the two series u_{2n} or u_{2n+1} is divergent, since all terms are positive. Then the above inequalities show that the series $\sum_{n=1}^{+\infty} c_n$ is divergent, which proves that the continued fraction is divergent by Seidel's criterion.

Exercise 3.13   Since a_n and b_n are natural integers, $\sqrt{b_{n-1}b_n/a_n} \geq 1$ for n sufficiently large. Hence $\sum_{n=1}^{+\infty}\sqrt{b_{n-1}b_n/a_n}$ is divergent, and the continued fraction is convergent by Pringsheim's criterion.

The proof of the irrationality is similar to the proof of theorem 3.8. Let

$$\alpha_n = \frac{a_n}{b_n +} \frac{a_{n+1}}{b_{n+1} +} \cdots .$$

Since all terms are positive, we have $\alpha_n < \dfrac{a_n}{b_n} \leq 1$ for every $n \geq N$.

If α_1 was rational, so would be α_N . Put $\alpha_N = A_1/A_0$, $(A_0, A_1) \in \mathbb{N}^2$. Then

$$\alpha_N = \frac{a_N}{b_N + \alpha_{N+1}} \quad \Rightarrow \quad \alpha_{N+1} = \frac{A_2}{A_1} .$$

Now an easy induction shows that $\alpha_{N+k} = A_{k+1}/A_k$, $(A_k, A_{k+1}) \in \mathbb{N}^2$. The sequence $A_n \in \mathbb{N}$ is decreasing since $\alpha_n < 1$ for $n \geq N$, contradiction.

This criterion and (3.6) imply that e is irrational (see also theorem 1.2).

Exercise 3.14   1) Replacing x by ix in (3.30) yields

$$\tan(ix) = i \tanh x = \frac{ix}{1 +} \frac{x^2}{3 +} \frac{x^2}{5 +} \cdots , \quad \text{whence } \tanh x = \frac{x}{1 +} \frac{x^2}{3 +} \frac{x^2}{5 +} \cdots \frac{x^2}{2n+1 +} \cdots .$$

Let $x = a/b \in \mathbb{Q}^*$. By arguing as in the proof of theorem 3.9, we obtain

$$\tanh x = \tanh\left(\frac{a}{b}\right) = \frac{a}{b +} \frac{a^2}{3b +} \frac{a^2}{5b +} \cdots \frac{a^2}{(2n+1)b +} \cdots .$$

Now theorem 3.8 applies, and $\tanh x$ is irrational for $x \in \mathbb{Q}^*$.

2) If e^x was rational for some $x \in \mathbb{Q}^*$, so would be $e^{2x} = (e^x)^2$, and therefore $\tanh x = (e^{2x} - 1)/(e^{2x} + 1)$, contradiction. Hence $e^x \notin \mathbb{Q}$ whenever $x \in \mathbb{Q}^*$.

Exercise 3.15   If $x \neq 0$ and $J_\nu(x) = 0$, we have, by (3.29),

$$\frac{x}{2\nu -} \frac{x^2}{2(\nu+1) -} \frac{x^2}{2(\nu+2) -} \frac{x^2}{2(\nu+3) -} \cdots = 0 .$$

Hence $\dfrac{x^2}{2(\nu+1)-}\dfrac{x^2}{2(\nu+2)-}\dfrac{x^2}{2(\nu+3)-\cdots}=\infty$.

Therefore $2(\nu+1)=\dfrac{x^2}{2(\nu+2)-}\dfrac{x^2}{2(\nu+3)-}\dfrac{x^2}{2(\nu+4)-\cdots}$

Assume that x^2 and ν are rational, $x^2=a/b$ and $\nu=c/d$. Then

$$2\left(\frac{c}{d}+1\right)=\frac{ad}{2b(c+2d)-}\frac{ad^2}{2(c+3d)-}\frac{ad^2}{2b(c+4d)-}\frac{ad^2}{2(c+5d)-\cdots}.$$

This equality contradicts theorem 3.8, which proves that x^2 is irrational if ν is rational. Especially, π^2 is irrational since π is a zero of $J_{1/2}(x)$.

Exercise 3.16   1) We have

$$
\begin{aligned}
F_q(qx)+qxF_q(q^2x) &= \sum_{n=0}^{+\infty}\frac{q^{n^2+n}}{(q;q)_n}x^n+\sum_{n=0}^{+\infty}\frac{q^{n^2+2n+1}}{(q;q)_n}x^{n+1}\\
&= \sum_{n=0}^{+\infty}\frac{q^{n^2+n}}{(q;q)_n}x^n+\sum_{n=0}^{+\infty}\frac{q^{(n+1)^2}\left(1-q^{n+1}\right)}{(q;q)_n}x^{n+1}\\
&= \sum_{n=0}^{+\infty}\frac{q^{n^2+n}}{(q;q)_n}x^n+\sum_{n=1}^{+\infty}\frac{q^{n^2}}{(q;q)_n}x^n-\sum_{n=1}^{+\infty}\frac{q^{n^2+n}}{(q;q)_n}x^n\\
&= 1+\sum_{n=1}^{+\infty}\frac{q^{n^2}}{(q;q)_n}x^n=\sum_{n=0}^{+\infty}\frac{q^{n^2}}{(q;q)_n}x^n=F_q(x).
\end{aligned}
$$

2) Replace x by $q^{n-1}x$ in the functional equation above.
We obtain, for every integer $n\geq 1$,

$$1+q^nx\frac{F_q(q^{n+1}x)}{F_q(q^nx)}=\frac{F_q(q^{n-1}x)}{F_q(q^nx)}.$$

Consequently, we have $\dfrac{F_q(q^nx)}{F_q(q^{n-1}x)}=\dfrac{1}{1+q^nx\dfrac{F_q(q^{n+1}x)}{F_q(q^nx)}}$ $(*)$.

When n is even, $(*)$ can be written as

$$\frac{F_q(q^{2n}x)}{F_q(q^{2n-1}x)}=\frac{1}{1+q^{2n}x\dfrac{F_q(q^{2n+1}x)}{F_q(q^{2n}x)}}\Leftrightarrow q^nx\frac{F_q(q^{2n}x)}{F_q(q^{2n-1}x)}=\frac{1}{\dfrac{1}{q^nx}+q^n\dfrac{F_q(q^{2n+1}x)}{F_q(q^{2n}x)}}.$$

On the other hand, when n is odd, $(*)$ becomes

$$\frac{F_q\left(q^{2n+1}x\right)}{F_q\left(q^{2n}x\right)} = \frac{1}{1+q^{2n+1}x\dfrac{F_q\left(q^{2n+2}x\right)}{F_q\left(q^{2n+1}x\right)}} \Leftrightarrow q^n\frac{F_q\left(q^{2n+1}x\right)}{F_q\left(q^{2n}x\right)} = \frac{1}{\dfrac{1}{q^n}+q^{n+1}x\dfrac{F_q\left(q^{2n+2}x\right)}{F_q\left(q^{2n+1}x\right)}}.$$

Put $c_{2n} = \dfrac{1}{q^n x}$ and $c_{2n+1} = \dfrac{1}{q^n}$. The above relations can be written as

$$\alpha_{2n} = \frac{1}{c_{2n}+\alpha_{2n+1}}, \quad \alpha_{2n+1} = \frac{1}{c_{2n+1}+\alpha_{2n+2}}.$$

3) Theorem 3.7 applies. Indeed:

a) First we have $\lim_{n\to+\infty} F_q\left(q^n x\right) = 1$, whence $\alpha_n \in \mathbb{C}$ for n sufficiently large.

b) Since $|q| < 1$, it is clear that $|c_n| \geq 2$ for n sufficiently large.

c) The series $\sum 1/c_n c_{n+1}$ is convergent because $|q| < 1$.

d) Finally, $\alpha_{2n} \sim q^n x$ and $\alpha_{2n+1} \sim q^n$ when $n \to +\infty$. Thus $\lim_{n\to+\infty} \alpha_{n+1}/c_n = 0$.

Therefore, we have for every $x \in \mathbb{C}^*$,

$$\frac{F_q(qx)}{F_q(x)} = \frac{1}{1+} \frac{1}{\dfrac{1}{qx}+} \frac{1}{\dfrac{1}{q}+} \frac{1}{\dfrac{1}{q^2x}+} \frac{1}{\dfrac{1}{q^2}+} \frac{1}{\dfrac{1}{q^3x}} +\cdots.$$

We transform this continued fraction as in the proof of theorem 3.5:

$$\frac{F_q(qx)}{F_q(x)} = \frac{1}{1+} \frac{qx}{1+} \frac{q^2x}{1+} \frac{q^3x}{1+} \frac{q^4x}{1+} \frac{q^5x}{1} +\cdots.$$

Clearly, this expression remains true for $x = 0$. Taking the inverse finally yields

$$\frac{F_q(x)}{F_q(qx)} = 1 + \frac{qx}{1+} \frac{q^2x}{1+} \frac{q^3x}{1+} \frac{q^4x}{1+} \frac{q^5x}{1} +\cdots \qquad (x \in \mathbb{C}).$$

Solutions to the exercises of chapter 4

Exercise 4.1 Let $\alpha_0 = [c_0, c_1, \cdots, c_n, \cdots]$ be the regular continued fraction expansion of α_0. Assume that $\alpha_0 = c_0' + \dfrac{1}{c_1'+} \cdots \dfrac{1}{c_n'} +\cdots$, with $c_n' \in \mathbb{N} - \{0\}$ for every $n \geq 1$. Then $0 < \dfrac{1}{c_1'+} \cdots \dfrac{1}{c_n'} +\cdots < 1$, and therefore $c_0' = [\alpha_0] = c_0$.

Hence $[c_1, c_2, ..., c_n, ...] = \dfrac{1}{c_1' +} \dfrac{1}{\cdots c_n' + \cdots}$, and similarly $c_1' = c_1$, $c_2' = c_2$, ...

Exercise 4.2   Suppose that, on the contrary,

$$\left| \alpha_0 - \frac{P_n}{Q_n} \right| \geq \frac{1}{2Q_n^2} \quad \text{and} \quad \left| \alpha_0 - \frac{P_{n+1}}{Q_{n+1}} \right| \geq \frac{1}{2Q_{n+1}^2}$$

for some $n \in \mathbb{N}$. Since $\alpha_0 \notin \mathbb{Q}$, the equality is impossible. Therefore

$$\left| \alpha_0 - \frac{P_n}{Q_n} \right| > \frac{1}{2Q_n^2} \quad \text{and} \quad \left| \alpha_0 - \frac{P_{n+1}}{Q_{n+1}} \right| > \frac{1}{2Q_{n+1}^2}.$$

Since α_0 lies between P_n/Q_n and P_{n+1}/Q_{n+1}, we have

$$\left| \frac{P_{n+1}}{Q_{n+1}} - \frac{P_n}{Q_n} \right| = \left| \frac{P_{n+1}}{Q_{n+1}} - \alpha_0 \right| + \left| \alpha_0 - \frac{P_n}{Q_n} \right| > \frac{1}{2} \left(\frac{1}{Q_n^2} + \frac{1}{Q_{n+1}^2} \right).$$

Hence $\dfrac{|P_{n+1}Q_n - P_n Q_{n+1}|}{Q_n Q_{n+1}} = \dfrac{1}{Q_n Q_{n+1}} > \dfrac{1}{2} \left(\dfrac{1}{Q_n^2} + \dfrac{1}{Q_{n+1}^2} \right).$

Consequently $0 > (Q_{n+1} - Q_n)^2$, which is impossible.

Exercise 4.3   Let $\alpha_N = c_N + \dfrac{1}{c_{N+1} +} \dfrac{1}{\cdots c_{N+T-1} +} \dfrac{1}{c_{N+T} +} \dfrac{1}{c_{N+T+1} + \cdots}$.

Since $c_{n+T} = c_n$ for every $n \geq N$, we have $\alpha_N = c_N + \dfrac{1}{c_{N+1} +} \dfrac{1}{\cdots c_{N+T-1} +} \dfrac{1}{\alpha_N}$.

By using theorem 3.1, we obtain $\alpha_N = \dfrac{P_{N+T-2} + \alpha_N P_{N+T-1}}{Q_{N+T-2} + \alpha_N Q_{N+T-1}}$.

Hence $Q_{N+T-1}\alpha_N^2 + (Q_{n+T-2} - P_{N+T-1})\alpha_N - P_{N+T-2} = 0$, which proves that α_N is quadratic irrational. Let $\omega = \sqrt{\Delta}$, where $\Delta = (Q_{N+T-2} - P_{N+T-1})^2 + 4Q_{N+T-1}P_{N+T-2}$. Then there exists $(p, q) \in \mathbb{Q}^2$ such that $\alpha_N = p + q\omega$.

But $\alpha_0 = c_0 + \dfrac{1}{c_1 +} \dfrac{1}{\cdots c_{N-1} +} \dfrac{1}{\alpha_N}$, whence

$$\alpha_0 = \frac{P_{N-2} + \alpha_N P_{N-1}}{Q_{N-2} + \alpha_N Q_{N-1}} = \frac{a + b\omega}{a' + b'\omega} = \frac{(a + b\omega)(a' - b'\omega)}{a'^2 + b'\omega^2}, \quad a, b, a', b' \in \mathbb{Q}.$$

Since $\omega^2 = \Delta$, we have $\alpha_0 = a'' + b''\omega$, with $a'', b'' \in \mathbb{Q}$, which proves that α_0 is quadratic irrational.

Exercise 4.4   Assume that $A_0 \alpha_0^2 + B_0 \alpha_0 + C_0 = 0$ (*), A_0, B_0, C_0 integers.

Then $A_0\left(c_0+\dfrac{1}{\alpha_1}\right)^2+B_0\left(c_0+\dfrac{1}{\alpha_1}\right)+C_0=0$, whence $A_1\alpha_1^2+B_1\alpha_1+C_1=0$ (**),

with $A_1=-(A_0c_0^2+B_0c_0+C_0)$, $B_1=-(2A_0c_0+B_0)$, $C_1=-A_0$.
Therefore α_1 is quadratic. The discriminant Δ_1 of (**) is

$$\Delta_1=(2A_0c_0+B_0)^2-4A_0(A_0c_0^2+B_0c_0+C_0)=B_0^2-4A_0C_0=\Delta_0,$$

the same as the discriminant of the equation (*) satisfied by α_0.

Now we can suppose that $A_0>0$ and $C_0<0$, since the two roots α_0 and α_0^* of (*) have differents signs by assumption. Since $\alpha_0>1$, we have $\alpha_0^*<[\alpha_0]<\alpha_0$. Hence c_0 lies between the roots of (*) and $A_1>0$, whereas $C_1<0$.

Therefore α_1 and α_1^* have different signs, and since $\alpha_1=c_1+\dfrac{1}{c_2+\dfrac{1}{c_3+\cdots}}>1$,

we have $\alpha_1^*<0$. Thus α_1 has the same properties as α_0. Put

$$\alpha_1=c_1+\frac{1}{\alpha_2},\quad \alpha_2=c_2+\frac{1}{\alpha_3},\quad \ldots$$

Then the α_n's satisfy an equation (E_n) of the form $A_n\alpha_n^2+B_n\alpha_n+C_n=0$, where A_n, B_n, C_n are integers, $A_n>0$, $C_n<0$, $\Delta_n=B_n^2-4A_nC_n=\Delta_0$. Since $A_nC_n<0$, we have $B_n<\sqrt{\Delta_0}$, $A_n<\Delta_0/4$, $|C_n|<\Delta_0/4$. Therefore the triple (A_n,B_n,C_n) takes only *finitely* many values. Hence there exists integers $N\ge 0$ and $T\ge 1$ such that $A_{N+T}=A_N$, $B_{N+T}=B_N$, $C_{N+T}=C_N$ by the pigeon-hole principle. Thus $\alpha_N=\alpha_{N+T}=\alpha_{N+2T}=\cdots=\alpha_{N+kT}=\cdots$. As the regular continued fraction expansion is unique, $c_{n+T}=c_n$ for every $n\ge N$.

Exercise 4.5   Let $c_0=[\sqrt{D}]$, and let $\alpha_0=c_0+\sqrt{D}$. Then α_0 is reduced quadratic irational, since $\alpha_0>1$ and $\alpha_0^*=-\left(\sqrt{D}-c_0\right)\in\,]-1,0[$. Hence the expansion of α_0 is purely periodic. Since $[\alpha_0]=2c_0$, it is can be written as $\alpha_0=[\overline{2c_0,c_1,\cdots,c_{T-1}}]$, whence $\sqrt{D}=\alpha_0-c_0=[c_0,\overline{c_1,\cdots,c_{T-1},2c_0}]$.

Exercise 4.6   Suppose that $x^2-Dy^2=L$, $1\le|L|<\sqrt{D}$. Then

$$(x-\sqrt{D}y)(x+\sqrt{D}y)=L \;\Rightarrow\; \left|x-\sqrt{D}y\right|<\frac{\sqrt{D}}{x+\sqrt{D}y}\;(*).$$

First assume that $L>0$. Then $x^2-Dy^2=L$ implies that $x>\sqrt{D}y$. Replacing in (*) yields

$$\left|\frac{x}{y}-\sqrt{D}\right|<\frac{1}{2y^2},$$

and x/y is one of the convergents to $\sqrt{D}$ by theorem 4.3.

Assume now that $L<0$. Then $\sqrt{D}y>x$. Replacing in (*) yields

$$\left|\frac{y}{x}-\frac{1}{\sqrt{D}}\right|<\frac{1}{2x^2},$$

whence y/x is a convergent to $1/\sqrt{D}$. Thus x/y is a convergent to $\sqrt{D}$.
In both cases, since x and y are coprime, we have $x=P_n$ and $y=Q_n$, where P_n/Q_n is one of the convergents to $\sqrt{D}$.

Exercise 4.7   11 and 43 are coprime. We have $\alpha_0=11/43=0+1/\alpha_1$, $\alpha_1=43/11=3+10/11$, $\alpha_2=11/10=1+1/10$, $\alpha_3=10$. Therefore

$$\frac{11}{43}=[0,3,1,10].$$

Thus $P_{-1}=1$, $Q_{-1}=0$, $P_0=0$, $Q_0=1$, $P_1=3P_0+P_{-1}=1$, $Q_1=3Q_0+Q_{-1}=3$, $P_2=P_1+P_0=1$, $Q_2=Q_1+Q_0=4$, $P_3=10P_2+P_1=11$, $Q_3=10Q_2+Q_1=43$ (as expected). Now $P_3Q_2-P_2Q_3=1$, whence $11Q_2-43P_2=1$. Substracting to the equation to solve, we get $x-10Q_2=43t$, $-y-10P_2=11t$, $t\in\mathbb{Z}$, and finally

$$x=40+43t,\ y=-10-11t,\ t\in\mathbb{Z}.$$

Exercise 4.8   We have $\alpha_0=\sqrt{34}=5+\sqrt{34}-5, \alpha_1=\dfrac{5+\sqrt{34}}{9}=1+\dfrac{\sqrt{34}-4}{9}$,

$\alpha_2=\dfrac{4+\sqrt{34}}{2}=4+\dfrac{\sqrt{34}-4}{2}$, $\alpha_3=\dfrac{\sqrt{34}+4}{9}=1+\dfrac{\sqrt{34}-5}{9}$, $\alpha_4=\sqrt{34}+5$

$=10+\sqrt{34}-5$, $\alpha_5=\alpha_1$. Therefore α_n is periodic with period $T=4$.
The sequence $(-1)^n B_n$ is also periodic with period 4. It takes the values $(-1)^1 B_1=-9$, $(-1)^2 B_2=2$, $(-1)^3 B_3=-9$, $(-1)^4 B_4=1$.

If $L\neq\pm4$, then the solutions x and y are coprime. Hence the equation $x^2-34y^2=L$ has no solution for $L=-5,\ -3,\ -2,\ -1,\ 3,\ 5$. For $L=1$, the fundamental solution is given by $x=P_3$, $y=Q_3$. Now $P_{-1}=1$, $Q_{-1}=0$, $P_0=5$, $Q_0=1$, $P_1=P_0+P_{-1}=6$, $Q_1=Q_0+Q_{-1}=1$, $P_2=4P_1+P_0=29$, $Q_2=4Q_1+Q_0=5$, $P_3=P_2+P_1=35$, $Q_3=Q_2+Q_1=6$.

Therefore the fundamental solution of the Pell equation $x^2 - 34y^2 = 1$ is $x = 35$, $y = 6$. And the fundamental solution of $x^2 - 34y^2 = 2$ is $x = P_1 = 6$, $y = Q_1 = 1$.

In the cases where $L = -4$ and $L = 4$, there can be solutions of the form $x = 2x'$, $y = 2y'$. Thus we see that the equation $x^2 - 34y^2 = -4$ has no solution, whereas the fundamental solution of $x^2 - 34y^2 = 4$ is $x = 70$, $y = 12$.

Exercise 4.9   We have $\alpha_0 = c_0 + \dfrac{1}{\alpha_1}$, $\alpha_1 = c_1 + \dfrac{1}{\alpha_2}$, ..., $\alpha_{T-2} = c_{T-2} + \dfrac{1}{\alpha_{T-1}}$,

and $\alpha_{T-1} = c_{T-1} + \dfrac{1}{\alpha_0}$. Indeed, $\alpha_0 = \alpha_T$ since α_0 is reduced quadratic.

Taking the conjugates yields $-1/\alpha_1^* = c_0 - \alpha_0^*$, $-1/\alpha_2^* = c_1 - \alpha_1^*$, $\cdots$, $-1/\alpha_{T-1}^* = c_{T-2} - \alpha_{T-2}^*$, $-1/\alpha_0^* = c_{T-1} - \alpha_{T-1}^*$. Starting from the end, we get

$$\frac{-1}{\alpha_0^*} = c_{T-1} + \frac{1}{c_{T-2} - \alpha_{T-2}^*} = c_{T-1} + \frac{1}{c_{T-2} +} \frac{1}{c_{T-3} - \alpha_{T-3}^*}$$

$$= \cdots = c_{T-1} + \frac{1}{c_{T-2} +} \cdots \frac{1}{c_1 +} \frac{1}{c_0 - \alpha_0^*}.$$

Therefore $\dfrac{-1}{\alpha_0^*} = \overline{[c_{T-1}, c_{T-2}, \cdots, c_0]}$. This is Galois theorem.

Now $\sqrt{D} = \overline{[c_0, c_1, c_2, ..., c_{T-1}, 2c_0]}$ by corollary 4.2.

Moroever, $\beta_0 = \dfrac{1}{\sqrt{D} - c_0} = \overline{[c_1, c_2, \cdots, 2c_0]}$ is clearly reduced quadratic.

Hence Galois theorem yields $\dfrac{-1}{\beta_0^*} = \sqrt{D} + c_0 = \overline{[2c_0, c_{T-1}, \cdots, c_2, c_1]}$.

But we also have, from the expansion of $\sqrt{D}$,

$$\sqrt{D} + c_0 = \overline{[2c_0, c_1, c_2, ..., c_{T-1}, 2c_0]} = \overline{[2c_0, c_1, c_2, \cdots, c_{T-1}]}.$$

As the expansion in regular continued fraction is unique, we get

$$c_k = c_{T-k} \quad \text{for} \quad k = 1, 2, ..., T-1.$$

Exercise 4.10   1) We have $\alpha_n = c_n + 1/\alpha_{n+1}$. Therefore

$$\frac{A_n + \sqrt{D}}{B_n} = c_n + \frac{B_{n+1}}{A_{n+1}^2 - D} \left(A_{n+1} - \sqrt{D} \right).$$

Since $\sqrt{D}$ is irrational, this implies that

$$\frac{A_n}{B_n} = c_n + \frac{B_{n+1}A_{n+1}}{A_{n+1}^2 - D} \quad \text{and} \quad \frac{1}{B_n} = -\frac{B_{n+1}}{A_{n+1}^2 - D}.$$

The second equality is equivalent to $D = A_{n+1}^2 + B_n B_{n+1}$. Replacing $A_{n+1}^2 - D$ by $-B_n B_{n+1}$ in the first one, we obtain $A_n + A_{n+1} = c_n B_n$.

2) We know that $\alpha_0 = \sqrt{D} = [c_0, \overline{c_1, c_2, \cdots, c_{T-1}, 2c_0}]$. Hence $B_0 = 1$. Now, since $\alpha_T = [\overline{2c_0, c_1, c_2, \cdots, c_{T-1}}]$, we have $\alpha_T = c_0 + \sqrt{D}$. Therefore $B_T = B_0 = 1$.

Now we show that $B_{T-1} = B_1$. We have $\alpha_0 = c_0 + 1/\alpha_1$, whence

$$\alpha_1 = \frac{1}{\alpha_0 - c_0} = \frac{1}{\sqrt{D} - c_0} = \frac{c_0 + \sqrt{D}}{D - c_0^2}.$$

Thus $B_1 = D - c_0^2$. Similarly

$$\alpha_{T-1} = c_{T-1} + \frac{1}{\alpha_T} = c_{T-1} + \frac{1}{c_0 + \sqrt{D}} = c_{T-1} + \frac{\sqrt{D} - c_0}{D - c_0^2}.$$

Hence $B_{T-1} = D - c_0^2 = B_1$.

Now we prove by induction that $B_k = B_{T-k}$ for $k = 0, 1, 2, \ldots, T$. By using question 1, we have $A_{k+1} = \sqrt{D - B_k B_{k+1}}$ and $A_k = \sqrt{D - B_k B_{k-1}}$ ($k \geq 1$). Replacing in $A_k + A_{k+1} = c_k B_k$ yields

$$\sqrt{D - B_k B_{k-1}} + \sqrt{D - B_k B_{k+1}} = c_k B_k \qquad (k \geq 1). \quad (*)$$

Suppose that $B_k = B_{T-k}$ and $B_{k-1} = B_{T-k+1}$, and replace k by $T-k$ in (*). Taking into account that $c_k = c_{T-k}$ (exercise 4.9), we obtain

$$\sqrt{D - B_k B_{T-k-1}} + \sqrt{D - B_k B_{k-1}} = c_k B_k \qquad (k \geq 1). \quad (**)$$

Comparing (*) and (**) shows that $B_{k+1} = B_{T-(k+1)}$, Q.E.D.

3) If $T = 2m + 1$ is odd, we have $B_m = B_{T-m} = B_{m+1}$. Replacing in the relation $D = A_{n+1}^2 + B_n B_{n+1}$ immediatly yields $D = A_{m+1}^2 + B_{m+1}^2$, which proves that D is a sum of two squares.

4) Practically, we will expand $\sqrt{D}$ in regular continued function until we find some integer m such that $B_m = B_{m+1}$. We have

$$\alpha_0 = \sqrt{697} = 26 + \sqrt{697} - 26 = 26 + \frac{21}{26 + \sqrt{697}},$$

$$\alpha_1 = \frac{26 + \sqrt{697}}{21} = 2 + \frac{\sqrt{697} - 16}{21} = 2 + \frac{21}{16 + \sqrt{697}}, \quad \alpha_2 = \frac{16 + \sqrt{697}}{21}.$$

Therefore $B_1 = B_2 = 21$, which yields $697 = A_2^2 + B_1 B_2 = 16^2 + 21^2$.

5) Suppose that there exists $k \in \{1, 2, \cdots, T-1\}$ such that $B_k = 1$. Then $\alpha_k = A_k + \sqrt{D} = A_k^* + c_0 + 1/\alpha_1$. Thus $A_k + c_0 = [\alpha_k]$ and $1/\alpha_1 = 1/\alpha_{k+1}$, whence

$$\alpha_0 = c_0 + \cfrac{1}{c_1 + \cdots c_k + \cfrac{1}{\alpha_1}}.$$

This means that $\alpha_0 = [c_0, c_1, \dots, c_k]$ and has period $k < T$, a contradiction. Consequently $B_k \geq 2$ if $k \in \{1, 2, \dots, T-1\}$, which implies, since B_k has period T, that $B_k = 1$ if, and only if, k is a multiple of T.

Now assume that the equation $x^2 - Dy^2 = -1$ has solutions. Then x and y are coprime. Therefore, by theorem 4.6, there exists n_0 such that $(-1)^{n_0} B_{n_0} = -1$. Hence n_0 is odd and $B_{n_0} = 1$. Since n_0 is a multiple of T, T is also odd. Consequently, D is a sum of two squares by question 3.

The converse is false. Indeed, $34 = 25 + 9 = 5^2 + 3^2$ is a sum of two squares, but the equation $x^2 - 34y^2 = -1$ has no solution (see exercise 4.8).

Exercise 4.11 1) If x_1 is even, $x_1 = 2k$, then $x_1^2 \equiv 0 \pmod 4$. Hence, reducing modulo 4 the equation $x_1^2 - py_1^2 = 1$ yields $y_1^2 \equiv -1$. This is impossible, since the only squares modulo 4 are $0^2 \equiv 0 \pmod 4$, $1^2 \equiv 1 \pmod 4$, $2^2 \equiv 0 \pmod 4$, $3^2 \equiv 1 \pmod 4$. Therefore x_1 is odd.

Now $x_1^2 - py_1^2 = 1 \Leftrightarrow (x_1 - 1)(x_1 + 1) = py_1^2$. But $\mathrm{GCD}(x_1 - 1, x_1 + 1) = 2$, since any divisor of $x_1 - 1$ and $x_1 + 1$ must divide their difference 2 and these two numbers are even. Therefore $x_1 - 1 = 2pu^2$ and $x_1 + 1 = 2v^2$, or $x_1 - 1 = 2u^2$ and $x_1 + 1 = 2pv^2$ (consider the prime divisors of y_1). In both cases, $y_1 = 2uv$.

Consider the first case. Then $v^2 - pu^2 = \frac{1}{2}((x_1 + 1) - (x_1 - 1)) = 1$.

But $2v^2 = x_1 + 1$ and $x_1 > 1 \Rightarrow v < x_1$. Since (x_1, y_1) is the fundamental solution of Pell's equation $x^2 - py^2 = 1$, this yields $v = 1$, $u = 0$, which is impossible.

Therefore $x_1 - 1 = 2u^2$ and $x_1 + 1 = 2pv^2$. Substracting yields $u^2 - pv^2 = -1$, which proves that the equation $x^2 - py^2 = -1$ has a solution.

2) If $p \equiv 1 \pmod 4$, the equation $x^2 - py^2 = -1$ has a solution (question 1). Now the result of exercise 4.10, 5), shows that p is a sum of two squares.

Conversely, assume that $p = x^2 + y^2$, $x, y \in \mathbb{N}$. Reduce modulo 4. Then $x^2 \equiv 0$ or 1 $\pmod 4$ and $y^2 \equiv 0$ or 1 $\pmod 4$, as we have seen in question 1. Hence

$x^2 + y^2 \equiv 0$, 1 or 2 (mod 4). The cases $p \equiv 0$ (mod 4) and $p \equiv 0$ (mod 2) must be rejected since p is an odd prime. Therefore $p \equiv 1$ (mod 4).

Exercise 4.12 We proceed the same way as for the continued fraction expansion of e. We start from (4.10) with $a = 1$, $b = 1$, which yields

$$\frac{1}{\tanh 1} = [1,3,5,7,\ldots] = [c_0', c_1', \ldots], \quad \text{with } c_n' = 2n+1.$$

Let P_n'/Q_n' be the convergents of this expansion. We compute easily

$$\frac{P_0'}{Q_0'} = 1, \quad \frac{P_1'}{Q_1'} = \frac{4}{3}, \quad \frac{P_2'}{Q_2'} = \frac{21}{16}, \quad \frac{P_3'}{Q_3'} = \frac{151}{115}, \quad \frac{P_4'}{Q_4'} = \frac{1380}{1051}, \quad \frac{P_5'}{Q_5'} = \frac{15331}{11676}.$$

Now let P_n/Q_n be the convergents of the expansion of

$$\alpha_0 = [7, \overline{3n+2, 1, 1, 3(n+1), 6(2n+3)}]_{n=0}^{+\infty}.$$

Again, an easy computation shows that

$$\frac{P_0}{Q_0} = \frac{7}{1}, \quad \frac{P_1}{Q_1} = \frac{15}{2}, \quad \frac{P_2}{Q_2} = \frac{22}{3}, \quad \frac{P_3}{Q_3} = \frac{37}{5}, \quad \frac{P_4}{Q_4} = \frac{133}{18}, \quad \frac{P_5}{Q_5} = \frac{2431}{329},$$

$$\frac{P_6}{Q_6} = \frac{12288}{1663}, \quad \frac{P_7}{Q_7} = \frac{14719}{1992}, \quad \frac{P_8}{Q_8} = \frac{27007}{3655}.$$

We observe that $P_0 = P_1' + Q_1'$, $Q_0 = P_1' - Q_1'$, $P_3 = P_2' + Q_2'$, $Q_3 = P_2' - Q_2'$, $P_5 = P_4' + Q_4'$, $Q_5 = P_4' - Q_4'$, $P_8 = P_5' + Q_5'$, $Q_8 = P_5' - Q_5'$. This suggests that, for every $n \in \mathbb{N}$,

$$\begin{cases} P_{5n} = P_{3n+1}' + Q_{3n+1}', \quad Q_{5n} = P_{3n+1}' - Q_{3n+1}', \\ P_{5n+3} = P_{3n+2}' + Q_{3n+2}', \quad Q_{5n+3} = P_{3n+2}' - Q_{3n+2}'. \end{cases}$$

Now we prove these formulas by induction. They are clearly true for $n = 0$ and $n = 1$. By using the recurrence relations (4.5), we obtain

(a) $P_{5n+8} = P_{5n+7} + P_{5n+6}$, (b) $P_{5n+7} = P_{5n+6} + P_{5n+5}$,

(c) $P_{5n+6} = (3n+5)P_{5n+5} + P_{5n+4}$, (d) $P_{5n+5} = 6(2n+3)P_{5n+4} + P_{5n+3}$,

(e) $P_{5n+4} = 3(n+1)P_{5n+3} + P_{5n+2}$, (f) $P_{5n+3} = P_{5n+2} + P_{5n+1}$, (g) $P_{5n+2} = P_{5n+1} + P_{5n}$.

Substracting (g) to (f) yields $2P_{5n+2} = P_{5n+3} + P_{5n}$. If we replace in (e), we get

(h) $2P_{5n+4} = (6n+7)P_{5n+3} + P_{5n}$.

Now replacing in (d) yields

$$P_{5n+5} = 4(3n+4)^2 P_{5n+3} + 3(2n+3)P_{5n}. \qquad (*)$$

Moreover (a) + (b) $\Rightarrow P_{5n+8} = 2P_{5n+6} + P_{5n+5}$. By using (c), we obtain

$$P_{5n+8} = (6n+11)P_{5n+5} + 2P_{5n+4}.$$

Hence, by using (h), we have

$$P_{5n+8} = (6n+11)P_{5n+5} + (6n+7)P_{5n+3} + P_{5n}. \qquad (**)$$

The relations (*) and (**) remain true with Q_n in place of P_n.

For P'_n and Q'_n, we have (a') $P'_{3n+5} = (6n+11)P'_{3n+4} + P'_{3n+3}$,

(b') $P'_{3n+4} = (6n+9)P'_{3n+3} + P'_{3n+2}$, (c') $P'_{3n+3} = (6n+7)P'_{3n+2} + P'_{3n+1}$.

Replacing P'_{3n+3} in (b') yields $P'_{3n+4} = 4(3n+4)^2 P'_{3n+2} + 3(2n+3)P'_{3n+1}$. $\hspace{1em}$ (□)

Similarly, if we replace P'_{3n+3} in (a'), we obtain

$$P'_{3n+5} = (6n+11)P'_{3n+4} + (6n+7)P'_{3n+2} + P'_{3n+1}. \hspace{2em} (□□)$$

Adding the relations (□) and (□□) for P'_n and Q'_n yields the two equalities

$$(P'_{3n+4} + Q'_{3n+4}) = 4(3n+4)^2(P'_{3n+2} + Q'_{3n+2}) + 3(2n+3)(P'_{3n+1} + Q'_{3n+1}) \hspace{1em} \text{and}$$

$$(P'_{3n+5} + Q'_{3n+5}) = (6n+11)(P'_{3n+4} + Q'_{3n+4}) + (6n+7)(P'_{3n+2} + Q'_{3n+2}) + (P'_{3n+1} + Q'_{3n+1}).$$

Comparing the first one with (*) yields $P_{10} = P'_7 + Q'_7$. Then, comparing the second one with (**) yields $P_{13} = P'_8 + Q'_8$, and so on...

By induction, we see that $P_{5n} = P'_{3n+1} + Q'_{3n+1}$ and $P_{5n+3} = P'_{3n+2} + Q'_{3n+2}$.

Similarly, we obtain $Q_{5n} = P'_{3n+1} - Q'_{3n+1}$ and $Q_{5n+3} = P'_{3n+2} - Q'_{3n+2}$.

Therefore $\alpha_0 = \lim\limits_{n \to +\infty} \dfrac{P_{5n}}{Q_{5n}} = \lim\limits_{n \to +\infty} \dfrac{\dfrac{P'_{3n+1}}{Q'_{3n+1}} + 1}{\dfrac{P'_{3n+1}}{Q'_{3n+1}} - 1} = \dfrac{\dfrac{e^2 + 1}{e^2 - 1} + 1}{\dfrac{e^2 + 1}{e^2 - 1} - 1} = e^2$.

Finally, since the expansion of e^2 in regular continued fraction is not ultimately periodic, e^2 is not quadratic by theorem 4.4. This means that $ae^4 + be^2 + c \neq 0$ if a, b, c are not all zero integers.

Exercise 4.13 $\hspace{1em}$ Let (x_1, y_1) be the fundamental solution of the Pell equation $x^2 - Dy^2 = 1$. Then $x_1^2 - 1 = Dy_1^2$. Define the sequence (x_n), for $n \geq 1$, by $x_{n+1} = 2x_n^2 - 1$. Then we know from exercise 2.8 that

$$x_1 - \sqrt{x_1^2 - 1} = x_1 - y_1\sqrt{D} = \sum_{n=1}^{+\infty} \frac{1}{2x_1 2x_2 \ldots 2x_n}.$$

Hence $\sqrt{D} = \dfrac{1}{y_1}\left(x_1 - \sum\limits_{n=1}^{+\infty} \dfrac{1}{2x_1 2x_2 \ldots 2x_n} \right)$.

Exercise 4.14 $\hspace{1em}$ Again, let (x_1, y_1) be the fundamental solution of the equation $x^2 - Dy^2 = 1$, and (x_n) defined by $x_{n+1} = 2x_n^2 - 1$ for $n \geq 1$. By formula (2.10),

$$\frac{y_1\sqrt{D}}{x_1 - 1} = \prod_{n=1}^{+\infty} \left(1 + \frac{1}{x_n} \right).$$

Therefore $\sqrt{D} = \dfrac{x_1 - 1}{y_1} \displaystyle\prod_{n=1}^{+\infty} \left(1 + \dfrac{1}{x_n}\right)$.

Remark: In exercise 4.13, as well as in exercise 4.14, the expression of $\sqrt{D}$ is not unique, since we can use any solution of Pell's equation in place of, (x_1, y_1).

Exercise 4.15   1) We have

$$\alpha - \frac{r_n}{s_n} = 3\left[\prod_{k=0}^{+\infty} \frac{10^{4^k}+1}{10^{4^k}} - \prod_{k=0}^{n-1} \frac{10^{4^k}+1}{10^{4^k}}\right] = 3\prod_{k=0}^{n-1} \frac{10^{4^k}+1}{10^{4^k}}\left(\prod_{k=n}^{+\infty}\left(1+10^{-4^k}\right)-1\right).$$

Now, similarly to exercise 2.10, question 3, we see that

$$\prod_{k=n}^{+\infty}\left(1+10^{-4^k}\right)-1 = \sum_{k\in E, k\geq 4^n} 10^{-k} \leq \sum_{k\geq 4^n} 10^{-k} = \frac{1}{9}10^{-4^n+1} = \frac{1}{9}s_n^{-3}.$$

Consequently $\left|\alpha - \dfrac{r_n}{s_n}\right| \leq 3\displaystyle\prod_{k=0}^{+\infty} \frac{10^{4^k}+1}{10^{4^k}}\,\frac{1}{9}s_n^{-3} = \frac{\alpha}{3}s_n^{-3}$.

Since $\alpha = \displaystyle\sum_{k\in E} 10^{-k} \leq \sum_{k\geq 0} 10^{-k} = \frac{10}{9}$, we have finally $\left|\alpha - \dfrac{r_n}{s_n}\right| < \dfrac{1}{2}s_n^3$.

In particular, for $n \geq 1$, we have $|\alpha - r_n/s_n| < 1/2s_n^2$, which proves, by theorem 4.3, that r_n/s_n is a convergent of the continued fraction expansion of α.

2) First we prove the formulas for p_{2n} and p_{2n-1} by induction. We have

$$p_1 = c_1 p_0 + p_{-1} = c_1 c_0 + 1 = 10,\ p_2 = c_2 p_1 + p_0 = 33,\ p_3 = c_3 p_2 + p_1 = 1000.$$

Hence the formulas for p_{2n} and p_{2n-1} are correct for $n=1$. Assuming that are true for a given $n \geq 1$, we have now

$$P_{2n+2} = c_{2n+2}P_{2n+1} + p_{2n} = 3.10^{(4^n-1)/3}\prod_{k=0}^{n-1}\left(10^{4^k}+1\right).10^{(2.4^n+1)/3} + 3\prod_{k=0}^{n-1}\left(10^{4^k}+1\right)$$

$$= 3\prod_{k=0}^{n-1}\left(10^{4^k}+1\right)\left(10^{4^n}+1\right) = 3\prod_{k=0}^{n}\left(10^{4^k}+1\right).$$

This shows that the formula is true at the order $n+1$ for even indices. Moreover

$$P_{2n+3} = c_{2n+3}P_{2n+2} + p_{2n+1} = 3.10^{(2.4^n+1)/3}\prod_{k=0}^{n-1}\left(10^{2.4^k}+1\right).3\prod_{k=0}^{n}\left(10^{4^k}+1\right)+10^{(2.4^n+1)/3}$$

$$= 10^{(2.4^n+1)/3}\left(9\left(10^{4^n}+1\right)\prod_{k=0}^{n-1}\left(1+10^{4^k}+10^{2.4^k}+10^{3.4^k}\right)+1\right)$$

$$= 10^{(2.4^n+1)/3}\left(9\left(10^{4^n}+1\right)\prod_{k=0}^{n-1}\frac{10^{4^{k+1}}-1}{10^{4^k}-1}+1\right) = 10^{(2.4^n+1)/3}\left(9\times\frac{10^{2.4^n}-1}{9}+1\right)$$

$$= 10^{(2.4^n+1)/3}\times 10^{2.4^n} = 10^{(2.4^{n+1}+1)/3}.$$

This completes the proof of the formulas for p_n. The proof of the formulas for q_n is similar and left to the reader.

3) We observe immediatly that, for every $n \geq 1$, $\dfrac{r_n}{s_n} = \dfrac{p_{2n}}{q_{2n}}$.

Therefore $\alpha = \lim\limits_{n \to +\infty} \dfrac{r_n}{s_n} = \lim\limits_{n \to +\infty} \dfrac{p_{2n}}{q_{2n}} = \beta$.

This proves that $\alpha = \beta = [c_0, c_1, \cdots, c_n, \cdots]$.

Exercise 4.16   1) This is a direct consequence of theorem 3.3.

2) We have $u_0 = Q_0$, $u_1 = Q_1$, and u_n and Q_n satisfy the same recurrence relation. Thus $u_n = Q_n$ for every $n \geq 0$, and the result follows from question 1.

3) The denominators Q'_n of the convergents of $[0, c_1 - 2, 1, c_2 - 2, 1, c_3 - 2, 1, \cdots]$ satisfy $Q'_{-1} = 0$, $Q'_0 = 1$, and the recurrence relations

$$Q'_{2p+1} = (c_{p+1} - 2)Q'_{2p} + Q'_{2p-1}, \quad Q'_{2p+2} = Q'_{2p+1} + Q'_{2p}.$$

Therefore $Q'_1 = c_1 - 2$ and $Q'_2 = c_1 - 1$.

The second relation can be written $Q'_{2p+1} = Q'_{2p+2} - Q'_{2p}$. Replacing in the first one yields $Q'_{2p+2} = c_{p+1}Q'_{2p} - Q'_{2p-2}$. Now we see that the sequences u_n and Q'_{2n} satisfy $u_0 = Q'_0$, $u_1 = Q'_2$, and the same recurrence relation. Therefore $u_n = Q'_{2n}$ for every $n \geq 0$. By using the result of question 1, we obtain

$$[0, c_1 - 2, 1, c_2 - 2, 1, c_3 - 2, 1, \cdots] = \sum_{n=0}^{+\infty} \frac{(-1)^n}{Q'_n Q'_{n+1}} = \sum_{n=0}^{+\infty} \left(\frac{1}{Q'_{2n}Q'_{2n+1}} - \frac{1}{Q'_{2n+1}Q'_{2n+2}} \right)$$

$$= \sum_{n=0}^{+\infty} \frac{Q'_{2n+2} - Q'_{2n}}{Q'_{2n}Q'_{2n+1}Q'_{2n+2}} = \sum_{n=0}^{+\infty} \frac{1}{Q'_{2n}Q'_{2n+2}} = \sum_{n=0}^{+\infty} \frac{1}{u_n u_{n+1}}.$$

4) a) In order to compute S, we apply the result of question 2 to the sequence $u_n = F_{n+1}$. We obtain $S = [0, 1, 1, \cdots, 1, \cdots] = \dfrac{1}{[1, 1, 1, \cdots, 1, \cdots]} = \dfrac{1}{1 + S}$.

Therefore $S^2 + S - 1 = 0$, whence $S = (\sqrt{5} - 1)/2$, the golden number.

b) To compute T, we observe that the sequence $u_n = F_{2n+2}$ satisfies $u_0 = 1$, $u_1 = 3 = c_1 - 1$, and the recurrence relation $u_{n+1} = 3u_n - u_{n-1}$.

Applying the result of question 3 yields $T = [0, 2, 1, 1, 1, 1, \cdots] = \dfrac{1}{2 + S} = \dfrac{3 - \sqrt{5}}{2}$.

c) Similarly, the sequence $u_n = F_{2n+1}$ satisfies $u_0 = 1$, $u_1 = 2 = c_1 - 1$, and $u_{n+1} = 3u_n - u_{n-1}$. Applying the result of question 3 again yields

$$U = [0,1,1,1,1,1,\cdots] = S = \frac{\sqrt{5}-1}{2}.$$

d) It is clear that $V = T + U = 1$.

5) We can write V as a telescoping series, namely

$$V = \sum_{n=1}^{+\infty} \frac{F_{n+2} - F_n}{F_n F_{n+1} F_{n+2}} = \sum_{n=1}^{+\infty} \left(\frac{1}{F_n F_{n+1}} - \frac{1}{F_{n+1} F_{n+2}} \right) = \frac{1}{F_1 F_2} = 1.$$

Exercise 4.17   1) The function $P(x) = x^3 - 2x - 5$ is increasing on $[2,3]$, $P(2) = -1$, $P(3) = 16$. Therefore P has one and only one zero α in $[2,3]$. Assume that $\alpha = p/q \in \mathbb{Q}$. Then we know (see exercise 1.6) that $q|1$ and $p|5$. Hence $q = 1$ and $p = 1$ or 5, which is impossible. Therefore α is irrational.

2) Put $\alpha = 2 + 1/\alpha_1$, and replace in the equation. We obtain

$$\frac{1}{\alpha_1^3} + \frac{6}{\alpha_1^2} + \frac{10}{\alpha_1} - 1 = 0.$$

Now, it is clear that the equation $x^3 + 6x^2 + 10x - 1 = 0$ has one and only one root $1/\alpha_1$ in $]0,1[$. Therefore the equation $x^3 - 10x^2 - 6x - 1 = 0$ (E_1) has one and only one root in $]1,+\infty[$. By trial and error, we find that $\alpha_1 \in]10,11[$. We put $\alpha_1 = 10 + 1/\alpha_2$ and replace in (E_1). We see that α_2 satisfies the equation

$$61x^3 - 94x^2 - 20x - 1 = 0 \quad (E_2).$$

Hence $\alpha_2 \in]1,2[$. If we put $\alpha_2 = 1 + 1/\alpha_3$, α_3 satisfies the equation

$$54x^3 + 25x^2 - 89x - 61 = 0 \quad (E_3).$$

Therefore $\alpha_3 \in]1,2[$, $\alpha_3 = 1 + 1/\alpha_4$, $\alpha = [2,10,1,1,2,\cdots]$, and the first five convergents to α are 2, $\dfrac{21}{10}$, $\dfrac{23}{11}$, $\dfrac{44}{21}$, $\dfrac{111}{53}$. Thus $\dfrac{111}{53} < \alpha < \dfrac{44}{21}$.

Solutions to the exercises of chapter 5

Exercise 5.1   First observe that the equality $\alpha + \beta\sqrt{d} = 0$, with $\alpha, \beta \in \mathbb{Q}$, implies $\alpha = \beta = 0$ since $\sqrt{d}$ is irrational. Now, to prove that $\mathbb{K}$ is a subfield of $\mathbb{C}$, it suffices to prove that $(\mathbb{K},+)$ is a subgroup of $(\mathbb{C},+)$ and that $(\mathbb{K}^*,\cdot)$ is a subgroup of $(\mathbb{C}^*, \cdot)$. First, if $x = a + b\sqrt{d}$ and $y = a' + b'\sqrt{d} \in \mathbb{K}$, with a, a', $b, b' \in \mathbb{Q}$, we have $x - y = (a - a') + (b - b')\sqrt{d}$ with $a - a' \in \mathbb{Q}$, $b - b' \in \mathbb{Q}$. Therefore $(\mathbb{K}, +)$ is a subgroup of $(\mathbb{C}, +)$.

Now, if $y \neq 0$, we have $y^* = a' - b'\sqrt{d} \neq 0$ since $\sqrt{d} \notin \mathbb{Q}$, whence

$$\frac{x}{y} = \frac{xy^*}{yy^*} = \frac{(a + b\sqrt{d})(a' - b'\sqrt{d})}{(a' + b'\sqrt{d})(a' - b'\sqrt{d})} = \frac{aa' - dbb'}{a'^2 - db'^2} + \frac{a'b - ab'}{a'^2 - db'^2}\sqrt{d} \in \mathbb{K},$$

and $(\mathbb{K}^*, \cdot)$ is a subgroup of $(\mathbb{C}^*, \cdot)$.

To prove that $\mathbb{K}$ is a vector space over $\mathbb{Q}$, it suffices to check that the multiplication of the elements of $\mathbb{K}$ by the elements of $\mathbb{Q}$ fullfills the axioms of the vector space structure, since we know already that $(\mathbb{K}, +)$ is a multiplicative group. Now, if $\lambda, \mu \in \mathbb{Q}$, we have immediatly for all $x, y \in \mathbb{K}$

$$\lambda x \in \mathbb{K}, \quad \lambda(x + y) = \lambda x + \lambda y, \quad (\lambda + \mu)x = \lambda x + \mu x, \quad (\lambda\mu)x = \lambda(\mu x), \quad 1 \cdot x = x.$$

Finally, the numbers 1 and $\sqrt{d}$ span the $\mathbb{Q}$-vector space $\mathbb{K}$ by definition of $\mathbb{K}$, and are linearly independent since $\sqrt{d} \notin \mathbb{Q}$. Hence they form a basis of $\mathbb{K}$.

Exercise 5.2 Let $x = \dfrac{a}{b} + \dfrac{c}{d}\sqrt{d}$, $a, b, c, d \in \mathbb{Z}$. If $c = 0$, x is rational.

If not, x is irrational since $\sqrt{d}$ is, and $bdx - ad = bc\sqrt{d}$. Squaring yields $b^2d^2x^2 - 2abd^2x + a^2d^2 - db^2c^2 = 0$, which proves that x is quadratic irrational.

Exercise 5.3 Let $x = \alpha + \beta\sqrt{d}$, $y = \alpha' + \beta'\sqrt{d}$, $\alpha, \beta, \alpha', \beta' \in \mathbb{Q}$. Then $x^* = \alpha - \beta\sqrt{d}$, $y^* = \alpha' - \beta'\sqrt{d}$, and

$$(x + y)^* = (\alpha + \alpha' + (\beta + \beta')\sqrt{d})^* = \alpha + \alpha' - (\beta + \beta')\sqrt{d} = x^* + y^*,$$

$$(xy)^* = (\alpha\alpha' + d\beta\beta' + (\alpha\beta' + \alpha'\beta)\sqrt{d})^* = \alpha\alpha' + d\beta\beta' - (\alpha\beta' + \alpha'\beta)\sqrt{d} = x^* y^*.$$

Finally $N(xy) = (xy)(xy)^* = xyx^* y^* = N(x)N(y)$.

Exercise 5.4 1) Assume first that $x \in \mathbb{Q}(\sqrt{d})$ is an integer. Then x and x^* are the roots (possibly equal if $x \in \mathbb{Z}$) of an equation of the form $X^2 + aX + b = 0$, with $a, b \in \mathbb{Z}$. Then $N(x) = xx^* = b$ and $\mathrm{Tr}(x) = x + x^* = -a$ belong to $\mathbb{Z}$. Conversely, if $N(x) \in \mathbb{Z}$ and $\mathrm{Tr}(x) \in \mathbb{Z}$, then x and x^* are the roots of $X^2 - \mathrm{Tr}(x)X + N(x) = 0$, which proves that x is an integer of $\mathbb{Q}(\sqrt{d})$

2) We have $\mathrm{Tr}(x) = 2\alpha/\beta \in \mathbb{Z}$.

Since α and β are coprime, $\beta \mid 2$, whence $\beta = 1$ or 2. If $\beta = 1$, then

$$N(x) = \alpha^2 - \frac{\gamma^2}{\delta^2} d \in \mathbb{Z},$$

whence $\delta^2 \mid d$. Since d is squarefree, this implies that $\delta = 1$, and therefore $x = \alpha + \gamma\sqrt{d} \in \mathbb{Z}(\sqrt{d})$. If $\beta = 2$, $N(x) = \frac{\alpha^2}{4} - \frac{\gamma^2}{\delta^2} d \in \mathbb{Z}$.

Hence $4\delta^2 \mid \alpha^2 \delta^2 - 4\gamma^2 d$. Thus $4 \mid \alpha^2 \delta^2$. Since α and β are coprime, α is odd and δ is even, and we put $\delta = 2\delta'$. Returning to $N(x)$, we have

$$N(x) = \frac{\alpha^2}{4} - \frac{\gamma^2 d}{4\delta'^2} \in \mathbb{Z},$$

whence $4\delta'^2 \mid \alpha^2 \delta'^2 - \gamma^2 d$. Thus $\delta'^2 \mid \gamma^2$. Since δ and γ are coprime, this implies $\delta' = 1$. Thus $\delta = 2$ and $4 \mid \alpha^2 - \gamma^2 d$. Hence, since α is odd, γ (and d) must be odd as well. Now, since α and γ are odd, we have $\alpha^2 \equiv 1 \pmod 4$ and $\gamma^2 \equiv 1 \pmod 4$. Thus $\alpha^2 - \gamma^2 d \equiv 0 \pmod 4$ yields $d \equiv 1 \pmod 4$.

3) As d is squarefree, then $d \equiv 1 \pmod 4$, $d \equiv 2 \pmod 4$, or $d \equiv 3 \pmod 4$. If $d \equiv 2$ or 3 (mod 4), the only possible integers of $\mathbb{Q}(\sqrt{d})$ are, by question 2, the numbers $x = \alpha + \gamma\sqrt{d}$, $\alpha, \gamma \in \mathbb{Z}$. In this case, we see that, conversely, $\mathrm{Tr}(x) \in \mathbb{Z}$ and $N(x) \in \mathbb{Z}$, whence $A_K = \mathbb{Z}(\sqrt{d})$.

Si $d \equiv 1 \pmod 4$, the only possible integers of $\mathbb{Q}(\sqrt{d})$ are the numbers

$$x = \frac{1}{2}\left(\alpha + \gamma\sqrt{d}\right), \text{ with } \alpha \text{ and } \gamma \text{ both even or both odd.}$$

If x can be expressed in this form, we have $\mathrm{Tr}(x) = \alpha \in \mathbb{Z}$ and

$$N(x) = \frac{1}{4}\left(\alpha^2 - d\gamma^2\right) \in \mathbb{Z} \quad \text{since } d \equiv 1 \pmod 4.$$

Therefore $A_{\mathbb{K}} = \left\{ x = \frac{1}{2}\left(\alpha + \gamma\sqrt{d}\right) \mid \alpha, \gamma \in \mathbb{Z}, \text{ of the same parity} \right\}$.

Notice that any $x \in A_{\mathbb{K}}$ can also be written $x = \frac{\alpha-\gamma}{2} + \gamma\frac{1+\sqrt{d}}{2}$, $\frac{\alpha-\gamma}{2} \in \mathbb{Z}$.

Therefore $A_{\mathbb{K}} = \mathbb{Z}\left(\frac{1+\sqrt{d}}{2}\right)$ if $d \equiv 1 \pmod 4$.

4) It suffices to check that $A_{\mathbb{K}}$ is a subring of $\mathbb{K}$, in other words that $x - y \in A_{\mathbb{K}}$ and $xy \in A_{\mathbb{K}}$ whenever $x, y \in A_{\mathbb{K}}$. Use theorem 3, parts a) and b).

Exercise 5.5   If ε and $\varepsilon' \in A^\times$, there exist $\eta, \eta' \in A$ such that $\varepsilon\eta = 1$ and $\varepsilon'\eta' = 1$. By definition of $A^\times$, we see that, in fact, $\eta \in A^\times$ and $\eta' \in A^\times$. Hence $1/\varepsilon \in A^\times$, and $(\varepsilon\varepsilon')(\eta\eta') = 1 \Rightarrow \varepsilon\varepsilon' \in A^\times$. Thus $A^\times$ is a subgroup of $(\mathbb{C}^*, .)$.

Exercise 5.6   Since $\varepsilon \in \mathbb{A}_\mathbb{K}$, we have $N(\varepsilon) \in \mathbb{Z}$. Therefore

$$|N(\varepsilon)| = 1 \implies N(\varepsilon) = \pm 1 \implies \varepsilon \varepsilon^* = \pm 1 \implies \varepsilon \text{ invertible.}$$

Conversely, $\varepsilon\eta = 1 \implies \varepsilon^*\eta^* = 1 \implies \varepsilon\varepsilon^*\eta\eta^* = 1 \implies |\varepsilon\varepsilon^*| = 1.$

Exercise 5.7   Since $\mathbb{K}$ is an imaginary quadratic field, we have

$$N(x) = xx^* = x\bar{x} = |x|^2.$$

Therefore the units of $\mathbb{A}_\mathbb{K}$ are the elements of $\mathbb{A}_\mathbb{K}$ which lie on the circle with centre 0 and radius 1 in the complex plane. Looking at figures 5.1 and 5.2 shows that $\mathbb{Z}(i)^\times = \{-1, 1, i, -i\}$, whereas $\mathbb{Z}(j)^\times = \{-1, 1, e^{i\pi 3}, e^{2i\pi 3}, e^{-i\pi 3}, e^{-2i\pi 3}\}$.

This can be proved by using theorem 5.4:

First case: $-d \equiv 2$ or $3 \pmod 4$. Then $\mathbb{A}_\mathbb{K} = \mathbb{Z}(i\sqrt{d})$ by theorem 5.3. Now let $\varepsilon = \alpha + i\beta\sqrt{d} \in \mathbb{A}_\mathbb{K}$. We have

$$\varepsilon \in \mathbb{A}_\mathbb{K}^\times \Leftrightarrow |N(\varepsilon)| = 1 \Leftrightarrow |\varepsilon|^2 = 1 \Leftrightarrow \alpha^2 + d\beta^2 = 1$$
$$\Leftrightarrow (\alpha^2 = 1 \text{ and } d\beta^2 = 0) \text{ or } (\alpha^2 = 0 \text{ and } d\beta^2 = 1).$$

The first condition yields the units 1 and -1. The second one can be fulfilled only if $d = 1$ and $\beta^2 = 1$, which yields the units i and $-i$ of $\mathbb{Z}(i)$.

Second case: $-d \equiv 1 \pmod 4$. Then $\mathbb{A}_\mathbb{K} = \mathbb{Z}\left(\frac{1}{2}\left(1 + i\sqrt{d}\right)\right)$.

Let $\varepsilon = \left(\alpha + i\beta\sqrt{d}\right)/2 \in \mathbb{A}_\mathbb{K}^\times$, with α and β having the same parity.

If α and β are both even, the same computation as in the first case shows that $\varepsilon = 1$ or $\varepsilon = -1$. If α and β are both odd,

$$|\varepsilon|^2 = 1 \Leftrightarrow \alpha^2 + d\beta^2 = 4 \Leftrightarrow [(\alpha^2 = 1 \text{ and } d\beta^2 = 3) \text{ or } (\alpha^2 = 3 \text{ and } d\beta^2 = 1)].$$

The first condition yields $\beta^2 = 1$ and $d = 3$, $\alpha^2 = 1$, which gives the units $\left(1 \pm i\sqrt{3}\right)/2 = e^{\pm i\pi/3}$ and $\left(-1 \pm i\sqrt{3}\right)/2 = e^{\pm 2i\pi/3}$ of $\mathbb{Z}(j)$.

The second condition cannot be fulfilled, because it implies $d = 1$.

Exercise 5.8   1) $\dfrac{a + ib\sqrt{3}}{2} \times \dfrac{1 + i\sqrt{3}}{2} = \dfrac{1}{4}[(a - 3b) + i\sqrt{3}(a + b)].$

Since $a \equiv 1 \pmod 4$ and $b \equiv -1 \pmod 4$, we have $a - 3b \equiv 0 \pmod 4$ and $a + b \equiv 0 \pmod 4$, whence

$$\frac{a + ib\sqrt{3}}{2} \times \frac{1 + i\sqrt{3}}{2} \in \mathbb{Z}(i\sqrt{3}).$$

2) Let $x \in \mathbb{Z}(j)$. Then $x = \dfrac{1}{2}\left(a + ib\sqrt{3}\right)$, where a and b have the same parity.

If a and b are both even, there is nothing to prove since $1 \cdot x \in \mathbb{Z}(i\sqrt{3})$.

If a and b are both odd, we distinguish four cases:

First case: $a \equiv 1 \pmod 4$ and $b \equiv -1 \pmod 4$. We have seen in question 1 that

$$\frac{1}{2}\left(a+ib\sqrt{3}\right)e^{i\pi/3} \in \mathbb{Z}(i\sqrt{3}).$$

Second case: $a \equiv -1 \pmod 4$ and $b \equiv 1 \pmod 4$. It is similar to the first case since $\frac{1}{2}\left(a+ib\sqrt{3}\right)e^{-2i\pi/3} = -\frac{1}{2}\left(a+ib\sqrt{3}\right)e^{i\pi/3} \in \mathbb{Z}(i\sqrt{3})$.

Third case: $a \equiv -1 \pmod 4$ and $b \equiv -1 \pmod 4$. Again it is similar to the first case since $\frac{1}{2}\left(a+ib\sqrt{3}\right)e^{-i\pi/3}$ is the conjugate of $\frac{1}{2}\left(a-ib\sqrt{3}\right)e^{i\pi/3} \in \mathbb{Z}(i\sqrt{3})$.

Forth case: $a \equiv 1 \pmod 4$ and $b \equiv 1 \pmod 4$. It is nothing more than the third case, since $\frac{1}{2}\left(a+ib\sqrt{3}\right)e^{2i\pi/3} = -\frac{1}{2}\left(a+ib\sqrt{3}\right)e^{-i\pi/3}$.

Exercise 5.9   1) If η and ε belong to G, then

$$\begin{cases} N(\varepsilon\eta) = N(\varepsilon)N(\eta) = 1, \text{ with } \varepsilon\eta > 0 \\ N(\varepsilon\varepsilon^{-1}) = 1 = N(\varepsilon)N(\varepsilon^{-1}) \Rightarrow N(\varepsilon^{-1}) = 1, \text{ with } \varepsilon^{-1} > 0. \end{cases}$$

Therefore $\varepsilon\eta \in G$ and $\varepsilon^{-1} \in G$, which proves that G is a subgroup of $\mathbb{R}_+^*$.

2) The least element of G which is greater than 1 is given by the fundamental solution (example 4.8, p. 52) $\omega = x_1 + y_1\sqrt{d}$. Thus $G = \left\{(x_1 + y_1\sqrt{d})^n / n \in \mathbb{Z}\right\}$.

The positive n correspond to $(x_1 + y_1\sqrt{d})^n = x_n + y_n\sqrt{d}$ with $x_n \geq 0$, $y_n \geq 0$, and the negative n correspond to $x_n \geq 0$, $y_n \leq 0$. Hence the solutions $x_n \geq 0$, $y_n \geq 0$ of $x^2 - dy^2 = 1$ satisfy $x_n + y_n\sqrt{d} = (x_1 + y_1\sqrt{d})^n$, with $n \in \mathbb{N}$.

3) Consequently $x_{n+1} + y_{n+1}\sqrt{d} = (x_1 + y_1\sqrt{d})(x_n + y_n\sqrt{d})$, whence

$$x_{n+1} = x_1 x_n + dy_1 y_n, \quad y_{n+1} = y_1 x_n + x_1 y_n.$$

Replacing n by $n+1$ in the first relation yields $x_{n+2} = x_1 x_{n+1} + dy_1 y_{n+1}$. Moreover, if we multiply the second one by dy_1, we obtain

$$x_{n+2} - x_1 x_{n+1} = dy_1^2 x_n + x_1(x_{n+1} - x_1 x_n).$$

Hence, taking into account that $x_1^2 - dy_1^2 = 1$, we get $x_{n+2} = 2x_1 x_{n+1} - x_n$. The proof is similar for y_n.

Exercise 5.10   Let $x \in \mathbb{A}_\mathbb{K}$. Assume that $x = yz$, with y, $z \in \mathbb{A}_\mathbb{K}$.

Then $N(x) = N(y)N(z)$, and $N(y)$ and $N(z)$ belong to $\mathbb{Z}$. If $N(x)$ is prime in $\mathbb{Z}$, then $N(y) = \pm 1$ or $N(z) = \pm 1$. Consequently y or z is invertible in $\mathbb{A}_\mathbb{K}$ by theorem 5.4. Therefore x is irreducible in $\mathbb{A}_\mathbb{K}$.

Exercise 5.11   Since x is not invertible and $N(x) \in \mathbb{Z}$, $|N(x)| > 1$. Let

$$D = \left\{ y \in \mathbb{A}_{\mathbb{K}} \ / \ |N(y)| > 1 \text{ and } y \mid x \right\}.$$

Then D is not empty, since it contains x. Therefore there exists $p_1 \in D$ such that $|N(p_1)|$ is *minimal*. If p_1 was reducible, then we could write $p_1 = yz$, with y and z non invertible. Hence we would have $1 < |N(y)| < |N(p_1)|$, since z non invertible $\Rightarrow |N(z)| > 1$. As $y \in D$, this contradicts the fact that $|N(p_1)|$ is minimal in D. Thus x has an irreducible divisor p_1. Now put $x = p_1 x_1$. We have $|N(x_1)| < |N(x)|$. If $|N(x_1)| > 1$, then x_1 has an irreducible divisor p_2, and $x = p_1 p_2 x_2$, with $|N(x_2)| < |N(x_1)|$... Since $N(x_k)$ is a sequence of positive integers, there exists n such that $|N(x_n)| = 1$, which means that x_n is a unit. Then $x = p_1 p_2 ... p_{n-1} p_n x_n$ is an expression of x as a product of irreducibles.

Exercise 5.12   1) In $\mathbb{Z}(i\sqrt{5})$, we have $N(a + ib\sqrt{5}) = a^2 + 5b^2$. Hence the equations $N(x) = 2$ and $N(x) = 3$ have no solutions in $\mathbb{Z}(i\sqrt{5})$. If one of the numbers 2, 3, $1 + i\sqrt{5}$ or $1 - i\sqrt{5}$ was reducible, there would exists an $y \in \mathbb{Z}(i\sqrt{5})$ such that $N(y) = 2$ or $N(y) = 3$, impossible.

2) If $1 + i\sqrt{5}$ was associated to 2, there would exist $\varepsilon \in \mathbb{A}_{\mathbb{K}}^{\times}$ with $1 + i\sqrt{5} = 2\varepsilon$. Taking the norms would yield $6 = 4N(\varepsilon) = 4$, contradiction.

Exercise 5.13   1) We have $|N(b)| > |N(r_1)| > |N(r_2)|$... As the euclidian algorithm can be used as long as $r_k \neq 0$, there exists k such that $N(r_k) = 0$, in other words $r_k = 0$. Then $r_{k-2} = r_{k-1} q_k$ and $r_{k-1} \mid r_{k-2}$. As $r_{k-3} = r_{k-2} q_{k-1} + r_{k-1}$, $r_{k-1} \mid r_{k-3}$,... By induction, we see that $r_{k-1} \mid r_n$ for every $n \leq k - 1$. In particular $r_{k-1} \mid r_1$, $r_{k-1} \mid r_0 = b$, and $r_{k-1} \mid a$ since $a = bq_1 + r_1$. Since a and b are coprime, we have $r_{k-1} \in \mathbb{A}_{\mathbb{K}}^{\times}$.

2) We have $a = bq_1 + r_1$, whence $r_1 = a - bq_1$. Replacing in $b = r_1 q_2 + r_2$ yields $r_2 = u_2 a + v_2 b$, with $u_2, v_2 \in \mathbb{A}_{\mathbb{K}}$. By induction, we see that $r_n = u_n a + v_n b$, with $u_n, v_n \in \mathbb{A}_{\mathbb{K}}$. Since r_{k-1} is invertible, we obtain $1 = u_{k-1} r_{k-1}^{-1} a + v_{k-1} r_{k-1}^{-1} b$. This is Bézout's identity.

Exercise 5.14   If GCD $(2a, a^2 + 3b^2) = 3$, put $a = 3c$. Then (5.10) becomes $18c(3c^2 + b^2) = -y^3$, $18c$ and $3c^2 + b^2$ coprime. By the unique factorization theorem in $\mathbb{Z}$, we have now $18c = r^3$ and $3c^2 + b^2 = s^3$, s odd. Arguing the same way as in the first part of the proof yields $b = u(u + 3v)(u - 3v)$ and $c = 3v(u - v)(u + v)$, u odd, v even, u and v coprime. Since $18c = r^3$, we get $r^3 = 54v(u - v)(u + v)$. Thus $3 | r$, $r = 3r'$ and $r'^3 = 2v(u - v)(u + v)$. Now, the numbers $2v$, $u - v$ and $u + v$ are pairwise coprime. Therefore there exist h, m, n such that $h^3 = 2v$, $m^3 = u - v$, $n^3 = u + v$ and $h^3 + m^3 + (-n)^3 = 0$, h even. Finally $|y^3| = |18c(3c^2 + b^2)| = |27h^3(u^2 - v^2)(3c^2 + b^2)| \geq 27|h^3| > |h^3|$. Contradiction with the fact that $|y|$ is minimal.

Exercise 5.15   We have $N(1 + i\sqrt{13}) = 14 = 2 \times 7$. Hence, if $1 + i\sqrt{13} = xy$, with x and y non invertible, $N(x) = 2$ or $N(x) = 7$. But the equations $a^2 + 13b^2 = 2$ and $a^2 + 13b^2 = 7$ have no solutions in $\mathbb{Z}^2$. Thus $1 + i\sqrt{13}$ is irreducible. However, it is not prime. Indeed, $(1 + i\sqrt{13})(1 - i\sqrt{13}) = 2 \times 7 \Rightarrow 1 + i\sqrt{13} | 2 \times 7$. But $1 + i\sqrt{13}$ does not divide 2 and $1 + i\sqrt{13}$ does not divide 7 (otherwise $N(1 + i\sqrt{13}) = 14$ would divide $N(2) = 4$ or $N(7) = 49$).
We conclude that $\mathbb{Z}(i\sqrt{13})$ is not a unique factorization domain, since $1 + i\sqrt{13}$ is irreducible but not prime in $\mathbb{Z}(i\sqrt{13})$.

Exercise 5.16   First we expand $\sqrt{87}$ in regular continued fraction:
$$\alpha_0 = \sqrt{87} = 9 + (\sqrt{87} - 9), \; \alpha_1 = \frac{1}{\sqrt{87} - 9} = \frac{\sqrt{87} + 9}{6} = 3 + \frac{\sqrt{87} - 9}{6},$$
$$\alpha_2 = \frac{6}{\sqrt{87} - 9} = \sqrt{87} + 9 = 18 + (\sqrt{87} - 9).$$
Hence B_n takes only the two values 1 and 6, and by theorem 4.6 the equations $x^2 - 87b^2 = 2$ and $x^2 - 87b^2 = -2$ have no solutions. Consequently,
$$\forall x \in \mathbb{Z}(\sqrt{87}), \; N(x) \neq 2.$$
Now let $y = 1 + \sqrt{87}$. Then y is irreducible in $\mathbb{Z}(\sqrt{87})$, since
$$y = xz \Rightarrow N(y) = N(x)N(z) \Rightarrow N(x)N(z) = 86 = 2 \times 43.$$
Therefore $N(x) = 2$ or $N(z) = 2$, impossible. Hence y is irreducible.

However, y is not prime since $y | 2 \times 43$ but y does not divide 2 nor 43. Indeed we have $N(y) = -86$, $N(2) = 4$, $N(43) = 1849$. Thus $\mathbb{Z}(\sqrt{87})$ is not a unique factorization domain, hence it is not euclidean.

Remark: Theorem 4.6 allows to find all elements of A_K with a given norm less than $\sqrt{d}$ when $\mathbb{Q}(\sqrt{d})$ is a real quadratic field.

Exercise 5.17   First observe that x is odd. Indeed, if it was even, then we would have $y^2 \equiv -2 \pmod{4}$, which is impossible because the only squares modulo 4 are 0 and 1. Now we factorize in $\mathbb{Z}(i\sqrt{2})$: $(y+i\sqrt{2})(y-i\sqrt{2}) = x^3$.

First we show that $y+i\sqrt{2}$ and $y-i\sqrt{2}$ are coprime. Let $d \in \mathbb{Z}(i\sqrt{2})$, such that $d \mid y+i\sqrt{2}$ and $d \mid y-i\sqrt{2}$. Then d divides their difference $2i\sqrt{2}$. Taking the norms yields $N(d) \mid 8$. If $N(d) \neq 1$, then $N(d)$ is even. But $d \mid x^3$, whence $N(d) \mid N(x^3) = x^6$. Hence x^6 is even, contradiction. Therefore $N(d) = 1$, d is invertible, and $y+i\sqrt{2}$ are $y-i\sqrt{2}$ coprime. As $\mathbb{Z}(i\sqrt{2})$ is a unique factorization domain, this imples $y+i\sqrt{2} = \varepsilon(u+iv\sqrt{2})^3$, where ε is a unit of $Z(i\sqrt{2})$. But the only units of $Z(i\sqrt{2})$ are 1 and -1 by theorem 5.5. They are both cubes, and therefore $y+i\sqrt{2} = (a+ib\sqrt{2})^3$, with $a,b \in \mathbb{Z}$. Thus

$$y+i\sqrt{2} = a^3 - 6ab^2 + (3a^2b - 2b^3)i\sqrt{2}.$$

This yields $y = a(a^2 - 6b^2)$ and $1 = b(3a^2 - 2b^2)$. The solutions of the second equation are $b=1$, $3a^2 - 2b^2 = 1$ or $b = -1$, $3a^2 - 2b^2 = -1$. Hence $(a,b) = (1,1)$ or $(-1,1)$ and $y = 5$ or $y = -5$.

The solutions (x, y) of the given equation are therefore $(3,5)$ and $(3,-5)$.

Exercise 5.18 x is odd. If not, we would have $y^2 \equiv -11 \equiv 5 \pmod 8$, which is impossible since the squares modulo 8 are sont 0, 1 and 4. We factorize in the ring $A_K = \mathbb{Z}\left(\frac{1}{2}(1+i\sqrt{11})\right)$ of integers of $\mathbb{K} = \mathbb{Q}(i\sqrt{11})$, which is euclidean (and hence a unique factorization domain) by theorem 5.16, and obtain

$$(y+i\sqrt{11})(y-i\sqrt{11}) = x^3.$$

Every common divisor d to $y+i\sqrt{11}$ and $y-i\sqrt{11}$ divides $2i\sqrt{11}$. Hence $N(d) \mid 44$. Since $d \mid x^3$, $N(d) \mid x^6$ and $N(d)$ cannot be even because x is odd. If $11 \mid N(d)$, then $11 \mid x$ and 11 also divides y since $y^2 = x^3 - 11$. Therefore $11^2 \mid 11$, which is impossible. Thus $N(d) = 1$ and $y+i\sqrt{11}$ and $y-i\sqrt{11}$ are coprime. Consequently $y+i\sqrt{11}$ is a cube in $A_K = \mathbb{Z}\left(\frac{1}{2}(1+i\sqrt{11})\right)$ and

$$y+i\sqrt{11} = ((a+ib\sqrt{11})/2)^3, \quad a \text{ and } b \text{ of the same parity.}$$

Identifying yields $8y = a(a^2 - 33b^2)$ and $8 = b(3a^2 - 11b^2)$. The couples (a,b) solutions of the second equation are $(1,-1)$, $(-1,-1)$, $(4,2)$, $(-4,2)$. Finally we find that 4 couples are solutions of Mordell's equation $y^2 = x^3 - 11 : (x = 3, y = 4)$, $(x = 3, y = -4)$, $(x = 15, y = 58)$, $(x = 15, y = -58)$.

Exercise 5.19   Since $y^2 \equiv 0$ or $1 \pmod 4$, x is odd. We factorize in the ring of Gauss integers $\mathbb{Z}(i)$ and obtain $x^5 = (y+i)(y-i)$. Every common divisor to $y+i$ and $y-i$ divides $2i$. Therefore these numbers are coprime (take the norms). Since $\mathbb{Z}(i)$ is euclidean, it has unique factorization. Thus there exists a unit $\varepsilon \in \mathbb{Z}(i)^\times$ and $u \in \mathbb{Z}(i)$ such that $y+i = \varepsilon u^5$. But the units in $\mathbb{Z}(i)$ are all fifth powers: $1 = 1^5$, $(-1) = (-1)^5$, $i = i^5$, $-i = (-i)^5$. Hence $y+i = (a+ib)^5$, $a,b \in \mathbb{Z}$. This yields $y = a^5 - 10a^3b^2 + 5ab^4$ and $1 = b(5a^4 - 10a^2b^2 + b^4)$. The solutions are $b = \pm 1$, $a = 0$. Thus $x = 1$, $y = 0$ is the only solution.

Exercise 5.20   We can assume that x, y and z are pairwise coprime. Moreover, x is odd. If not, we would have $z^2 \equiv -1 \pmod 4$, which is impossible. Assume first that z is odd, $z > 0$ *minimum*. Then x is odd, y even, z odd and z, y^2, x^2 is a pythagorean triple. Therefore there exist u, v coprime, of different parities, such that $z = u^2 - v^2$, $y^2 = 2uv$, $x^2 = u^2 + v^2$ (theorem 5.9). Put $y = 2y'$. We have $uv = 2y'^2$. If u is even, then $u = 2\ell^2$ and $v = m^2$. As $x^2 = u^2 + v^2$, applying again theorem 5.9 yields $u = 2\alpha\beta$ and $v = \alpha^2 - \beta^2$. Therefore $\alpha\beta = \ell^2 \Rightarrow \alpha = h^2$ and $\beta = k^2$, and finally $m^2 = h^4 - k^4$, with m odd, $m < z$, contradiction. We argue the same way if u is odd.

Now assume that z is even. Then y is odd and $(x-y)(x+y)(x^2 + y^2) = z^2$. Since $x-y$, $x+y$ and $x^2 + y^2$ are even, $4|z$, and 4 divides at least one of the numbers $x-y$ and $x+y$. Assume that $4|x+y$. We have

$$\frac{x-y}{2} \times \frac{x+y}{4} \times \frac{x^2 + y^2}{2} = \frac{z^2}{16}.$$

Since the three numbers in the left-hand side are pairwise coprime, we obtain

$$\frac{x-y}{2} = u^2, \quad \frac{x+y}{4} = v^2, \quad \frac{x^2 + y^2}{2} = w^2, \quad u,v,w \text{ pairwise coprime.}$$

Therefore $x = u^2 + 2v^2$, $y = -u^2 + 2v^2$. Replacing in the third equation above yields $u^4 + 4v^4 = w^2$. Thus $\left(u^2, 2v^2, w\right)$ is a pythagorean triple.

Hence $u^2 = \alpha^2 - \beta^2$, $2v^2 = 2\alpha\beta$, $w = \alpha^2 + \beta^2$, α and β coprime.
The second equation yields $\alpha = m^2$ and $\beta = h^2$, and replacing in the first one we get $u^2 = m^4 - h^4$ with $0 < u < z$, a contradiction.
We argue the same way if 4 divides $x - y$.

Exercise 5.21   1) When $d = 2$, the fundamental unit ω corresponds to the least solution of Pell's equations $x^2 - 2y^2 = 1$ or $x^2 - 2y^2 = -1$ (see theorem 5.4). Since $(x = 1, y = 1)$ satisfies the second one, we cannot find a smaller one and $\omega = 1 + \sqrt{2}$. Therefore $A_K^\times = \left\{ \pm(1 + \sqrt{2})^n, n \in \mathbb{Z} \right\}$.

2) When $d = 13$, the fundamental unit ω corresponds to the least solution of Pell's equations $x^2 - 13y^2 = 1$, or $x^2 - 13y^2 = -1$, or $x^2 - 13y^2 = 4$ (x and y odd), or $x^2 - 13y^2 = -4$ (x and y odd) since $\omega = (a + b\sqrt{13})/2$ with a and b having the same parity (theorems 5.3 and 5.4). We see at once that $x = 3$, $y = 1$ is a solution of the last one, and it is the least (note that we cannot use here theorem 4.6 since $4 > \sqrt{13}$). Indeed, the couples ($x = 1, y = 1$) and ($x = 2, y = 1$) are not solutions of the two last equations. Moreover, ω cannot be of the form $\omega' = x + y\sqrt{13}$, since the unit $(3 + \sqrt{13})/2$ cannot be a power of such an ω'.
Therefore $\omega = (3 + \sqrt{13})/2$ and $A_K^\times = \left\{ \pm \left((3 + \sqrt{13})/2 \right)^n \ / \ n \in \mathbb{Z} \right\}$.

Exercise 5.22   1) a) By d'Alembert's ratio test, we see that the radius of convergence of f_q is $+\infty$, whereas the radius of convergence of g_q si $|q|$.
b) We have

$$f_q(x) - f_q(xq) = \sum_{n=1}^{+\infty} \frac{q^n - 1}{(q-1)\ldots(q^n - 1)} \left(\frac{x}{q} \right)^n$$

$$= \frac{x}{q} \sum_{n=1}^{+\infty} \frac{1}{(q-1)\ldots(q^{n-1} - 1)} \left(\frac{x}{q} \right)^{n-1} = \frac{x}{q} f_q \left(\frac{x}{q} \right).$$

Therefore $f_q(x) = \left(1 + \frac{x}{q} \right) f_q \left(\frac{x}{q} \right)$. By induction, we see that, for every $n \in \mathbb{N}$,

$$f_q(x) = \left(1 + \frac{x}{q} \right) \left(1 + \frac{x}{q^2} \right) \ldots \left(1 + \frac{x}{q^n} \right) f_q \left(\frac{x}{q^n} \right).$$

When $n \to +\infty$, $f_q\left(x/q^n \right) \to 1$, and therefore $f_q(x) = \prod_{n=1}^{+\infty} \left(1 + x/q^n \right)$.
c) Similarly we have $g_q(x) = g_q\left(x/q \right) + x/(q-x)$. Hence, for every $n \in \mathbb{N}$,

$$g_q(x) = g_q\left(\frac{x}{q^n}\right) + \sum_{k=1}^{n} \frac{x}{q^k - x}.$$

Let $n \to +\infty$. We deduce that $g_q(x) = \sum_{n=1}^{+\infty} x/(q^n - x)$.

d) Using b) yields $\text{Log}(f_q(-x)) = \sum_{k=1}^{n} \text{Log}\left(1 - xq^{-k}\right) + \text{Log}\left(f_q\left(-xq^{-n}\right)\right)$.

If we derive with respect to x, we get

$$-\frac{f_q'(-x)}{f_q(-x)} = -\sum_{k=1}^{n} \frac{1}{q^k - x} + \frac{f_q'(-x/q^n)}{f_q(-x/q^n)} \cdot \frac{-1}{q^n}.$$

But $\lim_{n \to +\infty} f_q\left(-xq^{-n}\right) = 1$ and $\lim_{n \to +\infty} f_n'\left(-xq^{-n}\right) = 1/(q-1)$.

Hence $-\dfrac{f_q'(-x)}{f_q(-x)} = -\sum_{k=1}^{n} \dfrac{1}{q^k - x} = g_q(x)$ by question 1)c).

2) a) $g_{-\Phi^2}(-\Phi) = \sum_{n=1}^{+\infty} \dfrac{(-\Phi)^n}{(-\Phi^2)^n - 1} = \sum_{n=1}^{+\infty} \dfrac{1}{\Phi^n - \Psi^n}$.

However, $F_n = \dfrac{\Phi^n - \Psi^n}{\Phi - \Psi}$ (exercise 3.11). Thus $(\Phi - \Psi)g_{-\Phi^2}(-\Phi) = \sum_{n=1}^{+\infty} \dfrac{1}{F_n}$.

b) If $\theta = A/B$, by 2)a) and 1)d) we have $\Phi B(\Phi - \Psi)f'_{-\Phi^2}(\Phi) + Af_{-\Phi^2}(\Phi) = 0$.

Hence $A + \sum_{n=1}^{+\infty} \dfrac{A + Bn(\Phi - \Psi)}{(1 + \Phi^2)(1 - \Phi^4)...(1 - (-\Phi^2)^n)}(-\Phi)^n = 0$

c) We have $X_k = A(1 + \Phi^2)(1 - \Phi^4)...(1 - (-\Phi^2)^k)$

$$+ \sum_{n=1}^{k} (A + Bn(\Phi - \Psi))(-\Phi)^n (1 - (-\Phi^2)^{n+1})...(1 - (-\Phi^2)^k).$$

Now, since $A, B, \Phi \in \mathbb{Z}\left((1 + \sqrt{5})/2\right) = \mathbb{Z}(\Phi)$, we see that $X_k \in \mathbb{Z}(\Phi)$. Moreover, taking 2)b) into account, we have

$$|X_k| = \left|(1 + \Phi^2)(1 - \Phi^4)...(1 - (-\Phi^2)^k) \sum_{n=k+1}^{+\infty} \frac{A + Bn(\Phi - \Psi)}{(1 + \Phi^2)...(1 - (-\Phi^2)^n)}(-\Phi)^n\right|.$$

Hence $|X_k| \le \sum_{n=k+1}^{+\infty} \dfrac{|A + Bn(\Phi - \Psi)|\Phi^n}{(\Phi^{2k+2} - 1)...(\Phi^{2n} - 1)}$.

For k sufficiently large, $\Phi^{2j} - 1 \ge \left(\Phi^2 - 1/2\right)^j$ for $j \ge k+1$, whence

$$|X_k| \le \sum_{n=k+1}^{+\infty} k^2(n-k)\frac{\Phi^n}{(\Phi^2 - 1/2)^{k+1}...(\Phi^2 - 1/2)^n}.$$

But $\Phi^2 - 1/2 \ge 1$, and therefore $|X_k| \le k^2 \sum_{n=k+1}^{+\infty} (n-k)\left(\dfrac{2\Phi}{2\Phi^2 - 1}\right)^n$.

As $\Phi^2 = \Phi + 1$, we get

$$|X_k| \le k^2 \sum_{n=k+1}^{+\infty} (n-k)\left(\frac{2\Phi}{2\Phi+1}\right)^n \le k^2 \left(\frac{2\Phi}{2\Phi+1}\right)^k \sum_{m=1}^{+\infty} m\left(\frac{2\Phi}{2\Phi+1}\right)^m.$$

This completes the proof with $c = \sum_{m=1}^{+\infty} m\left(2\Phi/(2\Phi+1)\right)^m$.

d) Since $\Phi^* = \Psi$, we have

$$X_k^* = A^*(1+\Psi^2)(1-\Psi^4)\ldots(1-(-\Psi^2)^k)$$

$$+ \sum_{n=1}^{k}(A^* + Bn(\Psi - \Phi))(-\Psi)^n(1-(-\Psi^2)^{n+1})\ldots(1-(-\Psi^2)^k).$$

Hence $|X_k^*| \le |A^*| \prod_{n=1}^{k}(1+|\Psi^2|^n) + \sum_{n=1}^{k}\left(|A^*| + Bn(\Phi-\Psi)\right)\prod_{i=n+1}^{k}(1+|\Psi^2|^i)$.

By question 1)b) we know that $\prod_{n=1}^{+\infty}(1+|\Psi^2|^n)$ converges to $f_{|\Psi^{-2}|}(1)$. Thus

$$|X_k^*| \le f_{|\Psi^{-2}|}(1)\left(|A^*|(k+1) + B(\Phi-\Psi)\frac{k(k+1)}{2}\right),$$

which implies that $|X_k^*| \le c'k^2$ for k sufficiently large.

e) The above two questions yield $|N(X_k)| = |X_k X_k^*| \le cc'k^4\left(\frac{2\Phi}{2\Phi+1}\right)^k$.

Therefore $\lim_{k\to+\infty} N(X_k) = 0$. But $X_k \in \mathbb{Z}(\Phi)$, the ring of integers of $\mathbb{Q}(\sqrt{5})$, whence $N(X_k) \in \mathbb{Z}$. Thus for k sufficiently large, $N(X_k) = 0$, which implies $X_k = 0$. Now, for k sufficiently large, we have by definition of X_k

$$Y_k = A + \sum_{n=1}^{k}\frac{(A+Bn(\Phi-\Psi))(-\Phi)^n}{(1+\Phi^2)(1-\Phi^4)\ldots(1-(-\Phi^2)^n)} = 0.$$

Consequently, for k sufficiently large,

$$Y_k - Y_{k-1} = \frac{(A+Bk(\Phi-\Psi))(-\Phi)^k}{(1+\Phi^2)(1-\Phi^4)\ldots(1-(-\Phi^2)^k)} = 0.$$

This is impossible since $B \ne 0$, which proves that $\sum_{n=1}^{+\infty} 1/F_n \notin \mathbb{Q}(\sqrt{5})$.

Solutions to the exercises of chapter 6

Exercise 6.1   It is *reflexive*: $x \mathcal{R} x$ since $x - x = 0 \in H$.
It is *symmetric*: if $x \mathcal{R} y$, then $x - y \in H \Rightarrow y - x \in H \Rightarrow y \mathcal{R} x$.
It is *transitive*: if $x \mathcal{R} y$ and $y \mathcal{R} z$, then $x - y \in H$ and $y - z \in H$, whence $(x - y) + (y - z) = x - z \in H$ and $x \mathcal{R} z$.

Exercise 6.2   First we show that $\mathcal{R}$ is *compatible* with the addition. Assume
that $x \mathcal{R} y$ and $z \mathcal{R} t$. Then $(x-y) \in H$ and $(z-t) \in H$. Therefore

$$(x-y)+(z-t) = ((x+z)-(y+t)) \in H \text{ and } (x+z)\,\mathcal{R}(y+t).$$

Now we *define* the sum of two equivalence classes by $\hat{x}+\hat{y} = \widehat{x+y}$. First we
must check that $\hat{x}+\hat{y}$ does not depend on the choice of the representants of the
classes $\hat{x}$ and $\hat{y}$. Thus, let x' and y' be two other representants. Then $x'\mathcal{R}x$
and $y'\mathcal{R}y$, whence $(x'+y')\mathcal{R}(x+y)$, which means that $\widehat{x'+y'}=\widehat{x+y}$.
Therefore the addition depends only on the choice of the class, in other words on
the elements of $\hat{x}$ and $\hat{y}$ of G/H. It is *commutative*. Indeed

$$\hat{x}+\hat{y} = \widehat{x+y} = \widehat{y+x} = \hat{y}+\hat{x}$$

since the addition is commutative in G. It is *associative*. Indeed,

$$\hat{x}+(\hat{y}+\hat{z}) = \widehat{x+(y+z)} = \widehat{(x+y)+z} = (\hat{x}+\hat{y})+\hat{z}$$

since the addition is associative in G. The neutral element is $\hat{0}$, where 0 is the
neutral element of G, because $\hat{0}+\hat{x} = \widehat{0+x} = \hat{x}$. And the opposite of $\hat{x}$ is
$\widehat{(-x)}$, since $\hat{x}+\widehat{(-x)} = \widehat{x+(-x)} = \hat{0}$.

Exercise 6.3   We only have to prove that the multiplication is compatible
with $\mathcal{R}$. But $x \mathcal{R} y$ and $z \mathcal{R} t \Rightarrow (x-y) \in I$ and $(z-t) \in I$. Therefore
$z(x-y) = zx - zy \in I$ and $y(z-t) = yz - yt \in I$.
Consequently $(zx-zy)+(yz-yt) = zx - yt \in I$, whence $(zx)\,\mathcal{R}\,(yt)$, q.e.d.

Exercise 6.4   Let $x \in \mathbb{A}_K$. Then $x/a = \alpha_1 + \beta_1\sqrt{d}$, with $(\alpha_1, \beta_1) \in \mathbb{Q}^2$. We
can write $x/a = [\alpha_1]+\alpha_2 + ([\beta_1]+\beta_2)\sqrt{d}$, with $0 \le \alpha_2 < 1$ and $0 \le \beta_2 < 1$.
Thus $x = a([\alpha_1]+[\beta_1]\sqrt{d})+a(\alpha_2 + \beta_2\sqrt{d})$.
Now define $q = [\alpha_1]+[\beta_1]\sqrt{d} \in \mathbb{A}_K$ and $r = a(\alpha_2 + \beta_2\sqrt{d}) = \alpha + \beta\sqrt{d}$. Then
$x = aq + r$, whence $r = x - aq \in \mathbb{A}_K$, and $|\alpha| = |a\alpha_2| \le |a|$, $|\beta| = |a\beta_2| \le |a|$.
Now we prove theorem 6.3. For $x \in \mathbb{A}_K$, the equality $x = aq + r$ shows that
$x - r = aq \in a\mathbb{A}_K$. Thus any $x \in \mathbb{A}_K$ is equivalent to some $r = \alpha + \beta\sqrt{d} \in \mathbb{A}_K$,
with $|\alpha| \le |a|$, $|\beta| \le |a|$. But there exist only finitely many such elements r,
since α and β are *integers* or *half integers* by theorem 5.3. Therefore there
exist only a finite number of equivalence classes, and $\mathbb{A}_K / a\mathbb{A}_K$ is finite.

Exercise 6.5   First assume that p is prime, and let $\hat{a}$ and $\hat{b}$ two elements of $\mathbb{A}/p\mathbb{A}$ satisfying $\hat{a}\hat{b}=\hat{0}$. This is equivalent to $ab\equiv 0$ (mod p), that is $ab=kp$, $k\in\mathbb{A}$ (since $ab\in p\mathbb{A}$). Therefore $p\mid a$ or $p\mid b$ by definition 5.2, whence $a\in p\mathbb{A}$ or $b\in p\mathbb{A}$, in other words $\hat{a}=\hat{0}$ or $\hat{b}=\hat{0}$ and $\mathbb{A}/p\mathbb{A}$ is an integral domain.

Conversely, suppose that $\mathbb{A}/p\mathbb{A}$ is an integral domain. Then

$$(p\mid ab)\Rightarrow(ab\in p\mathbb{A})\Rightarrow(\widehat{ab}=\hat{0})\Rightarrow(\hat{a}\hat{b}=\hat{0})\Rightarrow(\hat{a}=\hat{0}\text{ or }\hat{b}=\hat{0})$$
$$\Rightarrow(a\in p\mathbb{A}\text{ or }b\in p\mathbb{A})\Rightarrow(p\mid a\text{ or }p\mid b).$$

Exercise 6.6   1) If $a\equiv 0$ (mod p), it is clear that $a^p\equiv 0\equiv a$ (mod p). Now assume that $a\not\equiv 0$ (mod p), in other words that $a\in\mathbb{F}_p^*$ (we denote by the same symbol a and its equivalence class mod p). Let k be the order of a in the multiplicative group $\mathbb{F}_p^*$. Then k divides $\left|\mathbb{F}_p^*\right|=p-1$ (example 6.3). Then we can put $p-1=km$ and $a^{p-1}\equiv(a^k)^m\equiv 1^m\equiv 1$ (mod p).

2) We have $\binom{p}{k}=\big(p(p-1)\cdots(p-k+1)\big)/k!$ for $1\le k\le p$. If $k<p$, $k!$ is not divisible by the prime p. Therefore the prime factor p remains in $\binom{p}{k}$ after division of $p(p-1)\cdots(p-k+1)$ by $k!$. Thus $\binom{p}{k}\equiv 0$ (mod p) if $1\le k\le p-1$. Now Fermat's theorem can be proved by induction on a. Simply observe that it is true for $a=0$, and that $(a+1)^p=\sum_{k=0}^{p}\binom{p}{k}a^k\equiv a^p+1$ (mod p).

3) It is easy to check that $561=3\times 11\times 17$. We have
$$a^{561}\equiv a^{3\times 11\times 17}\equiv a^{11\times 17}\equiv(a^{3\times 3+2})^{17}\equiv(a^3\cdot a^2)^{17}\equiv(a\cdot a^2)^{17}$$
$$\equiv a^{17}\equiv a^{3\times 5+2}\equiv a^5\cdot a^2\equiv a^3\equiv a\ (\text{mod}\,3),$$

since $a^3\equiv a$ (mod 3) by Fermat's theorem. Similarly $a^{561}\equiv a$ (mod 11) and $a^{561}\equiv a$ (mod 17). Therefore $a^{561}-a$ is a multiple of 3, 11 and 17. Hence it is a multiple of 561 and $a^{561}-a\equiv 0$ (mod 561).

Remark   The first Carmichael numbers are $561=3\cdot 11\cdot 17$, $1105=5\cdot 13\cdot 17$, $1729=7\cdot 13\cdot 19$, $2465=5\cdot 17\cdot 29$, $2821=7\cdot 13\cdot 31$. It is not known if there exist infinitely many Carmichael numbers.

Exercise 6.7   First we prove that f maps A into B. If $(x,y)\in A$, we have $0<x<\frac{1}{2}q$, $0<y<\frac{1}{2}p$, $px-qy\le-\frac{1}{2}q$. Hence $x'=\frac{1}{2}(q+1)-x$ yields

a) $x' > \frac{1}{2}(q+1) - \frac{1}{2}q > 0$;

b) $x' < \frac{1}{2}(q+1) < \frac{1}{2}q$ since x' et $\frac{1}{2}(q+1)$ are integers.

Similarly we have $0 < y' < \frac{1}{2}p$. Finally,

$$qy' - px' = \left[\frac{1}{2}(p+1) - y\right] - p\left[\frac{1}{2}(q+1) - x\right] = \frac{1}{2}q - \frac{1}{2}p + px - qy$$
$$\leq \frac{1}{2}q - \frac{1}{2}p - \frac{1}{2}q \leq -\frac{1}{2}p,$$

which proves that $(x', y') \in B$. Conversely, to every element $(x', y') \in B$ corresponds a unique $(x, y) \in A$ by the same computation. Indeed, (x, y) and (x', y') play symmetric roles.

Exercise 6.8   1) In the product $(p-1)!$, only 1 and $p-1 \equiv -1 \pmod{p}$ are their own inverses modulo p, since $x^2 \equiv 1 \pmod{p} \Leftrightarrow (x+1)(x-1) \equiv 0 \pmod{p} \Leftrightarrow x \equiv 1$ or $x \equiv -1$. Now associate each number $k \neq 1$ and $\neq p-1$ with its inverse k^{-1}. We get $(p-1)! \equiv 1 \cdot k_1 k_1^{-1} k_2 k_2^{-1} \cdots k_{(p-3)/2} k_{(p-3)/2}^{-1} \cdot (-1) \equiv -1 \pmod{p}$.

2) If $p \equiv 1 \pmod{4}$, $(p-1)/2$ is even and we have

$$(p-1)! \equiv 1 \cdot 2 \cdots \frac{p-1}{2} \cdot (-1)(-2) \cdots \frac{-p+1}{2} \equiv (-1)^{(p-1)/2}\left(\frac{p-1}{2}!\right)^2 \pmod{p}.$$

Exercise 6.9   1) In $\mathbb{F}_p$, we have $a^\alpha = 1$ and $b^\beta = 1$. Hence $a^{\alpha\beta} = 1^\beta = 1$ and $b^{\alpha\beta} = 1$. Consequently $(ab)^{\alpha\beta} = 1$ (in other words $(ab)^{\alpha\beta} \equiv 1 \pmod{p}$). Thus the order k of ab divides $\alpha\beta$. Put $k = \alpha_1\beta_1$, where $\alpha_1 \mid \alpha$ and $\beta_1 \mid \beta$. We have $(ab)^k = a^{\alpha_1\beta_1} b^{\alpha_1\beta_1} = 1$ (*). Now we write $\alpha = \alpha_1\alpha_2$, and raise (*) to the power α_2. We obtain $1 = a^{\alpha_1\alpha_2\beta_1} b^{\alpha_1\beta_1\alpha_2} = (a^\alpha)^{\beta_1} b^{k\alpha_2} = b^{k\alpha_2}$. Therefore β, which is the order of b, divides $k\alpha_2$. But $\alpha_2 \mid \alpha$ and α, and β are coprime. Hence $\beta \mid k$. Similarly $\alpha \mid k$. Finally $\alpha\beta \mid k$ since α and β are coprime. As $k \mid \alpha\beta$, we have $k = \alpha\beta$, Q.E.D.

2) We have $x^{p-1} - 1 = x^{dk} - 1 = (x^d - 1)(x^{d(k-1)} + x^{d(k-2)} + \cdots + x^d + 1)$. Now the equation $x^{p-1} - 1 = 0$ has *exactly* $p-1$ solutions in the field $\mathbb{F}_p$ by Fermat's theorem. The equation $x^{d(k-1)} + x^{d(k-2)} + \cdots + x^d + 1) = 0$ has *at most* $d(k-1)$ solutions in the field $\mathbb{F}_p$. Therefore the equation $x^d - 1 = 0$ has *at least* $(p-1) - d(k-1) = d$ solutions. As it has at most d solutions because it has degree d, it has exactly d solutions.

This proves that the equation $x^{p_k^{a_k}} - 1 = 0$ has exactly $p_k^{a_k}$ solutions.

Among these solutions, the ones that do not have order $p_k^{a_k}$ satisfy $x^{p_k^{a_k-1}} - 1 = 0$, since their order divides $p_k^{a_k}$. There are $p_k^{a_k-1}$ such solutions. Hence there exist exactly $p_k^{a_k} - p_k^{a_k-1}$ elements x in $\mathbb{F}_p$ with order $p_k^{a_k}$.

3) This results directly from questions 1 and 2. To obtain a primitive root g, one only has to multiply elements of orders $p_1^{a_1}, p_2^{a_2}, ..., p_j^{a_j}$.

4) Here $p - 1 = 40 = 2^3 \cdot 5$. First we look for an element of order 2^3, that is satisfying $x^8 = 1$, $x^4 \neq 1$. We have $x^8 - 1 = (x^4 - 1)(x^4 + 1)$, whence $x^4 + 1 = 0$. Put $X = x^2$. We can solve the equation $X^2 = -1$ by using theorem 6.9 (but it is rather long) or we can observe that $9^2 = 81 \equiv -1 \pmod{41}$. Therefore $X^2 = -1 \Leftrightarrow X = 9$ or -9, whence $x = 3$ or -3 or $9 \cdot 3 = 27$ or -27. Now we look for an element of order 5. We have $x^5 - 1 = 0 \Leftrightarrow x^4 + x^3 + x^2 + x + 1 = 0$. This symmetric equation can be solved, as usual, by putting $u = x + 1/x$, whence $u^2 = x^2 + 1/x^2 + 2$, and

$$x^4 + x^3 + x^2 + x + 1 = 0 \Leftrightarrow x^2 + x + 1 + \frac{1}{x} + \frac{1}{x^2} = 0 \Leftrightarrow u^2 + u - 1 = 0.$$

Here $\Delta = 5$. Now trial and error shows that $13^2 = 169 \equiv 5 \pmod{41}$. Hence $u_1 = (-1 + 13)/2 = 6$ and $u_2 = (-1 - 13)/2 = -7$ are the roots of $u^2 + u - 1 = 0$. Now $x + 1/x = 6 \Leftrightarrow x^2 - 6x + 1 = 0$. Here $\Delta' = 8$ and $7^2 = 49 = 8 \pmod{41}$. Thus the solutions of $x + 1/x = 6$ are $x_1 = 3 + 7 = 10$, $x_2 = 3 - 7 = -4$.

Finally $x + 1/x = -7 \Leftrightarrow x^2 + 7x + 1 = 0$. Here $\Delta = 45 = 9 \cdot 5$, and the solutions of $x + 1/x = -7$ are $x_3 = (-7 + 3 \cdot 13)/2 = 16$; $x_4 = (-7 - 3 \cdot 13)/2 = -23$.

Therefore x_1, x_2, x_3, x_4 are the four elements of $\mathbb{F}_{41}$ with order 5.

To obtain the primitive roots in $\mathbb{F}_{41}$, we multiply the solutions of $x^4 + 1 \equiv 0$ and $x^4 + x^3 + x^2 + x + 1 \equiv 0$ in pairs. We obtain $g = 6, 7, 11, 12, 13, 15, 17, 19, 22, 24, 26, 28, 29, 30, 34, 35$.

Exercise 6.10   1) Denote $M = \begin{pmatrix} a & \frac{1}{2}b \\ \frac{1}{2}b & c \end{pmatrix}$ the matrix of the quadratic form.

Then $\varphi(x, y) = (x \quad y) M^T \begin{pmatrix} x \\ y \end{pmatrix} = X^T M X$.

Since $\det M = ac - b^2/4$ we see that $\Delta = -4 \cdot \det M$.

2) If $X' = \begin{pmatrix} x' \\ y' \end{pmatrix}$, we clearly have $X = UX'$.

This corresponds to a change of basis in $\mathbb{R}^2$, with matrix U. Therefore $\varphi'(X') = (X')^T M'X'$, with $M' = U^T MU$. Thus φ and φ' are equivalent if and only if their matrices satisfy $M' = U^T MU$, where U is unimodular.

Hence the relation is *reflexive*: $\varphi \sim \varphi$ since $M = I^T MI$, where $I = \begin{pmatrix} 1 & 0 \\ 0 & 1 \end{pmatrix}$.

It is *symmetric*. Indeed,

$$M' = U^T MU \Leftrightarrow M = (U^{-1})^T M'U^{-1} \text{ and } U^{-1} = \begin{pmatrix} s & -q \\ -r & p \end{pmatrix} \text{ is unimodular.}$$

It is *transitive* since $M' = U^T MU$ and $M'' = V^T M'V$, with U and V unimodular, implies $M'' = (UV)^T M(UV)$, and UV is unimodular because it has integer coefficients and $\det(UV) = \det U \det V = 1$.

3) If φ and φ' are equivalent, their matrices satisfy $M' = U^T MU$, with U unimodular. Hence $\det M' = \det M$ and $\Delta' = \Delta$ by question 1. Clearly this relation could also be proved by a direct computation using (6.11).

Exercise 6.11   The squares modulo 8 are 0, 1 and 4. By trying systematically all possible combinations, we see that $a^2 + b^2 + c^2 \equiv 0, 1, 2, 3, 4, 5$ or $6 \pmod{8}$. Therefore it cannot be congruent to 7. This proves that the numbers of the form $8n + 7$ cannot be written as sums of 3 squares.

Exercise 6.12   Let $x, x' \in \{0, 1, ..., \frac{p-1}{2}\}$, such that $1 + x^2 \equiv 1 + x'^2 \pmod{p}$. Then $x \equiv x' \pmod{p}$ or $x \equiv -x' \pmod{p}$. Since $|x + x'| \leq p - 1$ and $|x - x'| \leq p - 1$, this implies $x = x'$. Hence we have $p + \frac{1}{2}$ different numbers $1 + x^2$ in $\mathbb{F}_p$ when $x \in \{0, 1, ..., \frac{p-1}{2}\}$. Similarly we have $p + \frac{1}{2}$ different numbers $-y^2$ when $y \in \{0, 1, ..., \frac{p-1}{2}\}$. Since $|F_p| = p$, the *pigeon-hole principle* shows that these two sets have at least one common element in $\mathbb{F}_p$. Hence there exists $(x, y) \in \mathbb{N}^2$ such that $1 + x^2 \equiv -y^2 \pmod{p}$.

Exercise 6.13   a) By corollary 6.1, (6.6), and the quadratic reciprocity law:

$$\left(\tfrac{26}{31}\right) = \left(\tfrac{2}{31}\right)\left(\tfrac{13}{31}\right) = (-1)^{120} \left(\tfrac{13}{31}\right) = (-1)^{\frac{12\cdot30}{4}} \left(\tfrac{31}{13}\right) = \left(\tfrac{5}{13}\right) = (-1)^{\frac{4\cdot12}{4}} \left(\tfrac{13}{5}\right) = \left(\tfrac{3}{5}\right) = -1.$$

b) $\left(\tfrac{33}{37}\right) = \left(\tfrac{3}{37}\right)\left(\tfrac{11}{37}\right) = (-1)^{\frac{2\cdot36}{4}} \left(\tfrac{37}{3}\right)(-1)^{\frac{10\cdot36}{4}} \left(\tfrac{37}{11}\right) = \left(\tfrac{1}{3}\right)\left(\tfrac{4}{11}\right) = 1$ since $1 = 1^2$ and $4 = 2^2$.

c) Let $x^2 + 7x - 2 \equiv 0 \pmod{31}$. Then $\Delta = 57 \equiv 26 \pmod{31}$.

Now 26 is not a square modulo 31 by a). Thus the equation has no solution.

d) Let $2x^2 + 5x - 1 \equiv 0$ (mod 37). Here $\Delta = 33$, which is a square mod 37 by b). Table 6.1 shows that the index of 33 relative to the primitive root 2 is 20. Thus $33 = (2^{10})^2$, and the table shows that $33 = (25)^2$. Finally the roots of the equation are $x_1 = 204 = 5$ and $x_2 = -304 = -152 = 222 = 11$.

Exercise 6.14   By using (6.14), (6.12), (6.13), (6.13), we obtain
$$(12,31,21) \sim (12,7,2) \sim (2,-7,12) \sim (2,-3,7) \sim (2,1,6),$$
and $(2,1,6)$ is reduced since $-2 < 1 \le 2 < 6$.

Exercise 6.15   a) $\Delta = -3$. By theorem 6.13, $a \le \sqrt{3/3}$, whence $a = 1$. Thus $b = 0$ or 1 by theorem 6.11. Finally $b^2 - 4ac = -3 \Rightarrow b = 1$, $c = 1$. The only reduced form is $\varphi = (1,1,1)$, and therefore $h(-3) = 1$.

b) $\Delta = -12$. Here $a \le \sqrt{4}$, whence $a = 1$ or 2.

If $a = 1$, $b = 0$ or 1 and $b^2 - 4ac = -12 \Rightarrow b = 0$, $c = 3$. Hence $\varphi_1 = (1,0,3)$.

If $a = 2$, $b = -1$, 0, 1 or 2, and $b^2 - 4ac = -12 \Rightarrow b = 2$, $c = 2$, $\varphi_2 = (2,2,2)$. There exist two reduced forms φ_1 and φ_2 and $h(-12) = 2$. The form φ_2 is *not primitive* because a, b and c are not coprime. In fact, any non primitive form is the product of an integer by a *primitive form* (a, b, c coprime). Here $\varphi_2 = 2\varphi$, where φ is the form in question a).

c) Three reduced forms: $\varphi_1 = (1,1,8)$, $\varphi_2 = (2,1,4)$, $\varphi_3 = (2,-1,4)$

Exercise 6.16   When $\Delta = -7$, $a \le \sqrt{7/2}$, whence $a = 1$ and there exists only one reduced form: $\varphi = (1,1,2)$.

The prime number p is represented (necessarily properly) by a form of discriminant -7 (in other words by φ, since any form of discriminant -7 is equivalent to φ) if and only if there exists $k \in \mathbb{Z}$ such that $-7 \equiv k^2$ (mod $4p$) (theorem 6.14). In other words, iff there exists $m \in \mathbb{Z}$ such that $-7 \equiv (2m+1)^2$ (mod $4p$) $\Leftrightarrow m^2 + m + 2 \equiv 0$ (mod p). The discriminant of this equation is $\Delta = -7$. If $p = 7$, $\Delta \equiv 0$ and the equation has solutions.

If $p \ne 7$, $\left(\frac{-7}{p}\right) = \left(\frac{-1}{p}\right)\left(\frac{7}{p}\right) = (-1)^{(p-1)/2}(-1)^{3(p-1)/2}\left(\frac{p}{7}\right) = \left(\frac{p}{7}\right)$. But the squares mod 7 are 1, 2, and 4. Thus p is represented by φ if and only if $p = 7$ or $p \equiv 1$, 2 or 4 (mod 7). For example, $211 \equiv 1$ (mod 7). Therefore 211 is represented by φ. Trial and error yields $211 = 1^2 + 1 \times 1 \times 10 + 2 \times 10^2$.

Exercise 6.17   1) We have $a = 1$ or 2. Similarly to the above exercise, we see that the only reduced form is $\varphi = (1, 0, 2)$, whence $h(-8) = 1$. The prime number p is represented by φ if and only if the equation $-8 \equiv k^2 \pmod{4p}$ has solutions, that is if and only if the equation $-2 \equiv m^2 \pmod{p}$ has solutions.

This is the case if $p = 2$. If p is odd, we have $\left(\frac{-2}{p} \right) = \left(\frac{-1}{p} \right)\left(\frac{-2}{p} \right) = (-1)^{\frac{p-1}{2} + \frac{p^2-1}{8}}$.

Hence p is represented by φ if and only if $p = 2$ or $p \equiv 1$ or $3 \pmod 8$.

2) The possibility of representing $n \in \mathbb{N}$ by the form $x^2 + 2y^2$ is connected to the ring of integers $\mathbb{A}_\mathbb{K} = \mathbb{Z}(i\sqrt{2})$ of the quadratic field $\mathbb{K} = \mathbb{Q}(i\sqrt{2})$. We have

$$N(\alpha\beta) = N(\alpha)N(\beta) \Leftrightarrow (a^2 + 2b^2)(c^2 + 2d^2) = (ac - 2bd)^2 + 2(ad + bc)^2,$$

as in Lagrange's identity (6.1). Since the ring $\mathbb{Z}(i\sqrt{2})$ is an *euclidean domain* by theorem 5.16, the proof of theorem 6.2 can be used almost word for word, and therefore *the natural integer n can be written as $x^2 + 2y^2$ if, and only if, its prime factors congruent to 5 or 7 (mod 8) have an even exponent.*

Exercise 6.18   1) We have $F_0 = 3$, $F_1 = 5$, $F_2 = 17$, $F_3 = 257$.

To know if $N \in \mathbb{N}$ is prime, it suffices to try to divide it by all primes $p \leq \sqrt{N}$. For F_3 for example, it suffices to try to divide it by all odd prime $p \leq 16$, that is 3, 5, 7, 11, 13. Thus we see easily that F_3 is prime. Checking that $F_4 = 65\,537$ is prime by this method is rather tedious. It becomes impossible for

$$F_5 = 4\,294\,967\,297 \text{ and } F_6 = 18\,446\,744\,073\,709\,551\,617.$$

2) a) We have $F_n \equiv 0 \pmod{p}$, whence $2^{2^n} \equiv -1 \pmod{p}$. Squaring yields $2^{2^{n+1}} \equiv 1 \pmod{p}$. Thus the order m of 2 in $\mathbb{F}_p$ is a divisor of 2^{n+1}. Therefore $m = 2^k$, $k \leq n+1$. But $k \leq n$ is impossible. If so, $2^{2^n} = (2^{2^k})^{2^{n-k}} \equiv 1 \pmod{p}$, contradiction. Thus the order of 2 in $\mathbb{F}_p$ is $m = 2^{n+1}$. Consequently $2^{n+1} \mid p - 1$.

b) In particular, since $n \geq 2$, we have $8 \mid p - 1$. Therefore

$$\left(\frac{2}{p} \right) = (-1)^{\frac{p^2-1}{8}} = (-1)^{k(p+1)} = 1.$$

Now by Euler's criterion we get $2^{(p-1)/2} \equiv 1 \pmod{p}$, whence $2^{n+1} \mid \frac{p-1}{2}$ and therefore there exists an integer k such that $p = k2^{n+2} + 1$.

c) We know from above that the only possible prime divisors of $F_4 = 65\,537$ can be written as $p = 64k + 1$.

Hence we only have to try the prime divisors of this form less than 256. Only $p = 193$ has to be checked. As $193 | F_4$, F_4 is prime.

The only possible prime divisors of $F_5 = 4\ 294\ 967\ 297$ can be written as $p = 128k + 1$. The first primes to be tried are 257, 641, 769, ... We observe easily that $F_5 = 641 \times 6700417$. Therefore F_5 is not prime.

A similar compuation, but a bit longer (use a computer with 20 significant figures) shows that $F_6 = 274177 \times 67\ 280\ 421\ 310\ 721$.

Remark: Fermat claimed in 1650 that all F_n are prime. However, in 1770, Euler succeeded in factorizing F_5, by using the criterion of question 2.

Exercise 6.19   We have $x^p + y^p + z^p = 0 \Leftrightarrow x^p + y^p = -z^p$, in other words
$$(x + y)(x^{p-1} - x^{p-2}y + x^{p-3}y^2 - \cdots + (-1)^{p-1}y^{p-1}) = -z^p.$$
The two factors in the left-hand side are coprime. Indeed, h divides both of them, then $x \equiv -y \pmod{h}$, whence $px^{p-1} \equiv 0 \pmod{h}$ if we replace in $x^{p-1} - x^{p-2}y + \cdots \equiv 0 \pmod{h}$. But h does not divide x, otherwise it would also divides y, and $h \neq p$, otherwise p would divide z. Since $\mathbb{Z}$ is a unique factorization domain, we now have $x + y = t^p$, $x^{p-1} - x^{p-2}y + \cdots = t_1^p$, $z = -tt_1$. As the problem is symmetric in x, y, z, Abel-Barlow relations are proved.

2) If $x^p + y^p + z^p = 0$, then $x^p + y^p + z^p \equiv 0 \pmod{q}$. Hence, by (P), $q \mid x$ or $q \mid y$ or $q \mid z$. Without loss of generality, we can assume that $q \mid x$ (in which case $q \nmid y$ and $q \nmid z$). Now the relations $x + y = t^p$, $y + z = r^p$, $z + x = s^p$ yield $2x = t^p - r^p + s^p$, whence $t^p - r^p + s^p \equiv 0 \pmod{q}$. Using again (P), we see that $q \mid t$ or $q \mid r$ or $q \mid s$. And the relations $z = -tt_1$, $x = -rr_1$, $y = -ss_1$ yield $q \mid r$, $q \nmid s$, $q \nmid z$. Therefore
$$y + z = r^p \Rightarrow y \equiv -z \pmod{q}. \tag{*}$$
Moreover, since $q \mid x$ and $q \nmid y$,
$$x^p + y^p = t_1^p(x + y) \Rightarrow y^p \equiv yt_1^p \pmod{q} \Rightarrow y^{p-1} \equiv t_1^p \pmod{q}. \tag{**}$$
Finally (*) and $y^{p-1} - zy^{p-2} + z^2 y^{p-3} - \cdots = r_1^p$ yield $py^{p-1} \equiv r_1^p \pmod{q}$.
Hence $pt_1^p \equiv r_1^p \pmod{q}$ by using (**). But $q \nmid t_1$ since $q \nmid z$, and therefore t_1 has an inverse in $\mathbb{F}_q^*$. Thus $p \equiv (r_1 t_1^{-1})^p \pmod{q}$, Q.E.D.

3) We only have to prove that, if $q = 2p + 1$, then (P) is satisfied and $p \not\equiv k^p$ $\pmod{q}$. First we prove that (P) is satisfied: if $q = 2p + 1$, $p = q - \frac{1}{2}$. Hence, if

$x \not\equiv 0$ (mod q), we have $x^{(p-1)/2} \equiv \pm 1$ (mod q) by Fermat's theorem. Therefore, if $x \not\equiv 0$, $y \not\equiv 0$, $z \not\equiv 0$ (mod q), $x^p + y^p + z^p \equiv 0$ (mod q) $\Rightarrow \pm 1 \pm 1 \pm 1 \equiv 0$ (mod q), which is impossible since this sum takes only the values $1, -1, 3, -3$, and $q = 2p + 1 \geq 7$.

Now we show that p is not a p-th power modulo q. Suppose, on the contrary, that $p \equiv k^p$ (mod q). Compute Legendre symbol $\left(\frac{k}{q}\right)$. By Euler's criterion,

$$\pm 1 \equiv \left(\frac{k}{q}\right) \equiv k^{(q-1)/2} \equiv k^p \equiv p \text{ (mod } q).).$$

Hence $p \equiv \pm 1$ (mod q), which is impossible since $q = 2p + 1$.

4) Now assume that $q = 4p + 1$ is prime. We first show that (P) is satisfied. If $x \not\equiv 0$, $y \not\equiv 0$, $z \not\equiv 0$ (mod q), then $x^{2p} \equiv \pm 1$, $y^{2p} \equiv \pm 1$, $z^{2p} \equiv \pm 1$ (mod q). Therefore $x^p + y^p + z^p \equiv 0$ (mod q) $\Rightarrow x^{2p} + 2x^p y^p + y^{2p} \equiv z^{2p}$ (mod q) $\Rightarrow 4x^{2p} y^{2p} \equiv (z^{2p} - x^{2p} - y^{2p})^2$ (mod q), whence $\pm 4 \equiv 1$ or 9 (mod q) if $x^{2p} y^{2p} \equiv -1$ or $4 \equiv 1$ or 9 (mod q) if $x^{2p} y^{2p} \equiv 1$ (mod q). In both cases, this is impossible since $q \geq 13$.

Now we prove that $p \not\equiv k^p$ (mod q). If $p \equiv k^p$ (mod q), then

$$\pm 1 \equiv \left(\frac{k}{q}\right) \equiv k^{(q-1)/2} \equiv k^{2p} \equiv p^2 \text{ (mod } q).$$

Now $q = 4p + 1 \Rightarrow 16p^2 = q^2 - 2q + 1 \Rightarrow \pm 16 \equiv 1$ (mod q). As $q \geq 13$, this implies $q = 17$, which is impossible since $(17 - 1)/4 = 4$ is not prime.

Exercise 6.20   1) The most important three examples are:

a) $\mathbb{A} = \mathbb{Z}$, with $\varphi(r) = |r|$.

b) $\mathbb{A} = \mathbb{A}_{\mathbb{K}}$, $\mathbb{K} = \mathbb{Q}(\sqrt{d})$, $d = -11, -7, -3, -2, -1, 2, 3, 5$ and 13 (theorem 5.16), with $\varphi(r) = |N(r)|$.

c) $\mathbb{A} = \mathbb{K}[x]$, the ring of polynomials over a commutative field $\mathbb{K}$, with $\varphi(P) = \deg P$ (polynomial long division).

2) We have to prove that every ideal of an euclidean domain $\mathbb{A}$ is principal. Let I an ideal of $\mathbb{A}$. If $I = \{0\}$, then $I = 0 \cdot \mathbb{A}$ and I is principal. Otherwise, let $\alpha \neq 0 \in I$, such that $\varphi(\alpha)$ is *minimal* in $\mathbb{N}$. Let $x \in I$. Dividing x by α yields $x = \alpha q + r$, with $\varphi(r) < \varphi(\alpha)$ or $r = 0$. But $r = x - \alpha q \in I$ since I is an ideal. Therefore $\varphi(r) < \varphi(\alpha)$ is impossible because $\varphi(\alpha)$ is minimal. Hence $r = 0$, $x = \alpha q$ and $I \subset \alpha \mathbb{A}$.

As evidently $\alpha \mathbb{A} \subset I$, we have $I = \alpha \mathbb{A}$ and I is principal, Q.E.D.

3) I is an additive subgroup of $\mathbb{A}$ since
$(ax + by) + (ax' + by') = a(x + x') + b(y + y') \in I$ and $a(-x) + b(-y) \in I$.
Moreover, $\forall z \in \mathbb{A}$, $z(ax + by) = a(xz) + b(yz) \in I$. Hence I is an ideal of $\mathbb{A}$.

Now we prove Bézout's identity. If $\mathbb{A}$ is a principal ideal domain, then I defined as above is principal and therefore $I = d\mathbb{A}$ for some $d \in \mathbb{A}$. Hence d divides all elements of I, in particular a (for $x = 0$, $y = 1$) and b (for $x = 1$, $y = 0$). If a and b are coprime, d is therefore invertible, $d = \varepsilon$. Since $d \in I$, there exist $x_0, y_0 \in \mathbb{A}$ such that $ax_0 + by_0 = d = \varepsilon$, whence

$$a(x_0 \varepsilon^{-1}) + b(y_0 \varepsilon^{-1}) = 1.$$

This is Bézout's identity.

4) Look closely at the proof of theorem 5.15. You will see that it rests only on Bézout's identity. Since Bézout's identity remains true in any principal ideal domain, any principal ideal domain is a unique factorization domain. Thus:

$\mathbb{A}$ is an euclidean domain $\Rightarrow \mathbb{A}$ is a principal ideal domain $\Rightarrow \mathbb{A}$ is a unique factorization domain.

Solutions to the exercises of chapter 7

Exercise 7.1 1) Suppose that

$$u_{n+p} = \alpha_{p-1} u_{n+p-1} + \alpha_{p-2} \mu_{n+p-2} + \cdots + \alpha_0 u_n, \alpha_0 \neq 0.$$

Then $f(x) = \sum_{n=0}^{+\infty} u_n x^n = \sum_{n=0}^{p-1} u_n x^n + \sum_{n=p}^{+\infty} u_n x^n = \sum_{n=0}^{p-1} u_n x^n + \sum_{k=1}^{p} \alpha_{p-k} \sum_{n=p}^{+\infty} u_{n-k} x^n$

$= \sum_{n=0}^{p-1} u_n x^n + \sum_{k=1}^{p} \alpha_{p-k} x^k \sum_{n=p-k}^{+\infty} u_n x^n = \sum_{n=0}^{p-1} u_n x^n + \sum_{k=1}^{p} \alpha_{p-k} x^k \left(f(x) - \sum_{n=0}^{p-k-1} u_n x^n \right)$,

whence $\left[1 - \sum_{k=1}^{p} \alpha_{p-k} x^k \right] f(x) = \sum_{n=0}^{p-1} u_n x^n - \sum_{k=1}^{p} \sum_{n=0}^{p-k-1} \alpha_{p-k} u_n x^{n+k}$.

Thus $f(x) = P(x)/Q(x)$ is a rational fraction, $\deg Q = p$ since $\alpha_0 \neq 0$, and $\deg P \leq p - 1$.

2) Conversely, assume that $f(x) = P(x)/Q(x)$, with P, Q coprime and $Q(x) = \sum_{k=0}^{p} q_k x^k$. Then $q_0 \neq 0$, otherwise f would not be a power series. Now

$$Q(x) f(x) = \sum_{k=0}^{p} q_k x^k \sum_{n=0}^{+\infty} u_n x^n = P(x).$$

By collecting the terms of degree n, we obtain $Q(x)f(x) = \sum_{n=0}^{+\infty} a_n x^n$, with $a_n = \sum_{k=0}^{p} q_k u_{n-k}$. Since $\deg P \leq p-1$, we have $a_n = 0$ for $n \geq p$.

Hence $q_0 u_n = -\sum_{k=1}^{p} q_k u_{n-k}$ for $n \geq p$ and (u_n) is a linear recurring sequence.

Remark: We see easily that the generating function of (u_n) is a rational fraction (without restriction on the degrees of P and Q) if and only if (u_n) is linear recurring from a given rank n_0.

Exercise 7.2   Put $v_n(x) = \prod_{k=0}^{n}(1+u_k(x))$. The computation of exercise 3.4 shows that $|v_{n+p}(x) - v_n(x)| \leq |w_{n+p}(x) - w_n(x)|$, with $w_n(x) = \prod_{k=0}^{n}(1+|u_k(x)|)$. We deduce that v_n converges uniformly to f on every compact K of Ω. Hence f is holomorphic in Ω by lemma 7.1. To prove (7.13), it suffices to apply to v_n the product rule, and obtain

$$v'_n(x) = \sum_{m=0}^{n} u'_m(x) \prod_{k \neq m}(1+u_k(x)) = v_n(x) \sum_{m=0}^{n} \frac{u'_m(x)}{1+u_m(x)}.$$

Then we let n tend to infinity.

Exercise 7.3   Put $x = -1$ in the triple product identity. We obtain

$$\sum_{n=-\infty}^{+\infty} (-1)^n q^{n^2} = \prod_{n=1}^{+\infty}(1-q^{2n})(1-q^{2n-1})^2$$

$$= \prod_{n=1}^{+\infty}(1-q^n)(1-q^{2n-1}) = \prod_{n=1}^{+\infty} \frac{(1-q^{2n})(1-q^{2n-1})}{1+q^n}.$$

Exercise 7.4   The computations are similar to those of exercise 7.3:

$$\prod_{n=1}^{+\infty}(1+q^n)^2(1-q^n) = \prod_{n=1}^{+\infty}(1+q^n)(1-q^{2n})$$

$$= \prod_{n=1}^{+\infty}(1+q^{2n})(1+q^{4n-3})(1+q^{4n-1})(1-q^{2n}) = \prod_{n=1}^{+\infty}(1+q^{4n-3})(1+q^{4n-1})(1-q^{4n}).$$

Exercise 7.5   1) We have

$$\sum_{n=1}^{+\infty} q^n/(1-q^n)^2 = \sum_{n=1}^{+\infty} q^n \sum_{k=1}^{+\infty} kq^{n(k-1)} = \sum_{n=1}^{+\infty} \sum_{k=1}^{+\infty} kq^{nk}.$$

Here we can exchange the summations in n and k since the double series $\sum_{n=1}^{+\infty} \sum_{k=1}^{+\infty} k|q|^{nk} = \sum_{n=1}^{+\infty} |q|^n/(1-|q|^n)^2$ is convergent. Therefore we have

$$\sum_{n=1}^{+\infty} \frac{q^n}{(1-q^n)^2} = \sum_{k=1}^{+\infty} k \sum_{n=1}^{+\infty} (q^k)^n = \sum_{k=1}^{+\infty} \frac{kq^k}{1-q^k}.$$

2) The *Euler operator* $D_x = x\frac{d}{dx}$ satisfies $D_x(x^k) = kx^k$ for every k. Thus
$$D_x^2\theta_3(q,x) = \sum_{n=-\infty}^{+\infty} n^2 x^n q^{n^2}. \text{ Similarly } D_q\theta_3(q,x) = \sum_{n=-\infty}^{+\infty} n^2 x^n q^{n^2}.$$

3) By using (7.13) in the triple product identity, we get
$$D_x\theta_3(q,x) = \theta_3(q,x)\left(\sum_{n=1}^{+\infty} \frac{xq^{2n-1}}{1+xq^{2n-1}} - \sum_{n=1}^{+\infty} \frac{x^{-1}q^{2n-1}}{1+x^{-1}q^{2n-1}}\right), \quad \text{whence}$$

$$D_x^2\theta_3(q,x) = D_x\theta_3(q,x)\left(\sum_{n=1}^{+\infty} \frac{xq^{2n-1}}{1+xq^{2n-1}} - \sum_{n=1}^{+\infty} \frac{x^{-1}q^{2n-1}}{1+x^{-1}q^{2n-1}}\right)$$

$$+ \theta_3(q,x)\left(\sum_{n=1}^{+\infty} \frac{xq^{2n-1}}{(1+xq^{2n-1})^2} + \sum_{n=1}^{+\infty} \frac{x^{-1}q^{2n-1}}{(1+x^{-1}q^{2n-1})^2}\right).$$

This proves the result for $D_x^2\theta_3(q,x)$. The computation is similar for $D_q\theta_3(q,x)$.

4) This result from questions 2 and 3 after simplification by $\theta_3(q,x)$.

5) We simplify the left-hand side of the above result. First we have
$$\sum_{n=1}^{+\infty}\left(\frac{q^{4n-3}}{(1-q^{4n-3})^2} + \frac{q^{4n-1}}{(1-q^{4n-1})^2}\right) = \sum_{n=0}^{+\infty} \frac{q^{2n+1}}{(1-q^{2n+1})^2} = \sum_{n=1}^{+\infty} \frac{q^n}{(1-q^n)^2} - \sum_{n=1}^{+\infty} \frac{q^{2n}}{(1-q^{2n})^2}.$$

Using question 1 yields
$$\sum_{n=1}^{+\infty}\left(\frac{q^{4n-3}}{(1-q^{4n-3})^2} + \frac{q^{4n-1}}{(1-q^{4n-1})^2}\right) = \sum_{n=1}^{+\infty} \frac{nq^n}{1-q^n} - \sum_{n=1}^{+\infty} \frac{nq^{2n}}{1-q^{2n}}. \tag{*}$$

On the other hand,
$$\sum_{n=1}^{+\infty} \frac{(2n-1)q^{4n-1}}{1-q^{4n-1}} + \sum_{n=1}^{+\infty} \frac{(2n-1)q^{4n-3}}{1-q^{4n-3}}$$

$$= \frac{1}{2}\sum_{n=1}^{+\infty}\left(\frac{(4n-1)q^{4n-1}}{1-q^{4n-1}} + \frac{(4n-3)q^{4n-3}}{1-q^{4n-3}}\right) - \frac{1}{2}\sum_{n=1}^{+\infty}\left(\frac{q^{4n-1}}{1-q^{4n-1}} - \frac{q^{4n-3}}{1-q^{4n-3}}\right)$$

$$= \frac{1}{2}\sum_{n=0}^{+\infty} \frac{(2n+1)q^{2n+1}}{1-q^{2n+1}} + \frac{1}{2}\sum_{n=0}^{+\infty} (-1)^n \frac{q^{2n+1}}{1-q^{2n+1}}.$$

Replacing in the result of question 4 and taking (*) into account yields
$$\left(\sum_{n=0}^{+\infty} \frac{(-1)^n q^{2n+1}}{1-q^{2n+1}}\right)^2 = \frac{1}{2}\left(\sum_{n=1}^{+\infty} \frac{nq^n}{1-q^n} - \sum_{n=1}^{+\infty} \frac{4nq^{4n}}{1-q^{4n}}\right) - \frac{1}{2}\sum_{n=0}^{+\infty} \frac{(-1)^n q^{2n+1}}{1-q^{2n+1}}.$$

This is equivalent to
$$\left(1 + 4\sum_{n=0}^{+\infty} (-1)^n \frac{q^{2n+1}}{1-q^{2n+1}}\right)^2 = 1 + 8\left(\sum_{n=1}^{+\infty} \frac{nq^n}{1-q^n} - \sum_{n=1}^{+\infty} \frac{4nq^{4n}}{1-q^{4n}}\right).$$

And this proves (7.20) by using (7.18).

Exercise 7.6 We prove the sieve formula by induction. First, for $n=2$,

$$|A_1 \cup A_2| = |A_1| + |A_2| - |A_1 \cap A_2|.$$

Indeed, when we add $|A_1|$ and $|A_2|$, we count the elements of $A_1 \cap A_2$ twice. Now assume that the formula is true for $n \geq 2$. We have

$$|A_1 \cup A_2 \cup \cdots \cup A_{n+1}| = |A_1 \cup \cdots \cup A_n| + |A_{n+1}| - |(A_1 \cap A_{n+1}) \cup \cdots \cup (A_n \cap A_{n+1})|,$$

since the formula is true for $n = 2$. Using the induction hypothesis yields

$$|A_1 \cup \cdots \cup A_{n+1}| = \sum_{i=1}^{n+1} |A_i| - \sum_{i<j} |A_i \cap A_j| + \cdots + (-1)^{n-1} |A_1 \cap \cdots \cap A_n|$$

$$- \left(\sum_{i=1}^{n} |A_i \cap A_{n+1}| - \sum_{i<j} |A_i \cap A_j \cap A_{n+1}| + \cdots + (-1)^{n-1} |A_1 \cap A_2 \cap \ldots \cap A_{n+1}| \right),$$

which proves that the formula is true for $n+1$.

Exercise 7.7   a) We have $a_{n+1} - 1 = a_n^2 - a_n$, whence

$$\frac{1}{a_{n+1} - 1} = \frac{1}{a_n(a_n - 1)} = \frac{1}{a_n - 1} - \frac{1}{a_n}.$$

Consequently $\dfrac{1}{a_n} = \dfrac{1}{a_n - 1} - \dfrac{1}{a_{n+1} - 1}$.

Hence $\sum_{n=1}^{+\infty} 1/a_n$ is a *telescoping series*, and we have

$$\sum_{n=1}^{N} \frac{1}{a_n} = \frac{1}{a_1 - 1} - \frac{1}{a_{N+1} - 1}, \text{ whence } \sum_{n=1}^{+\infty} \frac{1}{a_n} = \frac{1}{a_1 - 1} = 1.$$

b) Since $a_n \geq 2$, we have $a_n^2 = a_{n+1} + a_n - 1 \geq a_{n+1} + 1$ and $a_n^2 > a_{n+1}$. Moreover $a_{n+1} = a_n^2 - a_n + 1 > a_n^2 - 2a_n + 1$ and $a_{n+1} > (a_n - 1)^2$. To study the convergence of (u_n), we take its logarithm. We have

$$\mathrm{Log}\, u_n = \sum_{k=1}^{n} \mathrm{Log}\, a_k / a_k = \sum_{k=1}^{n} v_k.$$

Now we study the convergence of the series $\sum_{k=1}^{+\infty} v_k$ by d'Alembert's ratio test. Since $a_n^2 > a_{n+1} > (a_n - 1)^2$, we have

$$\frac{v_{k+1}}{v_k} = \frac{\mathrm{Log}\, a_{k+1}}{\mathrm{Log}\, a_k} \times \frac{a_k}{a_{k+1}} < \frac{2a_k}{(a_k - 1)^2} \leq \frac{7}{18} \text{ for } k \geq 3.$$

Hence the series is convergent, and $\sum_{k=1}^{+\infty} v_k = \sum_{k=1}^{5} v_k + \sum_{k=6}^{+\infty} v_k$, with

$$0 < \sum_{k=6}^{+\infty} v_k < v_6 \sum_{n=0}^{+\infty} \left(\frac{7}{18} \right)^n = \frac{18}{11} v_6.$$

Finally $\sum_{k=1}^{n} v_k < \sum_{k=1}^{5} \frac{\mathrm{Log}\, a_k}{a_k} + \frac{\mathrm{Log}\, a_6}{a_6} \times \frac{18}{11} < 1.0824$ and $u_n < e^{1.0824} < 2.952$.

c) Since $a_{n+1} > (a_n - 1)^2$, an easy induction shows that $a_n \geq 2^{2^{n-2}} + 1$ for $n \geq 2$.

Therefore $a_k \leq n \Rightarrow 2^{2^{k-2}} + 1 \leq n$, whence $2^{2^{k-2}} \leq n$, $2^{k-2} \leq \log_2 n$, and finally $k - 2 \leq \log_2(\log_2 n)$.

Exercise 7.8   This formula rests on the following elementary remark: if $a \geq 2$ is a natural integer, the sequence $1, 2, ..., n$ contains exactly $[n/a]$ multiples of a, namely the numbers $a, 2a, 3a, ..., [n/a]a$.

Now let p be any prime number. In the sequence $1, 2, ..., n$, there are exactly $[n/p]$ multiples of p. Among them, $[n/p^2]$ are multiples of p^2, $[n/p^3]$ are multiples of p^3, and so on. The process stops at k satisfying $p^k \leq n$, that is $k = [\log_p n] = [\mathrm{Log}\, n/\mathrm{Log}\, p])$. This yields the exponent of p in $n!$.

Exercise 7.9   1) We have $[x/m] \geq [[x]/m]$ since $x \geq [x]$ and the integer part is a non decreasing function. Moreover, $x < [x] + 1$, whence $\frac{x}{m} < \frac{[x]}{m} + \frac{1}{m}$.

Now divide $[x]$ by m. We have $[x] = m[[x]/m] + r$, with $0 \leq r \leq m - 1$. Thus

$$\frac{x}{m} < \left[\frac{[x]}{m}\right] + \frac{r+1}{m} \leq \left[\frac{[x]}{m}\right] + 1.$$

Consequently $\left[\frac{x}{m}\right] \leq \left[\frac{[x]}{m}\right]$, and finally $\left[\frac{x}{m}\right] = \left[\frac{[x]}{m}\right]$. This yields

$$\sum_{i=1}^{n} \left[\frac{x}{a_i}\right] = \sum_{i=1}^{n} \left[\frac{[x]}{a_i}\right] \leq \sum_{i=1}^{n} \frac{[x]}{a_i} < [x] \quad \text{since} \quad \sum_{i=1}^{+\infty} \frac{1}{a_i} = 1.$$

2) By using Legendre's formula (lemma 7.3), we see that

$$C(n) = \prod_{p \leq n} p^{\beta_p}, \quad \text{with} \quad \beta_p = \sum_{i=1}^{[\log_p n]} \left(\left[\frac{n}{p^i}\right] - \sum_{j=1}^{k}\left[\frac{n}{p^i a_j}\right]\right).$$

Now question 1) yields $\beta_p \geq \sum_{i=1}^{[\log_p n]} 1 = [\log_p n]$.

Therefore $\delta(n) \mid C(n)$ by (7.26), and consequently $\delta(n) \leq C(n)$.

3) We now that $\mathrm{Log}(1 + X) < X$ for every $X > 0$.

Therefore $\mathrm{Log}\left(1 + \frac{1}{x}\right)^x = x\,\mathrm{Log}\left(1 + \frac{1}{x}\right) < 1$, whence $\left(1 + \frac{1}{x}\right)^x < e$.

Now, we observe that the result we have to prove is evident if $n = a_i$. If $n > a_i$, we have $n = a_i q + r$, with $0 \leq r \leq a_i - 1$ and $q = [n/a_i]$.

Therefore $\frac{n-r}{a_i} = \left[\frac{n}{a_i}\right]$ and $\left[\frac{n}{a_i}\right] \geq \frac{n - a_i + 1}{a_i}$. Consequently,

$$\frac{(n/a_i)^{n/a_i}}{[n/a_i]^{[n/a_i]}} \leq \frac{(n/a_i)^{n/a_i}}{((n-a_i+1)/a_i)^{(n-a_i+1)/a_i}} = \left(\frac{n}{a_i}\right)^{n/a_i}\left(\frac{a_i}{n}\times\frac{n}{n-a_i+1}\right)^{(n-a_i+1)/a_i},$$

whence $\dfrac{(n/a_i)^{n/a_i}}{[n/a_i]^{[n/a_i]}} \leq \left(\dfrac{n}{a_i}\right)^{(a_i-1)/a_i}\left(1+\dfrac{a_i-1}{n-a_i+1}\right)^{\frac{n-a_i+1}{a_i-1}\times\frac{a_i-1}{a_i}} < \left(\dfrac{ne}{a_i}\right)^{(a_i-1)/a_i}$.

4) The coefficient of $x_1^{n_1}x_2^{n_2}...x_k^{n_k}$ in the expansion of $(x_1+x_2+\cdots+x_k)^N$ is the binomial coefficient $N!/(n_1!n_2!...n_k!)$. Replacing x_i by $n_i \in \mathbb{N}$ yields

$$N^N \geq \frac{N!}{n_1!n_2!...n_k!}n_1^{n_1}n_2^{n_2}...n_k^{n_k}, \text{ with } N = n_1+\cdots+n_k.$$

Now replace in (7.27), with $n_i = [n/a_i]$. Since $N < n$, we obtain

$$C(n) = \frac{N!(N+1)(N+2)...n}{n_1!n_2!...n_k!} < \frac{N^N n^{n-N}}{n_1^{n_1}...n_k^{n_k}} < \frac{n^n}{n_1^{n_1}...n_k^{n_k}}.$$

Using question 3 yields $n_i^{n_i} > (n/a_i)^{n/a_i}/(en/a_i)^{(a_i-1)/a_i}$, Q.E.D.

5) We know from exercise 7.7, question a), that

$$\frac{1}{a_1}+\frac{1}{a_2}+\cdots+\frac{1}{a_k} = 1-\frac{1}{a_{k+1}-1}.$$

Thus, by introducing the sequence u_k defined in exercise 7.7, we have

$$C(n) < n^n\left[n^{\frac{1}{a_1}+\cdots+\frac{1}{a_k}}\right]^{-n}(en)^{k-\frac{1}{a_1}-\cdots-\frac{1}{a_k}}\frac{\left[a_1^{\frac{1}{a_1}}a_2^{\frac{1}{a_2}}...a_k^{\frac{1}{a_k}}\right]^n}{a_1^{(a_1-1)/a_1}...a_k^{(a_k-1)/a_k}} < n^{\frac{n}{a_{k+1}-1}}(en)^{k-1+\frac{1}{a_{k+1}-1}}(u_k)^n.$$

But $a_{k+1} \geq n+1 \geq 2$. Thus the results of questions b and c, exercise 7.7 yield

$$C(n) < e^k n^{k+1}(2.952)^n < (en)^{\log_2(\log_2 n)+3}\times(2.952)^n$$

Exercise 7.10   By theorem 7.3, we know that $r_2(n) = 4(d_1(n)-d_3(n))$.

First suppose that all primes dividing $p_1, p_2,..., p_k$ satisfy $p_i \equiv 1$ (mod 4). Then $d_3(n)=0$, and we obtain $d_1(n)$ by counting all numbers of the form $p_1^{\alpha_1}p_2^{\alpha_2}...p_k^{\alpha_k}$, with $0 \leq \alpha_i \leq a_i$ and $n = p_1^{a_1}p_2^{a_2}...p_k^{a_k}$. We have clearly a_i+1 possible choices for α_i, whence $d_1(n) = \prod_{i=1}^{k}(a_i+1)$, which proves the formula for $r_2(n)$ in this case.

Now we argue by induction on the number k of distinct prime divisors of n satisfying $q_i \equiv 3$ (mod 4). If $k=0$, the formula is true as we have just seen. Assume it is true for $k-1$, and denote $n = mq_k^s$. If s is odd, then $r_2(n) = 0$ by theorem 6.1, and therefore (7.29) is satisfied. If s is even, we obtain the new divisors $d \equiv 1 \pmod 4$ of n by multiplying those of m by $q_k^2, q_k^4, \ldots, q_k^s$, and the new divisors $d \equiv 3 \pmod 4$ by multiplying those of m by $q_k, q_k^3, \ldots, q_k^{s-1}$. Hence, when we multiply m by q_k^s, the number of new divisors congruent to 1 mod 4 is the same than the number of new divisors congruent to 3 mod 4. Therefore $r_2(n) = r_2(m)$ by theorem 7.3, which proves (7.25).

Exercise 7.11   Replace q by $q^{3/2}$ and x by $-q^{1/2}$ in (7.12). We get

$$\sum\nolimits_{n=-\infty}^{+\infty} (-1)^n q^{n/2} q^{3n^2/2} = \prod\nolimits_{n=1}^{+\infty} (1-q^{3n})(1-q^{3n-1})(1-q^{3n-2}) = \prod\nolimits_{n=1}^{+\infty} (1-q^n).$$

Exercise 7.12   Derive (7.17) with respect to x. Then put $x = 1$, and note that

$$\sum_{n=-\infty}^{+\infty} (-1)^n q^{n(n+1)/2} (2n+1) = \sum_{n=0}^{+\infty} (-1)^n (2n+1) q^{n(n+1)/2} + \sum_{n=1}^{+\infty} (-1)^n (-2n+1) q^{n(n-1)/2}.$$

The conclusion follows by putting $m = n-1$ in the second sum.

Exercise 7.13   1) The series $\sum_{k=1}^{+\infty} q^k$ converges uniformly on every compact of $D = \{q \in \mathbb{C}/|q| < 1\}$. Therefore $g(q) = \prod_{k=1}^{+\infty} (1-q^k)$ is holomorphic in D by lemma 7.2 and does not vanish in D by exercise 3.4, question 2). Thus $f(q) = (g(q))^{-1}$ is holomorphic in D, and $f(q) = \prod_{k=1}^{+\infty} (1 + q^k + q^{2k} + \cdots)$ for $|q| < 1 <$ In other words,

$$f(q) = (1+q+q^2+q^3+\cdots)(1+q^2+q^4+q^6+\cdots)(1+q^3+q^6+q^9+\cdots)\cdots.$$

Now every partition of n adds 1 to the coefficient of q^n in the power series expansion of $f(q)$. For example, the partition $5+4+4+3+2+2+2+1$ of $n = 23$ corresponds to the product of

- q^5 from the fifth bracket of the infinite product,
- $q^8 = q^{4+4}$ from the fourth bracket,
- q^3 from the third bracket,
- $q^6 = q^{2+2+2}$ from the second bracket,
- q from the first bracket.

This proves that $f(q)$ is the generating function of $p(n)$.

2) By exercise 7.11, we have

$$g(q) = \sum\nolimits_{n=-\infty}^{+\infty} (-1)^n q^{n(3n+1)/2} = \sum\nolimits_{n=0}^{+\infty} (-1)^n q^{n(3n+1)/2} + \sum\nolimits_{n=1}^{+\infty} (-1)^{-n} q^{-n(-3n+1)/2}.$$

Thus $g(q) = \sum_{n=0}^{+\infty} \alpha(n)q^n$. As $f(q)g(q) = 1$, multiplying the two series yields

$\sum_{k=0}^{n} \alpha(k)p(n-k) = 0$ for $n \geq 1$, which proves the formula since $\alpha(0) = 1$.

3) It is easy to compute $\alpha(n)$ for $1 \leq n \leq 12$.

We obtain $\alpha(1) = \alpha(2) = \alpha(12) = -1$, $\alpha(5) = \alpha(7) = 1$, $\alpha(n) = 0$ otherwise.

Now, by induction: $p(1) = 1$, $p(2) = -\alpha(1)p(1) - \alpha(2)p(0) = 2$,

$p(3) = -\alpha(1)p(2) - \alpha(2)p(1) - \alpha(3)p(0) = 3$, $p(4) = p(3) + p(2) = 5$,

$p(5) = p(4) + p(3) - p(0) = 7$, $p(6) = p(5) + p(4) - p(1) = 11$,

$p(7) = p(6) + p(5) - p(2) - p(0) = 15$, $p(8) = p(7) + p(6) - p(3) - p(1) = 22$,

$p(9) = p(8) + p(7) - p(4) - p(2) = 30$, $p(10) = p(9) + p(8) - p(5) - p(3) = 42$,

$p(11) = p(10) + p(9) - p(6) - p(4) = 56$,

$p(12) = p(11) + p(10) - p(7) - p(5) + p(0) = 77$.

Exercise 7.14   1) Dividing 1 by $1 + t/2! + t^2/3! + t^4/4! + \cdots$ yields

$$B_0 = 1,\ B_1 = -\frac{1}{2},\ B_2 = \frac{1}{6},\ B_3 = 0,\ B_4 = -\frac{1}{30},\ B_5 = 0,\ B_6 = \frac{1}{42}.$$

2) a) We have

$$\frac{te^{tx}}{e^t - 1} = \frac{t + t^2 x + o(t^2)}{t + t^2/2 + o(t^2)} = \left(1 + tx + o(t)\right)\left(1 - \frac{t}{2} + o(t)\right) = 1 + \left(x - \frac{1}{2}\right)t + o(t).$$

Therefore $B_0(x) = 1$ and $B_1(x) = x - \frac{1}{2}$.

b) Take the partial derivative of (*) with respect to x. We obtain

$$\frac{t^2 e^{tx}}{e^t - 1} = \sum_{n=1}^{+\infty} B_n'(x)\frac{t^n}{n!} = t\sum_{n=0}^{+\infty} B_{n+1}'(x)\frac{t^n}{(n+1)!} = t\sum_{n=0}^{+\infty} B_n(x)\frac{t^n}{n!},\quad \text{Q.E.D.}$$

c) We have $f(t) = \dfrac{t}{e^t - 1} - 1 + \dfrac{1}{2}t$ and $f(-t) = \dfrac{te^t}{e^t - 1} - 1 - \dfrac{1}{2}t$.

Hence $f(t) - f(-t) = -t + t = 0$, which proves that f is even.

Since $f(t) = \sum_{n=2}^{+\infty} B_n t^n/n!$, this proves that $B_{2p+1} = 0$ for $p \geq 1$.

d) Replacing x by 1 and t by $-t$ in (*) yields $\dfrac{t}{e^t - 1} = \sum_{n=0}^{+\infty} (-1)^n B_n(1)\dfrac{t^n}{n!}$.

Comparing to (*) with $x = 0$ yields $B_n(0) = B_n(1)$ for n even. This remains

true for n odd ≥ 3 by question c. Finally $B_n(0) = B_n$ by definition of B_n.

e) We prove it by induction. The formula is true for $n = 0$. Now assume that

$B_n(x) = \sum_{k=0}^{n} \binom{n}{k} B_k x^{n-k}$. Integrating and taking question b into account yields

$$B_{n+1}(x) = (n+1)\sum_{k=0}^{n}\binom{n}{k}B_k\frac{x^{n+1-k}}{n+1-k} + B_{n+1}(0).$$

But $\binom{n+1}{k} = \frac{n+1}{n+1-k}\binom{n}{k}$ and $B_{n+1}(0) = B_{n+1}$, which proves the formula.

f) We argue again by induction. The formula is true for $n = 0$. The induction hypothesis is $B_n(x+1) - B_n(x) = nx^{n-1}$. Integrating yields

$$B_{n+1}(x+1) - B_{n+1}(x) + A = (n+1)x^n, \text{ where } A \text{ is a constant.}$$

For $x = 0$ we obtain $A = 0$ by question d, Q.E.D.

Now summing for $x = 1, 2, ..., k$ yields $\displaystyle\sum_{i=1}^{k}i^n = \frac{1}{n+1}\left(B_{n+1}(k+1) - B_{n+1}\right).$

Exercise 7.15   1) $\mu(4) = \mu(8) = \mu(9) = \mu(12) = 0$,

$\mu(2) = \mu(3) = \mu(5) = \mu(7) = \mu(11) = -1$, $\mu(6) = \mu(10) = 1$.

2) In $\sum_{d\backslash n}\mu(d)/d$, the only divisors of $n = p_1^{\alpha_1}p_2^{\alpha_2}...p_m^{\alpha_m}$ (p_i prime) which count are 1 and those of the form $p_{i_1}p_{i_2}...p_{i_j}$, with $p_a \neq p_b$ if $a \neq b$. Therefore, by the proof of (7.21),

$$n\sum_{d\backslash n}\frac{\mu(d)}{d} = n\left(1 - \sum\frac{1}{p_i} + \sum_{i<j}\frac{1}{p_ip_j} - \cdots + (-1)^m\frac{1}{p_1p_2\cdots p_m}\right) = \varphi(n).$$

3) Similarly, if n has at least one prime divisor, that is if $n > 1$,

$$\sum_{d\backslash n}\mu(d) = 1 - \sum_{i=1}^{m}1 + \sum_{1<j}1 - \cdots = \binom{m}{0} - \binom{m}{1} + \binom{m}{2} - \cdots = (1-1)^m = 0.$$

4) We have

$$\sum_{d\backslash n}\mu\left(\tfrac{n}{d}\right)g(d) = \sum_{d\backslash n}\mu(d)g\left(\tfrac{n}{d}\right) = \sum_{d\backslash n}\mu(d)\sum_{c\backslash\frac{n}{d}}f(c)$$

$$= \sum_{cd\backslash n}\mu(d)f(c) = \sum_{c\backslash n}f(c)\sum_{d\backslash\frac{n}{c}}\mu(d).$$

Now, if $c < n$, we have $n/c > 1$ and $\sum_{d\backslash(n/c)}\mu(d) = 0$ by question 3. The only divisor c of n which counts in the sum is therefore $c = n$, whence

$$\sum_{d\backslash n}\mu\left(\tfrac{n}{d}\right)g(d) = f(n)\sum_{d\backslash 1}\mu(d) = f(n).$$

Exercise 7.16   The divisors of n are the abscissas of the lattice points which lie on the hyperbola $xy = n$. Thus $\sum_{k=1}^{n}d(k) = d(1) + d(2) + \cdots + d(n)$ is the number of lattice points $(x, y) \in \mathbb{N}^{*2}$ which lie under, or on, this hyperbola.

Among these points, n have the x-coordinate 1, $[n/2]$ have the x-coordinate 2, $[n/3]$ have the x-coordinate 3, and so on. Therefore

$$\sum_{k=1}^{n} d(k) = n + \left[\frac{n}{2}\right] + \cdots + \left[\frac{n}{n-1}\right] + 1 = n + \frac{n}{2} + \cdots + \frac{n}{n-1} + \frac{n}{n} + 0(n),$$

since $\frac{n}{k} - \left[\frac{n}{k}\right] \in [0,1[$ for $k = 1, 2,..., n$.

Hence $\sum_{k=1}^{n} d(k) = n\left(1 + \frac{1}{2} + \cdots + \frac{1}{n}\right) + 0(n)$. But we know that

$$\lim_{n \to +\infty} \left(1 + \frac{1}{2} + \cdots + \frac{1}{n} - \text{Log}\, n\right) = \gamma, \text{ where } \gamma = 0.577... \text{ is Euler's constant.}$$

Therefore $\sum_{k=1}^{n} d(k) = n\,\text{Log}\, n + 0(n)$.

See [13], page 264, for a more precise estimate.

Exercise 7.17   1) We have $n^s = e^{s\,\text{Log}\, n}$. Hence, if $s = x + iy$, $n^s = e^{x\,\text{Log}\, n}e^{iy\,\text{Log}\, n}$ and $|n^s| = e^{x\,\text{Log}\, n} = n^{\text{Re}\, s}$. Therefore $\sum_{n=1}^{+\infty} 1/n^s$ is absolutely convergent if $\text{Re}\, s > 1$ and defines a function $\zeta(s)$. To prove that ζ is holomorphic in the open half-plane Ω, consider a compact subset K of Ω. Then there exists $s_0 \in \mathbb{R}$, $s_0 > 1$, such that $\text{Re}\, s \geq s_0 > 1$ for every $s \in K$. Hence $1/|n^s| \leq 1/n^{s_0}$, which is a convergent series. This proves that the series $\sum_{n=1}^{+\infty} 1/n^s$ is uniformly convergent on K. Therefore ζ is holomorphic in Ω by lemma 7.1.

2) Put $B_{2k}(x) = a_0 + \sum_{n=1}^{+\infty} a_n(k)\cos(2\pi nx) + b_n(k)\sin(2\pi nx)$.

Then $a_0 = \int_0^1 B_{2k}(x)dx = \frac{1}{2k+1}\left[B_{2k+1}(x)\right]_0^1 = 0$ for $k \geq 1$ (exercise 7.14, 2), c)).

If $n \geq 1$, $a_n(k) = 2\int_0^1 B_{2k}(x)\cos(2\pi nx)dx$. Integrating by parts twice and taking into account that $B'_n(x) = nB_{n-1}(x)$ and $B_{2k-1}(0) = B_{2k-1}(1) = 0$ for $k \geq 2$ yields

$$a_n(k) = -\frac{(2k)(2k-1)}{(2\pi n)^2}a_n(k-1) = (-1)^{k-1}\frac{2k(2k-1)...4 \cdot 3}{(2\pi n)^{2(k-1)}}a_n(1).$$

Now it is easy to compute $a_n(1)$. Again we integrate by parts twice, here taking into account that $B_1(0) = -\frac{1}{2}$ and $B_1(1) = \frac{1}{2}$). We obtain $a_n(1) = 2 \cdot \frac{2}{(2\pi n)^2}$.

Therefore $a_n(k) = (-1)^{k-1}\frac{2(2k)!}{(2\pi n)^{2k}}$.

Similarly, it can be shown that $b_n(k) = 2\int_0^1 B_{2k}(x)\sin(2\pi nx)dx = 0$.

Therefore, for $k \geq 1$ and $x \in [0,1]$, $B_{2k}(x) = (-1)^{k-1} \dfrac{2(2k)!}{(2\pi)^{2k}} \displaystyle\sum_{n=1}^{+\infty} \dfrac{1}{n^{2k}} \cos(2\pi nx)$.

For $x = 0$, we obtain $\zeta(2k) = (-1)^{k-1} \dfrac{(2\pi)^{2k}}{(2k)!} \dfrac{B_{2k}}{2}$.

In particular, $\zeta(2) = \displaystyle\sum_{n=1}^{+\infty} \dfrac{1}{n^2} = \dfrac{\pi^2}{6}$, $\zeta(4) = \displaystyle\sum_{n=1}^{+\infty} \dfrac{1}{n^4} = \dfrac{\pi^4}{90}$, $\zeta(6) = \displaystyle\sum_{n=1}^{+\infty} \dfrac{1}{n^6} = \dfrac{\pi^6}{945}$.

3) Since $p \geq 2$, we have $\dfrac{1}{1 - 1/p^s} = \displaystyle\sum_{k=0}^{+\infty} \dfrac{1}{p^{ks}}$. Hence we observe that

$$\dfrac{1}{1 - 1/2^s} \times \dfrac{1}{1 - 1/3^s} = \sum_{k=0}^{+\infty} \dfrac{1}{2^{ks}} \times \sum_{k=0}^{+\infty} \dfrac{1}{3^{ks}} = \sum_{i=1}^{+\infty} \dfrac{1}{n_i^s},$$

where the n_i's are all integers of the form $2^a 3^b$. More generally, denote by S_m the set of all integers of the form $2^{a_1} 3^{a_2} \dots p_m^{a_m}$, where $2, 3, \dots, p_m$ are the m first primes. Then we clearly have

$$\prod_{k=1}^{m} \dfrac{1}{1 - 1/p_k^s} = \sum_{n \in S_n} \dfrac{1}{n^s}.$$

But $\displaystyle\lim_{m \to +\infty} \sum_{n \in S_m} \dfrac{1}{n^s} = \sum_{n=1}^{+\infty} \dfrac{1}{n^s}$ since $\left| \displaystyle\sum_{n=1}^{+\infty} \dfrac{1}{n^s} - \sum_{n \in S_m} \dfrac{1}{n^s} \right| \leq \displaystyle\sum_{n \in S_m} \dfrac{1}{n^{\operatorname{Re} s}} \leq \sum_{n = p_{m+1}}^{+\infty} \dfrac{1}{n^{\operatorname{Re} s}}$.

Therefore $\displaystyle\prod_{k=1}^{+\infty} \dfrac{1}{1 - 1/p_k^s} = \zeta(s)$.

Remark: This result has been proved by Euler. It is at the beginning of analytic number theory. See for example [1].

Exercise 7.18   1) This is the same as exercise 7.17, since

$$\left| \dfrac{a(n)}{n^s} \right| \leq \dfrac{1}{|n^{s-\alpha}|} = \dfrac{1}{n^{\operatorname{Re} s - \alpha}}.$$

2) We have $F(s)G(s) = \displaystyle\sum_{d=1}^{+\infty} \sum_{h=1}^{+\infty} \dfrac{a(d)b(h)}{(dh)^s}$.

We collect the terms of this sum with respect to the $1/n^s$ (compare to Lambert series, proof of formula (7.9)). We obtain

$$F(s)G(s) = \sum_{n=1}^{+\infty} \left(\sum_{d|n} a(d)b(n/d) \right) \dfrac{1}{n^s}.$$

3) We have $\zeta(s) \displaystyle\sum_{n=1}^{+\infty} \mu(n)/n^s = \sum_{n=1}^{+\infty} \left(\sum_{d|n} \mu(d) \cdot 1 \right) \Big/ n^s$ by question 2).

And by question 3, exercise 7.15, this is equal to 1.

4) $\zeta^2(s) = \sum_{n=1}^{+\infty} \left(\sum_{d|n} 1 \cdot 1 \right) / n^s = \sum_{n=1}^{+\infty} d(n) / n^s$.

5) By (7.23), $\sum_{n=1}^{+\infty} \varphi(n) / n^s \times \sum_{n=1}^{+\infty} 1/n^s = \sum_{n=1}^{+\infty} \left(\sum_{d|n} \varphi(d) \right) / n^s = \sum_{n=1}^{+\infty} n / n^s$.

Therefore $\sum_{n=1}^{+\infty} \varphi(n) / n^s \times \zeta(s) = \zeta(s-1)$.

6) $\zeta(s-1)\zeta(s) = \sum_{n=1}^{+\infty} n / n^s \times \sum_{n=1}^{+\infty} 1/n^s = \sum_{n=1}^{+\infty} \left(\sum_{d|n} d \right) / n^s = \sum_{n=1}^{+\infty} \sigma(n) / n^s$.

Exercise 7.19   By using (7.28), we see that, for $n \geq 2$,

$$\delta(2n) = \left(\prod_{p \leq \lfloor \sqrt{2n} \rfloor} p^{\lceil \mathrm{Log}(2n)/\mathrm{Log}\, p \rceil} \right) \left(\prod_{\lfloor \sqrt{2n} \rfloor < p \leq n} p^{\lceil \mathrm{Log}(2n)/\mathrm{Log}\, p \rceil} \right) \left(\prod_{n < p < 2n} p^{\lceil \mathrm{Log}(2n)/\mathrm{Log}\, p \rceil} \right)$$

$$\leq \left(\prod_{p \leq \lfloor \sqrt{2n} \rfloor} p^{\mathrm{Log}\, n / \mathrm{Log}\, p} \right) \left(\prod_{\lfloor \sqrt{2n} \rfloor < p \leq n} p \right) \left(\prod_{n < p < 2n} p \right) \leq (2n)^{\pi(\lfloor \sqrt{2n} \rfloor)} \delta(n) \left(\prod_{n < p < 2n} p \right).$$

Now it is easy to check that $\pi(n) \leq n/2$ for every $n \geq 8$. Indeed, the sequence 9, 10, $\cdots$, n contains at most $\frac{n-8}{2}$ prime numbers, since nor 9 nor even numbers are primes, and there are only four primes ≤ 7. Hence, for every $n \geq 8$, we have by theorem 7.8

$$4^n \leq \delta(2n) \leq e^{\frac{1}{2}\sqrt{2n}\,\mathrm{Log}(2n)} \delta(n) \left(\prod_{n < p < 2n} p \right) \leq e^{n\,\mathrm{Log}\, 3 + \frac{1}{2}\sqrt{2n}\,\mathrm{Log}(2n)} \left(\prod_{n < p < 2n} p \right).$$

Therefore $\prod_{n < p < 2n} p \geq e^{(\mathrm{Log}\, 4 - \mathrm{Log}\, 3)n - \frac{1}{2}\sqrt{2n}\,\mathrm{Log}(2n)} > 1$ as soon as $n \geq 226$.

This proves that there exists at least one prime p between n and $2n$ if $n \geq 226$. It can be checked directly that this remains true for $n = 2, 3, \cdots, 225$. Bertrand's postulate has been proved by Chebyshev in 1850.

Exercise 7.20   1) Assume that p and q are coprime. Then they have no common prime factors p_i, and we have

$$p = p_1^{a_1} p_2^{a_2} \cdots p_k^{a_k}, \quad q = p_{k+1}^{a_{k+1}} p_{k+2}^{a_{k+2}} \cdots p_{k+m}^{a_{k+m}}, \quad p_i \neq p_j \text{ if } i \neq j.$$

Therefore, by formula (7.21),

$$\varphi(pq) = pq \prod_{i=1}^{k+m} \left(1 - \frac{1}{p_i} \right) = p \prod_{i=1}^{k} \left(1 - \frac{1}{p_i} \right) \times q \prod_{i=k+1}^{k+m} \left(1 - \frac{1}{p_i} \right) = \varphi(p)\varphi(q).$$

2) If p and q are coprime, the divisors of pq are exactly the numbers hm, where h divides p and m divides q. Hence $d(pq) = d(p)d(q)$. But it is easy to see that, for any prime p, $d(p^\alpha) = \alpha + 1$. Therefore

$$d\left(p_1^{a_1} p_2^{a_2} \cdots p_k^{a_k}\right) = (a_1 + 1)(a_2 + 1) \cdots (a_k + 1).$$

3) With the same notations as in question 2), we have

$$\sigma(pq) = \sum_{h|p} \sum_{m|q} hm = \sum_{h|p} h \sum_{m|q} m = \sigma(p)\sigma(q).$$

Moreover, for any prime p, $\sigma\left(p^\alpha\right) = 1 + p + p^2 + \cdots + p^\alpha = \dfrac{p^{\alpha+1} - 1}{p - 1}$.

Therefore $\sigma\left(p_1^{a_1} p_2^{a_2} \cdots p_k^{a_k}\right) = \displaystyle\prod_{i=1}^{k} \dfrac{p_i^{a_i+1} - 1}{p_i - 1}$.

Solutions to the exercises of chapter 8

Exercise 8.1   1) We look for a power series solution of (8.4):

$$y = \sum_{n=0}^{+\infty} a_n x^n, \quad y' = \sum_{n=0}^{+\infty} n a_n x^{n-1}, \quad y'' = \sum_{n=0}^{+\infty} n(n-1) a_n x^{n-2}.$$

Replacing in (8.4) yields

$$\sum_{n=1}^{+\infty} n(n-1) a_n x^{n-1} - \sum_{n=0}^{+\infty} n(n-1) a_n x^n + c \sum_{n=1}^{+\infty} n a_n x^{n-1}$$
$$- (a + b + 1) \sum_{n=0}^{+\infty} n a_n x^n - ab \sum_{n=0}^{+\infty} a_n x^n = 0.$$

By changing n into $n-1$ in the first and third sums, we obtain the recurrence relation $(n+1)(c+n)a_{n+1} = (n^2 + an + bn + ab)a_n = (n+a)(n+b)a_n$, whence

$$a_n = \frac{(a)_n (b)_n}{(c)_n n!} a_0,$$

which proves that the hypergeometric series is a solution of (8.4).

2) The function $x^{c-1} f(x) = {}_2F_1\left(\genfrac{}{}{0pt}{}{a-c+1, b-c+1}{2-c} \middle| x\right)$ is a solution of

$$x(1-x)y'' + (2 - c - (a + b - 2c + 3)x) y' - (a - c + 1)(b - c + 1)y = 0.$$

Therefore

$$x(1-x)x^{c-1}[(c-1)(c-2)x^{-2}f(x) + 2(c-1)x^{-1}f'(x) + f''(x)]$$
$$+ (2 - c - (a + b - 2c + 3)x)x^{c-1}[(c-1)x^{-1}f(x) + f'(x)]$$
$$- (a - c + 1)(b - c + 1)x^{c-1}f(x) = 0.$$

Simplify by x^{c-1}. Then an easy computation shows that f is solution of (8.4).

3) Since $f(x) \underset{0}{\sim} x^{1-c}$ with $c \notin \mathbb{Z}$, $\lim_{x\to 0^+} f(x) = 0$ or $\lim_{x\to 0^+} f(x) = +\infty$.

Thus f and ${}_2F_1\left(\genfrac{}{}{0pt}{}{a,b}{c} \middle| x\right)$ are linearly independent, which proves that the general solution of (8.4) is given by (8.5).

Exercise 8.2   Put $f(x) = {}_2F_1\left({c-a,c-b \atop c}\Big|x\right)(1-x)^{c-a-b}$. By (8.4) we have

$$x(1-x)\left(f(x)(1-x)^{a+b-c}\right)'' + \left(c-(2c-a-b+1)x\right)\left(f(x)(1-x)^{a+b-c}\right)'$$
$$-(c-a)(c-b)f(x)(1-x)^{a+b-c} = 0.$$

This yields $x(1-x)f''(x) + (c-(a+b-1)x)f'(x) - abf(x) = 0$.

Hence, by theorem 8.1, $f(x) = A\,{}_2F_1\left({a,b \atop c}\Big|x\right) + Bx^{1-c}\,{}_2F_1\left({c-a,c-b \atop c}\Big|x\right)$.

Since f is analytic for $|x|<1$, this implies $B=0$ if $c \notin \mathbb{Z}$.

This remains true if $c \notin \mathbb{Z}^-$ by continuity. For $x=0$, we obtain $A=1$, which proves (8.6).

Exercise 8.3   a) For $|x| < b$, we have

$$\frac{x}{b}\left(1-\frac{x}{b}\right){}_2F_1''\left({a,b \atop c}\Big|\frac{x}{b}\right) + \left(c-(a+b+1)\frac{x}{b}\right){}_2F_1'\left({a,b \atop c}\Big|\frac{x}{b}\right) - ab\,{}_2F_1\left({a,b \atop c}\Big|\frac{x}{b}\right) = 0.$$

Dividing by b yields

$$x\left(1-\frac{x}{b}\right)\frac{1}{b^2}\,{}_2F_1''\left({a,b \atop c}\Big|\frac{x}{b}\right) + \left(c-\left(\frac{a}{b}+1+\frac{1}{b}\right)b\right)\frac{1}{b}\,{}_2F_1'\left({a,b \atop c}\Big|\frac{x}{b}\right) - a\,{}_2F_1\left({a,b \atop c}\Big|\frac{x}{b}\right) = 0.$$

Now we observe that $\displaystyle\lim_{b\to+\infty}\frac{1}{b^2}\,{}_2F_1''\left({a,b \atop c}\Big|\frac{x}{b}\right) = {}_1F_1''\left({a \atop c}\Big|x\right)$, $\displaystyle\lim_{b\to+\infty}\frac{1}{b}\,{}_2F_1'\left({a,b \atop c}\Big|\frac{x}{b}\right) = {}_1F_1'\left({a \atop c}\Big|x\right)$,

and $\displaystyle\lim_{b\to+\infty}{}_2F_1\left({a,b \atop c}\Big|\frac{x}{b}\right) = {}_1F_1\left({a \atop c}\Big|x\right)$. This proves (8.11).

b) To prove (8.12), we proceed the same way as for (8.5).

Exercise 8.4   1) We have $t^{\alpha-1}(1-t)^{\beta-1} \underset{0}{\sim} t^{\alpha-1}$ and $t^{\alpha-1}(1-t)^{\beta-1} \underset{1}{\sim} (1-t)^{\beta-1}$.

This proves that the integral which defines $B(\alpha,\beta)$ is convergent.

Moreover, the change of variable $u = 1-t$ yields $B(\beta,\alpha) = B(\alpha,\beta)$.

2) $\Gamma(\alpha)\Gamma(\beta) = \int_0^{+\infty} e^{-x}x^{\alpha-1}dx \int_0^{+\infty} e^{-y}y^{\beta-1}dy$. Putting $x=u^2$, $y=v^2$ yields

$$\Gamma(\alpha)\Gamma(\beta) = 4\int_0^{+\infty} e^{-u^2}u^{2\alpha-1}du \int_0^{+\infty} e^{-v}v^{2\beta-1}dv = 4\int_{u=0}^{+\infty}\int_{v=0}^{+\infty} e^{-(u^2+v^2)}u^{2\alpha-1}v^{2\beta-1}dudv.$$

By using polar coordinates, we now have

$$\Gamma(\alpha)\Gamma(\beta) = 4\int_{\theta=0}^{\pi/2}\int_{r=0}^{+\infty} e^{-r^2}r^{2\alpha+2\beta-1}\cos^{2\alpha-1}\theta\sin^{2\beta-1}\theta\,d\theta dr.$$

Put $x = r^2$. We obtain $\Gamma(\alpha)\Gamma(\beta)\, 2\int_{x=0}^{+\infty} e^{-x}x^{\alpha+\beta-1}dx \int_{\theta=0}^{\pi/2}\cos^{2\alpha-1}\theta\sin^{2\alpha-1}\theta\,d\theta$.

Hence $\Gamma(\alpha)\Gamma(\beta) = 2\Gamma(\alpha+\beta)\int_{\theta=0}^{\pi/2}(\cos^2\theta)^{\alpha-1}(\sin^2\theta)^{\beta-1}\sin\theta\cos\theta d\theta$.

Finally, putting $t = \sin^2 \theta$ yields $\Gamma(\alpha)\Gamma(\beta) = \Gamma(\alpha + \beta)B(\alpha, \beta)$.

3) Using the functional relation for Γ repeatedly yields

$$\Gamma(a+n) = (a+n-1)(a+n-2)...a\Gamma(a),$$

whence $(a)_n = \Gamma(a+n)/\Gamma(a)$. Consequently

$$_1F_1\left(\begin{matrix}a\\c\end{matrix}\middle|x\right) = \frac{\Gamma(c)}{\Gamma(a)}\sum_{n=0}^{+\infty}\frac{\Gamma(a+n)}{n!\Gamma(c+n)}x^n$$

$$= \frac{\Gamma(c)}{\Gamma(a)\Gamma(c-a)}\sum_{n=0}^{+\infty}\frac{\Gamma(a+n)\Gamma\big((c+n)-(a+n)\big)}{\Gamma(c+n)}\frac{x^n}{n!}.$$

By using question 2, we now obtain

$$_1F_1\left(\begin{matrix}a\\c\end{matrix}\middle|x\right) = \frac{\Gamma(c)}{\Gamma(a)\Gamma(c-a)}\sum_{n=0}^{+\infty}\left(\int_0^1 t^{a+n-1}(1-t)^{c-a-1}dt\right)\frac{x^n}{n!}.$$

We can permute the integral and the series. Indeed, $\sum_{n=0}^{+\infty} t^{a-1}(1-t)^{c-a-1}(tx)^n/n!$

converges uniformly on $[0,1]$, since $\left|t^{a-1}(1-t)^{c-a-1}(tx)^n/n!\right| \le |x|^n/n!$. Thus

$$_1F_1\left(\begin{matrix}a\\c\end{matrix}\middle|x\right) = \frac{\Gamma(c)}{\Gamma(a)\Gamma(c-a)}\left(\int_0^1 t^{a-1}(1-t)^{c-a-1}\sum_{n=0}^{+\infty}\frac{(tx)^n}{n!}dt\right), \text{ Q.E.D.}$$

4) Similarly,

$$_2F_1\left(\begin{matrix}a,b\\c\end{matrix}\middle|x\right) = \frac{\Gamma(c)}{\Gamma(a)\Gamma(c-a)}\sum_{n=0}^{+\infty}\frac{\Gamma(a+n)\Gamma(c-a)}{\Gamma(c+n)}\frac{b(b+1)\cdots(b+n-1)}{n!}x^n$$

$$= \frac{\Gamma(c)}{\Gamma(a)\Gamma(c-a)}\sum_{n=0}^{+\infty}\left(\int_0^1 t^{a+n-1}(1-t)^{c-a-1}dt\right)\frac{-b(-b-1)...(-b-n+1)}{n!}(-x)^n$$

$$= \frac{\Gamma(c)}{\Gamma(a)\Gamma(c-a)}\int_0^1 t^{a-1}(1-t)^{c-a-1}\left(\sum_{n=0}^{+\infty}\frac{-b(-b-1)...(-b-n+1)}{n!}(-tx)^n\right)dt$$

$$= \frac{\Gamma(c)}{\Gamma(a)\Gamma(c-a)}\int_0^1 t^{a-1}(1-t)^{c-a-1}(1-tx)^{-b}dt.$$

Exercise 8.5   Lemma 8.3 yields $\left|_1F_1\left(\begin{matrix}a\\c\end{matrix}\middle|\alpha\right)\right| \le \dfrac{\Gamma(c)}{\Gamma(a)\Gamma(c-a)}\int_0^1 \left|e^{\alpha t}\right|dt.$

Now $\left|e^{\alpha t}\right| = e^{t\,\text{Re}\,\alpha}$ by exercise 7.17, question 1). Hence $e^{t\,\text{Re}\,\alpha} \le 1$ if $\text{Re}\,\alpha \le 0$ and $e^{t\,\text{Re}\,\alpha} \le e^{\text{Re}\,\alpha}$ if $\text{Re}\,\alpha \ge 0$. This proves lemma 8.4.

Exercise 8.6   If $n = -k+1, -k+2, ...,-1, 0$, then $(n)_k = 0$ and the result is obvious.

If $n > 0$, $\dfrac{(n)_k}{k!} = \dfrac{n(n+1)...(n+k-1)}{k!} = \binom{n+k-1}{k} \in \mathbb{N}$.

If $n \le -k$, $\dfrac{(n)_k}{k!} = (-1)^k \dfrac{-n(-n-1)...(-n-k+1)}{k!} = (-1)^k \binom{-n}{k} \in \mathbb{Z}$.

Finally, $n(n-1)...(n-k+1) = (-1)^k (-n)_k$.

Exercise 8.7   1) Take $m = n$, $\alpha = 1/3$, $x = 1/18^3$ in theorem 8.2. Then

$$\left(1 - \frac{1}{18^3}\right)^{1/3} = \left(\frac{5831}{18^3}\right)^{1/3} = \left(\frac{17 \cdot 7^3}{18^3}\right)^{1/3} = \frac{7}{18} \sqrt[3]{17}.$$

This proves the result after multiplication by 18.

2) Let p be any prime different from 3. Since 3 is invertible in $\mathbb{Z}/p\mathbb{Z}$, there exists m such that $p \mid u_m$. Since $u_{m+p} \equiv u_m \pmod{p}$, the multiples of p in the sequence u_n follow every p terms. Consequently, there exist exactly $[k/p]$ multiples of p among the numbers $u_1, u_2, ..., u_k$. Similarly, 3 is invertible in $\mathbb{Z}/p^2\mathbb{Z}$, and there exist $\left[k/p^2\right]$ multiples of p^2 among the numbers u_1, $u_2, ..., u_k$, and so on. Hence the exponant of p in the product $u_1 u_2 ... u_k$ is $[k/p] + \left[k/p^2\right] + \cdots$, that is the same as in $k!$ (lemma 7.3).

Therefore $u_1 u_2 ... u_k$ is divisible by $k!/3^{\alpha(k)}$.

Finally, $\alpha(k) = \left[\dfrac{k}{3}\right] + \left[\dfrac{k}{3^2}\right] + \cdots \le \dfrac{k}{3} + \dfrac{k}{3^2} + \cdots = k \cdot \dfrac{1}{3} \cdot \dfrac{1}{1 - 1/3}$, whence $\alpha(k) \le k/2$.

3) $_2F_1\left(\begin{array}{c} -n, -n+1/3 \\ -2n \end{array} \middle| \dfrac{1}{18^3}\right) = \sum_{i=0}^{n} \dfrac{(-n)_i (-n+1/3)_i}{(-2n)_i\, i!} \dfrac{1}{18^{3i}}$

$$= \sum_{i=0}^{n} \dfrac{(-n)_i}{i!} \dfrac{(-3n+1)(-3n+4)...(-3n+3i-2)}{(-1)^i (2n)(2n-1)...(2n-i+1)} \dfrac{1}{(3 \cdot 18^3)^i}.$$

By question 2, $i!/3^{\alpha(i)}$ divides $(-3n+1)(-3n+4)...(-3n+3i-2)$. Therefore there exist integers $\delta(i)$ such that

$$_2F_1\left(\begin{array}{c} -n, -n+1/3 \\ -2n \end{array} \middle| \dfrac{1}{18^3}\right) = \sum_{i=0}^{n} \dfrac{\delta(i) n(n-1)\cdots(n-i+1)}{3^{\alpha(i)} (2n)(2n-1)...(2n-i+1)} \dfrac{1}{(3 \cdot 18^3)^i}.$$

Moreover $\binom{2n}{n} \dfrac{n(n-1)...(n-i+1)}{2n(2n-1)...(2n-i+1)} = \dfrac{(2n-i)...(2n-i-1)...(n+1)}{(n-i)!} = \binom{2n-i}{n-i} \in \mathbb{N}$.

Therefore $q_n \in \mathbb{Z}$. Similarly $p_n \in \mathbb{Z}$.

4) By exercise 8.4, we can write

$$\left| {}_2F_1\!\left(\begin{matrix} n+1, n+2/3 \\ 2n+2 \end{matrix}\middle|\; \frac{1}{18^3}\right)\right| \le \frac{\Gamma(2n+2)}{(\Gamma(n+1))^2}\int_0^1 t^n(1-t)^n\left(1-\frac{t}{18^3}\right)^{-n-2/3} dt.$$

But the graph of $t(1-t)$ on $[0,1]$ shows that $t(1-t)\le\frac{1}{4}$. Moreover

$$1-\frac{t}{18^3}\ge 1-\frac{1}{18^3}\;\Rightarrow\; \left(1-\frac{t}{18^3}\right)^{-n-2/3}\le\left(\frac{18^3}{18^3-1}\right)^{n+2/3}=\frac{18^2}{7^2\cdot 17^{2/3}}\left(\frac{18^3}{18^3-1}\right)^n.$$

This yields the desired estimate.

5) We have $r_n = \dfrac{D_n\cdot(-1)^n\left(-n-1/3\right)_{2n+1}}{18^{6n+2}\binom{2n}{n}(2n+1)!}\;{}_2F_1\!\left(\begin{matrix} n+1, n+2/3 \\ 2n+2 \end{matrix}\middle|\; \frac{1}{18^3}\right).$

Now $D_n\le(\sqrt{3})^n\cdot(3\cdot 18^3)^n\binom{2n}{n}$. Moreover,

$$\left|\left(-n-\frac{1}{3}\right)_{2n+1}\right|=\prod_{k=0}^{n-1}\left(n+\frac{1}{3}-k\right)\times\frac{1}{3}\times\prod_{k=0}^{n-1}\left(n-\frac{1}{3}-k\right)=\frac{1}{3}\prod_{k=0}^{n-1}\left((n-k)^2-\frac{1}{9}\right)\le\frac{1}{3}(n!)^2.$$

Using the result of question 4) finally yields $|r_n|\le\dfrac{1}{3\cdot 7^2\cdot 17^{2/3}}\left(\dfrac{3\sqrt{3}}{4(18^3-1)}\right)^n.$

6) For every n, we have

$$\begin{vmatrix} q_n & p_n \\ q_{n+1} & p_{n+1} \end{vmatrix}=D_n D_{n+1}\begin{vmatrix} Q_n(18^{-3}) & P_n(18^{-3}) \\ Q_{n+1}(18^{-3}) & P_{n+1}(18^{-3}) \end{vmatrix}\neq 0$$

by lemma 8.1. Consequently at least one of the two numbers $r_n=q_n\sqrt[3]{17}-p_n$ or $r_{n+1}=q_{n+1}\sqrt[3]{17}-p_{n+1}$ is not zero. Otherwise, the homogeneous system whose matrix is $\begin{pmatrix} q_n & -p_n \\ q_{n+1} & p_{n+1} \end{pmatrix}$ would have a non-zero solution, and its determinant should be zero. Hence the irrationality of $\sqrt[3]{17}$ results from theorem 1.5.

Remark: Of course, this proof of the irrationality of $\sqrt[3]{17}$ is very complicated. We can give a much simpler proof by arguing as in section 1.1. However, we have obtained here an excellent diophantine approximation. It will enable us to obtain, in chapter 9, exercise 9.13, an *irrationality measure* of $\sqrt[3]{17}$.

Exercise 8.8   1) We have $\frac{\partial}{\partial\alpha}\left((1-x)^\alpha\right)=\frac{\partial}{\partial\alpha}\left(e^{\alpha\mathrm{Log}(1-x)}\right)=\mathrm{Log}(1-x)(1-x)^\alpha.$

Therefore $\mathrm{Log}(1-x)=\left[\frac{\partial}{\partial\alpha}(1-x)^\alpha\right]_{\alpha=0}.$

2) For $m=n$, we have by theorem 8.2

$$_2F_1\left(\begin{matrix}-n,-n+\alpha\\-2n\end{matrix}\Big|x\right)(1-x)^\alpha - {}_2F_1\left(\begin{matrix}-n,-n-\alpha\\-2n\end{matrix}\Big|x\right) = x^{2n+1}\frac{(-1)^n(-n-\alpha)_{2n+1}}{\binom{2n}{n}(2n+1)!}\,_2F_1\left(\begin{matrix}n+1-\alpha,n+1\\2n+2\end{matrix}\Big|x\right).$$

Now we take derivatives with respect to α by using logarithmic derivatives for the computation of $\frac{\partial}{\partial\alpha}[(-n+\alpha)_k]$, and we obtain

$$\left[\frac{\partial}{\partial\alpha}\,_2F_1\left(\begin{matrix}-n,-n+\alpha\\-2n\end{matrix}\Big|x\right)\right]_{\alpha=0} = \left[\sum_{k=0}^n\frac{(-n)_k\frac{\partial}{\partial\alpha}[(-n+\alpha)_k]}{(-2n)_k\,k!}x^k\right]_{\alpha=0}$$

$$= \left[\sum_{k=1}^n\frac{(-n)_k(-n+\alpha)_k}{(-2n)_k\,k!}\left(\frac{1}{-n+\alpha}+\frac{1}{-n+1+\alpha}+\cdots+\frac{1}{-n+k-1+\alpha}\right)x^k\right]_{\alpha=0}$$

$$= \sum_{k=1}^n\frac{(-n)_k^2}{(-2n)_k\,k!}\left(\sum_{i=0}^{k-1}\frac{1}{-n+i}\right)x^k\ .$$

Similarly $\left[\dfrac{\partial}{\partial\alpha}\,_2F_1\left(\begin{matrix}-n,-n-\alpha\\-2n\end{matrix}\Big|x\right)\right]_{\alpha=0} = \sum_{k=1}^n\dfrac{(-n)_k^2}{(-2n)_k\,k!}\left(\sum_{i=0}^{k-1}\dfrac{1}{n-i}\right)x^k.$

It is not necessary to compute the derivative of the $_2F_1$ in the right-hand side. Indeed, the product $(-n-\alpha)_{2n+1}$ contains the factor $(-\alpha)$. Therefore it vanishes for $\alpha=0$. For the same reason, it suffices to compute, in the derivative of $(-n-\alpha)_{2n+1}$, the term corresponding to the derivative of $(-\alpha)$, that is

$$\left[\frac{\partial}{\partial\alpha}(-n-\alpha)_{2n+1}\right]_{\alpha=0} = -[(-n-\alpha)(-n-\alpha+1)...(-\alpha-1)(-\alpha+1)...(-\alpha+n)]_{\alpha=0}$$

$$= (-1)^{n+1}(n!)^2.$$

Finally, we obtain $Q_n(x)(1-x) - P_n(x) = x^{2n+1}R_n(x)$, with

$$Q_n(x) = \,_2F_1\left(\begin{matrix}-n,-n\\-2n\end{matrix}\Big|x\right) = \sum_{k=0}^n\frac{(-n)_k^2}{(-2n)_k\,k!}x^k, \quad P_n(x) = 2\sum_{k=1}^n\frac{(-n)_k^2}{(-2n)_k\,k!}\left(\sum_{i=0}^{k-1}\frac{1}{n-i}\right)x^k,$$

$$R_n(x) = \frac{-(n!)^2}{\binom{2n}{n}(2n+1)!}\,_2F_1\left(\begin{matrix}n+1,n+1\\2n+2\end{matrix}\Big|x\right).$$

3) First observe that, for $k\ge 1$,

$$(-n)_k = (-n)(-n+1)...(-n+k-1) = \frac{(-1)^k\,n!}{(n-k)!}, \quad (-2n)_k = \frac{(-1)^k(2n)!}{(2n-k)!}.$$

Therefore $\dfrac{(-n)_k^2}{(-2n)_k\,k!} = \dfrac{(-1)^k(n!)^2(2n-k)!}{(2n)!\,k!(n-k)!^2}$

$$= \frac{(-1)^k(2n-k)...(2n-2k+1)\binom{2n-2k}{n-k}}{\binom{2n}{n}k!} = \frac{(-2n+k)_k\binom{2n-2k}{n-k}}{\binom{2n}{n}k!}.$$

By lemma 8.5, $\dfrac{(-2n+k)_k}{k!} \in \mathbb{Z}$, whence $\binom{2n}{n} \dfrac{(-n)_k^2}{(-2n)_k \, k!} \in \mathbb{Z}$.

Now, $\mathrm{LCM}(1,2,\cdots,n) \times \sum_{i=0}^{k-1} \dfrac{1}{n-i} \in \mathbb{N}$. Hence, if $D_n = 2^n \binom{2n}{n} \mathrm{LCM}(1,2,...,n)$,

then $p_n = D_n P_n\left(\frac{1}{2}\right) \in \mathbb{Z}$ and $q_n = D_n Q_n\left(\frac{1}{2}\right) \in \mathbb{Z}$.

4) We have $r_n = \dfrac{-D_n \, (n!)^2}{2^{2n+1} \binom{2n}{n} (2n+1)!} \, {}_2F_1\!\left(\begin{array}{c} n+1, n+1 \\ 2n+2 \end{array} \middle| \frac{1}{2} \right)$.

But the integral representation of ${}_2F_1$ (exercise 8.4, question 4) yields

$$\left| {}_2F_1\!\left(\begin{array}{c} n+1, n+1 \\ 2n+2 \end{array} \middle| \frac{1}{2} \right) \right| \le \frac{(2n+1)!}{(n!)^2} \int_0^1 t^n (1-t)^n \left(1 - \frac{t}{2} \right)^{-n-1} dt.$$

Since $t(1-t) \le \frac{1}{4}$ and $1 - \frac{t}{2} \ge \frac{1}{2}$ for $t \in [0,1]$, we see that

$$\left| {}_2F_1\!\left(\begin{array}{c} n+1, n+1 \\ 2n+2 \end{array} \middle| \frac{1}{2} \right) \right| \le \frac{(2n+1)!}{(n!)^2} \cdot 2 \cdot \left(\frac{1}{2} \right)^n.$$

As $D_n \le 6^n \binom{2n}{n}$ by theorem 7.8, we obtain finally $|r_n| \le \left(\frac{3}{4} \right)^n$.

5) For every n, we have

$$\begin{vmatrix} q_n & p_n \\ q_{n+1} & p_{n+1} \end{vmatrix} = D_n D_{n+1} \begin{vmatrix} Q_n(1/2) & P_n(1/2) \\ Q_{n+1}(1/2) & P_{n+1}(1/2) \end{vmatrix} \ne 0$$

by lemma 8.1. Consequently, for every n, at least one of the two numbers $r_n = q_n 2 + p_n$ or $r_{n+1} = q_{n+1} 2 + p_{n+1}$ is not zero. Since $\lim_{n \to +\infty} r_n = 0$, theorem 1.5 implies that $\mathrm{Log}\,2$ is irrational.

Remark: As in exercise 8.7, there is a much simpler proof of the irrationality of $\mathrm{Log}\,2$: if $\mathrm{Log}\,2$ was rational, then $e^{\mathrm{Log}\,2} = 2$ would be irrational by theorem 8.5, which is clearly absurd. However, the good diophantine approximations we have obtained here by using Padé approximants will enable us to compute an irrationality measure of $\mathrm{Log}\,2$ in exercise 9.14.

Exercise 8.9   1) If $|x| \le k < 1$, we have for n sufficiently large

$$\left| \frac{x^{2^n}}{1 + a x^{2^n}} \right| \le 2 \left| x^{2^n} \right| \le 2 k^{2^n}.$$

Hence the series which defines $f(x)$ converges uniformly on every compact subset of D and f is analytic in D by lemma 7.1. Moreover, for $|x| < 1$,

$$f(x^2) = \sum_{n=0}^{+\infty} \frac{x^{2^{n+1}}}{1+ax^{2^{n+1}}} = \sum_{n=1}^{+\infty} \frac{x^{2^n}}{1+ax^{2^n}} = f(x) - \frac{x}{1+ax}.$$

2) An easy induction shows that, for any $n \geq 1$, $f(x^{2^n}) = f(x) - \sum_{k=0}^{n-1} \frac{x^{2^k}}{1+ax^{2^k}}$.

Suppose that $f(1/2) = p/q \in \mathbb{Q}$. Then $f\left(1/2^{2^n}\right) = p/q - \sum_{k=0}^{n-1} 1/\left(2^{2^k} + a\right)$.

Therefore $f\left(1/2^{2^n}\right) \in \mathbb{Q}$, and $D_n f\left(1/2^{2^n}\right) \in \mathbb{Z}$, with

$$D_n = q\prod_{k=0}^{n-1}(2^{2^k}+a) = q\prod_{k=0}^{n-1} 2^{2^k} \prod_{k=0}^{n-1}\left(1+a2^{-2^k}\right).$$

Now the infinite product $\prod_{k=0}^{+\infty}\left(1+a2^{-2^k}\right)$ converges to some limit $L \neq 0$ by

exercise 3.4. Therefore $D_n \underset{+\infty}{\sim} qL \cdot 2^{1+2+\cdots+2^{n-1}} = \frac{qL}{2} \cdot 2^{2^n}$, Q.E.D.

3) The fourth order Taylor approximation of f is

$$f(x) = \frac{x}{1+ax} + \frac{x^2}{1+ax}^2 + 0(x^4) = x + (1-a)x^2 + a^2 x^3 + 0(x^4).$$

We look for $Q(x) = \alpha x + \beta$ and $P(x) = \gamma x + \delta$, such that

$$(\alpha x + \beta)f(x) - \gamma x - \delta = 0(x^3).$$

We expand, and write that the terms of degrees 0,1 and 2 must vanish, whence

$$-\delta = 0, \quad \beta - \gamma = 0, \quad (1-a)\beta + \alpha = 0.$$

The solution is $\alpha = a-1$, $\beta = 1$, $\gamma = 1$, $\delta = 0$. Replacing in the fourth order Taylor approximation of f finally yields

$$[(a-1)x+1]f(x) - x = (2a-1)x^3 + 0(x^4).$$

4) If $f(1/2) \in \mathbb{Q}$, then $D_n f(1/2^{2^n}) \in \mathbb{Z}$ by question 2). Now we replace x by 2^{-2^n} in the result of question 3) and obtain

$$\left[(a-1)\frac{1}{2^{2^n}}+1\right]f\left(\frac{1}{2^{2^n}}\right) - \frac{1}{2^{2^n}} = (2a-1)\frac{1}{2^{3 \cdot 2^n}} + 0\left(\frac{1}{2^{4 \cdot 2^n}}\right).$$

Multiply by $2^{2^n} D_n$. The left-hand side is now an integer A_n and

$$A_n \underset{+\infty}{\sim} (2a-1)D_n 2^{-2 \cdot 2^n} \underset{+\infty}{\sim} C(2a-1)2^{-2^n}.$$

This proves that $\lim_{n \to +\infty} A_n = 0$, and that $A_n \neq 0$ if n is sufficiently large. Since A_n is an integer, this is impossible and we conclude that $f(1/2) \notin \mathbb{Q}$.

Remark: In fact, it can be proved that this number is transcendental (see exercise 12.13).

5) a) For every $r \in [0,1[$, choose $\eta = 2/(r+1)$. Then, for every $x \in \mathbb{C}$ with $|x| \le r$, we have for fixed h and n sufficiently large

$$\left| a_{n+h} x^{2^n} \right| \le \eta^{2^n} (\eta|x|)^{2^n} \le \frac{1}{2} \Rightarrow \left| \frac{x^{2^n}}{1+a_{n+h}x^{2^n}} \right| \le 2r^{2^n}.$$

Hence f_h is analytic in D. Moreover,

$$f_h(x) = \sum_{n=0}^{+\infty} \frac{x^{2^n}}{1+a_{n+h}x^{2^n}} = \sum_{d=1}^{+\infty} \frac{b_d x^d}{1-c_d x^d},$$

with $b_d = 1$ if $d = 2^n$ and 0 otherwise, $c_d = -a_{n+h}$ if $d = 2^n$ and 0 otherwise. Thus f_h looks almost like a Lambert series. Hence proceeding as in the proof of (7.9) yields

$$f_h(x) = \sum_{d=1}^{+\infty} \sum_{m=1}^{+\infty} b_d c_d^{m-1} x^{md} = \sum_{n=1}^{+\infty} \left(\sum_{d|n} b_d c_d^{n/d-1} \right) x^n = \sum_{n=1}^{+\infty} p_{n,h} x^n.$$

Since $b_d = 0$ if $d \ne 2^k$, the following estimate of $|p_{n,h}|$ holds for large h:

$$|p_{n,h}| \le \sum_{k=0}^{\text{Log}_2 n} |a_{k+h}|^{\frac{n}{2^k}} \le \sum_{k=0}^{n-1} \eta^{2^{k+h} \times \frac{n}{2^k}} \le n \left(\eta^{2^h} \right)^n.$$

b) First observe that

$$f_h(x^2) = \sum_{n=0}^{+\infty} \frac{x^{2^{n+1}}}{1+a_{n+h}x^{2^{n+1}}} = \sum_{n=1}^{+\infty} \frac{x^{2^n}}{1+a_{n+h-1}x^{2^n}} = f_{h-1}(x) - \frac{x}{1+a_{h-1}x}.$$

Consequently, $$f_h(x^2) = f_0(x) - \sum_{k=0}^{h-1} x^{2^k} / \left(1 + a_k x^{2^k} \right). \qquad (*)$$

Moreover, we have as in question 3

$$f_h(x) = \frac{x}{1+a_h x} + \frac{x^2}{1+a_{h+1}x^2} + O(x^4) = x + (1-a_h)x^2 + a_h^2 x^3 + O(x^4),$$

and therefore the Padé approximants we have computed in question 3) remain valid if we replace a by a_h, whence

$$\left[(a_h - 1)x + 1 \right] f_h(x) - x = (2a_h - 1)x^3 + O(x^4) = x^3 g_h(x).$$

Since $a_h \in \mathbb{Z}$, we have $|g_h(0)| \ge 1$.

Moreover, we can estimate the coefficients of the Taylor series expansion of g_h. Indeed, if we denote

$$x^3 g_h(x) = \sum_{n=3}^{+\infty} q_{n,h} x^n, \qquad (**)$$

we have for every fixed $\eta > 1$, for $n \ge 4$ and h sufficiently large,

$$|q_{n,h}| = |(a_h - 1)p_{n-1,h} + p_{n,h}| \le 2\eta^{2^h}(n-1)\left(\eta^{2^h} \right)^{n-1} + n\left(\eta^{2^h} \right)^n \le 3n\left(\eta^{2^h} \right)^n.$$

Now choose $\eta = 2^{1/5}$. For h sufficiently large, replace x by $1/2^{2^h}$ in $(**)$:

$$\left(\frac{1}{2}\right)^{3.2^h} g_h\left(\left(\frac{1}{2}\right)^{2^h}\right) = \left(\frac{1}{2}\right)^{3.2^h}\left(q_{3,h} + \sum_{n=4}^{+\infty} q_{n,h}\left(\frac{1}{2}\right)^{(n-3).2^h}\right).$$

But we see that

$$\left|\sum_{n=4}^{+\infty} q_{n,h}\left(\frac{1}{2}\right)^{(n-3).2^h}\right| \le \sum_{n=4}^{+\infty} 3n\eta^{n2^h}\left(\frac{1}{2}\right)^{(n-3).2^h} \le 3\eta^{n2^h}\sum_{n=4}^{+\infty} n\left(\frac{\eta}{2}\right)^{(n-3).2^h} \le C_1\left(\frac{\eta^4}{2}\right)^{2^h} \le \frac{C_1}{\eta^{2^h}}.$$

Since $\left|q_{3,h}\right| = \left|2a_h - 1\right| \ge 1$, we deduce at once that, for $h \to +\infty$,

$$\left(\frac{1}{2}\right)^{3.2^h} g_h\left(\left(\frac{1}{2}\right)^{2^h}\right) \underset{+\infty}{\sim} (2a_h - 1)\left(\frac{1}{2}\right)^{3.2^h} \ne 0. \qquad (***)$$

Assume now that $f_0(1/2) = \alpha/\beta$ is rational.

Then, by $(*)$, $f_h\left[\left(\frac{1}{2}\right)^{2^h}\right] = f_0\left(\frac{1}{2}\right) - \sum_{k=0}^{h-1}\frac{1}{2^{2^k} + a_k}$ is also rational.

Denote $B_h = \prod_{k=0}^{h-1}\left(2^{2^k} + a_k\right) = 2^{2^h - 1}\prod_{k=0}^{h-1}\left(1 + a_k 2^{-2^k}\right)$.

Then $\beta B_h f_h\left((1/2)^{2^h}\right) \in \mathbb{Z}$ and $B_h \le C_2 2^{2^h}$. Now multiply the equality

$$\left[(a_h - 1)\left(\frac{1}{2}\right)^{2^h} + 1\right]f_h\left(\left(\frac{1}{2}\right)^{2^h}\right) - \left(\frac{1}{2}\right)^{2^h} = \left(\frac{1}{2}\right)^{3.2^h} g_h\left(\left(\frac{1}{2}\right)^{2^h}\right)$$

by $\beta B_h 2^{2^h}$. We obtain $E_h = \beta B_h 2^{2^h}\left(\frac{1}{2}\right)^{3.2^h} g_h\left(\left(\frac{1}{2}\right)^{2^h}\right) \in \mathbb{Z}$.

But $(***)$ shows first that $E_h \ne 0$. Moreover, for h sufficiently large,

$$|E_h| \le |\beta| C_2 2^{2.2^h}.2(2a_h - 1)\left(\frac{1}{2}\right)^{3.2^h} \le 4|\beta| C_2\left(\frac{\eta}{2}\right)^{2^h}.$$

Thus $\lim_{h \to +\infty} E_h = 0$, contradiction.

Exercise 8.10   1) We have

$$g_n(qz) = \frac{1}{(qz - 1)(qz - q)\cdots(qz - q^n)} = \frac{1}{q^n}\frac{z - q^n}{qz - 1}g_n(z).$$

Put $g_n(z) = \sum_{k=0}^{+\infty} a_n(k)z^k$, $|z| < 1$. Then the above equality becomes

$$q^n(qz - 1)\sum_{k=0}^{+\infty} q^k a_n(k)z^k = (z - q^n)\sum_{k=0}^{+\infty} a_n(k)z^k.$$

Expanding and changing indices yields

$$\sum_{k=1}^{+\infty} q^{n+k}\left(a_n(k-1) - a_n(k)\right)z^k = \sum_{k=1}^{+\infty}\left(a_n(k-1) - q^n a_n(k)\right)z^k.$$

Hence, for $k \geq 1$, $a_n(k) = a_n(k-1)(q^{n+k}-1)/(q^n(q^k-1))$.

Since $a_n(0) = (-1)^{n+1}q^{-n(n+1)/2}$, we obtain finally

$$a_n(k) = \frac{(-1)^{n+1}}{q^{n(n+1)/2+nk}} \frac{(q^{n+k}-1)(q^{n+k-1}-1)\cdots(q^{n+1}-1)}{(q^k-1)(q^{k-1}-1)\cdots(q-1)} = \frac{(-1)^{n+1}}{q^{n(n+1)/2+nk}} \frac{(n+k)_q!}{k_q!n_q!}.$$

2) a) Denote $h_q(z) = \dfrac{E_q(xz)}{z^{n+1}(z-1)(z-q)\cdots(z-q^n)}$.

We have $I_n(x) = \sum_{k=0}^{n} \mathrm{Res}\left(h_q, q^k\right) + \mathrm{Res}\left(h_q, 0\right)$. It is clear that

$$\mathrm{Res}\left(h_q, q^k\right) = \frac{E_q\left(xq^k\right)}{q^{(n+1)k}\left(q^k-1\right)\left(q^k-q\right)\cdots\left(q^k-q^{k-1}\right)\left(q^k-q^{k+1}\right)\cdots\left(q^k-q^n\right)}$$

$$= \frac{(1+x)(1+qx)\cdots\left(1+q^{k-1}x\right)E_q(x)}{q^{(n+1)k}\left(q^k-1\right)\left(q^k-q\right)\cdots\left(q^k-q^{k-1}\right)\left(q^k-q^{k+1}\right)\cdots\left(q^k-q^n\right)}.$$

$$\mathrm{Res}\left(h_q, q^k\right) = \frac{(1+x)(1+qx)\cdots\left(1+q^{k-1}x\right)E_q(x)}{q^{(n+1)k}q^{k(k-1)/2}\left(q^k-1\right)\left(q^{k-1}-1\right)\cdots(q-1)q^{k(n-k)}(1-q)\cdots\left(1-q^{n-k}\right)}$$

$$= \frac{(-1)^{n-k}(1+x)(1+qx)\cdots\left(1+q^{k-1}x\right)}{q^{2nk-k(k-1)/2}k_q!(n-k)_q!}E_q(x).$$

The residue of h_q at 0 is the coefficient of z^n in the Taylor series expansion of

$E_q(xz)g_n(z)$. Therefore $\mathrm{Res}\left(h_q, 0\right) = \sum_{k=0}^{n} a_n(k)x^{n-k}/(n-k)_q!$

Hence $I_n(x) = Q_n(x)E_q(x) + P_n(x)$, where P_n and Q_n are given by (8.19).

2) b) We have immediately

$$I_n(x) = \frac{1}{2i\pi}\sum_{k=0}^{+\infty}\left[\int_{C_n}\frac{z^k\,dz}{z^{n+1}(z-1)(z-q)\cdots\left(z-q^n\right)}\right]\frac{x^k}{k_q!} = \frac{1}{2i\pi}\sum_{k=0}^{+\infty}J_k\frac{x^k}{k_q!}.$$

Now denote by Γ_n the positively oriented circle with center O and radius $2/\left(|q|^n+|q|^{n+1}\right)$. The change of variable $t = 1/z$ in J_k shows that

$$J_k = \int_{C_n}\frac{z^{k-n-1}\,dz}{(z-1)(z-q)\cdots\left(z-q^n\right)} = \int_{\Gamma_n}\frac{t^{2n-k}\,dt}{(1-t)(1-qt)\cdots\left(1-q^n t\right)}.$$

The only pole inside Γ_n is 0, and only if $k \geq 2n+1$. Thus, if $k \leq 2n$, $J_k = 0$.

Therefore $I_n(x) = x^{2n+1} f_n(x)$, where $f_n(x) = \dfrac{1}{2i\pi} \displaystyle\sum_{k=2n+1}^{+\infty} J_k \dfrac{x^{k-2n-1}}{k_q!}$.

We conclude that Q_n and P_n given by (8.19) are the $[n/n]$ Padé approximants of the q-exponential function. Moreover it is clear that

$$f_n(0) = \frac{1}{2i\pi} \frac{1}{(2n+1)_q!} \int_{\Gamma_n} \frac{dt}{t(1-t)(1-qt)\cdots(1-q^n t)} = \frac{1}{(2n+1)_q!} \neq 0. \qquad (*)$$

c) If $z \in C_n$, we have $|z| - |q|^m \geq |q|^{n+1} - |q|^n \geq |q|^n$ for any $m \leq n$.

Thus, if $k \geq 2n+1$, $|J_k| \leq 2\pi |q|^{n+1} \dfrac{|q|^{(n+1)(k-n-1)}}{|q|^{n(n+1)}} \leq 2\pi |q|^{n+1} |q|^{(n+1)(k-2n-1)}$.

Consequently $|f_n(x)| \leq |q|^{n+1} \displaystyle\sum_{k=2n+1}^{+\infty} \dfrac{\left| xq^{n+1}\right|^{k-2n-1}}{k_{|q|}!}$

$$= |q|^{n+1} \sum_{k=0}^{+\infty} \frac{\left| xq^{n+1}\right|^k}{(k+2n+1)_{|q|}!} \leq \frac{|q|^{n+1}}{(2n+1)_{|q|}!} E_{|q|}\left(\left| xq^{n+1}\right|\right).$$

3) a) As in the case of the classical binomial coefficients (which are a limit case when $q \to 1$), we have

$$q^k \binom{n-1}{k}_q + \binom{n-1}{k-1}_q = q^k \frac{(n-1)_q!}{k_q!(n-1-k)_q!} + \frac{(n-1)_q!}{(k-1)_q!(n-k)_q!}$$

$$= \frac{(n-1)_q!}{k_q!(n-k)_q!}\left[q^k\left(q^{n-k}-1\right)+\left(q^k-1\right)\right] = \frac{(n-1)_q!\left(q^n-1\right)}{k_q!(n-k)_q!} = \binom{n}{k}_q.$$

Since $\binom{n}{0}_q = \binom{n}{n}_q = 1$ for every n, this proves that $\binom{n}{k}_q \in \mathbb{Z}$.

b) We have $B_n(x) = q^{n(3n+1)/2} \displaystyle\sum_{k=0}^{n} \frac{(-1)^{n-k}}{q^{2nk-k(k-1)/2}} \binom{n}{k}_q (1+x)(1+qx)\cdots(1+q^{k-1}x)$.

But for $k = 0, 1, \cdots, n$, $2nk - k(k-1)/2$ is maximum for $k = n$, and in this case its value is $n(3n+1)/2$. Therefore $B_n \in \mathbb{Z}[x]$. Similarly

$$A_n(x) = q^{n(3n+1)/2} \sum_{k=0}^{n} \frac{(-1)^{n+1}}{q^{nk+n(n+1)/2}} \binom{n+k}{k}_q \frac{n_q!}{(n-k)_q!} x^{n-k} \in \mathbb{Z}[x].$$

Now question 2)c) yields, for every $x \neq 0$,

$$\left| B_n(x) E_q(x) + A_n(x)\right| \leq n_q! |q|^{\frac{n(3n+1)}{2}} \frac{|q|^{n+1}}{(2n+1)_{|q|}!} E_{|q|}\left(\left| xq^{n+1}\right|\right)|x|^{2n+1}.$$

But we have $|n_q!| \le (|q|+1)(|q|^2+1)\cdots(|q|^n+1) \le \left(\prod_{k=1}^{+\infty}(1+|q|^{-k})\right)|q|^{n(n+1)/2}$.

And also, for n sufficiently large,

$$(2n+1)_{|q|}! = (|q|-1)(|q|^2-1)\cdots(|q|^{2n+1}-1) \ge \frac{1}{2}|q|^{(2n+2)(2n+1)/2}\prod_{k=1}^{+\infty}(1-|q|^{-k}).$$

This proves the desired result.

c) Assume that $E_q(r/s) = \alpha/\beta \in \mathbb{Q}$, with $\beta > 0$. Then $E_q\left(r/(sq^n)\right) \in \mathbb{Q}$ since

$$E_q\left(\frac{r}{sq^n}\right) = \prod_{k=1}^{+\infty}\left(1+\frac{r}{sq^{n+k}}\right) = \frac{\prod_{k=1}^{+\infty}\left(1+r/sq^k\right)}{\prod_{k=1}^{+\infty}\left(1+r/sq^k\right)} = \frac{s^n q^{n(n+1)/2}}{\prod_{k=1}^{+\infty}\left(r+sq^k\right)}E_q\left(\frac{r}{s}\right).$$

Now put $D_n = \beta\prod_{k=1}^{n}\left(r+sq^k\right)$. It is clear that $D_n E_q\left(r/sq^n\right) \in \mathbb{Z}$ and that

$$|D_n| \le s^n \beta|q|^{n(n+1)/2}\prod_{k=1}^{+\infty}\left(1+|r||q|^{-k}/s\right) \le C's^n\beta|q|^{n(n+1)/2}. \qquad (**)$$

Put $K_n = s^n q^{n^2} D_n\left(B_n\left(\frac{r}{sq^n}\right)E_q\left(\frac{r}{sq^n}\right) + A_n\left(\frac{r}{sq^n}\right)\right),$

$$L_n = s^{n+1}q^{n(n+1)}D_n\left(B_{n+1}\left(\frac{r}{sq^n}\right)E_q\left(\frac{r}{sq^n}\right) + A_{n+1}\left(\frac{r}{sq^n}\right)\right).$$

Then $K_n \in \mathbb{Z}$ and $L_n \in \mathbb{Z}$. By question 3)b), we have for n sufficiently large

$$|K_n| \le Cs^n|q|^{n^2}D_n E_{|q|}\left(\left|\frac{rq}{s}\right|\right)\left|\frac{r}{sq^n}\right|^{2n+1} \quad ; \quad |L_n| \le Cs^{n+1}|q|^{n(n+1)}D_n E_{|q|}\left(\left|\frac{rq^2}{s}\right|\right)\left|\frac{r}{sq^n}\right|^{2n+3}.$$

Taking (**) into account, in both cases the exponent of $|q|$ contains $-n^2/2$. Therefore K_n and L_n tend to 0 when $n \to \infty$. Since they are integers, this implies $K_n = L_n = 0$ for n sufficiently large. We conclude that

$$\begin{vmatrix} B_n\left(r/sq^n\right) & A_n\left(r/sq^n\right) \\ B_{n+1}\left(r/sq^n\right) & A_{n+1}\left(r/sq^n\right) \end{vmatrix} = 0$$

for n sufficiently large. As (*) holds, this contradicts lemma 8.1. Here it should be noted that $Q_n(0) \ne 0$ since $P_n(0) \ne 0$ and $E_q(0) \ne 0$.

Solutions to the exercises of chapter 9

Exercise 9.1   Assume that the algebraic number α has two minimal polynomials P_α and Q_α. Then $P_\alpha \ne 0$, $Q_\alpha \ne 0$, $P_\alpha(\alpha) = 0$, $Q_\alpha(\alpha) = 0$, and since P_α and Q_α have minimal degree, $\deg P_\alpha \le \deg Q_\alpha$ and $\deg Q_\alpha \le \deg P_\alpha$, which implies that $\deg P_\alpha = \deg Q_\alpha$. Put $P_\alpha(x) = x^d + a_{d-1}x^{d-1} + \cdots + a_0$ and

$Q_\alpha(x) = x^d + b_{d-1}x^{d-1} + \cdots + b_0$. Then $R(x) = (a_{d-1} - b_{d-1})x^{d-1} + \cdots + (a_0 - b_0)$ is not zero since $P_\alpha \neq Q_\alpha$, and $R(\alpha) = 0$ and $\deg R < \deg P_\alpha$. Contradiction.

Exercise 9.2   1) If P_α was reducible, we would have $P_\alpha(x) = Q(x)R(x)$, with $Q, R \in \mathbb{Q}[x]$, $\deg Q \geq 1$, $\deg R \geq 1$. Hence $Q(\alpha)$ or $R(\alpha) = 0$, which contradicts the fact that P_α has minimal degree.

2) Divide P par P_α. We have $P(x) = P_\alpha(x)Q(x) + R(x)$, with $\deg R < \deg P_\alpha$. Hence $R(\alpha) = 0$ and $R = 0$ since P_α has minimal degree. Thus $P(x) = P_\alpha(x)Q(x)$. Since P is irreducible in $\mathbb{Q}[x]$, this yields $Q(x)$ = constant $= q$. Indeed, P_α cannot be constant since $P_\alpha \neq 0$ and $P_\alpha(\alpha) = 0$. Now $P(x) = qP_\alpha(x)$. Looking at the terms of higher degree yields $q = 1$, Q.E.D.

Exercise 9.3   1)   Assume   that $P(x) = Q(x)R(x)$,   with $Q, R \in \mathbb{Q}[x]$, $\deg Q \geq 1, \deg R \geq 1$. Put $d = \deg P$ and

$$Q(x) = \frac{a}{b}x^q + \cdots, \qquad R(x) = \frac{a'}{b'}x^{d-q} + \cdots.$$

Then $aa'/(bb') \in \mathbb{Z}$, and therefore we can write $P(x) = Q_0(x)R_0(x)$, with

$$Q_0(x) = \frac{aa'}{bb'}x^q + \cdots, \quad R_0(x) = x^{d-q} + \cdots,$$

which means that the leading coefficients of Q_0 and R_0 are rational integers. Let $\alpha_1, \ldots, \alpha_q$ be the zeros of Q_0 in $\mathbb{C}$, $\alpha_{q+1}, \ldots, \alpha_d$ those of R_0. Then $\alpha_1, \ldots, \alpha_q$ are roots of $P(x)$, therefore they are *algebraic integers*. Hence the coefficients of $Q_0(x) = \frac{aa'}{bb'}(x - \alpha_1)\ldots(x - \alpha_q)$ are algebraic integers by theorem 9.4. Since they are also rational, we have $Q_0 \in \mathbb{Z}[x]$. Similarly $R_0 \in \mathbb{Z}[x]$.

2) Suppose that P is reducible in $\mathbb{Q}[x]$. By question 1), there exist $Q, R \in \mathbb{Z}[x]$ such that $P(x) = Q(x)R(x)$ with $\deg Q \geq 1$, $\deg R \geq 1$. Denote by $\hat{n}$ the residue class of n modulo p, $\hat{Q}(x) = \hat{a}_0 + \hat{a}_1 x + \cdots + x^q$, and $\hat{R}(x) = \hat{b}_0 + \hat{b}_1 x + \cdots + x^{d-q}$. Then $x^d = \hat{Q}(x)\hat{R}(x)$ in the ring $\mathbb{F}_p[x]$ of polynomials over the field $\mathbb{F}_p$. This yields $\hat{Q}(x) = x^q$, $\hat{R}(x) = x^{d-q}$. Therefore $a_0 \equiv b_0 \equiv 0 \pmod{p}$, $a_0 b_0$ is a multiple of p^2, contradiction.

Exercise 9.4   1) Let a and b be the denominators of α and β, and let $m = \text{LCM}(a, b)$. Then $m\alpha = m'(a\alpha)$ and $m\beta = m''(b\alpha)$ are algebraic integers by theorem 9.4 (products of algebraic integers). Therefore $m(\alpha + \beta)$ is an algebraic integer (sum of algebraic integers), which shows that $\alpha+\beta$ is algebraic. Similarly $ab\alpha\beta$ is an algebraic integer and $\alpha\beta$ is algebraic.

2) If α and β are algebraic, $\alpha+\beta$ and $\alpha\beta$ are algebraic. Moreover, $-\alpha$ is algebraic since $a_0 + a_1\alpha + .. + a_d\alpha^d = 0 \Rightarrow a_0 - a_1(-\alpha) + .. + (-1)^d a_d(-\alpha)^d = 0$.

Finally, if $\alpha \neq 0$, then $1/\alpha$ is algebraic since $a_0(1/\alpha)^d + a_1(1/\alpha)^{d-1} + \cdots + a_d = 0$. Thus $\overline{\mathbb{Q}}$ is a subfield of $\mathbb{C}$.

Exercise 9.5 Assume that α has irrationality measure μ and is a Liouville number. For every n sufficiently large, there exists $p/q \in \mathbb{Q}$ with $q \geq 2$ and

$$\frac{C}{q^\mu} \leq \left| \alpha - \frac{p}{q} \right| < \frac{1}{q^n}.$$

This yields $0 < C \leq q^{\mu-n}$. Contradiction when $n \to +\infty$.

Exercise 9.6 Assume that $\mu(\alpha) < 2$. Fix $\mu \in]\mu(\alpha), 2[$. Since $\mu(\alpha)$ is the lowest bound of all irrationality measures of α , we have $\left| \alpha - \frac{p}{q} \right| \geq \frac{C}{q^\mu}$, where C is a constant. Take $p = P_n$, $q = Q_n$, where P_n/Q_n is the n -th convergent of the expansion of α in regular continued fraction. Using (4.8) yields

$$\frac{C}{Q_n^\mu} \leq \left| \alpha - \frac{P_n}{Q_n} \right| < \frac{1}{Q_n^2}.$$

Hence $0 < C \leq Q_n^{\mu-2}$. But $\mu - 2 < 0$ and $Q_n \to +\infty$, contradiction.

Exercise 9.7 We have to show that $n+1 \leq 3\sqrt{n!}$ for every $n \in \mathbb{N}$. Clearly this is true for $n = 0$. Assume that $n+1 \leq 3\sqrt{n!}$ for some $n \geq 1$, then

$$n+2 \leq 3\sqrt{n!} + 1 \leq 4\sqrt{n!} \leq \frac{4}{\sqrt{n+1}}\sqrt{(n+1)!} \leq \frac{4}{\sqrt{2}}\sqrt{(n+1)!} \leq 3\sqrt{(n+1)!},$$

which proves the estimate by induction.

Exercise 9.8 As $(1+x)^{2n} = \sum_{k=0}^{2n} \binom{2n}{k} x^k$, we have $\sum_{k=0}^{2n} \binom{2n}{k} = 2^{2n} = 4^n$.

Consequently $\binom{2n}{n} \leq 4^n$. Now, since $|q_n| \leq e \cdot \binom{2n}{n} n!$, we have $|q_n| \leq e \cdot 4^n n!$, and it remains to prove that $4^n e \leq 30.n!$ for every $n \in \mathbb{N}$. This is true for $n = 0, 1, 2, 3$. For $n \geq 3$, we have $4^{n+1}e \leq 4 \cdot 4^n e \leq 4 \cdot 30 \cdot n! \leq (n+1) \cdot 30n! \leq 30(n+1)!$, which proves the desired result by induction.

Exercise 9.9   1) Let $f(n) = \dfrac{(n+1)!}{(n!)^{1+\eta}} = \dfrac{n+1}{(n!)^{\eta}}$. Then $\lim_{n\to+\infty} f(n) = 0$. Therefore $f(n)$ is bounded over $\mathbb{N}$ and there exists $C_1 > 0$ such that $f(n) \le C_1$, $\forall n \in \mathbb{N}$.

2) We have seen in exercise 9.8 that $|q_n| \le e \cdot 4^n n!$. Let $\varphi(n) = \dfrac{e \cdot 4^n n!}{(n!)^{1+\eta}} = \dfrac{e \cdot 4^n}{(n!)^{\eta}}$.

Since $\lim_{n\to+\infty} \varphi(n) = 0$, $\varphi(n)$ is bounded over $\mathbb{N}$ and $e \cdot 4^n n! \le C_2 (n!)^{1+\eta}$.

3) Apply theorem 9.7 with $g(n) = n!$, $a = 1+\eta$, $b = C_1$, $k = C_2$, $\ell = e$ (see 9.14), $h = 1+\eta$. Then $|e - p/q| \ge C/q^{\mu}$ with $\mu = (1+\eta)^3 + 1$. For any given $\varepsilon > 0$, choose η in such a way that $\mu \le 2+\varepsilon$ (this is possible since $\lim_{\eta\to 0}\mu = 2$). Then $|e - p/q| \ge C(\varepsilon)/q^{2+\varepsilon}$ for every rational p/q with $q > 0$.

Remark: Notice that $C(\varepsilon)$ can be *explicitly* computed. Indeed, one can compute a numerical value of η for a given $\varepsilon > 0$. This being done, C_1 and C_2 can be computed, as well $C(\varepsilon)$, by using a program on *Maple* or *Mathematica*.

Exercise 9.10   Assume that the equation (9.25) has infinitely many solutions (x_n, y_n). Then $\lim_{n\to+\infty} y_n = +\infty$. Indeed, if not, the sequence (y_n) would take only finitely many values $b_1, b_2, ..., b_N$, and (9.25) would reduce to a finite number of equations $\sum_{i=0}^{d} a_i b_j^{d-i} x^i = k$, each of them having only finitely many solutions x, contradiction. Thus $\lim_{n\to+\infty} y_n = +\infty$.

But (9.25) can be written $\sum_{i=0}^{d} a_i (x_n/y_n)^i = k/y_n^d$. Thus $\lim_{n\to+\infty} P(x_n/y_n) = 0$, and we can find two subsequences (p_m) and (q_m) of (x_n) and (y_n), and a zero α of P, with $\lim_{m\to+\infty} p_m/q_m = \alpha$, $\lim_{m\to+\infty} |q_m| = +\infty$. Then α is algebraic of degree $d \ge 3$, since P is irreducible in $\mathbb{Q}[X]$ and $P'(\alpha) \ne 0$ (otherwise α would be algebraic of degree $\le d-1$). Hence, for m sufficiently large,

$$|P'(p_m/q_m)| \ge |P'(\alpha)/2| > 0, \text{ and}$$

$$P\left(\frac{p_m}{q_m}\right) = \frac{k}{q_m^d} \Rightarrow P\left(\frac{p_m}{q_m}\right) - P(\alpha) = \frac{k}{q_m^d} \Rightarrow \left(\frac{p_m}{q_m} - \alpha\right)\sum_{i=0}^{d} a_i \sum_{j=0}^{i-1}\left(\frac{p_m}{q_m}\right)^j \alpha^{i-1-j} = \frac{k}{q_m^d}.$$

When $m \to +\infty$, $\sum_{i=0}^{d} a_i \sum_{j=0}^{i-1}(p_m/q_m)^j \alpha^{i-1-j} \to \sum_{i=1}^{d} a_i i \alpha^{i-1} = P'(\alpha)$.

Therefore, for m sufficiently large, $\left|\dfrac{k}{q_m^d}\right| \ge \left|\dfrac{p_m}{q_m} - \alpha\right| \cdot \left|\dfrac{P'(\alpha)}{2}\right|$.

In other words $\left|\dfrac{p_m}{q_m}-\alpha\right|\le\dfrac{2k}{|P'(\alpha)|}\left|\dfrac{1}{q_m^d}\right|$.

Now Thue's theorem 9.9 yields, for m sufficiently large,

$$\frac{C}{|q_m|^{1+d/2}}\le\frac{2k}{|P'(\alpha)||q_m^d|}, \qquad \text{whence} \quad 0<\left|\frac{CP'(\alpha)}{2k}\right|\le\frac{1}{|q_m|^{d/2-1}}.$$

But $d/2-1\ge 1/2$ since $d\ge 3$, and $\lim_{m\to+\infty}|q_m|=+\infty$. Contradiction when m tends to $+\infty$. Thus (9.25) has only finitely many solutions.

Exercise 9.11   1) We can use elementary means. Let $\gamma=\alpha+\beta$. We have $(\gamma-\sqrt{2})^3=3$, whence $(\gamma^3+6\gamma-3)=\sqrt{2}(3\gamma^2+2)$. Squaring yields $P(\gamma)=0$, with $P(x)=x^6-6x^4-6x^3+12x^2-36x+1$. Now we have to prove that P is irreducible in $\mathbb{Q}[x]$. Here Eisenstein's criterion does not apply, but we can argue the following way. It is clear, by construction, that the six complex roots of P are $x_1=\sqrt{2}+\sqrt[3]{3}$, $x_2=\sqrt{2}+j\sqrt[3]{3}$, $x_3=\sqrt{2}+j^2\sqrt[3]{3}$, $x_4=-\sqrt{2}+\sqrt[3]{3}$, $x_5=-\sqrt{2}+j\sqrt[3]{3}$, $x_6=-\sqrt{2}+j^2\sqrt[3]{3}$, with $j=e^{2i\pi/3}$. And we have

$$P(x)=(x-x_1)(x-x_2)(x-x_3)(x-x_4)(x-x_5)(x-x_6).$$

If P was reducible in $\mathbb{Q}[x]$, we could group one, two, three, four or five factors in such a sort that their product would belong to $\mathbb{Q}[x]$, as well as the product of the remaining factors. For example, we would have $P(x)=Q(x)R(x)$, with $Q(x)=(x-x_1)(x-x_3)$ and $R(x)=(x-x_2)(x-x_4)(x-x_5)(x-x_6)$. In this example, we would have $x_1+x_3\in\mathbb{Q}$ and $x_2+x_4+x_5+x_6\in\mathbb{Q}$. But it is easy to see that these sums of roots cannot be *real* except if we group them in order to eliminate j, for example $x_a+x_b+x_c$ with $x_a=\pm\sqrt{2}+\sqrt[3]{3}$, $x_b=\pm\sqrt{2}+j\sqrt[3]{3}$, $x_c\pm\sqrt{2}+j^2\sqrt[3]{3}$. But in all cases the sum is irrational, which proves that P is irreducible in $\mathbb{Q}[x]$. Therefore it is the minimal polynomial of $\alpha+\beta$.

2) By repeating the proof of theorem 9.4, we see that $\alpha+\beta$ is a root of

$$P(x)=\big((x-i)^2+(x-i)+1\big)\big((x+i)^2+(x+i)+1\big)=x^4+2x^3+5x^2+4x+1.$$

It is easy to see that P is irreducible in $\mathbb{Q}[x]$ (or equivalently in $\mathbb{Z}[x]$ by exercise 9.3, 1). If P was reducible, only two cases could occur:
$P(x)=(x+a)(x^3+bx^2+cx+d)$ (*)   or $P(x)=(x^2+ax+b)(x^2+cx+d)$ (**),
with $a,\ b,\ c,\ d\in\mathbb{Z}$. Case (*) is impossible. Indeed, it implies $ad=1$, whence $a=1$ or $a=-1$. However, 1 and -1 are no roots of P.

Case (**) is impossible: it yields $bd=1$, $bc+ad=4$, $a+c=2$. The first equation yields $b=d=1$ or $b=d=-1$. In both cases, this leads to a contradiction when replacing in the second equation.

Therefore P is irreducible in $\mathbb{Q}[x]$ and it is the minimal polynomial of $\alpha+\beta$.

3) Both polynomials $P_1(x)=x^3+x+1$ and $P_2(x)=x^3+x^2+1$ have only one real root. Therefore $\beta=1/\alpha$. But $\alpha^3+\alpha+1=0 \Rightarrow \dfrac{1}{\alpha}=-1-\alpha^2$, whence

$$\gamma=\alpha+\beta=\alpha-1-\alpha^2. \quad (\#)$$

Squaring yields $\gamma^2=\alpha^4-2\alpha^3+3\alpha^2-2\alpha+1$. But $\alpha^3=-\alpha-1$, $\alpha^4=-\alpha^2-\alpha$.
Replacing yields

$$\gamma^2=2\alpha^2-\alpha+3. \quad (\#\#)$$

Using (#) and (##) yields $\alpha=-1+2\gamma+\gamma^2$ and $\alpha^2=-2+\gamma+\gamma^2$.
Finally, if we multiply (#) and (##), we obtain

$$\gamma^3=-2\alpha^4+3\alpha^3-6\alpha^2+4\alpha-3=-4\alpha^2+3\alpha-6,$$

and replacing α and α^2 yields $\gamma^3+\gamma^2-2\gamma+1=0$.
The polynomial $P(x)=x^3+x^2-2x+1$ is irreducible in $\mathbb{Q}[x]$ since it has no root in $\mathbb{Z}$. Therefore it is the minimal polynomial of $\alpha+\beta$.

Remark: This is an example of computation in the field $\mathbb{Q}(\alpha)$ (see chapter 10).

Exercise 9.12   We have to prove:
   a) $P\in A$ and $Q\in A \Rightarrow P-Q\in A$,
   b) $P\in A$ and $Q\in \mathbb{Q}[x] \Rightarrow PQ\in A$.
This is obvious. Therefore A is an ideal of $\mathbb{Q}[x]$.
Now $\mathbb{Q}[x]$ is euclidean, and therefore it is a principal ideal domain (exercise 6.20). Hence every element of A can be written as $Q(x)P_0(x)$, and we can assume that P_0 is monic. Moreover, P_0 is irreducible in $\mathbb{Q}(x)$, since it has minimal degree (see exercise 6.20). Therefore $P_0=P_\alpha$. Thus, if α is algebraic, every polynomial P satisfying $P(\alpha)=0$ is a multiple (in $\mathbb{Q}[x]$) of the minimal polynomial of α.

Exercise 9.13   We have seen in exercise 8.7, 3), that

$$q_n=7D_n\,{}_2F_1\left(\begin{matrix}-n,-n+1/3\\-2n\end{matrix}\bigg|\frac{1}{18^3}\right), \text{ with } 0<D_n\le(3\sqrt{3}\cdot18^3)^n\binom{2n}{n}\le(12\sqrt{3}\cdot18^3)^n$$

by (9.18). Moreover,

$$\left| {}_2F_1\left(\begin{matrix} -n,-n+1/3 \\ -2n \end{matrix} \middle| \frac{1}{18^3} \right) \right| \le \sum_{i=0}^n \frac{n(n-1)...(n-i+1)(n-\frac{1}{3})(n-\frac{1}{3}+1)...(n-\frac{1}{3}-i+1)}{2n(2n-1)...(2n-i+1)\cdot i!} \cdot \frac{1}{18^{3i}}$$

$$= \sum_{i=0}^n \binom{n}{i} \frac{(n-\frac{1}{3})(n-\frac{1}{3}+1)...(n-\frac{1}{3}-i+1)}{2n(2n-1)...(2n-i+1)\cdot i!} \cdot \frac{1}{18^{3i}}.$$

As each of the coefficients of $\binom{n}{i}\dfrac{1}{18^{3i}}$ is less than 1, we obtain

$$\left| {}_2F_1\left(\begin{matrix} -n,-n+1/3 \\ -2n \end{matrix} \middle| \frac{1}{18^3} \right) \right| \le \sum_{i=0}^n \binom{n}{i} \frac{1}{18^{3i}} = \left(1+\frac{1}{18^3}\right)^n.$$

Hence $q_n \le 7(12\sqrt{3}\cdot 5833)^n$, Q.E.D.

2) By exercise 8.7, questions 5) and 6), we have

$$\left| q_n \sqrt[3]{17} - p_n \right| \le 1.03\times 10^{-3}(4488)^{-n}, \text{ with } \left| \begin{matrix} q_n & p_n \\ q_{n+1} & p_{n+1} \end{matrix} \right| \ne 0 \text{ for every } n\in \mathbb{N}.$$

But we have just seen that $\left| q_n \right| \le 7\cdot(4488)^{1.392n}$. We apply theorem 9.7 with $g(n) = (4488)^n$, $k = 7$, $a = 1.392$, $\ell = 0.5$, $b = 4488$, $h = 1$. This yields the desired result, obtained by Baker as soon as 1966. The irrationality measure given by Bennet (table 9.1, page 146) is better.

Exercise 9.14 1) We have seen in exercise 8.8 that $q_n = D_n\, {}_2F_1\left(\begin{matrix} -n,-n \\ -2n \end{matrix} \middle| \frac{1}{2} \right)$,

with $D_n \le 6^n \binom{2n}{n} \le 24^n$. Proceeding as in exercise 9.13, question 1), yields

$$\left| q_n \right| \le 24^n \sum_{i=0}^n \binom{n}{i}\left(\frac{1}{2}\right)^i = 36^n.$$

2) In exercise 8.8, we have proved that $\left| q_n \operatorname{Log} 2 - p_n \right| \le \left(\frac{4}{3}\right)^{-n}$ and $\left| \begin{matrix} q_n & p_n \\ q_{n+1} & p_{n+1} \end{matrix} \right| \ne 0$

for every n. Moreover, it results from question 1) that $\left| q_n \right| \le \left(\frac{4}{3}\right)^{12.46n}$.

We apply theorem 9.7 with $g(n) = \left(\frac{4}{3}\right)^n$, $k = 1$, $\ell = 1$, $a = 12.46$, $b = \frac{4}{3}$, $h = 1$.

This yields the desired result, which has been obtained by Baker in 1964. This implies that $\operatorname{Log} 2$ is not a Liouville number.

Remark: We know now that $\mu(\operatorname{Log} 2) \le 3.9$ (E.A. Rukhadze, 1987).

Exercise 9.15 We may assume that $x > 0$, $y > 0$, x and y coprime, since 11 is not divisible by any prime at a fourth power. We use theorem 9.8. The only solution (x, y) satisfying $y \le 22$ is $x = 2$, $y = 1$. In order to find other solutions, we use the irrationality measure of $\sqrt[4]{5}$ given in table 9.1. We also have to compute the convergents P_n/Q_n of $\sqrt[4]{5}$ satisfying $Q_n \le (11/0.03)^{1/1.23}$, that is $Q_n \le 121$. From a numerical value of $\sqrt[4]{5}$ we obtain the continued fraction expansion $\sqrt[4]{5} = [1, 2, 53, 4, \ldots]$, whence $P_1 = 1$, $Q_1 = 1$, $P_2 = 3$, $Q_2 = 2$, $P_3 = 160$, $Q_3 = 107$, $P_4 = 643$, $Q_4 = 430$. This yields no solution of the equation. Hence its only solution is $x = \pm 2$, $y = \pm 1$.

Exercise 9.16   1) If the polynomial $P(x) = ax^3 - b$ is irreducible in $\mathbb{Q}[x]$, then the equation $ax^3 - by^3 = c$ has only finitely many solutions by corollary 9.2. If $P(x) = ax^3 - b$ is reducible, it has a rational root p/q, with p and q coprime. Then $a/b = q^3/p^3$, and there exists $d \in \mathbb{N}$ such that $a = dq^3, b = dp^3$. If $d \mid c$, the equation $ax^3 - by^3 = c$ has no solution. Otherwise, it can be written $X^3 - Y^3 = c'$, $X = qx$, $Y = py$, $c' = c/d$. Then $(X - Y)(X^2 + XY + Y^2) = c'$, whence $X^2 \le X^2 + XY + Y^2 \le c'$, the equation has only finitely many solutions.
2) Put $a_n = 2^{\alpha_1} 3^{\alpha_2}$, $a_{n+1} = 2^{\beta_1} 3^{\beta_2}$, with
$$\alpha_i = 3u_i + v_i, \quad \beta_i = 3t_i + w_i, \quad 0 \le v_i \le 2, \quad 0 \le w_i \le 2.$$
The equation $a_{n+1} - a_n = c$ can be written $2^{w_1} 3^{w_2} (2^{t_1} 3^{t_2})^3 - 2^{v_1} 3^{v_2} (2^{u_1} 3^{u_2})^3 = c$.
For each value of (w_1, w_2, v_1, v_2), this equation has only finitely many solutions $x = 2^{t_1} 3^{t_2}$, $y = 2^{u_1} 3^{u_2}$. Since (w_1, w_2, v_1, v_2) can take only 3^4 values, the equation $a_{n+1} - a_n = c$ has only finitely many solutions.
3) Let $m \in \mathbb{N}^*$ be given. The inequation $a_{n+1} - a_n \le m$ has only finitely many solutions since it is equivalent to $(a_{n+1} - a_n = 1)$ or $(a_{n+1} - a_n = 2)$ ... or $(a_{n+1} - a_n = m)$, and that each of these equations has only finitely many solutions. Therefore $\forall m \in \mathbb{N}^*, \exists N \in \mathbb{N}$ such that $n \ge N \Rightarrow a_{n+1} - a_n > m$. This means that $\lim_{n \to +\infty} (a_{n+1} - a_n) = +\infty$.

Exercise 9.17   1) Since $\lim_{n \to +\infty} |q_n \alpha - p_n| = 0$ by (**), n exists.
Since $|q_0 \alpha - p_0| \ge \ell_1$ by (**), we have $n \ge 1$.
Now $q_n = 0 \Rightarrow |p_n| < \ell_1/q \le 1/2 \Rightarrow p_n = 0 \Rightarrow \ell_1 E_1^{-n} \le 0$, which is impossible.
Therefore $q_n \ne 0$. Finally, since n is minimal, we have
$$\ell_1/q \le |q_{n-1} \alpha - p_{n-1}| \le \ell_0 E_0^{-n+1},$$
whence $E_0^{n-1} \le \ell_0 q/\ell_1$, which yields $n \le \mathrm{Log}(\ell_0 q/\ell_1)/\mathrm{Log}\, E_0 + 1$.

2) As in the proof of theorem 9.7, we write

$$0 = q_n p - p_n q = q_n(p - q\alpha) + q(q_n \alpha - p_n),$$

whence $\left|\alpha - \frac{p}{q}\right| = \frac{1}{|q_n|}\left|q_n\alpha - p_n\right| \geq \ell_1 E_1^{-n}/|q_n|$. Using (*), we get

$$\left|\alpha - \frac{p}{q}\right| \geq \frac{\ell_1}{k_0}(QE_1)^{-n} \geq \frac{\ell_1}{k_0}(QE_1)^{-\mathrm{Log}(\ell_0 q/\ell_1)/\mathrm{Log}\,E_0 - 1}$$

$$\geq \frac{\ell_1}{k_0}(QE_1)^{-\mathrm{Log}(\ell_0/\ell_1)/\mathrm{Log}\,E_0 - 1} q^{-\mathrm{Log}(QE_1)/\mathrm{Log}\,E_0}.$$

3) If $q_n p - p_n q \neq 0$, we have $q_n p - p_n q \geq 1$, whence

$$1 \leq |q_n||p - q\alpha| + q|q_n\alpha - p_n| \leq |q_n||p - q\alpha| + \frac{1}{2}$$

by definition of n. Consequently $|p - q\alpha| \geq 1/2|q_n| \geq 1/2k_0 Q^n$. Now using the

method of question 2, we obtain $\left|\alpha - \frac{p}{q}\right| \geq \frac{1}{2k_0}Q^{-\mathrm{Log}(\ell_0/\ell_1)/\mathrm{Log}\,E_0 - 1} q^{-\mathrm{Log}\,Q/\mathrm{Log}\,E_0 - 1}$.

But $\frac{1}{2} \geq \ell_1$ and $QE_1 \geq Q$. Thus $\frac{1}{2k_0}Q^{-\mathrm{Log}(\ell_0/\ell_1)/\mathrm{Log}\,E_0 - 1} \geq \frac{\ell_1}{k_0}(QE_1)^{-\mathrm{Log}(\ell_0/\ell_1)/\mathrm{Log}\,E_0 - 1}$.

Moreover $\dfrac{\mathrm{Log}(QE_1)}{\mathrm{Log}\,E_0} = \dfrac{\mathrm{Log}\,Q}{\mathrm{Log}\,E_0} + \dfrac{\mathrm{Log}\,E_1}{\mathrm{Log}\,E_0} \geq \dfrac{\mathrm{Log}\,Q}{\mathrm{Log}\,E_0} + 1$.

Hence $q^{-\mathrm{Log}\,Q/\mathrm{Log}\,E_0 - 1} \geq q^{-\mathrm{Log}(QE_1)/\mathrm{Log}\,E_0}$. Therefore the result of question 2 remains valid if $q_n p - q p_n \neq 0$.

Exercise 9.18   1) We have a convergence problem at $x = 1$, $y = 1$.
For $a \in\,]0,1[$, $b \in\,]0,1[$, we have

$$\int_0^a \int_0^b \frac{x^r y^s dx dy}{1 - xy} = \int_0^a \int_0^b \sum_{n=0}^{+\infty} x^{r+n} y^{s+n} dx dy = \sum_{n=0}^{+\infty} \frac{a^{r+n+1} b^{s+n+1}}{(r+n+1)(s+n+1)}.$$

Therefore $I_{r,s} = \displaystyle\sum_{n=0}^{+\infty} \frac{1}{(r+n+1)(s+n+1)}$.

For $s = r$, we have $I_{r,r} = \displaystyle\sum_{n=0}^{+\infty} \frac{1}{(r+n+1)^2} = \zeta(2) - \sum_{n=0}^{r} \frac{1}{n^2}$.

For $s < r$, we expand the rational function $\frac{1}{(r+n+1)(s+n+1)}$ in partial fractions:

$$I_{r,s} = \lim_{N \to +\infty} \sum_{n=0}^{N} \frac{1}{r-s}\left(\frac{1}{s+n+1} - \frac{1}{r+n+1}\right)$$

$$= \frac{1}{r-s} \lim_{N \to +\infty} \left(\frac{1}{s+1} + \frac{1}{s+2} + \cdots + \frac{1}{r} - \frac{1}{s+N+2} - \cdots - \frac{1}{r+N+1}\right).$$

Therefore $I_{r,s} = \dfrac{1}{r-s}\left(\dfrac{1}{s+1} + \dfrac{1}{s+2} + \cdots + \dfrac{1}{r}\right).$

2) By using the binomial theorem, we obtain

$$P_n(x) = \frac{1}{n!}\frac{d^n}{dx^n}\sum_{k=0}^{n}\binom{n}{k}(-1)^k x^{n+k} = \sum_{k=0}^{n}\binom{n}{k}\frac{(n+k)(n+k-1)\ldots(k+1)}{n!}(-1)^k x^k.$$

Therefore $P_n(x) = \sum_{k=0}^{n}\binom{n}{k}\binom{n+k}{n}(-1)^k x^k \in \mathbb{Z}[x].$

3) We have

$$K_n = \int_0^1\int_0^1 \frac{\sum_{k=0}^{n}\sum_{\ell=0}^{n}\binom{n}{k}\binom{n+k}{n}\binom{n}{\ell}(-1)^{k+\ell}x^k y^\ell}{1-xy}\,dxdy = \sum_{k=0}^{n}\sum_{\ell=0}^{n}\binom{n}{k}\binom{n+k}{n}\binom{n}{\ell}(-1)^{k+\ell}I_{k,\ell}.$$

If we distinguish the cases $k = \ell$ and $k \neq \ell$, we obtain

$$K_n = \sum_{k=0}^{n}\binom{n}{k}^2\binom{n+k}{n}I_{k,k} + \sum_{k=0}^{n}\sum_{\ell<k}\binom{n}{k}\binom{n+k}{n}\binom{n}{\ell}(-1)^{k+\ell}I_{k,\ell}$$

$$+ \sum_{k=0}^{n}\sum_{\ell>k}\binom{n}{k}\binom{n+k}{n}\binom{n}{\ell}(-1)^{k+\ell}I_{k,\ell}.$$

Hence question 1) yields $K_n = b_n\zeta(2) + a_n$, with $b_n = \sum_{k=0}^{n}\binom{n}{k}^2\binom{n+k}{n} \in \mathbb{N}$,

$$a_n = -\sum_{k=0}^{n}\binom{n}{k}^2\binom{n+k}{n}\sum_{i=1}^{k}\frac{1}{i^2} + \sum_{k=0}^{n}\sum_{\ell<k}\binom{n}{k}\binom{n+k}{n}\binom{n}{\ell}(-1)^{k+\ell}I_{k,\ell}$$

$$+ \sum_{k=0}^{n}\sum_{\ell>k}\binom{n}{k}\binom{n+k}{n}\binom{n}{\ell}(-1)^{k+\ell}I_{k,\ell}.$$

Since $I_{k,\ell} \in \mathbb{Q}$ if $k \neq \ell$, it is clear that $a_n \in \mathbb{Q}$. Moreover, since $k \leq n$ and $\ell \leq n$, $(\delta(n))^2 I_{k,\ell} \in \mathbb{Z}$, as well as $(\delta(n))^2 \sum_{i=1}^{k} i^{-2}$ for $k \leq n$. Therefore we have $p_n = (\delta(n))^2 a_n \in \mathbb{Z}$ and $q_n = (\delta(n))^2 b_n \in \mathbb{Z}$.

4) For every $k = 0,\ 1,\ 2,\ldots,\ n$, $\binom{n+k}{n} \leq \binom{2n}{n}$. Hence $b_n \leq \binom{2n}{n}\sum_{k=0}^{n}\binom{n}{k}^2$.

This sum can be computed by looking at the coefficient of x^n in the polynomial $(1+x)^{2n} = (1+x)^n(1+x)^n$. This yields $\sum_{k=0}^{n}\binom{n}{k}^2 = \binom{2n}{n}$.

Therefore $q_n \leq (\delta(n))^2\binom{2n}{n}^2 \leq (3^2 4^2)^n \leq 144^n$ by (9.18) and theorem 7.5.

5) We have $K_n = \dfrac{1}{n!}\int_0^1\int_0^1\dfrac{\frac{d^n}{dx^n}\left(x^n(1-x)^n\right)(1-y)^n}{1-xy}\,dxdy$.

Since 0 and 1 are repeated roots with multiplicity n of $x^n(1-x)^n$, the polynomial $\frac{d^{n-1}}{dx^{n-1}}\left(x^n(1-x)^n\right)$ vanishes for $x = 0$ and $x = 1$.

Integrating by parts with respect to x yields

$$K_n = -\frac{1}{n!} \int_0^1 \int_0^1 \frac{y \frac{d^{n-1}}{dx^{n-1}}\left(x^n(1-x)^n\right)(1-y)^n}{(1-xy)^2} \, dxdy.$$

Doing this n times yields the desired result.

6) It is well known that, for $x \geq 0$ and $y \geq 0$, $\sqrt{xy} \leq (x+y)/2$ (inequality between the arithmetic and the geometric means). Therefore

$$(1-\sqrt{xy})^2 = 1 - 2\sqrt{xy} + xy \geq 1 - x - y + xy = (1-x)(1-y).$$

Hence $\displaystyle |K_n| \leq \int_0^1 \int_0^1 \frac{(xy)^n(1-\sqrt{xy})^{2n}}{(1-xy)^{n+1}} \, dxdy = \int_0^1 \int_0^1 \frac{1}{1-xy}\left(\frac{xy(1-\sqrt{xy})}{1+\sqrt{xy}}\right)^n dxdy$.

Now it is easy to check that the function $f(u) = u^2(1-u)/(1+u)$ is maximum on $[0,1]$ for $u = (-1+\sqrt{5})/2$. This maximum is $\left((\sqrt{5}-1)/2\right)^5$. Hence

$$|K_n| \leq I_{0,0}\left(\frac{\sqrt{5}-1}{2}\right)^{5n} = \frac{\pi^2}{6}\left(\frac{\sqrt{5}-1}{2}\right)^{5n}.$$

7) For $x \in [\varepsilon, 1-\varepsilon]$ and $y \in [\varepsilon, 1-\varepsilon]$, we have $x \geq \varepsilon$, $y \geq \varepsilon$, $1-x \geq \varepsilon$, $1-y \geq \varepsilon$,

$1-xy \leq 1-\varepsilon^2$, whence $|K_n| \geq \frac{(1-2\varepsilon)^2}{1-\varepsilon^2}\left(\frac{\varepsilon^4}{1-\varepsilon^2}\right)^n$. It is easy to see that $\varepsilon^4/(1-\varepsilon^2)$ is minimum for $\varepsilon = 1/2$. However we cannot use this value since it is a zero of $1-2\varepsilon$. We take a close value instead, namely $\varepsilon = 0.4999$, and we obtain $|K_n| \geq 5 \cdot 10^{-8}(0.0832)^n$.

8) We have $q_n\pi^2 + 6p_n = 6(D(n))^2 K(n) = r_n$, with

$$3 \cdot 10^{-7}(0.0832)^n \leq r_n \leq 9.9 \times (0.8118)^n$$

from questions 6 and 7 and theorem 7.5. We apply the result of exercise 9.17 with $k_0 = 1$, $Q = 144$, $\ell_1 = 3 \cdot 10^{-7}$, $E_1 = 12.0193$, $\ell_0 = 9.9$, $E_0 = 1.2318$.

We obtain $\left|\pi^2 - \dfrac{p}{q}\right| \geq 10^{-279} q^{35.77}$ for every $(p,q) \in \mathbb{Z} \times \mathbb{N}^*$.

This yields $\left|\pi^2 - \dfrac{p^2}{q^2}\right| \geq \dfrac{10^{-279}}{q^{71.54}}$, whence $\left|\pi - \dfrac{p}{q}\right|\left|\pi + \dfrac{p}{q}\right| \geq \dfrac{10^{-279}}{q^{71.54}}$.

Therefore $\left|\pi - \dfrac{p}{q}\right| \geq \dfrac{10^{-280}}{q^{71.54}}$ for every $(p,q) \in \mathbb{Z} \times \mathbb{N}^*$.

Indeed, if $\left|\pi - \dfrac{p}{q}\right| \geq \dfrac{\pi}{2}$, it is true. If not, we can write $\dfrac{\pi}{2} < \dfrac{p}{q} < \dfrac{3\pi}{2}$, whence

$\dfrac{10^{-279}}{q^{71.54}} \leq \left|\pi - \dfrac{p}{q}\right|\left(\pi + \dfrac{p}{q}\right) \leq \left|\pi - \dfrac{p}{q}\right| \cdot \dfrac{5\pi}{2}$ and the result.

Remark: This proves that π *is not a Liouville number*. The irrationality measure we have obtained here ($\mu(\pi) \geq 71.54$) is not very good. By using a different method, one can obtain $\mu(\pi) \geq 8.0161$ (Hata 1993).

Exercise 9.19   Let $\alpha_1 = \alpha$, $\alpha_2, \ldots,$ α_n be the roots of P_α in $\mathbb{C}$. then P_α is the minimal polynomial of α_i for every i (theorem 9.1). If one of the α_i's was a repeated root of P_α, we would have $P'_\alpha(\alpha_i) = 0$, contradiction with the definition of the minimal polynomial.

Exercise 9.20   We have $r_n = 3 \prod\limits_{k=0}^{n-1} \left(10^{4^k} + 1\right)$, $s_n = 10^{(4^n - 1)/3}$, $\left|s_n \alpha - r_n\right| < 1/2 s_n^2$.

An easy computation shows that $s_n r_{n+1} - s_{n+1} r_n = s_n r_n \neq 0$.

Theorem 9.7 applies with $g(n) = s_n^2$, $k = 1$, $a = 2$, $\ell = 1/2$, $b = 100$, $h = 4$.

We obtain $\left|\alpha - \dfrac{p}{q}\right| \geq \dfrac{C}{q^{33}}$ for every $\dfrac{p}{q} \in \mathbb{Q}$ with $q > 0$, and $C = 0.5 . 10^{-20}$.

Exercise 9.21   The partial sums of the series converge fast and allow to obtain an irrationality measure. Indeed, for every $n \in \mathbb{N}$ we can write

$$\left|q^{2^n} \beta - \sum_{k=0}^{n} q^{2^n - 2^k}\right| \leq |q|^{2^n} \sum_{k=n+1}^{+\infty} |q|^{-2^k} \leq |q|^{2^n} \sum_{k=2^{n+1}}^{+\infty} |q|^{-k} \leq \frac{|q|}{|q|-1} \frac{1}{|q|^{2^n}} \leq \frac{|q|}{|q|-1} \frac{|q|}{|q|^{2^n}}.$$

Put $q_n = q^{2^n}$ and $p_n = \sum_{k=0}^{n} q^{2^n - 2^k}$. Then $q_n p_{n+1} - q_{n+1} p_n = q_n \neq 0$. Theorem 9.7 applies with $g(n) = |q|^{2^n} / |q|$, $k = |q|$, $a = 1$, $\ell = |q|/(|q|-1)$, $b = |q|$, $h = 2$.

We obtain $\left|\beta - \dfrac{r}{s}\right| \geq \dfrac{C}{s^5}$ for every $\dfrac{r}{s} \in \mathbb{Q}$ with $s > 0$, and $C = \dfrac{1}{2}\left(\dfrac{|q|-1}{2|q|^2}\right)^4$.

Exercise 9.22   The same computation as above shows that

$$\left|q^{r^n} \gamma - \sum_{k=0}^{n} q^{r^n - r^k}\right| \leq \frac{|q|}{|q|-1} \frac{|q|^{r-1}}{|q|^{(r-1)r^n}}.$$

If we put $q_n = q^{r^n}$ and $p_n = \sum_{k=0}^{n} q^{r^n - r^k}$, we see that $\left|\gamma - \dfrac{p_n}{q_n}\right| \leq \dfrac{|q|}{|q|-1} \dfrac{1}{q_n^r}$.

Since $r \geq 3$, Roth's theorem shows that γ is transcendental.

Theorem 9.7 applies with $g(n) = |q|^{(r-1)r^n}/|q|^{r-1}$, $k = |q|$, $a = 1/(r-1)$, $\ell = |q|/(|q|-1)$, $b = |q|^{(r-1)^2}$, $h = r$.

We get $\left|\gamma - \dfrac{u}{v}\right| \geq \dfrac{C}{v^{r^2/(r-1)+1}}$ for every $\dfrac{u}{v} \in \mathbb{Q}$, $v > 0$, and $C = \dfrac{1}{2}\left(\dfrac{|q|-1}{2|q|^r}\right)^{r^2/(r-1)}$.

Solutions to the exercises of chapter 10

Exercise 10.1 $\mathbb{K}$ is a subfield of $\mathbb{C}$. Hence $(\mathbb{K}, +)$ is a commutative field. Since $\mathbb{Q} \subset \mathbb{K}$, we have $\lambda x \in \mathbb{K}$ for every $\lambda \in \mathbb{Q}$, $x \in \mathbb{K}$, which defines the scalar multiplication. Finally,
$$\lambda(x + y) = \lambda x + \lambda y, \ (\lambda + \mu)x = \lambda x + \mu x, \ (\lambda\mu)x = \lambda(\mu x), \ 1 \cdot x = x$$
for every $\lambda, \mu \in \mathbb{Q}$, $x, y \in \mathbb{K}$. Therefore $\mathbb{K}$ is a vector space over $\mathbb{Q}$.

Exercise 10.2 $\mathbb{K}$ is the $\mathbb{Q}$-vector space spanned by $(1, \alpha, ..., \alpha^{d-1})$. These vectors are linearly independent since $a_0 + \cdots + a_{d-1}\alpha^{d-1} = 0$ ($a_i \in \mathbb{Q}$) implies $P(\alpha) = 0$, with $P(x) = a_0 + \cdots + a_{d-1}x^{d-1}$. Therefore $P = 0$, because the minimal polynomial of α has degree d. Hence $a_0 = a_1 = \cdots = a_{d-1} = 0$. This proves that $(1, \alpha, ..., \alpha^{d-1})$ is a basis of the $\mathbb{Q}$-vector space $\mathbb{K}$, and $\dim K = d$. To show that $\mathbb{K}$ is a subfield of $\mathbb{C}$, it suffices to show that $(\mathbb{K}^*, \cdot)$ is a subfield of $(\mathbb{C}^*, \cdot)$. First we observe that $\alpha^n \in \mathbb{K}$ for every $n \in \mathbb{N}$. Indeed, if we denote by $P_\alpha(x) = x^d + a_{d-1}x^{d-1} + \cdots + a_0$ the minimum polynomial of α, we have $\alpha^d = -a_{d-1}\alpha^{d-1} - \cdots - a_0$, whence $\alpha^d \in \mathbb{K}$, and the result follows from an easy induction. This yields immediately that $xy \in \mathbb{K}$ for every $(x, y) \in \mathbb{K}^2$. It remains to prove that $x \in \mathbb{K}^* \Rightarrow 1/x \in \mathbb{K}$. For a given $x \in \mathbb{K}^*$ put $\varphi(y) = xy$ for every $y \in \mathbb{K}$. Then φ is an endomorphism of the $\mathbb{Q}$-vector space $\mathbb{K}$ and clearly $\mathrm{Ker}\,\varphi = \{0\}$. Since $\mathbb{K}$ is finite-dimensional, φ est bijective. Therefore there exists $y \in \mathbb{K}$ such that $xy = 1$, Q.E.D.

Exercise 10.3 Let $\omega = e^{2i\pi/5}$. We know that $1 + \omega + \omega^2 + \omega^3 + \omega^4 = 0$. Hence $1 + 2\cos(2\pi/5) + 2\cos(4\pi/5) = 0$. But $\cos(4\pi/5) = 2\cos^2(4\pi/5) - 1$, and therefore $4\cos^2(2\pi/5) + 2\cos(2\pi/5) - 1 = 0$. This yields $\cos(2\pi/5) = \left(-1 + \sqrt{5}\right)/4$.

Exercise 10.4   Since α is algebraic (meaning: over $\mathbb{Q}$), it is algebraic over $\mathbb{K}$. Let P be its *minimal polynomial* over $\mathbb{K}$. Observe that P is not necessarily equal to P_α: for example, if $\alpha = \sqrt{2}$, $P_\alpha(x) = x^2 - 2$, but the minimal polynomial of α over $\mathbb{K} = \mathbb{Q}(\sqrt{2})$ is $P(x) = x - \sqrt{2}$. Now, it is easy to see that $\mathbb{L} = \mathbb{K}(\alpha)$ is a number field by arguing the same way as in exercise 10.2.

Exercise 10.5   It is clear that $\sigma_i(x+y) = \sigma_i(x) + \sigma_i(y)$, for any $x, y \in \mathbb{K}$, and that $\sigma_i(r) = r$ if $r \in \mathbb{Q}$. It remains to prove that $\sigma_i(xy) = \sigma_i(x)\sigma_i(y)$ for every $x, y \in \mathbb{K}$. To do this, we have to notice the important following point:
If $x = P(\theta)$ and $y = Q(\theta)$, then $xy = R(\theta)$, where $R(X)$ is the remainder of the division of $P(X)Q(X)$ by $P_\theta(X)$, minimal polynomial of θ.
Indeed we have $P(X)Q(X) = P_\theta(X)H(X) + R(X)$, with $\deg R \le d - 1$.
We also observe that $\sigma_i(x) = \sigma_i(P(\theta)) = P(\theta_i)$ by definition. Therefore $\sigma_i(xy) = \sigma_i(R(\theta)) = R(\theta_i) = P_\theta(\theta_i)H(\theta_i) + R(\theta_i)$ since θ_i is a zero of P_θ. Thus $\sigma_i(xy) = P(\theta_i)Q(\theta_i) = \sigma_i(x)\sigma_i(y)$. Finally, $\sigma_i \ne \sigma_j$ if $i \ne j$ because the roots of P_θ are distinct from each other by exercise 9.19.

Exercise 10.6   1) We have
$$T(x+y) = \sum_{i=1}^d \sigma_i(x+y) = \sum_{i=1}^d \sigma_i(x) + \sum_{i=1}^d \sigma_i(y) = T(x) + T(y).$$
2) $N(xy) = \sigma_1(xy)...\sigma_d(xy) = \sigma_1(x)\sigma_1(y)...\sigma_d(x)\sigma_d(y) = N(x)N(y)$.
3) $N(x) = 0$ if and only if there exists $i \in \{1,2,...,d\}$ such that $\sigma_i(x) = 0$. Put $x = a_0 + a_1\theta + \cdots + a_{d-1}\theta^{d-1} = P(\theta)$. Then we have $P(\theta_i) = 0$. But P_θ is the minimal polynomial of θ_i as well as of θ (theorem 9.1), and θ_i is algebraic of degree d. Hence $\sigma_i(x) = 0 \Leftrightarrow P = 0 \Leftrightarrow x = 0$ (in other words, σ_i is *injective*).

Exercise 10.7   Let $x \in \mathbb{A}_\mathbb{K}$. Then $P_x \in \mathbb{Z}(x)$, and therefore there exist $a_0, a_1, ..., a_{k-1} \in \mathbb{Z}$ such that $a_0 + a_1 x + \cdots + a_{k-1}x^{k-1} + x^k = 0$. Consequently we have $\sigma_i(a_0 + a_1 x + \cdots + x^k) = \sigma_i(0) = 0$, whence $a_0 + a_1\sigma_i(x) + \cdots + \sigma_i(x)^k = 0$. Thus $\sigma_i(x)$ is an algebraic integer for every monomorphism σ_i of $\mathbb{K}$. Hence, by theorem 9.4, $\sigma_2(x)...\sigma_d(x)$ is an algebraic integer. Moreover $x\sigma_2(x)...\sigma_d(x) = N(x)$, which yields $\sigma_2(x)...\sigma_d(x) = N(x)/x \in \mathbb{K}$. Therefore $\sigma_2(x)...\sigma_d(x)$ is an algebraic integer of $\mathbb{K}$, that is an element of $\mathbb{A}_\mathbb{K}$.

Exercise 10.8   We have $(N(\varepsilon) = \pm 1) \Rightarrow (\varepsilon \sigma_2(\varepsilon)...\sigma_d(\varepsilon) = \pm 1)$
$\Rightarrow (\varepsilon \times (\pm \sigma_2(\varepsilon)...\sigma_d(\varepsilon)) = 1) \Rightarrow \varepsilon$ invertible in $\mathbb{A}_\mathbb{K}$ by lemma 10.1.
Conversely, $\varepsilon \varepsilon' = 1$ with $\varepsilon, \varepsilon' \in \mathbb{A}_\mathbb{K}$ yields $N(\varepsilon)N(\varepsilon') = 1$, whence $N(\varepsilon) = 1$
since $N(\varepsilon) \in \mathbb{Z}$ and $N(\varepsilon') \in \mathbb{Z}$.

Exercise 10.9   We have $P(x)P(x^{p-1}) = \sum_{i=0}^{p-1} \sum_{j=0}^{p-1} b_i b_j x^{pj+i-j}$. Let $r_{i,j}$ be the
remainder of the division of $pj + i - j$ by p. Then $i - j \equiv r_{i,j} \pmod{p}$, and

$$P(x)P(x^{p-1}) = \sum_{i=0}^{p-1} \sum_{j=0}^{p-1} b_i b_j x^{pq_{i,j} + r_{i,j}} = \sum_{i=0}^{p-1} \sum_{j=0}^{p-1} b_i b_j x^{r_{i,j}} (x^{pq_{i,j}} - 1) + \sum_{i=0}^{p-1} \sum_{j=0}^{p-1} b_i b_j x^{r_{i,j}}$$

$$= (x^p - 1)Q(x) + \sum_{r=0}^{p-1} \left(\sum_{(i,j) \in E_r} b_i b_j \right) x^r, \quad \text{Q.E.D.}$$

Exercise 10.10   1) We have $1 + \omega + \omega^2 + \omega^3 + \omega^4 + \omega^5 + \omega^6 = 0$, and
$$\eta = \omega + \omega^{-1} = \omega + \omega^6, \quad \eta^2 = \omega^2 + \omega^{-2} + 2 = \omega^2 + \omega^5 + 2,$$
$$\eta^3 = \omega^3 + 3\omega + 3\omega^{-1} + \omega^{-3} = \omega^3 + \omega^4 + 3\eta.$$
Consequently $1 + \eta + (\eta^2 - 2) + (\eta^3 - 3\eta) = 0$, whence $\eta^3 + \eta^2 - 2\eta - 1 = 0$.
Now the polynomial $P(x) = x^3 + x^2 - 2x - 1$ has no zero in $\mathbb{Q}$ by exercise 1.6.
Hence it is irreducible in $\mathbb{Q}[x]$ and it is the minimal polynomial of η. Let σ be
the monomorphism of $\mathbb{K}$ defined by $\sigma(\omega) = \omega^2$. Then $\nu = \sigma(\eta)$, $\mu = \sigma^2(\eta)$,
and $\eta^3 + \eta^2 - 2\eta - 1 = 0 \Rightarrow \nu^3 + \nu^2 - 2\nu - 1 = 0$ and $\mu^3 + \mu^2 - 2\mu - 1 = 0$. Thus
η, μ and ν are the three roots of P. Consequently $\eta \mu \nu = 1$, which proves that
η, μ and ν are units of $\mathbb{A}_\mathbb{K}$. Finally $\eta = 2\cos(2\pi/7)$, $\mu = 2\cos(6\pi/7)$, and
$\nu = 2\cos(4\pi/7)$. They are real units.
2) Assume that $\eta^r \mu^s = \pm 1$, with $r, s \in \mathbb{Z}$. By using the monomorphism σ
defined in question 1, we obtain $\sigma(\eta)^r \sigma(\mu)^s = \pm 1$, whence $\nu^r \eta^s = \pm 1$. Taking
the logarithms yields the homogeneous system
$$r \operatorname{Log}|\eta| + s \operatorname{Log}|\mu| = 0, \quad r \operatorname{Log}|\nu| + s \operatorname{Log}|\eta| = 0.$$
Its determinant is
$$\Delta = \left(\operatorname{Log}|2\cos(2\pi/7)| \right)^2 - \operatorname{Log}|2\cos(4\pi/7)| \operatorname{Log}|2\cos(6\pi/7)| = 0.525...,$$
whence $\Delta \neq 0$. Therefore $r = s = 0$.
Now, if $\varepsilon = \pm \eta^r \mu^s = \pm \eta^{r'} \mu^{s'}$, then $\eta^{r-r'} \mu^{s-s'} = \pm 1$, whence $r = r'$ and $s = s'$.
Therefore all units of the form $\pm \eta^r \mu^s$ are different.

3) We use the basis $\left(\omega, \omega^2, ..., \omega^6\right)$. Let $\varepsilon = E(\omega) = \sum_{i=1}^{6} a_i \omega^i$ $(a_i \in \mathbb{Z})$ be a real unit of $\mathbb{A}_K$. Then $\bar{\varepsilon} = E(\bar{\omega}) = \varepsilon = \sum_{i=1}^{6} a_i \bar{\omega}^i = \sum_{i=1}^{6} a_i \omega^{-i}$.

Hence $a_i = a_{7-i}$ for every i, $2\varepsilon = \sum_{i=1}^{6} a_i (\omega^i + \omega^{-i})$ and $\varepsilon = a_1 \eta + a_2 v + a_3 \mu$.

Drawing inspiration from question 2, we consider the system

$$r\log|\eta| + s\log|\mu| = \log|E(\omega)|, \quad r\log|v| + s\log|\eta| = \log|E(\omega^2)|. \quad (*)$$

This system has a unique solution $(r,s) \in \mathbb{R}^2$ since $\Delta = 0.525... \neq 0$. Then we have $|\eta|^r |\mu|^s = |E(\omega)|$, whence $E(\omega) = \alpha\eta^r \mu^s$, where α is a complex number and $|\alpha| = 1$. Since α is real, $\alpha = \pm1$, Q.E.D.

4) a) The system $(*)$ can be written $|E(\omega)| = |\eta|^r |\mu|^s$, $|E(\omega^2)| = |v|^r |\eta|^s$. Since $|\eta| = \eta = 1.24697...$, $|\mu| = -\mu = 1.80193...$, $|v| = -v = 0.44504...$, we have

$$|\eta|^{-1/2} |\mu|^{-1/2} \leq |E(\omega)| \leq |\eta|^{1/2} |\mu|^{1/2}, \text{ whence } 0,667 \leq |E(\omega)| \leq 1,499,$$

$$|v|^{1/2} |\eta|^{-1/2} \leq |E(\omega^2)| \leq |v|^{-1/2} |\eta|^{1/2}, \text{ whence } 0.597 \leq |E(\omega^2)| \leq 1.674.$$

Finally, $E(\omega)E(\omega^2)E(\omega^3) = \pm\sqrt{N(E(\omega))} = \pm1$ yields $0.398 \leq |E(\omega^3)| \leq 2.512$.

b) We have $(\eta - 2)E(\omega) + (\mu - 2)E(\omega^3) + (v - 2)E(\omega^2)$

$= (\eta - 2)(a\eta + b\mu + cv) + (\mu - 2)(a\mu + bv + c\eta) + (v - 2)(av + b\eta + c\mu)$

$= a(\eta^2 + \mu^2 + v^2) + (b + c)(\eta\mu + \eta v + \mu v) - 2(a + b + c)(\mu + v + \eta).$

But we know by question 1 that $\mu + v + \eta = -1$, $\eta\mu + \eta v + \mu v = -2$,

$\eta^2 + \mu^2 + v^2 = (\eta + \mu + v)^2 - 2(\eta\mu + \eta v + \mu v) = 5$.

This proves the first equality. The others can be obtained similarly. We deduce from the first one $|7a| \leq |\eta - 2||E(\omega)| + |\mu - 2||E(\omega^3)| + |v - 2||E(\omega^2)| \leq 14.78$.

From the second one $|7b| \leq 13.11$. From the third one $|7c| \leq 11.93$. Since $a, b, c \in \mathbb{Z}$, we have $|a| \leq 2$, $|b| \leq 1$, $|c| \leq 1$.

c) We have $N(E(\omega)) = 1 = [E(\omega)E(\omega^2)E(\omega^3)]^2$

$= [(a\eta + b\mu + cv)(av + b\eta + c\mu)(a\mu + bv + c\eta)]^2$

$= [(abc)(\eta^3 + \mu^3 + v^3) + (a^2b + ac^2 + b^2c)(\eta^2\mu + \mu^2v + v^2\eta)$

$\quad + (ab^2 + a^2c + bc^2)(\eta\mu^2 + \mu v^2 + v\eta^2) + (a^3 + b^3 + c^3 + 3abc)\eta\mu v]^2.$

But we have seen in question 1 that $\eta^2 = v + 2$. Applying σ yields $v^2 = \mu + 2$ and $\mu^2 = \eta + 2$. Hence $\eta^2\mu + \mu^2v + v^2\eta = 2(\mu + v + \eta) + \eta\mu + \eta v + \mu v = -4$.

Similarly $\eta\mu^2 + v^2\mu + v\eta^2 = \eta^2 + \mu^2 + v^2 + 2(\eta + \mu + v) = 5 - 2 = 3$.

Finally $\eta^3 + \mu^3 + v^3 = -(\eta^2 + \mu^2 + v^2) + 2(\eta + \mu + v) + 3 = -5 - 2 + 3 = -4$.

Therefore $1 = [-abc - 4(a^2b + ac^2 + b^2c) + 3(ab^2 + a^2c + bc^2) + (a^3 + b^3 + c^3)]^2$, whence $[(a+b+c)^3 - 7(a^2b + b^2c + c^2a + abc)]^2 = 1$.

d) Using an elementary program on a computer shows that the only (a,b,c), satisfying $|a| \le 2$, $|b| \le 1$, $|c| \le 1$ and the above equation are $(-2,-1,-1)$, $(-2,-1,0)$, $(-1,-1,-1)$, $(-1,-1,0)$, $(-1,0,-1)$, $(-1,0,0)$, $(-1,1,-1)$, $(-1,1,1)$, $(0,-1,-1)$, $(0,-1,0)$, $(0,0,-1)$, $(0,0,1)$, $(0,1,0)$, $(0,1,1)$, $(1,-1,-1)$, $(1,-1,1)$, $(1,0,0)$, $(1,0,1)$, $(1,1,-1)$, $(1,1,0)$, $(1,1,1)$, $(2,1,0)$, $(2,1,1)$.
The numerical values to 2 decimal places of the corresponding units $E(\omega) = a\eta + b\mu + c\nu$ are -0.25, -0.69, 1, 0.55, 0.11, -0.80, -1.24, -2.60, -3.49, 2.24, 1.80, 0.45, -0.45, -1.80, -2.24, 3.49, 2.60, 1.24, 0.80, -0.11, -0.55, -1, -0.69, 0.25. Comparing to the bounds for $|E(\omega)|$ obtained in question 4) a), we see that $E(\omega) = -2\eta - \mu$, $-\eta - \mu - \nu = 1$, $-\eta - \nu$, $-\eta$, η, $\eta + \nu$, $\eta + \mu + \nu = -1$, $2\eta + \mu$. The corresponding possible values of $E(\omega^2)$ are, up to two decimal places, $E(\omega^2) = -2\nu - \eta = -0.35$, $-\eta - \mu - \nu = 1$, $-\nu - \mu = 2.24$, $-\nu = 0.45$, $\nu = -0.45$, $\nu + \mu = -2.24$, $\eta + \mu + \nu = -1$, $2\nu + \eta = 0.35$. Using the bounds for $|E(\omega^2)|$ yields $E(\omega) = 1$ or -1, and consequently $r = s = 0$.

5) Let $E(\omega)$ be a real unit of $\mathbb{A}_K$. Then $E(\omega) = \pm \eta^r \mu^s$, with $(r,s) \in \mathbb{R}^2$ by question 3). Let m and n be integers such that $|r - m| \le 1/2$, $|s - n| \le 1/2$. Then $|\eta|^m |\mu|^n$ is a unit, and so is $E(\omega)|\eta|^{-m} |\mu|^{-n} = \pm |\eta|^{r-m} |\mu|^{s-n}$. Using question 4) yields $r - m = s - n = 0$ and $E(\omega) = \pm |\eta|^m |\mu|^n = \pm(-1)^n \eta^m \mu^n = \pm \eta^m \mu^n$. Thus we conclude from Kummer's lemma 10.2 that the units of $\mathbb{A}_K$ are $\pm \omega^k \eta^m \mu^n$.

Exercise 10.11 Δ is a polynomial in the indeterminates $\theta_1, \theta_2, \dots, \theta_j$ with real coefficients, and it is zero as soon as $\theta_i = \theta_j$ $(i \ne j)$. Therefore

$$\Delta = \prod_{1 \le j < i \le d} (\theta_i - \theta_j) Q(\theta_1, \theta_2, \dots, \theta_d).$$

Taking the degrees into account yields $Q(\theta_1, \theta_2, \dots, \theta_d) = C$. Comparing the leading coefficients in the indeterminate θ_1 yields $C = 1$.

Exercise 10.12 With the usual notations for squares matrices, we have

$$(\sigma_i(\alpha_j)) = \left(\sigma_i \left(\sum_{k=1}^d c_{kj} \beta_k \right) \right) = \left(\sum_{k=1}^d c_{kj} \sigma_i(\beta_k) \right) = (\sigma_i(\beta_j)) \times (c_{ij}).$$

This yields the desired result by taking the determinants.

Exercise 10.13 We have $\Delta[1,...,\theta^{d-1}] \in \mathbb{Q}$ by lemma 10.3. Therefore $\Delta[\alpha_1,...,\alpha_d] \in \mathbb{Q}$ for every basis $(\alpha_1, \alpha_2,...,\alpha_d)$ of $\mathbb{K}$ by lemma 10.4. Moreover, if all the α_i's are integers, $\Delta[\alpha_1, \alpha_2,...,\alpha_d]$ is an algebraic integer since it can be written as sums of products of the $\sigma_i(\alpha_j)$'s, which are algebraic integers. Hence $\Delta[\alpha_1, \alpha_2,...,\alpha_d]$ is an integer of $\mathbb{Q}$, that is $\Delta[\alpha_1, \alpha_2,...,\alpha_d] \in \mathbb{Z}$.

Exercise 10.14 1) Let $P(x) = a_0 + a_1 x + a_2 x^2 + a_3 x^3$. Then
$$N(\alpha) = P(\omega)P(\overline{\omega})P(\omega^2)P(\overline{\omega}^2),$$
since the four monomorphisms of $\mathbb{K}$ are defined by $\sigma_1(\omega) = \omega$, $\sigma_2(\omega) = \omega^2$, $\sigma_3(\omega) = \omega^3 = \overline{\omega}^2$, $\sigma_4(\omega) = \omega^4 = \overline{\omega}$. But
$$P(\omega)P(\overline{\omega}) = (a_0^2 + a_1^2 + a_2^2 + a_3^2) + (a_0 a_1 + a_1 a_2 + a_2 a_3)(\omega + \overline{\omega})$$
$$+ (a_0 a_2 + a_1 a_3 + a_0 a_3)(\omega^2 + \overline{\omega}^2).$$
Now $\omega + \overline{\omega} = 2\cos(2\pi/5) = (-1+\sqrt{5})/2$, $\omega^2 + \overline{\omega}^2 = 2\cos(4\pi/5) = (-1-\sqrt{5})/2$ (exercise 10.3), whence $P(\omega)P(\overline{\omega}) = \sum_{i=0}^3 a_i^2 - \frac{1}{2}\sum_{0 \le i < j \le 3} a_i a_j + B\sqrt{5}$, $B \in \mathbb{Q}$.

But $\sum_{0 \le i < j \le 3} a_i a_j = \frac{1}{2}\left[\left(\sum_{i=0}^3 a_i\right)^2 - \sum_{i=0}^3 a_i^2\right]$ yields $P(\omega)P(\overline{\omega}) = A + B\sqrt{5}$, with $A = \frac{5}{4}\sum_{i=0}^3 a_i^2 - \frac{1}{4}\left(\sum_{i=0}^3 a_i\right)^2$. Replacing ω by ω^2 reverses the roles of $\cos(2\pi/5)$ and $\cos(4\pi/5)$. Thus $P(\omega^2)P(\overline{\omega}^2) = A - B\sqrt{5}$ and $N(\alpha) = A^2 - 5B^2$.

2) We have $b_i \in [0,1[$ for $i = 0,1,2,3,4$. Divide $[0,1]$ in five parts of same length. We see by the pigeon-hole principle that at least one of the $b_i - b_j$ satisfies $|b_i - b_j| \le 1/5$ ($i \ne j$). However, we have
$$\alpha\omega = -a_3 + (a_0 - a_3)\omega + (a_1 - a_3)\omega^2 + (a_2 - a_3)\omega^3$$
$$\alpha\omega^2 = (a_3 - a_2) - a_2\omega + (a_0 - a_2)\omega^2 + (a_1 - a_2)\omega^3$$
$$\alpha\omega^3 = (a_2 - a_1) + (a_3 - a_1)\omega - a_1\omega^2 + (a_0 - a_1)\omega^3$$
$$\alpha\omega^4 = (a_1 - a_0) + (a_2 - a_0)\omega + (a_3 - a_0)\omega^2 - a_0\omega^3.$$
Therefore there exists $q \in \{1,2,3,4\}$ such that $\alpha\omega^q = A_0 + A_1\omega + A_2\omega^2 + A_3\omega^3$ and that, for one of the A_j's, there exists $B_j \in \mathbb{Z}$ satisfying $|A_j - B_j| \le 1/5$. For the three remaining A_i's, there exists an integer B_i such that $|A_i - B_i| \le 1/2$. Thus there exist $q \in \{1,2,3,4\}$ and $\beta = \sum_{i=0}^3 B_i\omega^i$ such that $\alpha\omega^q - \beta = \sum_{i=0}^3 b_i\omega^i$, with $\sum_{i=0}^3 b_i^2 \le \frac{1}{25} + 3\frac{1}{4} = 0.79$, $\left|\sum_{i=0}^3 b_i\right| \le \frac{1}{5} + \frac{3}{2} = 1.7$.

Now $\left|N(\alpha\omega^q - \beta)\right| = N(\alpha\omega^q - \beta) \leq A^2$ by question 1), with $-\frac{1}{4}\left(\sum_{i=0}^{3} b_i\right)^2 \leq A$

$\leq \frac{5}{4}\sum_{i=0}^{3} b_i^2$, whence $-0.7225 \leq A \leq 0.9875$. Thus $|A| < 1$ and $\left|N(\alpha\omega^q - \beta)\right| < 1$.

3) We deduce from above that $\left|N(\alpha - \beta\omega^{-q})\right| = \left|N(\omega^{-q})\right|\left|N(\alpha\omega^q - \beta)\right| < 1$ since $N(\omega^{-q}) = 1$. Therefore, for every $\alpha \in \mathbb{K}$, there exists $\gamma \in \mathbb{A}_\mathbb{K}$ such that $\left|N(\alpha - \gamma)\right| < 1$. Proceeding as in the proof of theorem 5.16, we see that for every $(a,b) \in \mathbb{A}_\mathbb{K} \times \mathbb{A}_\mathbb{K}^*$, there exists $\gamma \in \mathbb{A}_\mathbb{K}$ such that $\left|N(a/b - \gamma)\right| < 1$. If we put $r = a - b\gamma$, we have $\left|N(r)\right| = \left|N(b)\right|\left|N(a/b - \gamma)\right| < \left|N(b)\right|$. Thus $\mathbb{A}_\mathbb{K}$ is euclidean.

Exercise 10.15 Let $\varepsilon \in \mathbb{A}_\mathbb{K}^\times$. We know by example 10.9 that $\varepsilon = \pm\omega^n\Phi^k$. Suppose that $\varepsilon = \pm\omega^n\Phi^k \equiv a$ (mod 5), with $a \in \mathbb{Z}$. Then $\bar{\varepsilon} = \pm\bar{\omega}^n\Phi^k \equiv a$ (mod 5) since $\alpha \in \mathbb{A}_\mathbb{K} \Rightarrow \bar{\alpha} \in \mathbb{A}_\mathbb{K}$. Hence $\varepsilon\bar{\varepsilon} \equiv a^2$ (mod 5), that is $\Phi^{2k} \equiv a^2$ (mod 5). Since $\Phi = (1 + \sqrt{5})/2$, there exists $c \in \mathbb{Z}$ such that $(1 + \sqrt{5})^{2k} = c$ (mod 5), whence $1 + 2k\sqrt{5} \equiv c$ (mod 5). Taking the conjugates in the ring of integers of $\mathbb{Q}(\sqrt{5})$ yields $1 - 2k\sqrt{5} \equiv c$ (mod 5), whence $4k\sqrt{5} \equiv 0$ (mod 5), that is $4k\sqrt{5} = 5q$, $q \in \mathbb{Z}(\Phi)$. Taking the norms in $\mathbb{Z}(\Phi)$ yields $16k^2 \cdot 5 = 25N(q)$, whence $k = 5h$, $h \in \mathbb{Z}$. Therefore we have $\varepsilon = \pm\omega^n\Phi^{5h}$. Now, noticing that $\Phi = 1 + (-1 + \sqrt{5})/2$ and that $(-1 + \sqrt{5})/2$ is an integer of $\mathbb{K}$, we see that $\Phi^5 \equiv 1$ (mod 5), whence $\pm\omega^n \equiv a$ (mod 5). Thus $\pm\bar{\omega}^n \equiv a$ (mod 5) and $\cos(2n\pi/5) = \pm 2a$ (mod 5), $a \in \mathbb{Z}$. If n is not a multiple of 5, the only possible values for $\cos(2n\pi/5)$ are $(-1 + \sqrt{5})/4$ and $(-1 - \sqrt{5})/4$. Arguing as before, we see that the congruences $-1 + \sqrt{5} \equiv c$ (mod 5) and $1 + \sqrt{5} \equiv c$ (mod 5) have no solutions if $c \in \mathbb{Z}$. Therefore $n = 5m$, $\varepsilon = \pm\omega^{5m}\Phi^{5h} = (\pm\omega^m\Phi^h)^5$, Q.E.D.

Exercise 10.16 Let $d \in \mathbb{A}_\mathbb{K}$ be a prime dividing $x + \omega^i y$ and $x + \omega^j y$, with $i > j$. Then d divides their difference $y(\omega^i - \omega^j)$. Since d is prime in the euclidean domain $\mathbb{A}_\mathbb{K}$, $d \mid y$ or $d \mid \omega^i - \omega^j$. But d cannot divide y since x and y are coprime. Therefore $d \mid \omega^j(\omega^{i-j} - 1) = \omega^j(\omega - 1)(1 + \omega + \cdots + \omega^{i-j-1})$. Now the possible values of $1 + \omega + \cdots + \omega^{i-j-1}$ are 1, $1 + \omega$, $1 + \omega + \omega^2 = -\omega^3 - \omega^4 = -\omega^3(1 + \omega)$ and $1 + \omega + \omega^2 + \omega^3 = -\omega^4$. But $1 + \omega \in \mathbb{A}_\mathbb{K}^\times$ since $(1 + \omega)(1 + \omega^2 + \omega^4) = 1$. Hence $\omega^j(\omega^{i-j} - 1) = \varepsilon\lambda$, $\varepsilon \in \mathbb{A}_\mathbb{K}^\times$. Since λ is prime in

$\mathbb{A}_\mathbb{K}$, we have $d = \eta\lambda$, $\eta \in \mathbb{A}_\mathbb{K}^\times$, which proves that the numbers $(x + \omega^i y)/\lambda$ are pairwise coprime.

Exercise 10.17   1) Let $(\alpha_1, \alpha_2, ..., \alpha_n)$ be a basis of the $\mathbb{K}$ - vector space $\mathbb{L}$, and $(\beta_1, \beta_2, ..., \beta_m)$ a basis of the $\mathbb{L}$ - vector space $\mathbb{M}$. Then $[\mathbb{M}:\mathbb{L}] = m$ and $[\mathbb{L}:\mathbb{K}] = n$. We prove that $F = (\alpha_1\beta_1, ..., \alpha_1\beta_m, \alpha_2\beta_1, ..., \alpha_2\beta_m, ..., \alpha_n\beta_1, ..., \alpha_n\beta_m)$ is a basis of the $\mathbb{K}$ - vector space $\mathbb{M}$. First, we see that F spans $\mathbb{M}$ since $\forall x \in \mathbb{M}$, $x = \sum_{i=1}^m x_i\beta_i$ $(x_i \in \mathbb{L})$, whence $x = \sum_{i=1}^m \sum_{j=1}^n y_{ij}\alpha_j\beta_i$ $(y_{ij} \in \mathbb{K})$. And the vectors of F are linearly independent over $\mathbb{K}$ since $\sum_{i=1}^m \sum_{j=1}^n y_{ij}\alpha_j\beta_i = 0$

$(y_{ij} \in \mathbb{K}) \Rightarrow \sum_{i=1}^m \left(\sum_{j=1}^n y_{ij}\alpha_j\right)\beta_i = 0 \Rightarrow \sum_{j=1}^n y_{ij}\alpha_j = 0$ since $(\beta_1, \beta_2, ..., \beta_m)$ is a basis of $\mathbb{M}$ over $\mathbb{L} \Rightarrow y_{ij} = 0$ since $(\alpha_1, ..., \alpha_n)$ is a basis of $\mathbb{L}$ over $\mathbb{K}$. Therefore $[\mathbb{M}:\mathbb{K}] = mn = [\mathbb{M}:\mathbb{L}] \times [\mathbb{L}:\mathbb{K}]$.

2) Let $\alpha \in \mathbb{M}$. Then α is algebraic of degree ≤ 3 since $1, \alpha, \alpha^2, \alpha^3$ are linearly independent over $\mathbb{Q}$ $([\mathbb{M}:\mathbb{Q}] = 3)$. If α was algebraic of degree 2, $\mathbb{L} = \mathbb{Q}(\alpha)$ would be of degree 2. Then we would have $[\mathbb{M}:\mathbb{Q}] = 3 = [\mathbb{M}:\mathbb{L}] \times [\mathbb{L}:\mathbb{Q}]$, whence $2|3$, contradiction. Thus α is algebraic of degree 1 (rational) or algebraic of exact degree 3.

Exercise 10.18   Let $P \in \mathbb{Q}[x]$. Then P is equivalent, modulo the ideal $(P_\theta(x))$, to the remainder of the division of P by P_θ, since $P(x) = Q(x)P_\theta(x) + R(x) \Rightarrow P - R \in (P_\theta(x))$ (see section 6.2). Let $\alpha \in \mathbb{K}$. Then there exists $E \in \mathbb{Q}[x]$ such that $\alpha = E(\theta)$. Let $\varphi : \mathbb{K} \longrightarrow \mathbb{Q}[x]/(P_\theta(x))$, which maps any $\alpha \in \mathbb{K}$ to the class of its minimal polynomial E modulo $(P_\theta(x))$. It is clear that $\varphi(\alpha + \beta) = \varphi(\alpha) + \varphi(\beta)$. It remains to prove that $\varphi(\alpha\beta) = \varphi(\alpha)\varphi(\beta)$. Put $\alpha = E(\theta)$, $\beta = F(\theta)$, and let R be the remainder of the division of $E(x)F(x)$ by P_θ. We have $\alpha\beta = R(\theta)$. Therefore $\varphi(\alpha\beta)$ is the class of R modulo $(P_\theta(x))$. As $E(x)F(x)$ is équivalent to $R(x)$ modulo $(P_\theta(x))$, we have $\varphi(\alpha\beta) = \varphi(\alpha)\varphi(\beta)$. Finally, φ is clearly surjective, and it is injective since $\alpha \in \mathrm{Ker}\,\varphi \Rightarrow E(x) = 0$. Therefore $\mathbb{K} \cong \mathbb{Q}[x]/(P_\theta(x))$.

Exercise 10.19   1) If $x \in \mathbb{A}_\mathbb{K}$, then $\theta x \in \mathbb{A}_\mathbb{K}$ and $\theta^2 x \in \mathbb{A}_\mathbb{K}$. Therefore $T(x), T(\theta x), T(\theta^2 x) \in \mathbb{Z}$. Using (10.8) yields $3a \in \mathbb{Z}$, $6b \in \mathbb{Z}$, $6c \in \mathbb{Z}$. Now $N(x) \in \mathbb{Z}$. Hence, by putting $3a = a'$, $6b = b''$, $6c = c''$ and using (10.7), we see that $b''^3 + 2c''^3 - 6a'b''c''$ is a multiple of 4, whence $b'' = 2b'$. Replacing yields $c'' = 2c'$. Therefore $a' = 3a$, $b' = 3b$, $c' = 3c \in \mathbb{Z}$.

2) Let $x \in \mathbb{A}_\mathbb{K}$. Then x is rational or algebraic of degree 3 by exercise 10.17, question 2). If x is rational, then $x \in \mathbb{Z}$, whence $\sigma_i(x) \in \mathbb{Z}$ for every i. If x has degree 3, let $P(X) = X^3 + \alpha X^2 + \beta X + \gamma$, with α, β, $\gamma \in \mathbb{Z}$ be its minimal polynomial. The roots of $P(X)$ are $\sigma_1(x)$, $\sigma_2(x)$, $\sigma_3(x)$, whence $\sigma_1(x)\sigma_2(x) + \sigma_2(x)\sigma_3(x) + \sigma_1(x)\sigma_3(x) = \beta \in \mathbb{Z}$. But

$$\sigma_1(x)\sigma_2(x) + \sigma_2(x)\sigma_3(x) + \sigma_1(x)\sigma_3(x) = (a + b\theta + c\theta^2)(a + bj\theta + cj^2\theta^2) +$$
$$(a + bj\theta + cj^2\theta^2)(a + bj^2\theta + cj\theta^2) + (a + b\theta + c\theta^2)(a + bj^2\theta + cj\theta^2)$$
$$= ((a^2 - 2bc) + (2c^2 - ab)j^2\theta + (b^2 - ac)j\theta^2) + ((a^2 - 2bc) + (2c^2 - ab)\theta +$$
$$(b^2 - ac)\theta^2) + ((a^2 - 2bc) + (2c^2 - ab)j\theta + (b^2 - ac)j^2\theta^2)$$

(observe that the second term of the sum is obtained by replacing θ by $j\theta$, and the third one by taking the complex conjugate). Hence

$$\sigma_1(x)\sigma_2(x) + \sigma_2(x)\sigma_3(x) + \sigma_1(x)\sigma_3(x) = 3(a^2 - 2bc).$$

Therefore, by using the norm,

$$a'^2 - 2b'c' \in 3\mathbb{Z}, \quad a'^3 + 2b'^3 + 4c'^3 - 6a'b'c' \in 27\mathbb{Z}.$$

Modulo 3, the second condition yields $a'^3 - b'^3 + c'^3 \equiv 0$ (mod 3) $\Leftrightarrow a'^3 \equiv (b' - c')^3$ (mod 3) $\Leftrightarrow a' \equiv b' - c'$ (mod 3). Replacing in the first one yields $b' + c' \equiv 0$ (mod 3). Therefore $c' \equiv -b'$ and $a' \equiv -b'$ (mod 3). Put $a' = 3p + s$, $b' = 3q - s$, $c' = 3r + s$ and replace in the second condition. An easy computation shows that $9s^3 \equiv 0$ (mod 27), whence $s^3 \equiv 0$ (mod 3) and s is a multiple of 3. Thus a', b', c' are multiples of 3, $a, b, c \in \mathbb{Z}$ and $\mathbb{A}_\mathbb{K} = \mathbb{Z}(\theta)$.

3) By using lemma 10.3, we have $\Delta(\mathbb{K}) = \Delta[1, \theta, \theta^2] = -N(3\theta^2)$, since $(1, \theta, \theta^2)$ is an integral basis of $\mathbb{K}$. Therefore

$$\Delta(\mathbb{K}) = -N(3)(N(\theta))^2 = -27 \cdot (\theta j\theta j^2\theta)^2 = -108.$$

Exercise 10.20 Since $(\omega, \omega^2, ..., \omega^6)$ is a basis of $\mathbb{L}$ over $\mathbb{Q}$, we can write $x = \sum_{i=1}^6 a_i\omega^i$ ($a_i \in \mathbb{Q}$) for every $x \in \mathbb{L}$. This yields

$$x = (a_1 + a_6)\cos(2\pi/7) + (a_2 + a_5)\cos(4\pi/7) + (a_3 + a_4)\cos(6\pi/7)$$
$$+ i[(a_1 - a_6)\sin(2\pi/7) + (a_2 - a_5)\sin(4\pi/7) + (a_3 - a_4)\sin(6\pi/7)].$$

But $\eta = 2\cos(2\pi/7)$, whence $\cos(4\pi/7) = \eta^2/2 - 1$, $\cos(6\pi/7) = (\eta^3 - 3\eta)/2$, $\sin(4\pi/7) = \eta\sin(2\pi/7)$, $\sin(6\pi/7) = (\eta^2 - 1)\sin(2\pi/7)$. Thus $\mathbb{L} = \mathbb{K}(i\sin(2\pi/7))$ is an extension of degree 2 of $\mathbb{K}$ (observe that $\cos(4\pi/7) = 1 + 2(i\sin(2\pi/7))^2$

implies that $i\sin(2\pi/7)$ is algebraic of degree 2 over $\mathbb{K}$, or use the relation $[\mathbb{L}:\mathbb{Q}] = [\mathbb{L}:\mathbb{K}]\times[\mathbb{K}:\mathbb{Q}])$.

Now let $\alpha = a + b\eta + c\eta^2$ (a, b, $c \in \mathbb{Q}$) be an integer of $\mathbb{K}$. Then α is an integer of $\mathbb{L}$. But $\alpha = a + b(\omega + \omega^6) + c(\omega^2 + 2 + \omega^5) = a - b + 2c + (c-b)\omega^2 - b\omega^3 - b\omega^4 + (c-b)\omega^5$. This yields $a - b + 2c$, $c - b$, $-b \in \mathbb{Z}$. Therefore a, b, $c \in \mathbb{Z}$ and $\left(1, \eta, \eta^2\right)$ is an integral basis of $\mathbb{K}$.

Now, we know that the minimal polynomial of η is $P(x) = x^3 + x^2 - 2x - 1$ (exercise 10.10, question 1). With the notations of exercise 10.10, we have
$$N(P'(\eta)) = P'(\eta)P'(\nu)P'(\mu) = (3\eta^2 + 2\eta - 2)(3\nu^2 + 2\nu - 2)(3\mu^2 + 2\mu - 2).$$
But we have seen in exercise 10.10, question 4c), that $\eta^2 = \nu + 2$, $\nu^2 = \mu + 2$, $\mu^2 = \eta + 2$. Therefore
$$N\left(P'(\eta)\right) = (3\nu + 2\eta + 4)(3\mu + 2\nu + 4)(3\eta + 2\mu + 4)$$
$$= 35\eta\nu\mu + 12(\nu^2\mu + \mu^2\eta + \eta^2\nu) + 18(\nu\mu^2 + \mu\eta^2 + \eta\nu^2) + 76(\nu\eta + \mu\nu + \eta\mu)$$
$$+ 24(\mu^2 + \nu^2 + \eta^2) + 80(\eta + \mu + \nu) + 64.$$
By using the results of exercise 10.10, question 4c), we obtain
$$N(P'(\eta)) = 35 + 12\cdot3 - 18\cdot4 - 2\cdot76 + 5\cdot24 - 80 + 64 = -49.$$
Using lemma 10.3 finally yields $\Delta(\mathbb{K}) = 49$.

Solutions to the exercises of chapter 11

Exercise 11.1   First let $A = \langle\alpha_1, \alpha_2, ..., \alpha_n\rangle$ be an (integral) ideal of $\mathbb{A}_\mathbb{K}$ (which means that $\alpha_i \in \mathbb{A}_\mathbb{K}$ for every i). Let $x \in A$. Then $x = x_1\alpha_1 + \cdots + x_n\alpha_n$, $x_i \in \mathbb{A}_\mathbb{K}$. Therefore $x - y \in A$ for every $(x, y) \in A^2$ and $xy \in A$ for every $x \in A$ and $y \in \mathbb{A}_\mathbb{K}$. This proves that A is an ideal in the algebraic sense. Conversely, let $A \subset \mathbb{A}_\mathbb{K}$ be an ideal in the algebraic sense, that is satisfying (11.2). By theorem 10.8, any $x \in A$ can be written $x = x_1\alpha_1 + \cdots + x_d\alpha_d$, $x_i \in \mathbb{Z}$, $\alpha_i \in A$, with $d = [\mathbb{K}:\mathbb{Q}]$. Thus $A \subset \langle\alpha_1, \alpha_2, ..., \alpha_d\rangle$. But also $\langle\alpha_1, \alpha_2, ..., \alpha_d\rangle \subset A$ since $\alpha_1, \alpha_2, ..., \alpha_d \in A$ and A fulfils (11.2). Consequently $A = \langle\alpha_1, ..., \alpha_d\rangle$.

Remark: Be careful not to confuse the expressions of the ideal A as a $\mathbb{Z}$-*module* and as a $\mathbb{A}_\mathbb{K}$ -*module*.

Exercise 11.2   a) Let $A = \langle\alpha_1, \alpha_2, ..., \alpha_n\rangle$ and $B = \langle\beta_1, \beta_2, ..., \beta_m\rangle$. Then
$$x \in A + B \iff x = x_1\alpha_1 + \cdots + x_n\alpha_n + y_1\beta_1 + \cdots + y_m\beta_m \quad (x_i, y_i \in \mathbb{A}_\mathbb{K}).$$

Hence $x \in A + B \Leftrightarrow x \in \langle \alpha_1, \ldots, \alpha_n, \beta_1, \ldots, \beta_m \rangle$ and $A + B = \langle \alpha_1, \ldots, \alpha_n, \beta_1, \ldots, \beta_m \rangle$.

b) Let $x \in AB$. Then, by (11.3), x can be written as a finite sum of products of an element of A and an element of B, whence

$$x = \sum_{k=1}^{q} a_k b_k = \sum_{k=1}^{q} \left(\sum_{i=1}^{n} x_{ki} \alpha_i \right) \left(\sum_{j=1}^{m} y_{kj} \beta_j \right) (x_{ki}, y_{kj} \in \mathbb{A}_{\mathbb{K}}) = \sum_{i=1}^{n} \sum_{j=1}^{m} \left(\sum_{k=1}^{q} x_{ki} y_{kj} \right) \alpha_i \beta_j,$$

and $x \in \langle \alpha_1 \beta_1, \ldots, \alpha_1 \beta_m, \ldots, \alpha_n \beta_1, \ldots, \alpha_n \beta_m \rangle$. Conversely,

$$x \in \langle \alpha_1 \beta_1, \ldots, \alpha_1 \beta_m, \ldots, \alpha_n \beta_1, \ldots, \alpha_n \beta_m \rangle \Rightarrow x = \sum_{i=1}^{n} \sum_{j=1}^{m} x_{ij} \alpha_i \beta_j \ (x_{ij} \in \mathbb{A}_{\mathbb{K}}),$$

and therefore $x \in AB$ since $x_{ij} \alpha_i \in A$ for every i.

Exercise 11.3 1) If $n = 1$, then $P(x) = a_1(x - \rho)$ and the result is true. Assume that it holds for a given n, and let $P(x) = a_0 + a_1 x + \cdots + a_{n+1} x^{n+1}$. Consider $Q(x) = P(x) - a_{n+1} x^n (x - \rho)$. Since ρ is a root of P, we have $a_0 a_{n+1}^n + a_1 a_{n+1}^{n-1} a_{n+1} \rho + \cdots + (a_{n+1} \rho)^{n+1} = 0$, and therefore $a_{n+1} \rho$ is an algebraic integer. Thus $\deg Q \leq n$, the coefficients of Q are algebraic integers, and ρ is a root of Q. By the induction hypothesis, the coefficients of $Q(x)/(x - \rho)$ $= P(x)/(x - \rho) - a_{n+1} x^n$ are algebraic integers, and so are those of $P(x)/(x - \rho)$.

2) Put $A(x) = a_n(x - \rho_1) \ldots (x - \rho_n)$, $B(x) = b_m(x - \sigma_1) \ldots (x - \sigma_m)$. Then the coefficients of $\dfrac{C(x)}{\delta} = \dfrac{a_n b_m}{\delta} (x - \rho_1) \ldots (x - \rho_n)(x - \sigma_1) \ldots (x - \sigma_m)$ are algebraic integers by hypothesis. From question 1), we deduce that all products of the form $\dfrac{a_n b_m}{\delta} (x - \rho_{i_1}) \ldots (x - \rho_{i_k})(x - \sigma_{j_1}) \ldots (x - \sigma_{j_\ell})$ have algebraic integer coefficients. Therefore, in particular, all products $\dfrac{a_n b_m}{\delta} \rho_{i_1} \ldots \rho_{i_k} \sigma_{j_1} \ldots \sigma_{j_\ell}$ are algebraic integers for $1 \leq i_1 < \cdots < i_k \leq n$, $1 \leq j_1 < \cdots < j_\ell \leq m$. However we have, for $0 \leq k \leq n$ and $0 \leq \ell \leq m$,

$$\frac{a_{n-k}}{a_n} = \pm \sum_{1 \leq i_1 < \cdots < i_k \leq n} \rho_{i_1} \ldots \rho_{i_k}, \qquad \frac{b_{m-\ell}}{b_m} = \pm \sum_{1 \leq j_1 < \cdots < j_\ell \leq m} \sigma_{j_1} \ldots \sigma_{j_\ell}.$$

Therefore $a_{n-k} b_{m-\ell} = \pm \sum \sum a_n b_m \rho_{i_1} \ldots \rho_{i_k} \sigma_{j_1} \ldots \sigma_{j_\ell} = \delta \left(\sum \sum \alpha_{i_1, \ldots, i_k, j_1, \ldots, j_\ell} \right)$, where the $\alpha_{i_1, \ldots, i_k, j_1, \ldots, j_\ell}$ are algebraic integers. Since any sum of algebraic integers is an algebraic integer, so is $a_{n-k} b_{m-\ell} / \delta$.

3) a) The γ_i's are symmetric functions with rational coefficients of the $\sigma_j(\alpha_k)$'s (for every permutation of j).

Therefore they are symmetric functions of $\theta_1 = \theta, \theta_2, ..., \theta_d$, with $\mathbb{K} = \mathbb{Q}(\theta)$.
Hence the γ_i's are rational numbers by lemma 9.2. Moreover, the γ_i's are algebraic integers since the $\sigma_j(\alpha_k)$'s are. Therefore $\gamma_i \in \mathbb{Z}$.

Now we show that $\beta_i \in \mathbb{A}_{\mathbb{K}}$: we obtain the β_i's by dividing $f(x)g(x) \in \mathbb{Z}[x]$ by $f(x) \in \mathbb{A}_{\mathbb{K}}[x]$. Therefore $\beta_i \in \mathbb{K}$. Since the $\sigma_j(\alpha_k)$'s are algebraic integers, so are the β_i's and $\beta_i \in \mathbb{A}_{\mathbb{K}}$.

b) Since $a = \mathrm{GCD}(\gamma_0, \gamma_1, ..., \gamma_{nd})$, $a = \sum_{i=0}^{nd} n_i \gamma_i$, $n_i \in \mathbb{Z}$ by Bézout's identity.

Therefore $a = \sum_{i=0}^{nd} \sum_{k=0}^{i} n_i \alpha_k \beta_{i-k} \in AB$, and $\langle a \rangle \subset AB$.

c) Since $a \mid \gamma_i$ for every i, we know by question 2) that $a \mid \alpha_i \beta_j$ for every (i, j). Thus $\alpha_i \beta_j \in \langle a \rangle$, whence $AB \subset \langle a \rangle$. Since $\langle a \rangle \subset AB$, $\langle a \rangle = AB$.

4) Applying the above results, we see that the ideal $\langle a^{-1} \rangle B$ is the inverse of A.

Exercise 11.4   Put $A = \langle \alpha_1, ..., \alpha_n \rangle$, $B = \langle \beta_1, ..., \beta_m \rangle$, $C = \langle \gamma_1, ..., \gamma_q \rangle$. Then

$$A(B+C) = \langle \alpha_1\beta_1, ..., \alpha_1\beta_m, \alpha_1\gamma_1, ..., \alpha_1\gamma_q, ..., \alpha_n\beta_1, ..., \alpha_n\beta_m, \alpha_n\gamma_1, ..., \alpha_n\gamma_q \rangle$$
$$= \langle \alpha_1\beta_1, ..., \alpha_1\beta_m, ..., \alpha_n\beta_1, ..., \alpha_n\beta_m \rangle + \langle \alpha_1\gamma_1, ..., \alpha_1\gamma_q, ..., \alpha_n\gamma_1, ..., \alpha_n\gamma_q \rangle = AB + AC.$$

Exercise 11.5   Since A and B are coprime, $A + B = \langle 1 \rangle$ by theorem 11.3. Hence $AC + BC = C$. But $A \mid BC$, whence $BC = AD$, where $D \in \mathfrak{I}(\mathbb{A}_{\mathbb{K}})$. Thus $C = A(C + D)$ and $A \mid C$.

Exercise 11.6   Let $(\alpha_1, \alpha_2, ..., \alpha_d)$ and $(\beta_1, \beta_2, ..., \beta_d)$ be two basis of the $\mathbb{Z}$-modules A and B, with $d = [\mathbb{K} : \mathbb{Q}]$. Since $B \mid A$, we have $A \subset B$ by theorem 11.2, whence $\alpha_j = \sum_{i=1}^{d} c_{ij} \beta_i$, $c_{ij} \in \mathbb{Z}$. By lemma 10.4, $\Delta(A) = (\det(c_{ij}))^2 \Delta(B)$, whence $\Delta(B) \mid \Delta(A)$. If $\Delta(A) = \Delta(B)$, then $\det(c_{ij}) = \pm 1$, and therefore $\alpha_j = \sum_{i=1}^{d} c_{ij} \beta_i \Rightarrow \beta_i = \sum_{j=1}^{d} c'_{ij} \alpha_j$, $c'_{ij} \in \mathbb{Z}$ by Cramer's formulas. Thus $B \subset A$, whence $A = B$.

Exercise 11.7   1) We have $AP \neq A$. Indeed, $AP = A \Leftrightarrow P = \langle 1 \rangle$ since $\mathfrak{I}(\mathbb{K})^*$ is a group. But $P \neq 1$ because P is prime. Since $AP \neq A$ and $AP \subset A$, there exists $\alpha \in A$ such that $\alpha \notin AP$. Since $\langle \alpha \rangle \subset A$, $A \mid \langle \alpha \rangle$. Therefore there exists an (integral) ideal B such that $\langle \alpha \rangle = AB$. Now suppose that $P \mid B$. Then AP

divides $\langle \alpha \rangle$, contradiction. Therefore $P \mid B$ which shows that P and B are coprime since P is prime. Thus $P + B = \langle 1 \rangle$ and $A = AB + AP = \langle \alpha \rangle + AP$.

2) First we show that every $x \in \mathbb{A}_\mathbb{K}$ is congruent modulo (AP) to one of the $a_i + \alpha p_j$. To begin with, we observe that there exists an index i and an element y in A such that $x = a_i + y$. But $A = \langle \alpha \rangle + AP$. Hence there exists $\mu \in \mathbb{A}_\mathbb{K}$ and $z \in AP$ such that $y = \mu \alpha + z$, whence $x = a_i + \mu \alpha + z$. Finally, there exists an index j and an element t in P such that $\mu = p_j + t$. Thus $x = a_i + \alpha p_j + t \alpha + z$. Since $\alpha \in A$, $t \alpha \in AP$, whence $x \equiv a_i + \alpha p_j \pmod{AP}$.

Now we show that $a_i + \alpha p_j \equiv a_k + \alpha p_\ell \pmod{AP} \Rightarrow i = k$ and $j = \ell$. We have $AP \subset A$, whence $a_i + \alpha p_j \equiv a_k + \alpha p_\ell \pmod{AP} \Rightarrow a_i + \alpha p_j \equiv a_k + \alpha p_\ell$ $\pmod{A} \Rightarrow a_i \equiv a_k \pmod{A}$ because $\alpha \in A$. Therefore $i = k$, since $\{a_i, ..., a_{N(A)}\}$ is a system of residue classes modulo A. Thus $\alpha p_j \equiv \alpha p_\ell \pmod{AP}$. As $AP \subset P$, we have $\alpha p_j \equiv \alpha p_\ell \pmod{P}$. But $\alpha \notin P$ (otherwise α would belong to AP), and P is prime, whence $\mathbb{A}_\mathbb{K}/P$ is a field by corollary 11.3. Thus α is invertible mod P and $p_j \equiv p_\ell \pmod{P}$. This yields $j = \ell$.

3) Let $B = P_1 P_2 ... P_k$, with P_i prime. We have seen in question 2) that $N(AP) = N(A)N(P)$ for every prime ideal P. Therefore

$$N(AB) = N(AP_1 ... P_{k-1})N(P_k) = N(AP_1 ... P_{k-2})N(P_{k-1})N(P_k)$$
$$= N(AP_1 ... P_{k-2})N(P_{k-1}P_k) = N(A)N(P_1 ... P_k) = N(A)N(B).$$

Exercise 11.8   Suppose that $A = BC$, $B, C \in \mathfrak{I}(\mathbb{A}_\mathbb{K})$. Then
$$N(A) = N(B)N(C) \Rightarrow (N(B) = 1 \text{ or } N(C) = 1).$$
Hence $B = \langle 1 \rangle$ or $C = \langle 1 \rangle$) by (11.6), and A is prime.

Exercise 11.9   Let $a, b \in \text{Ker} \varphi$. Then
$$\varphi(a - b) = \varphi(a) - \varphi(b) = 0 \Rightarrow (a - b) \in \text{Ker} \varphi.$$
Let $x \in \mathbb{A}$. Then $\varphi(ax) = \varphi(a)\varphi(x) = 0$, whence $ax \in \text{Ker} \varphi$ and $\text{Ker} \varphi$ is an ideal of $\mathbb{A}$.

Exercise 11.10   Let $X = \bar{P}(\alpha_i)$, with $\bar{P}(x) \in \mathbb{F}_p[x]$, be an element of $\mathbb{F}_p(\alpha_i)$. Put $\bar{P}(x) = \sum_{k=0}^{m} \bar{a}_k x^k$, and let $a_0, a_1, ..., a_m \in \mathbb{Z}$ be representatives of $\bar{a}_0, ..., \bar{a}_m$

modulo p. If we put $P(x) = \sum_{k=0}^{m} a_k x^k \in \mathbb{Z}[x]$, it is clear that $v_i(P(\theta)) = \bar{P}(\alpha_i)$ by definition of v_i, and therefore v_i is surjective

Exercise 11.11   We start from $\rho_d = \sum_{j=1}^{d} c_{dj}\alpha_j$, and we substract ρ_{d-1} to it as many times as necessary to obtain $\rho_d^{(1)} = \sum_{j=1}^{d-1} c_{dj}^{(1)}\alpha_j + c_{dd}\alpha_d$, with $0 \le c_{d,d-1}^{(1)} < c_{d-1,d-1}$. Then we substract ρ_{d-2} as many times as necessary to obtain $\rho_d^{(2)} = \sum_{j=1}^{d-2} c_{dj}^{(2)} + c_{d,d-1}^{(1)}\alpha_{d-1} + c_{dd}\alpha_d$, with $0 \le c_{d,d-2}^{(2)} < c_{d-2,d-2}$. Applying the same process finally yields $\omega_d = \sum_{j=1}^{d-1} c'_{dj}\alpha_j + c_{dd}\alpha_d$, with $0 \le c'_{dj} < c_{jj}$ for $j = 1, 2, ..., d-1$. By construction, we have
$$\Delta(\rho_1, ..., \rho_{d-1}, \omega_d) = \Delta(\rho_1, ..., \rho_{d-1}, \rho_d).$$
Doing the same with $\rho_{d-1}, ..., \rho_1$ finally yields a set $(\omega_1, \omega_2, ..., \omega_d)$ with the desired property, and $\Delta(\omega_1, ..., \omega_d) = \Delta(\rho_1, ..., \rho_d)$. Therefore $(\omega_1 ..., \omega_d)$ is a basis of the $\mathbb{Z}$ - module A (theorem 10.8).

But we have seen in the proof of theorem 11.6 that $N(A) = c_{11}c_{22}...c_{dd}$. If $N(A)$ is fixed, $c_{11}, c_{22}, ..., c_{dd}$ can take only finitely many values, and so do $\omega_1, \omega_2, ..., \omega_d$. Hence, there exist only a finite number of ideals with a given norm.

Exercise 11.12   Let A and B be two fractional ideals. Then there exist a and $b \in \mathbb{N}$ such that $aA = A'$ and $bB = B'$ are (integral) ideals. And
$$A \sim B \iff \exists p \in \mathbb{A}_K, \exists q \in \mathbb{N} \text{ such that } A = \left\langle \frac{p}{q} \right\rangle B$$
$$\iff \langle q \rangle A = \langle p \rangle B \iff \langle bq \rangle A' = \langle ap \rangle B' \iff A' \sim B'.$$

Exercise 11.13   1) Let (α, β) be a basis of the $\mathbb{Z}$ - module A. Put $\varphi = \beta/\alpha$. First we observe that $\text{Im}\,\varphi \neq 0$; indeed, if φ was real, we would have
$$\varphi\alpha - \beta = 0 \implies \varphi(a + ib\sqrt{d}) - (a' + ib'\sqrt{d}) = 0 \implies \varphi \text{ rational}.$$
Hence α, β would be linearly dependent over $\mathbb{Q}$, contradiction. Now, if $\text{Im}\,\varphi > 0$, there is nothing to prove. If $\text{Im}\,\varphi < 0$, we see at once that $(\alpha, -\beta)$ is a basis of A, and we put $\theta = -\beta/\alpha$, $\text{Im}\,\theta > 0$.

2) a) We have $p = \alpha\bar{\alpha} = N(\alpha) \in \mathbb{Z}$ since $\alpha \in \mathbb{A}_K$, $r = N(\theta\alpha) \in \mathbb{Z}$, and $q = \alpha(\overline{\theta\alpha}) + \bar{\alpha}(\theta\alpha) = T(\alpha(\theta\alpha)) \in \mathbb{Z}$.

b) By using corollary 11.1, we have

$$q^2 - 4pr = (\alpha(\overline{\theta\alpha}) + \overline{\alpha}(\theta\alpha))^2 - 4(\alpha\overline{\alpha})(\theta\alpha)(\overline{\theta\alpha}) = (\alpha(\overline{\theta\alpha}) - \overline{\alpha}(\theta\alpha))^2$$

$$= \begin{vmatrix} \alpha & \overline{\alpha} \\ \alpha\theta & \overline{\theta}\alpha \end{vmatrix}^2 = \Delta(A) = N(A)^2 \Delta(\mathbb{K}).$$

c) Since $\alpha \in A$, $\langle\alpha\rangle \subset A$. Hence $A \mid \langle\alpha\rangle$ and $N(A) \mid N(\langle\alpha\rangle)$. Thus $N(A) \mid p$.

Similarly $N(A) \mid r$. Now question 2)b) yields $N(A)^2 \mid q^2$, whence $N(A) \mid q$.

3)a) $b^2 - 4ac = (q^2 - 4pr)/N(A)^2 = \Delta$ by 2) b) . Moreover,

$$Q(x, y) = \alpha\overline{\alpha}(x + \theta y)(x + \overline{\theta} y) = p(x + \theta y)(x + \overline{\theta} y).$$

b) Since $A_1 \sim A_2$, there exist $\lambda_1, \lambda_2 \in \mathbb{A}_{\mathbb{K}}$ such that $\langle\lambda_1\rangle A_1 = \langle\lambda_2\rangle A_2$. Hence there exist s, t, u, $v \in \mathbb{Z}$ such that

$$\lambda_1\alpha_1 = \lambda_2(s\alpha_2 + t\theta_2\alpha_2), \quad \lambda_1\theta_1\alpha_1 = \lambda_2(u\alpha_2 + v\theta_2\alpha_2). \tag{*}$$

Similarly there exist s', t', u', $v' \in \mathbb{Z}$ such that

$$\lambda_2\alpha_2 = \lambda_1(s'\alpha_1 + t'\theta_1\alpha_1), \quad \lambda_2\theta_2\alpha_2 = \lambda_1(u'\alpha_1 + v'\theta_1\alpha_1). \tag{**}$$

Therefore $\begin{pmatrix} s & t \\ u & v \end{pmatrix}\begin{pmatrix} s' & t' \\ u' & v' \end{pmatrix} = \begin{pmatrix} 1 & 0 \\ 0 & 1 \end{pmatrix}$. Taking the determinants yields

$$(sv - ut)(s'v' - u't') = 1 \iff sv - ut = \pm 1.$$

But we have by (*) $\theta_1 = \dfrac{u + v\theta_2}{s + t\theta_2} = \dfrac{(u + v\theta_2)(s + t\overline{\theta}_2)}{|s + t\theta_2|^2}$, whence

$$\theta_1 = \frac{1}{|s + t\theta|^2}(us + vt|\theta_2|^2 + sv\theta_2 + ut\overline{\theta}_2). \tag{***}$$

Since $\mathrm{Im}(\theta_1) > 0$, $\mathrm{Im}(\theta_2) > 0$, this yields $sv - ut > 0$, whence $sv - ut = 1$.

Now we show that $(a_1, b_1, c_1) \sim (a_2, b_2, c_2)$. By using the substitution $X = sx + uy$, $Y = tx + vy$, we obtain

$$a_2 X^2 + b_2 XY + c_2 Y^2 = a_2(X + \theta_2 Y)(X + \overline{\theta}_2 Y)$$

$$= a_2[(s + t\theta_2)x + (u + v\theta_2)y][(s + t\overline{\theta}_2)x + (u + v\overline{\theta}_2)y]$$

$$= a_2|s + t\theta_2|^2(x + \theta_1 y)(x + \overline{\theta}_1 y) \quad \text{(by (***))}$$

$$= \frac{N(\alpha_2)}{N(A_2)}N(s + t\theta_2)(x + \theta_1 y)(x + \overline{\theta}_1 y) \quad \text{(by question 2)a)}$$

$$= \frac{N(\lambda_1)}{N(\lambda_2)}\frac{N(\alpha_1)}{N(A_2)}(x + \theta_1 y)(x + \overline{\theta}_1 y) \quad \text{(by taking the norms in (*))}$$

$$= \frac{N(\alpha_1)}{N(A_1)}(x + \theta_1 y)(x + \overline{\theta}_1 y) \quad \text{(since } \langle\lambda_1\rangle A_1 = \langle\lambda_2\rangle A_2\text{)}$$

$$= a_1(x + \theta_1 y)(x + \overline{\theta}_1 y) = a_1 x^2 + b_1 xy + c_1 y^2.$$

Thus, if $A_1 \sim A_2$, we have $R_1(x, y) \sim R_2(x, y)$ (see subsection 6.5.2). Hence $\Psi : A \to R(x, y)$ defines a mapping from the set of the (integral) ideal classes of $\mathbb{K}$ into the set of all positive definite quadratic forms with discriminant Δ.

4) Let $R(x, y) = ax^2 + bxy + cy^2$ be a definite quadratic form with discriminant $\Delta = \Delta(\mathbb{K}) < 0$, $\mathbb{K} = \mathbb{Q}(\sqrt{d})$. We have $\Delta = 4d$ if $d \equiv 2$ ou $3 \pmod 4$, $\Delta = d$ if $d \equiv 1 \pmod 4$ (example 10.11). Put $\theta = (b + i\sqrt{-\Delta})/2a$, in such a way that $R(x, y) = a(x + \theta y)(x + \bar\theta y)$. Then clearly $\theta \in \mathbb{K}$ and θa is an integer, since $a\theta^2 + b\theta + c = 0 \Rightarrow (a\theta)^2 + b(a\theta) + ca = 0$. Thus $A = \mathbb{Z}(a, \theta a)$ is an integral ideal of $\mathbb{K}$. Its discriminant is

$$\Delta(A) = \begin{vmatrix} a & a \\ \theta a & \bar\theta a \end{vmatrix}^2 = a^4 (\bar\theta - \theta)^2 = a^2 \Delta.$$

Therefore $N(A) = a$ since $a > 0$. The quadratic form $Q(x, y) = \Psi(A)$ (see question 2) is $Q(x, y) = (ax + a\theta y)(ax + a\bar\theta y) = aR(x, y)$.

Hence $R(x, y) = Q(x, y)/N(A)$, and the equivalence class of R is the image of the equivalence class of A by Ψ. Therefore Ψ is surjective.

5) Assume that $R_1 = \Psi(A_1) = R_2 = \Psi(A_2)$. Put

$$A_1 = \mathbb{Z}(\alpha_1, \theta_1 \alpha_1), \quad A_2 = \mathbb{Z}(\alpha_2, \theta_2 \alpha_2), \quad \text{Im}(\theta_1) > 0, \quad \text{Im}(\theta_2) > 0,$$

in such a way that the forms $R_1(x, y) = a_1 x^2 + b_1 xy + c_1 y^2 = a_1(x + \theta_1 y)(x + \bar\theta_1 y)$ and $R_2(X, Y) = a_2(X + \theta_2 Y)(X + \bar\theta_2 Y)$ are equivalent. Hence there exist s, t, u, $v \in \mathbb{Z}$, with $\begin{vmatrix} s & u \\ t & v \end{vmatrix} = 1$, such that the substitution $X = sx + uy$, $Y = tx + vy$, transforms R_2 to R_1. Now we have

$$R_2(sx + uy, tx + vy) = a_2((s + t\theta_2)x + (u + v\theta_2)y)((s + t\bar\theta_2)x + (u + v\bar\theta_2)y)$$

$$= a_2 N(s + t\theta_2)\left(x + \frac{u + v\theta_2}{s + t\theta_2} y\right)\left(x + \frac{u + v\bar\theta_2}{s + t\bar\theta_2} y\right) = a_1(x + \theta_1 y)(x + \bar\theta_1 y).$$

This yields
$$\theta_1 = \frac{u + v\theta_2}{s + t\theta_2}. \tag{****}$$

(note that $\theta_1 = (u + v\bar\theta_2)/(s + t\bar\theta_2)$ is impossible since $\text{Im}(\theta_1) > 0$, $\text{Im}(\theta_2) > 0$, $sv - ut = 1$).

Now we choose λ_1 and $\lambda_2 \in \mathbb{A}_{\mathbb{K}}$ in order to satisfy the first equation of (*). For example, we can take $\lambda_1 = \alpha_2(s + t\theta_2)$, $\lambda_2 = \alpha_1$. Thus $\lambda_1 \alpha_1 = \lambda_2(s\alpha_2 + t\theta_2 \alpha_2)$. By using (****), we obtain $\theta_1 \lambda_1 \alpha_1 = \lambda_2(u\alpha_2 + v\theta_2 \alpha_2)$, and the system (*), which means that $\langle \lambda_1 \rangle A_1 \subset \langle \lambda_2 \rangle A_2$. But we can invert this system in (**) with

s', t', u', $v' \in \mathbb{Z}$ since $sv - ut = 1$. Therefore we have also $\langle \lambda_2 \rangle A_2 \subset \langle \lambda_1 \rangle A_1$. Thus $\langle \lambda_1 \rangle A_1 = \langle \lambda_2 \rangle A_2$, $A_1 \sim A_2$ and Ψ is injective. Hence Ψ is bijective by question 4), which proves theorem 11.9.

Exercise 11.14 By using theorem 6.13, we have to solve the equation $b^2 - 4ac = -163$, with $(a,b,c) \in \mathbb{N}^* \times \mathbb{Z} \times \mathbb{N}^*$ and $a \leq 7$ (since $-163 \equiv 1 \pmod 4$ $\Rightarrow \Delta(\mathbb{Q}(i\sqrt{163})) = -163$). By (6.15), since $b^2 + 163$ must be divisible by $4a$, the only possible value is $a = 1$, $b = 1$, $c = 41$. Finally, theorems 6.13 and 11.9 yield the desired result.

Exercise 11.15 $\langle \alpha \rangle = \langle \beta \rangle \Rightarrow (\langle \alpha \rangle \subset \langle \beta \rangle$ and $\langle \beta \rangle \subset \langle \alpha \rangle) \Rightarrow (\exists u, v \in \mathbb{A}_{\mathbb{K}}$ such that $\alpha = u\beta$ and $\beta = v\alpha) \Rightarrow (\alpha\beta = (uv)\alpha\beta) \Rightarrow (uv = 1) \Rightarrow (u$ and v inversible$) \Rightarrow (\exists \varepsilon \in \mathbb{A}_{\mathbb{K}}^{\times}$, such that $\alpha = \varepsilon\beta)$. Conversely, if $\alpha = \varepsilon\beta$, then $\alpha \in \langle \beta \rangle$ and $\beta \in \langle \alpha \rangle$ since $\beta = \varepsilon^{-1}\alpha$, whence $\langle \alpha \rangle = \langle \beta \rangle$.

Exercise 11.16 First we factorize the ideal $\langle 2 \rangle$ by using theorem 11.7. Here $\theta = i\sqrt{13}$ since $-13 \equiv 3 \pmod 4 \Rightarrow \mathbb{A}_{\mathbb{K}} = \mathbb{Z}(i\sqrt{13})$. Hence $P_\theta(x) = x^2 + 13$ $\equiv x^2 - 1 \pmod 2$. Thus $P_\theta(x) = (x+1)(x-1) \Rightarrow \langle 2 \rangle = \langle 2, 1+i\sqrt{13} \rangle \langle 2, 1-i\sqrt{13} \rangle$. Similarly $P_\theta(x) = x^2 + 13 \equiv x^2 + 3 \pmod 5$. This polynomial is irreducible (mod 5) because it has no root in $\mathbb{F}_5$ (one only has to try 0, 1 and 2). Therefore $\langle 5 \rangle$ is prime in $\mathbb{Z}(i\sqrt{13})$, and $\langle 10 \rangle = \langle 2, 1+i\sqrt{13} \rangle \langle 2, 1-i\sqrt{13} \rangle \langle 5 \rangle$.

Now denote $A = \langle 2, 1+i\sqrt{13} \rangle \langle 2, 1-i\sqrt{13} \rangle$. We check easily that

$$A = \langle 2, 1+i\sqrt{13} \rangle \langle 2, 1-i\sqrt{13} \rangle = \langle 4, 2(1-i\sqrt{13}), 2(1+i\sqrt{13}), 14 \rangle.$$

Therefore $2 = 14 - 3 \cdot 4 \in A$. Moreover, for every $x \in A$,

$$x = 4x_1 + 2(1-i\sqrt{13})x_2 + 2(1+i\sqrt{13})x_3 + 14x_3$$
$$= 2(2x_1 + x_2(1-i\sqrt{3}) + x_3(1+i\sqrt{13}) + 7x_3) \in \langle 2 \rangle.$$

Hence $A = \langle 2 \rangle$ and $\langle 10 \rangle = \langle 5 \rangle A = \langle 5 \rangle \langle 2, 1+i\sqrt{13} \rangle \langle 2, 1-i\sqrt{13} \rangle$.

Exercise 11.17 1) Let us compute the product

$$B = \langle 2, 1+\sqrt{51} \rangle \langle 2, 1-\sqrt{51} \rangle = \langle 4, 2(1-\sqrt{51}), 2(1+\sqrt{51}), -50 \rangle.$$

We have $-50 + 13 \cdot 4 = 2 \in B$, whence $\langle 2 \rangle \subset B$.

Moreover $B \subset \langle 2 \rangle$ since 4, $2(1-\sqrt{51})$, $2(1+\sqrt{51})$, -50 are multiples of 2.

Thus $B = \langle 2 \rangle$, $A^{-1} = \left\langle 1, \frac{1-\sqrt{51}}{2} \right\rangle$.

2) We factorize $\langle 2 \rangle$ as a product of prime ideals in $\mathbb{A}_\mathbb{K} = \mathbb{Z}(\sqrt{51})$. We have $P_\theta(x) = x^2 - 51 \equiv x^2 - 1 \pmod 2$. Hence $P_\theta(x) = (x-1)(x+1)$ in $\mathbb{F}_2[x]$ and $\langle 2 \rangle = \langle 2, \sqrt{51}-1 \rangle \langle 2, \sqrt{51}+1 \rangle$ and A is prime by theorem 11.7.

3) Denoting $C = \langle 2, 1-\sqrt{51} \rangle$ and taking the norms yields

$$N(\langle 2 \rangle) = |N(2)| = 4 = N(A)N(C).$$

Hence $N(A) = 1$, 2 or 4. But $N(A) \neq 1$ since $N(A) = 1 \Rightarrow A = \langle 1 \rangle \Rightarrow C = \langle 2 \rangle$. However, this is impossible because $N(2) = 4$ does not divide $N(1-\sqrt{51}) = -50$. Similarly $N(A) \neq 4$ since $C \neq \langle 1 \rangle$. Therefore $N(A) = 2$. This results yields a second proof of the fact that A is prime (see remark 11.3).

Exercise 11.18   Soit $(a_1, a_2, ..., a_{N(A)})$ be a complete system of residues modulo A. Then $a_1 + 1$, $a_2 + 1$, ..., $a_{N(A)} + 1$ is also a complete system of residues modulo A, since $a_i + 1 \equiv a_j + 1 \pmod A \Rightarrow a_i \equiv a_j \pmod A$ and, for every $x \in \mathbb{A}_\mathbb{K}$, there exists un index i such that $x - 1 \equiv a_i \pmod A$, whence $x \equiv a_i + 1 \pmod A$. Thus $a_1 + \cdots + a_{N(A)} \equiv (a_1 + 1) + \cdots + (a_{N(A)} + 1) \pmod A$. Hence $N(A) \equiv 0 \pmod A$, and therefore $N(A) \in A$.

Exercise 11.19   We have $P_\omega(x) = 1 + x + x^2 + x^3 + x^4$.

a) Modulo 3, $P_\omega(x)$ has no root (try 0, 1, -1). Hence, if it is reducible, we have
$$P_\omega(x) = (x^2 + ax + b)(x^2 + cx + d).$$
This yields $a + c \equiv 1$, $b + ac + d \equiv 1$, $bc + ad \equiv 1$, $bd \equiv 1$. Therefore $b \equiv d \equiv 1$, since $b \equiv d \equiv -1 \Rightarrow a + c \equiv -1$, impossible. Hence $a + c \equiv 1$, $ac \equiv -1$, which is again impossible. Thus P_ω is irreducible modulo 3. Now theorem 11.7 shows that $\langle 3 \rangle$ is prime in $\mathbb{Z}(\omega)$.

b) Modulo 5, it is clear that $P_\omega(1) = 0$. As a matter of fact, we see that, modulo 5, $P_\omega(x) = x^4 - 4x^3 + 6x^2 - 4x + 1 = (x-1)^4$. Consequently $\langle 5 \rangle = (\langle 5, \omega-1 \rangle)^4$.

Exercise 11.20   By lemma 11.7, every ideal A of $\mathbb{A}_\mathbb{K}$ is equivalent to an ideal B satisfying $N(B) \leq M$.

Write $N(B) = p_1 p_2 \ldots p_k$ as a product of primes in $\mathbb{N}$. We have $p_i \le M$ for every i. Hence by hypothesis every prime ideal dividing one of the $\langle p_i \rangle$'s is principal. But we know by exercise 11.18 that $N(B) \in B$, whence $B \mid \langle N(B) \rangle$. Every prime divisor Q_j of B divides one of the $\langle p_i \rangle$, and therefore it is principal. Thus $B = Q_1 Q_2 \ldots Q_m$ is principal. Now every ideal A is equivalent to a principal ideal. Hence A is principal, and $\mathbb{A}_{\mathbb{K}}$ is a principal ideal domain.

Exercise 11.21   Here we take $\omega_1 = 1$, $\omega_2 = \sqrt{7}$, and $M = (1 + \sqrt{7})^2 \approx 13.2915$. By using the result of exercise 11.20, we only have to study the ideals $\langle 2 \rangle$, $\langle 3 \rangle, \langle 5 \rangle, \langle 7 \rangle, \langle 11 \rangle, \langle 13 \rangle$.

a) $x^2 - 7 \equiv (x-1)(x+1) \equiv (x+1)^2$ (mod 2). Hence $\langle 2 \rangle = \langle 2, \sqrt{7} + 1 \rangle^2$. But $\langle 2, \sqrt{7} + 1 \rangle$ is principal. To show this, observe that its norm is 2, and look for an element with norm ± 2. We see that $(\sqrt{7} + 1) - 4 = \sqrt{7} - 3 \in \langle 2, \sqrt{7} + 1 \rangle$. Hence $\langle 2, \sqrt{7} + 1 \rangle$ divides $\langle \sqrt{7} - 3 \rangle$. Put $\langle \sqrt{7} - 3 \rangle = A \langle 2, \sqrt{7} + 1 \rangle$. Taking the norms yields $2 = N(A) \times 2$, whence $N(A) = 1$, $\langle A \rangle = 1$. Thus $\langle 2, \sqrt{7} + 1 \rangle = \langle \sqrt{7} - 3 \rangle$ is principal.

b) $x^2 - 7 \equiv (x-1)(x+1)$ (mod 3). Hence $\langle 3 \rangle = \langle 3, \sqrt{7} + 1 \rangle \langle 3, \sqrt{7} - 1 \rangle$. By arguing as in a), we see that $\langle 3, \sqrt{7} + 1 \rangle = \langle \sqrt{7} - 2 \rangle$ and that $\langle 3, \sqrt{7} - 1 \rangle = \langle \sqrt{7} + 2 \rangle$.

c) $x^2 - 7$ has no root modulo 5, hence it is irreducible modulo 5. Consequently $\langle 5 \rangle$ is prime, and it is evidently principal.

d) $\langle 7 \rangle = \langle \sqrt{7} \rangle^2$, and $\langle \sqrt{7} \rangle$ is prime because its norm is 7.

e) $x^2 - 7$ has no root modulo 11. Therefore $\langle 11 \rangle$ is prime.

f) $x^2 - 7$ has no root modulo 13 and $\langle 13 \rangle$ is prime.

Now for $\langle p \rangle = \langle 2 \rangle, \langle 3 \rangle, \langle 5 \rangle, \langle 7 \rangle, \langle 11 \rangle, \langle 13 \rangle$, every prime ideal dividing $\langle p \rangle$ is principal. Therefore $\mathbb{Z}(\sqrt{7})$ is principal, which means that $h\left(\mathbb{Q}(\sqrt{7}) \right) = 1$.

Exercise 11.22   Here we have $k = -13 \equiv 3 \pmod 4$.

Morover, by formula (10.19), we have $\Delta \left(\mathbb{Q}\left(i\sqrt{13} \right) \right) = -52$.

Now theorem 11.9 and table 6.2 show that $h\left(\mathbb{Q}\left(i\sqrt{13}\right)\right)=2$. Therefore theorem 11.12 applies. The equation has solutions since $k=-1-3a^2$, with $a=2$, The solutions are $x=a^2-k=17$, $y=\pm a\left(a^2+3k\right)=\pm70$.

Solutions to the exercises of chapter 12

Exercise 12.1   Let $x=\sum_{j=0}^{n-1}a_j\theta^j$, $a_j\in\mathbb{L}$, be an element of $\mathbb{K}$.

Then we can write $x=\sum_{j=0}^{n-1}\sum_{i=0}^{d-1}x_{ij}\alpha^i\theta^j$, $x_{ij}\in\mathbb{Q}$.

Put $\varphi_k(x)=\sum_{j=0}^{n-1}\sum_{i=0}^{d-1}x_{ij}\alpha_k^i\theta^j=\sum_{j=0}^{n-1}\Psi_k(a_j)\theta^j$, where Ψ_k is one of the monomorphisms of $\mathbb{L}$. We have immediatly $\varphi_k(x+y)=\varphi_k(x)+\varphi_k(y)$, $\varphi_k(xy)=\varphi_k(x)\varphi_k(y)$, $\varphi_k(r)=r$ if $r\in\mathbb{Q}$. Thus φ_k is a monomorphism of $\mathbb{K}$. In particular, $\varphi_k(\alpha)=\alpha_k$, whence $\max_k|\alpha_k|=\overline{|\alpha|}\le\max_i|\sigma_i(\alpha)|$.

Moreover, if we denote by P_α the minimal polynomial of α, we have $P_\alpha(\alpha)=0$, whence $P_\alpha(\sigma_k(\alpha))=0$ for every monomorphism σ_k of $\mathbb{K}$. Therefore there exists i with $\sigma_k(\alpha)=\alpha_i$ and $\overline{|\alpha|}=\max_i|\alpha_i|=\max_k|\sigma_k(\alpha)|$.

Exercise 12.2   Let $\sigma_1,...\sigma_d$ be the monomorphisms of $\mathbb{K}=\mathbb{Q}(\alpha,\beta)$. Then
$$\overline{|\alpha+\beta|}=\max_i|\sigma_i(\alpha+\beta)|\le\max_i|\sigma_i(\alpha)|+\max_i|\sigma_i(\beta)|,$$
whence $\overline{|\alpha+\beta|}\le\overline{|\alpha|}+\overline{|\beta|}$ by theorem 12.1. The proof of (12.12) is similar.

Exercise 12.3   1) Since α is an algebraic integer, its minimal polynomial P_α has integer coefficients. Hence there exist rational integers $a_0,a_1,...,a_{d-1}$ such that $a_0+a_1\alpha+\cdots+a_{d-1}\alpha^{d-1}+\alpha^d=0$, with $a_0\ne0$ since P_α is irreducible in $\mathbb{Q}[x]$. Thus $|\alpha_1\alpha_2...\alpha_d|=|a_0|\ge1$. Therefore at least one of the α_i's satisfies $|\alpha_i|\ge1$, which proves that $\overline{|\alpha|}=\max_i|\alpha_i|\ge1$.

2) Let $\beta_1=\alpha^i$, β_2, $\cdots$, β_{d_i} be the conjugates of α^i. Since $\beta_1=\alpha^i\in\mathbb{Q}(\alpha)$, for every k there exists a conjugate γ of α such that $\beta_k=\gamma^i$. Hence we have $|\beta_k|\le\overline{|\alpha|}^i=1$ for every $k=1,2,\cdots,d_i$. Therefore
$$|a_{j,i}|=\left|(-1)^{d_i-j}\sum_{k_1<k_2<\cdots<k_{d_i-j}}\beta_{k_1}\beta_{k_2}\cdots\beta_{k_{d_i-j}}\right|\le\sum_{k_1<k_2<\cdots<k_{d_i-j}}1=\binom{d_i}{d_i-j}=\binom{d_i}{j}.$$

Now $\alpha^i \in \mathbb{Q}(\alpha)$, whence $\deg(\alpha^i) = d_i \leq n$. This proves the desired result.

3) Let α be an algebraic integer satisfying $|\alpha| = 1$. From the above question we know that there are only finitely many polynomials P_i. Each of them has only finitely many roots since $\deg P_i \leq n$. Hence the sequence α^i can take only finitely many values. Therefore there exist $i \neq m$ such that $\alpha^i = \alpha^m$. This proves that α is a root of unity.

Exercise 12.4   Put $\alpha = \beta/a$, $a = \operatorname{den}(\alpha)$, $\beta \in \mathbb{A}_\mathbb{K}$. Since β is a *non zero integer* of $\mathbb{K}$, we have $|N(\beta)| \geq 1$ (see (10.11) and the beginning of section 10.3). As $N(\alpha) = \alpha_1 \alpha_2 ... \alpha_d$, we can write

$$|N(\alpha)| = |\sigma_1(\beta/a)...\sigma_d(\beta/a)| = |\sigma_1(\beta)...\sigma_d(\beta)|/a^d = |N(\beta)|/a^d \geq a^{-d}.$$

But $|\alpha_i| \leq |\overline{\alpha}|$ for $i = 2,...,d$. Hence $|\alpha| \cdot |\overline{\alpha}|^{d-1} \geq a^{-d}$, Q.E.D.

Exercise 12.5   Using the first estimate of $|\overline{B_n}|$, we have, since $c_3 \geq 1$,

$$|\overline{B_n}| \leq c_2 \sum_{i=0}^m \left(\sum_{j=0}^m c_3^{j2^n} \right) c_3^{i2^n} \leq c_2 \left(\sum_{i=0}^m c_3^{i2^n} \right)^2 \leq c_2 (m+1)^2 c_3^{m2^{n+1}}.$$

Exercise 12.6   Let $(\omega_1, \omega_2,..., \omega_d)$ be an integral basis of $\mathbb{K}$. Since the A_{ij}'s are integers, we have $A_{ij} = \sum_{k=1}^d a_{ijk} \omega_k$, $a_{ijk} \in \mathbb{Z}$.   (*)
Thus we can write (12.26) as

$$\sum_{j=1}^n a_{ijk} x_j = 0, \quad i = 1, 2,..., m, \quad k = 1,..., d.$$

This is a system with dm equations and n unknowns. If $n > dm$, we can apply Siegel's first lemma. The system has a solution $(x_1,..., x_n) \in \mathbb{Z}^n$ satisfying

$$0 < \max_i |x_i| \leq \left(n \max_{i,j,k} |a_{ijk}| \right)^{dm/(n-dm)}. \tag{**}$$

However, transforming (*) by the monomorphisms $\sigma_1,..., \sigma_d$ of K à (*), yields the following system with d equations:

$$\begin{cases} a_{ij1} \sigma_1(\omega_1) + \cdots + a_{ijd} \sigma_1(\omega_d) = \sigma_1(A_{ij}) \\ a_{ij1} \sigma_2(\omega_1) + \cdots + a_{ijd} \sigma_2(\omega_d) = \sigma_2(A_{ij}) \\ \vdots \\ a_{ij1} \sigma_d(\omega_1) + \cdots + a_{ijd} \sigma_d(\omega_d) = \sigma_d(A_{ij}) \end{cases}$$

The determinant D of this system satisfies $D^2 = \Delta[\omega_1, \omega_2,..., \omega_d] \neq 0$ (see section 10.5). Let H be the maximal value of the moduli of the minors of D.

Now we apply Cramer's rule. We obtain $\left|a_{ijk}\right| \le \dfrac{dH}{|D|} \max_{k} \left|\sigma_{k}(A_{ij})\right| \le \dfrac{dH}{|D|} A.$

Replacing in (**) yields the desired result.

Exercise 12.7 We have

$$F_{n}^{(k)}(x) = \sum_{i=0}^{p} \sum_{j=0}^{q} a_{ij} \sum_{\ell=0}^{k} \binom{k}{\ell} i(i-1)...(i-\ell+1) j^{k-\ell} x^{i-\ell} e^{jx},$$

$$F_{n}^{(k)}(\alpha) = \sum_{i=0}^{p} \sum_{j=0}^{q} a_{ij} \sum_{\ell=0}^{k} \binom{k}{\ell} i(i-1)...(i-\ell+1) j^{k-\ell} \alpha^{i-\ell} (e^{\alpha})^{j}.$$

In equations (12.29), we have to multiply $F_{n}^{(k)}(\alpha)$ by the denominators $a^{i+j-\ell}$, in order that the coefficients of the a_{ij}'s become algebraic integers. For (12.28) as well as for (12.29), an upper bound of the houses of the coefficients is

$$\sum_{\ell=0}^{k} \binom{k}{\ell} i(i-1)...(i-\ell+1) j^{k-\ell} \overline{|a\alpha|}^{i-\ell} \overline{|ae^{\alpha}|}^{j}$$

$$\le \sum_{\ell=0}^{k} \binom{k}{\ell} p! q^{\kappa} c_{1}^{p+q}, \quad \text{with } c_{1} = \max\left(\overline{|a\alpha|}, \overline{|ae^{\alpha}|}\right)$$

$$\le 2^{n} e^{n} (\text{Log } n)^{2n} c_{1}^{n/\text{Log } n + (\text{Log } n)^{2}} \quad (\text{since } p! \le p^{p} \le e^{n})$$

$$\le e^{3n \text{ LogLog } n} \quad \text{for } n \text{ sufficiently large.}$$

Thus the system (12.28) - (12.29) has N unknowns, with $N = (p+1)(q+1)$, $N \ge n \text{Log } n$, and $2n$ equations. By applying Siegel's lemma 12.4, we see that it has a solution $a_{ij} \in \mathbb{Z}$ such that, for n sufficiently large

$$0 < \max_{i,j}\left|a_{ij}\right| \le (2Mn \text{Log } ne^{3n \text{LogLog } n})^{2dn/(n \text{Log } n - 2dn)}.$$

Therefore $0 < \max_{i,j}\left|a_{ij}\right| \le e^{n}$ for n sufficiently large.

Exercise 12.8 Two cases can occur: $\beta = F_{n}^{(m)}(0)$ or $\beta = F_{n}^{(m)}(\alpha)$.

In the first case, we write $x^{m}(x-\alpha)^{m} G_{n}(x) = F_{n}(x) = \dfrac{\beta}{m!} x^{m} + \sum_{h=m+1}^{+\infty} a_{h} x^{h}$.

In the second case, $x^{m}(x-\alpha)^{m} G_{n}(x) = F_{n}(x) = \dfrac{\beta}{m!}(x-\alpha)^{m} + \sum_{h=m+1}^{+\infty} a_{h}'(x-\alpha)$.

Hence $\beta = (-\alpha)^{m} m! G_{n}(0)$ in the first case, and $\beta = \alpha^{m} m! G_{n}(\alpha)$ in the second case. In both cases, applying the maximum modulus principle yields

$$|\beta| \le |\alpha|^{m} m! \max_{|z|=m^{2/3}} |G_{n}(z)|.$$

Now we estimate $|G_{n}(z)|$. First we look for an upper bound for $|F_{n}(z)|$. By (12.27) and (12.30), we have $\max_{|z|=m^{2/3}} |F_{n}(z)| \le \sum_{i=0}^{p} \sum_{j=0}^{q} e^{n} m^{2i/3} e^{jm^{2/3}}$, since $|e^{z}| = e^{\text{Re} z}$. Now $m^{2i/3} \le m^{2p/3}$, $e^{jm^{2/3}} \le e^{qm^{2/3}}$, and $n \le m$.

Hence $e^n \le e^m$, $p \le (m/\mathrm{Log}\,m)+1$, $q \le (\mathrm{Log}\,m)^2 +1$. Thus, for n large,

$$\max_{|z|=m^{2/3}} |F_n(z)| \le \left(\frac{m}{\mathrm{Log}\,m}+1\right)\left((\mathrm{Log}\,m)^2 +1\right)e^m m^{\frac{2}{3}\left(\frac{m}{\mathrm{Log}\,m}+1\right)} e^{((\mathrm{Log}\,m)^2+1)m^{2/3}} \le e^{2m}.$$

Therefore $|\beta| \le |\alpha|^m m! \dfrac{e^{2m}}{m^{2m/3}\left(m^{2/3}-|\alpha|\right)^m} \le (2|\alpha|)^m m! \dfrac{e^{2m}}{m^{4m/3}}$.

As $m! \le m^m$, we obtain, for m sufficiently large, $|\beta| \le (2|\alpha|)^m e^{2m} m^{-\frac{m}{3}} \le m^{-\frac{m}{6}}$.

Exercise 12.9   We use the expressions of $F_n^{(k)}(0)$ and $F_n^{(k)}(\alpha)$ from exercise 12.7. If $\beta = F_n^{(m)}(0)$, we see that β is an integer of $\mathbb{K}$. Hence (12.35) evidently holds and, for m sufficiently large,

$$|\beta| \le \sum_{i=0}^{p}\sum_{j=0}^{q} |a_{ij}| \left|i!\binom{m}{i}\right| j^m \le (p+1)(q+1)e^n\, p! 2^m (\mathrm{Log}\,m)^{2m}$$

$$\le \left([m/\mathrm{Log}\,m]+1\right)!\left(\left[(\mathrm{Log}\,m)^2\right]+1\right)(2e)^m e^{2m\,\mathrm{LogLog}\,m} \le e^{3m\,\mathrm{LogLog}\,m}.$$

If $\beta = F_n^{(m)}(\alpha)$, then $\mathrm{den}\,\beta \le a^{p+q}$ and (12.35) holds. Moreover, for any m sufficiently large,

$$|\beta| \le \sum_{i=0}^{p}\sum_{j=0}^{q} |a_{ij}| \left|\sum_{\ell=0}^{\inf(i,m)}\binom{m}{\ell} i(i-1)\dots(i-\ell+1) j^{m-\ell}\right| |\alpha|^{i\ell} \left|e^\alpha\right|^j$$

$$\le \sum_{i=0}^{p}\sum_{j=0}^{q} e^n\, p! q^m |\alpha|^p \left|e^\alpha\right|^q 2^m.$$

This is almost the estimate of $|\beta|$ in the case where $\beta = F_n^{(m)}(0)$, and therefore we have $|\beta| \le e^{3m\,\mathrm{LogLog}\,m}$.

Exercise 12.10   We know by remark 3.2 that $[1,2,\dots,n,\dots] = i\dfrac{J_1(-2i)}{J_0(-2i)}$.

But $J_0'(x) = \left(\displaystyle\sum_{n=0}^{+\infty}\frac{(-1)^n x^{2n}}{(n!)^2 4^n}\right)' = 2\sum_{n=1}^{+\infty}\frac{(-1)^n x^{2n-1}}{n!(n-1)!4^n} = -\sum_{m=0}^{+\infty}\frac{(-1)^m x^{2m+1}}{m!(m+1)!2^{2m+1}} = -J_1(x)$.

Hence $[1,2,\dots,n,\dots] = -iJ_0'(-2i)/J_0(-2i)$ is transcendental by theorem 12.8.

Exercise 12.11   We proceed as in lemma 12.2. The Tschakaloff function T_q satisfies the functional equation (1.5), $T_q(qx) = 1+xT_q(x)$. Assume that T_q is algebraic, and write $(T_q(x))^d + a_{d-1}(x)(T_q(x))^{d-1} +\cdots+ a_0(x) = 0$, where $a_0,\dots,$ $a_{d-1} \in \mathbb{C}(x)$, d minimum. Replacing x by qx yields

$$(1+xT_q(x))^d + a_{d-1}(qx)(1+xT_q(x))^{d-1} + \cdots = 0.$$

Divide by x^d and look at the terms with degree $d-1$ in $T_q(x)$. We obtain

$$xa_{d-1}(x) = d + a_{d-1}(qx).$$

Put $a_{d-1}(x) = A(x)/B(x)$, with $A, B \in \mathbb{C}[x]$, A and B coprime. We get

$$xA(x)B(qx) = dB(x)B(qx) + A(qx)B(x).$$

Thus $B(qx) \mid A(qx)B(x)$. As A and B are coprime, this implies $B(qx) \mid B(x)$, whence $B(x) = b \in \mathbb{C}^*$. Contradiction because $\deg xA(x) > \deg A(qx)$.

Exercise 12.12   For $|x| < 1$, we have $\displaystyle\sum_{n=1}^{+\infty} \frac{x^{2^n}}{1-x^{2^{n+1}}} = \sum_{n=1}^{+\infty}\left(\frac{x^{2^n}}{1-x^{2^n}} - \frac{x^{2^{n+1}}}{1-x^{2^{n+1}}} \right).$

Indeed, $1 - x^{2^{n+1}} = (1 - x^{2^n})(1 + x^{2^n})$. Therefore we have a telescoping series:

$$\sum_{n=1}^{+\infty} \frac{x^{2^n}}{1-x^{2^{n+1}}} = \sum_{n=1}^{+\infty} \frac{x^{2^n}}{1-x^{2^n}} - \sum_{n=2}^{+\infty} \frac{x^{2^n}}{1-x^{2^n}} = \frac{x^2}{1-x^2}.$$

But we know that $F_n = (\Phi^n - \Psi^n)/\sqrt{5}$, where $\Phi = (1+\sqrt{5})/2$ is the golden number and $\Psi = -1/\Phi$ (exercise 3.11). Replacing x by $1/\Phi$ yields

$$\sum_{n=1}^{+\infty} \frac{\Phi^{-2^n}}{1-\Phi^{-2^{n+1}}} = \sum_{n=1}^{+\infty} \frac{1}{\Phi^{2^n} - \Phi^{-2^n}} = \frac{1}{\sqrt{5}} \sum_{n=1}^{+\infty} \frac{1}{F_{2^n}} = \frac{1}{\Phi^2 - 1}.$$

Now a short computation yields Lucas result.

Remark: This is an example of series whose sum is quadratic. Mahler's method does not apply to $f(x) = \sum_{n=0}^{+\infty} x^{2^n}/\left(1 - x^{2^{n+1}}\right)$. Indeed, f is a rational function, hence not transcendental. However, Mahler's method allows to prove, for example, that $\sum_{n=1}^{+\infty} 1/F_{2^n+1}$ is transcendental. See [19], example 1.3.3.

Exercise 12.13   Observe that $\theta = f(1/2)$, with $f_a(x) = \displaystyle\sum_{n=0}^{+\infty} \frac{x^{2^n}}{1+ax^{2^n}}$, $|x| < 1$.

1) It is easy to check that $f_a(x^2) = f_a(x) - x/(1+ax)$.

2) Now we show that f_a is transcendental. Suppose that

$$(f_a(x))^d + a_{d-1}(x)(f_a(x))^{d-1} + \cdots + a_0(x) = 0, \qquad a_i \in \mathbb{C}(x).$$

We assume that d is minimum. Replacing x by x^2 and using the functional relation 1) yields, by substracting and comparing the terms of degree $d-1$,

$$a_{d-1}(x) = a_{d-1}(x^2) - \frac{dx}{1+ax}.$$

Put $a_{d-1}(x) = A(x)/B(x)$, with A and B coprime. We obtain

$$(1+ax)A(x)B(x^2) = (1+ax)A(x^2)B(x) - dxB(x)B(x^2). \qquad (*)$$

First we have $B(0) \neq 0$. Indeed, $B(0) = 0 \Rightarrow A(0) \neq 0$ since A and B are coprime, and $(*)$ yields a contradiction if $B(x) = x^k C(x)$, $k \geq 1$, $C(0) \neq 0$. Moreover, by $(*)$, $B(x^2) \mid (1+ax)A(x^2)B(x)$. Since $B(x^2)$ and $A(x^2)$ are coprime, $B(x^2) \mid (1+ax)B(x)$. Hence, by looking at the degrees, $B(x) = \alpha x + \beta$ and $B(x^2) = \gamma(1+ax)B(x)$, $\gamma \in \mathbb{C}^*$. This yields $\beta = \gamma\beta$, whence $\gamma = 1$ since $\beta \neq 0$, and also $a\beta + \alpha = 0$ and $a\alpha = \alpha$. Thus $\alpha = -a\beta \neq 0$, whence $a = 1$. Therefore, if $a \neq 1$, f is transcendental.

If $a = 1$, we can observe, by using exercise 12.12, that

$$\sum_{n=0}^{+\infty} \frac{x^{2^n}}{1+x^{2^n}} + \sum_{n=0}^{+\infty} \frac{x^{2^n}}{1-x^{2^n}} = 2\sum_{n=0}^{+\infty} \frac{x^{2^n}}{1-x^{2^{n+1}}} = \frac{2x}{1-x}.$$

Thus, if f_1 was algebraic, so would be f_{-1}, which is not. Therefore f is transcendental for every $a \neq 0$.

3) It can be proved that θ is transcendental by using Mahler's method. The détails are left to the reader (see the proof of theorem 12.5, it is the same).

Exercise 12.14   1) Put $f(x) = \sum_{i=0}^{d} a_i x^i$, $a_i \in \mathbb{Z}$. Then

$$f^{(n)}(x) = \sum_{i=n}^{d} a_i i(i-1)...(i-n+1)x^{i-n} \text{ if } n \leq d, \ f^{(n)}(x) = 0 \text{ if } n > d.$$

Hence the coefficients of $f^{(n)}$ are divisible by $n!$ by lemma 8.5.

2) As $f^{(N+1)}(t) = 0$, $(e^{-t}F(t))' = -e^{-t}f(t)$, and we integrate from 0 to x.

3) Assume that e is algebraic of degree d, that is

$$a_d e^d + a_{d-1}e^{d-1} + \cdots + a_0 = 0, \quad a_i \in \mathbb{Z}, \quad a_0 \neq 0.$$

Replace x by k in Hermite's formula, multiply both sides by a_k, and add for $k = 0, 1,..., d$. We obtain

$$A_n = -\sum_{k=0}^{d} a_k F(k) = \sum_{k=0}^{d} a_k e^k \int_0^k f(t)e^{-t}dt.$$

with $f(t) = \frac{1}{(n-1)!}t^{n-1}\left((t-1)...(t-d)\right)^n$.

Now $f(t)$ satisfies $f^{(i)}(0) = 0$ if $i = 0, 1,..., n-2$, $f^{(n-1)}(0) = (-1)^{dn}(d!)^n$, $f^{(i)}(k) = 0$ if $i = 0, 1,..., n-1$, $k = 1, 2,..., d$.

Hence we have, since $\deg(f) = dn + n - 1 = N$,

$$F(0) = \sum_{i=n-1}^{dn+n-1} f^{(i)}(0) = (-1)^{dn}(d!)^n + nA, \qquad A \in \mathbb{Z}.$$

Indeed, for $i \geq n$ the coefficients of $(n-1)!f^{(i)}(t)$ are divisible by $i!$ (question 1) and therefore by $n!$. Similarly for $k \geq 1$, $F(k) = \sum_{i=n}^{dn+n-1} f^{(i)}(k) = nB$, $B \in \mathbb{Z}$. Thus, if n is a prime greater than d and a_0, we see that $n \nmid F(0)$ and $n \mid F(k)$ for $k \geq 1$. Therefore A_n *is a non-zero integer*.

Now, we estimate $|A_n|$ in order to prove that it vanishes when n tends to infinity. For $t \in [0,k]$, with $k \leq d$, we have $|f(t)| \leq d^{dn+n-1}/(n-1)!$, whence

$$|A_n| \leq \frac{d^{dn+n-1}}{(n-1)!} \sum_{k=0}^{d} |a_k| \int_0^k e^{k-t} dt \leq \frac{d^{dn+n-1}}{(n-1)!} e^d \sum_{k=0}^{d} |a_k| = K \frac{C^n}{(n-1)!},$$

where K and C do not depend on n. Thus $\lim_{n \to +\infty} A_n = 0$, and this contradiction proves that e is transcendental. It should be noted that this proof is relatively elementary, and does not refer to the algebraic theory of numbers.

Exercise 12.15 1) a) We show how to construct $x = p/q$, $p \in \mathbb{N}$, $q \in \mathbb{N}^*$. From the points $O(0,0)$ and $A(0,1)$ on the x-axis it is easy to obtain the point $P(p,0)$ by using the compass. Then we draw some straight line (OK), and using the compass again we construct the point Q satisfying $OQ = q$. Now constructing the parallel (CM) to (PQ) yields $M(p/q,0)$.

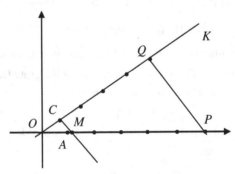

b) Starting from rational coordinates, it is only possible to construct algebraic coordinates of degree 2 (intersections of straight lines and circles), then algebraic coordinates of degree 4, and so on.

2) Let a be the length of side of the square. If squarring the circle was possible, than the number a satisfying $a^2 = \pi$ would be constructible. Hence π would be algebraic, contradiction.

Exercise 12.16 The system of equations $F(k + \beta m) = 0$ can be written as

$$\sum\nolimits_{i=0}^{N^{8}-1} \sum\nolimits_{j=0}^{N^{3}-1} a_{ij}(k+\beta m)^{i} \alpha^{jk} \gamma^{jm} = 0 \quad \left(0 \le k,\ m < N^{5}\right).$$

After multiplication by $(\mathrm{den}\,\alpha)^{N^{8}} (\mathrm{den}\,\beta)^{N^{8}} (\mathrm{den}\,\gamma)^{N^{8}} = \lambda^{N^{8}}$, we obtain a system with N^{10} equations and N^{11} unknowns a_{ij}, whose coefficients are integers of $\mathbb{K}$, which can be estimated by

$$\left|\lambda^{N^{8}}(k+\beta m)^{i} \alpha^{jk} \gamma^{jm}\right| \le \lambda^{N^{8}} \left(1+\overline{|\beta|}\right)^{N^{8}} N^{5N^{8}} \left(\overline{|\alpha||\gamma|}\right)^{N^{8}} \le (\lambda\mu)^{N^{8}} N^{5N^{8}} \le N^{6N^{8}},$$

considering our choice of N. Now Siegel's lemma 12.4 allows us to say that there exist $a_{ij} \in \mathbb{Z}$, solutions of the system and satisfying

$$0 < \max_{i,j}\left|a_{ij}\right| \le \left(N^{11} M . N^{6N^{8}}\right)^{d/(N-d)} \le e^{N^{8}}.$$

Exercise 12.17 1) We show that theorem 9.9 is equivalent to statement 1. First suppose there exists $C > 0$ such that $|\alpha - p/q| \ge Cq^{-d/2-1}$ for every $p/q \in \mathbb{Q}$, with $q > 0$. Let $\varepsilon > 0$. If inequation (I) has infinitely many solutions p/q, for each of them we have

$$Cq^{-\frac{d}{2}-1} \le |\alpha - p/q| < q^{-\frac{d}{2}-1-\varepsilon} \implies C < q^{-\varepsilon}.$$

Contradiction when $q \to +\infty$. Conversely, assume that, for every $\varepsilon > 0$, the inequation (I) has only finitely many solutions. Fix ε and denote by p_{1}/q_{1}, $p_{2}/q_{2},\ \cdots,\ p_{n}/q_{n}$ these solutions.

For every $q > Q = \max(q_{i})$, we have $|\alpha - p/q| \ge C_{0} q^{-d/2-1}$, with $C_{0} = q^{-\varepsilon}$.

Put $C_{1} = \min_{q \le Q}\left|\alpha - \dfrac{p}{q}\right| q^{\frac{d}{2}-1} > 0$ and $C = \min(C_{0}, C_{1})$. We have $\left|\alpha - \dfrac{p}{q}\right| \ge C q^{\frac{d}{2}+1}$.

Now it is clear that statement 1 implies statement 2. We show that the converse is true: if p/q is a solution of (I), then its irreducible expression p_{0}/q_{0} satisfies $|\alpha - p_{0}/q_{0}| < q^{-d/2-1-\varepsilon} \le q_{0}^{-d/2-1-\varepsilon}$. Hence there exist only finitely many associated irreducible rationals. Moreover, if p/q satisfies (I), we have $0 < |\alpha - p_{0}/q_{0}| < q^{-d/2-1-\varepsilon}$, which means that q is bounded, and that there are only finitely many rational p/q associated to this irreducible fraction. Hence statements 1 and 2 are equivalent.

2) We have to solve the system of equations ($h = 0,1,\cdots,m-1$):

$$\sum_{i=h}^{n} a_{i} i(i-1)\cdots(i-h+1)\alpha^{i-h} - \sum_{i=h}^{n} b_{i} i(i-1)\cdots(i-h+1)\alpha^{i-h+1} = 0.$$

We divide each equation by $h!$ and multiply by $\alpha^h (\mathrm{den}\,\alpha)^{n+1}$. The system is equivalent to the following one, with integer coefficients in $\mathbb{K} = \mathbb{Q}(\alpha)$.

$$\sum_{i=h}^{n} \binom{i}{h}(\mathrm{den}\,\alpha)^{n+1}\alpha^i a_i - \sum_{i=h}^{n}\binom{i}{h}(\mathrm{den}\,\alpha)^{n+1}\alpha^{i+1}b_i = 0, \quad h = 0,1,\cdots,m-1.$$

Since $\binom{i}{h} \le \binom{n}{h} \le (1+1)^n = 2^n$, the house of its coefficients is less than c^{n+1}, where $c > 1$ depends only on α. As $2^n \ge n$ for every $n \in \mathbb{N}$, Siegel's lemma 12.4 shows that this system has a solution satisfying

$$0 < \max_{i,j}\left(|a_i|,|b_j|\right) \le \left[2(n+1)Mc^{n+1}\right]^{\frac{dm}{2(n+1)-dm}} \le (4Mc)^{\frac{(n+1)dm}{2(n+1)-dm}}.$$

However, by definition of n, we have

$$n \ge \frac{1+\lambda}{2}dm - 1, \quad n \le \frac{1+\lambda}{2}dm < dm. \tag{*}$$

Thus $2(n+1) \ge (1+\lambda)dm$ and $n+1 \le dm$, Q.E.D. with $\beta = (4Mc)^d > 1$.

3) a) We observe that $W(x) = \begin{vmatrix} P(x) & Q(x) \\ P'(x) & Q'(x) \end{vmatrix}$ is a wronskian.

An easy induction shows that there exist natural integers p_{ij} such that

$$W^{(h)}(x) = \sum_{0 \le i < j \le h+1} p_{ij}\begin{vmatrix} P^{(i)}(x) & Q^{(i)}(x) \\ P^{(j)}(x) & Q^{(j)}(x) \end{vmatrix} \tag{**}$$

However, by definition of P and Q, we have for $0 \le i < j \le m-1$

$$P^{(i)}(\alpha) - Q^{(i)}(\alpha)\alpha = 0, \quad P^{(j)}(\alpha) - Q^{(j)}(\alpha)\alpha = 0$$

Therefore $\begin{vmatrix} P^{(i)}(\alpha) & Q^{(i)}(\alpha) \\ P^{(j)}(\alpha) & Q^{(j)}(\alpha) \end{vmatrix} = 0$ if $0 \le i < j \le h+1 \le m-1$.

Consequently $W(\alpha) = W'(\alpha) = \cdots = W^{(m-2)}(\alpha) = 0$.

b) If $W = 0$, P and Q are linearly dependent over $\mathbb{R}$ and there exists a constant $k \in \mathbb{R}^*$ such that $P = kQ$, say. Since P and Q have integer coefficients, $k \in \mathbb{Q}^*$. Now $P^{(h)}(\alpha) = kQ^{(h)}(\alpha)$ for every h. By definition of P and Q, this yields $(\alpha - k)Q^{(h)}(\alpha) = 0$ for every $h = 0, 1, \cdots, m-1$. Since $\alpha \notin \mathbb{Q}$, we have $Q^{(h)}(\alpha) = 0 = P^{(h)}(\alpha)$ for every $h = 0, 1, \cdots, m-1$. But P and $Q \in \mathbb{Z}[x]$. Hence for any conjugate $\alpha_1 = \alpha$, α_2, $\cdots, \alpha_d$, we have also $Q^{(h)}(\alpha_i) = 0 = P^{(h)}(\alpha_i)$. Thus P and Q are divisible by $(x - \alpha_1)^m .. (x - \alpha_d)^m$, contradiction because at least one of the polynomials is not zero and of degree at most $n < dm$ by $(*)$. Therefore $W \ne 0$.

c) We have $W(x) = \left(\sum_{i=0}^{n} a_i x^i \right) \left(\sum_{i=0}^{n} i b_i x^{i-1} \right) - \left(\sum_{i=0}^{n} i a_i x^{i-1} \right) \left(\sum_{i=0}^{n} b_i x^i \right)$.

Therefore $|c_i| \leq 2(n+1) \max|a_i| \max|b_i| \leq 2 dm \beta^{2m/\lambda} \leq 2 \dfrac{dm}{\lambda} \beta^{2dm/\lambda} \leq \beta^{4dm/\lambda}$.

This yields the desired result with $\gamma = 4d \operatorname{Log} \beta$.

4) If p_0/q_0 is a repeated zero with multiplicity h of W, we know by exercise 1.6 that q_0^h divides the leading coefficient of W, which is not zero. Thus $h \operatorname{Log} q_0 \leq \gamma m / \lambda$. Therefore the multiplicity of p_0/q_0 is at most equal to

$$w = \frac{\gamma m}{\lambda \operatorname{Log} q_0} \leq \lambda m \text{ thanks to the choice of } q_0.$$

5) Assume that, for every $0 \leq j \leq w+1$, $P^{(j)}\left(\dfrac{p_0}{q_0} \right) - \dfrac{p_k}{q_k} Q^{(j)}\left(\dfrac{p_0}{q_0} \right) = 0$.

Then $\begin{vmatrix} P^{(i)}\left(\dfrac{p_0}{q_0} \right) & Q^{(i)}\left(\dfrac{p_0}{q_0} \right) \\ P^{(j)}\left(\dfrac{p_0}{q_0} \right) & Q^{(j)}\left(\dfrac{p_0}{q_0} \right) \end{vmatrix} = 0$ if $0 \leq i < j \leq w+1$.

This yields $W\left(\dfrac{p_0}{q_0} \right) = W'\left(\dfrac{p_0}{q_0} \right) = \cdots = W^{(w)}\left(\dfrac{p_0}{q_0} \right) = 0$ by $(**)$, contradiction.

6) We have $|v_k| \leq \dfrac{1}{j!}\left(\left| P^{(j)}\left(\dfrac{p_0}{q_0} \right) - \alpha Q^{(j)}\left(\dfrac{p_0}{q_0} \right) \right| + \left| Q^{(j)}\left(\dfrac{p_0}{q_0} \right) \right| \cdot \left| \alpha - \dfrac{p_k}{q_k} \right| \right)$

$\leq \dfrac{1}{j!}\left(\left| \dfrac{\partial^j A}{\partial x^j}\left(\dfrac{p_0}{q_0}, \alpha \right) \right| + \left| Q^{(j)}\left(\dfrac{p_0}{q_0} \right) \right| \cdot \left| \alpha - \dfrac{p_k}{q_k} \right| \right)$

$\leq \dfrac{1}{j!}\left(\sum_{i=0}^{n-j} \dfrac{1}{i!}\left| \dfrac{\partial^{j+i} A}{\partial x^{j+i}}(\alpha, \alpha) \right| \cdot \left| \alpha - \dfrac{p_0}{q_0} \right|^i + \left| Q^{(j)}\left(\dfrac{p_0}{q_0} \right) \right| \cdot \left| \alpha - \dfrac{p_k}{q_k} \right| \right)$.

By the change of index $h = j+i$ and by taking into account the definition of $A(x, y)$ and the estimates $j \leq w \leq \lambda m < m$, we obtain

$$|v_k| \leq \sum_{h=m}^{n} \binom{h}{j} \dfrac{1}{h!}\left| \dfrac{\partial^h A}{\partial x^h}(\alpha, \alpha) \right| \left| \alpha - \dfrac{p_0}{q_0} \right|^{h-j} + \dfrac{1}{j!}\left| Q^{(j)}\left(\dfrac{p_0}{q_0} \right) \right| \cdot \left| \alpha - \dfrac{p_k}{q_k} \right|.$$

Now denote $\theta = 4^d (|\alpha|+1)^d \beta$. We have, as in question 2, the following estimates (note that $n+1 \leq dm/\lambda$ by $(*)$):

$$\dfrac{1}{j!}\left| Q^{(j)}(p_0/q_0) \right| \leq \sum_{i=j}^{n} \binom{i}{j} |b_i| \left| p_0/q_0 \right|^{i-j} \leq (n+1) 2^n \beta^{m/\lambda} (|\alpha|+1)^n \leq \theta^{m/\lambda},$$

$$\frac{1}{h!}\left|\frac{\partial^h A}{\partial x^h}(\alpha,\alpha)\right| \le \sum_{i=h}^{n}\binom{i}{h}|\alpha^i a_i| + \sum_{i=h}^{n}\binom{i}{h}|\alpha^{i+1}b_i| \le (2n+2)2^n \beta^{\frac{m}{\lambda}}(|\alpha|+1)^n \le (2^d \theta)^{\frac{m}{\lambda}}.$$

Consequently we obtain, by denoting $\delta = (8^d+1)\theta$,

$$|v_k| \le (n+1)2^n (2^d\theta)^{\frac{m}{\lambda}}\left|\alpha-\frac{p_0}{q_0}\right|^{m-j} + \theta^{\frac{m}{\lambda}}\left|\alpha-\frac{p_k}{q_k}\right| \le (8^d\theta)^{\frac{m}{\lambda}}q_0^{\rho(j-m)} + \theta^{\frac{m}{\lambda}}q_k^{-\rho}$$

$$\le (8^d\theta)^{\frac{m}{\lambda}}q_0^{\rho(\lambda-1)m} + \theta^{\frac{m}{\lambda}}q_0^{-\rho m} \le \left[(8^d\theta)^{\frac{m}{\lambda}}+\theta^{\frac{m}{\lambda}}\right]q_0^{\rho(\lambda-1)m} \le \delta^{\frac{m}{\lambda}}q_0^{\rho(\lambda-1)m}.$$

7) Since the coefficients of $\dfrac{1}{j!}P^{(j)}\left(\dfrac{p_0}{q_0}\right)$ and $\dfrac{1}{j!}Q^{(j)}\left(\dfrac{p_0}{q_0}\right)$ are rational integers,

we see that $q_0^{n-j}q_k v_k$ is also a rational integer. Since $v_k \ne 0$ and $m \ge \dfrac{\mathrm{Log}\, q_k}{\mathrm{Log}\, q_0}-1$,

we now obtain $|v_k| \ge q_0^{j-n}q_k^{-1} \ge q_0^{j-n}q_0^{-m-1} = q_0^{j-n-m-1} \ge q_0^{\frac{1+\lambda}{2}dm-m-1}.$

We deduce immediatly from the upper and lower bounds for $|v_k|$ that

$$-\left[\left(\frac{1+\lambda}{2}d+1\right)m+1\right]\mathrm{Log}\, q_0 \le \frac{m}{\lambda}\mathrm{Log}\,\delta + (\lambda-1)\rho m \,\mathrm{Log}\, q_0.$$

Thus $\quad \rho \le \dfrac{\mathrm{Log}\,\delta}{\lambda(1-\lambda)\mathrm{Log}\, q_0} + \dfrac{(1+\lambda)\frac{d}{2}+1}{1-\lambda} + \dfrac{1}{(1-\lambda)m}.$

Now we perform the following operations:

a) We choose a "small" λ, such that $\dfrac{(1+\lambda)\frac{d}{2}+1}{1-\lambda} \le \dfrac{d}{2}+1+\dfrac{\varepsilon}{6}.$

b) We choose q_0 in such a way that $\dfrac{\mathrm{Log}\,\delta}{\lambda(1-\lambda)\mathrm{Log}\, q_0} \le \dfrac{\varepsilon}{6}.$

c) We choose k such that $\dfrac{1}{(1-\lambda)m} \le \dfrac{\varepsilon}{6}.$

We obtain $\rho \le \dfrac{d}{2}+1+\dfrac{\varepsilon}{2}$. This contradiction proves Thue's theorem.

Exercise 12.18   1) If $\theta = p/q \in \mathbb{Q}$, $q>0$, then $a^q = b^p$, which is impossible since a and b are coprime.

2) If θ was algebraic, then b^θ would be transcendental by Gelfond-Schneider theorem, since $\theta \notin \mathbb{Q}$. Contradiction because $b^\theta = e^{\mathrm{Log}\,a} = a \in \mathbb{N}$.

Bibliography

[1] Apostol (T.), *Introduction to Analytic Number Theory*, Springer, 1976.

[2] Baker (A.), *Transcendental Number Theory*, Cambridge, 1974.

[3] Bertrand (D.) et al, *Les Nombres Transcendants*, Mémoire de la SMF, 1984.

[4] Borevich (Z.I.) and Shafarevich (I.R), *Number Theory*, Academic Press, 1986.

[5] De Koninck (J.M.) and Mercier (A.), *Approche élémentaire de l'étude des fonctions arithmétiques*, Presses de l'Université Laval, 1982.

[6] Descombes (R.), *Eléments de Théorie des Nombres*, PUF, 1986.

[7] Edwards (H.), *Fermat's Last Theorem*, Springer, 1977.

[8] Faisant (A.), *L'équation diophantienne du second degré*, Hermann, 1991.

[9] Galambos (J.), *Representation of Real Numbers by Infinite Series*, Springer, 1976.

[10] Gel'fond (A.) and Linnik (Yu.), *Elementary methods in the analytic theory of numbers*, Pergamon Press, 1966.

[11] Goldman (J.R.), *The Queen of Mathematics*, A K Peters, 1998.

[12] Grosswald (E.), *Representations of Integers as Sums of Squares*, Springer, 1985.

[13] Hardy (G.H.) and Wright (E.M), *An Introduction to the Theory of Numbers*, Oxford, 1938.

[14] Hellegouarch (Y.), *Invitation aux mathématiques de Fermat-Wiles*, Masson, 1997.

[15] Ireland (K.) and Rosen (M.), *A Classical Introduction to Modern Number Theory*, Springer, 1990.

[16] Landau (E.), *Vorlesungen über Zahlentheorie*, Hirzel, 1927.

[17] Lang (S.), *Transcendental Numbers*, Addison-Wesley, 1966.

[18] LeVeque (W.J.), *Topics in Number Theory*, Addison-Wesley, 1965.

[19] Nishioka (Ku.), *Mahler Functions and Transcendence*, Springer, 1996.

[20] Perron (O.), *Die Lehre von den Kettenbrüchen*, Teubner, 1913.

[21] Perron (O.), *Irrationalzahlen*, Chelsea, 1948.

[22] Ribenboim (P.), *The Book of Prime Numbers Records*, Springer, 1988.

[23] Ribenboim (P.), *Lectures on Fermat's Last theorem*, Springer, 1979.

[24] Samuel (P.), *Théorie Algébrique des Nombres*, Hermann, 1967.

[25] Serre (J.P.), *Cours d'Arithmétique*, PUF, 1970.

[26] Shidlovski (A.), *Transcendental Numbers*, de Gruyter, 1989.

[27] Stewart (I.) and Tall (D.), *Algebraic Number Theory*, Chapman and Hall, 1979.

[28] Waldschmidt (M.), *Nombres Transcendants*, Springer, 1974.

Index